Plant-Environment Interactions

BOOKS IN SOILS, PLANTS, AND THE ENVIRONMENT

Soil Biochemistry, Volume 1, edited by A. D. McLaren and G. H. Peterson
Soil Biochemistry, Volume 2, edited by A. D. McLaren and J. Skujiņš
Soil Biochemistry, Volume 3, edited by E. A. Paul and A. D. McLaren
Soil Biochemistry, Volume 4, edited by E. A. Paul and A. D. McLaren
Soil Biochemistry, Volume 5, edited by E. A. Paul and J. N. Ladd
Soil Biochemistry, Volume 6, edited by Jean-Marc Bollag and G. Stotzky
Soil Biochemistry, Volume 7, edited by G. Stotzky and Jean-Marc Bollag
Soil Biochemistry, Volume 8, edited by Jean-Marc Bollag and G. Stotzky

Organic Chemicals in the Soil Environment, Volumes 1 and 2, edited by C. A. I. Goring and J. W. Hamaker
Humic Substances in the Environment, M. Schnitzer and S. U. Khan
Microbial Life in the Soil: An Introduction, T. Hattori
Principles of Soil Chemistry, Kim H. Tan
Soil Analysis: Instrumental Techniques and Related Procedures, edited by Keith A. Smith
Soil Reclamation Processes: Microbiological Analyses and Applications, edited by Robert L. Tate III and Donald A. Klein
Symbiotic Nitrogen Fixation Technology, edited by Gerald H. Elkan
Soil-Water Interactions: Mechanisms and Applications, edited by Shingo Iwata, Toshio Tabuchi, and Benno P. Warkentin
Soil Analysis: Modern Instrumental Techniques, Second Edition, edited by Keith A. Smith
Soil Analysis: Physical Methods, edited by Keith A. Smith and Chris E. Mullins
Growth and Mineral Nutrition of Field Crops, N. K. Fageria, V. C. Baligar, and Charles Allan Jones
Semiarid Lands and Deserts: Soil Resource and Reclamation, edited by J. Skujiņš
Plant Roots: The Hidden Half, edited by Yoav Waisel, Amram Eshel, and Uzi Kafkafi
Plant Biochemical Regulators, edited by Harold W. Gausman
Maximizing Crop Yields, N. K. Fageria
Transgenic Plants: Fundamentals and Applications, edited by Andrew Hiatt
Soil Microbial Ecology: Applications in Agricultural and Environmental Management, edited by F. Blaine Metting, Jr.
Principles of Soil Chemistry: Second Edition, Kim H. Tan

Water Flow in Soils, edited by Tsuyoshi Miyazaki
Handbook of Plant and Crop Stress, edited by Mohammad Pessarakli
Genetic Improvement of Field Crops, edited by Gustavo A. Slafer
Agricultural Field Experiments: Design and Analysis, Roger G. Petersen
Environmental Soil Science, Kim H. Tan
Plant–Environment Interactions, edited by Robert E. Wilkinson

Additional Volumes in Preparation

Mechanisms of Plant Growth and Improved Productivity: Modern Approaches, edited by Amarjit S. Basra

Selenium in the Environment, edited by W. T. Frankenberger, Jr., and Sally Benson

Handbook of Plant and Crop Physiology, edited by Mohammad Pessarakli

Handbook of Phytoalexin Metabolism and Action, edited by M. Daniel and R. P. Purkayastha

Seed Development and Germination, edited by Jaime Kigel and Gad Galili

Stored Grain Ecosystems, edited by D. S. Jayas and N. D. G. White

Plant-Environment Interactions

edited by

Robert E. Wilkinson
The University of Georgia
Griffin, Georgia

Marcel Dekker, Inc.   New York • Basel • Hong Kong

Library of Congress Cataloging-in-Publication Data

Plant—environment interactions / edited by Robert E. Wilkinson
    p. cm. — (Books in soils, plants, and the environment)
    Includes bibliographical references and index.
    ISBN 0-8247-8940-7
    1. Plant ecophysiology. I. Wilkinson, R. E. (Robert E.). II. Series.
QK905.P56 1994
581.5—dc20   94-11104
                                                       CIP

The publisher offers discounts on this book when ordered in bulk quantities. For more information, write to Special Sales/Professional Marketing at the address below.

This book is printed on acid-free paper.

Marcel Dekker, Inc.
270 Madison Avenue, New York, New York 10016

Current printing (last digit):
10 9 8 7 6 5 4 3 2 1

PRINTED IN THE UNITED STATES OF AMERICA

Preface

Plant response to environment has been a paradigm for millenia. But the total ecosystem influencing plant growth and development has multiparameters. Science has attempted to isolate individual environmental factors so that the plant response to a single stimulus can be quantitated. The success of this attack on apparently insoluble problems can best be evaluated by the increased understanding of plant growth and development that has accumulated in the last century.

But, increasingly, evidence has shown that plant response to a single stimulus is not uniform during the life of the plant, and that a plant is an integrated whole biological entity whereby a change at one level can have a profound influence at a second tissue, organ, or process separated in time or location by some distance from the original stimulus. Thus, this text is an attempt to correlate some of these variables. And, because so many of the environmental parameters produce concomitant responses, those environmental influences produce interactions in the plant.

Basically, a large percentage of the interactions that have been reported have been studied in agricultural systems. Since agriculture is only applied ecology, the relationships between mineral nutrition, plant growth and development, plant–water relations, photoperiod, light intensity, temperature, pesticides, and plant biochemistry are closely interwoven.

Also, there is a natural progression of study and understanding that proceeds from (a) description of biological responses to (b) biochemical and biophysical mechanisms that produce the responses, to (c) genetic manipulation of DNA to understand and create new biological responses. Each portion must correlate with the other types of study.

Concepts of natural food production have evolved to an absence of pesticides that are only synthetic plant growth regulators (PGRs). True, the pesticides may not necessarily be hormone-type PGRs. But, as the various genetic mutants have shown, loss of the ability to produce a requisite component (i.e., chlorophyll) places a severe restraint on continued plant growth and development. This includes eukaryotic and prokaryotic plants. Thus, utilization of one pesticide (i.e., alar) may produce excellent apples that have a degradation product that may possibly induce cancer in x numbers of humans over a two-decade span. At the same time, that particular PGRs inhibits the growth and development of "natural" pathogens whose "natural" products induce lethal responses in $200x$ humans over the same period. These applied ecology problems are the province of a vast array of biologists, chemists, biochemists, etc., and, factually, a large proportion of the biochemical and biological knowledge that has accreted in the last few decades about plant growth and development has been a direct result of the development of herbicides, fungicides, and PGRs for use in agriculture.

Definition of the mode of action of these various chemicals has permitted scientists to further isolate specific processes in plants that control plant growth and development. Examples of this progression are seen in the study of diuron as a herbicide that inhibits photosystem II (PSII). And the ability to selectively inhibit specific portions of the entire photosynthetic process by certain diuron concentrations has led to major advances in the study of photosynthesis as a biochemical process. Currently, this study is focusing on the amino acid constituents of DNA involved in the production of specific proteins utilized in PSII. Similar progressions in the development of plant biochemistry, etc., are occurring in many other areas. However, there is always the consideration that conditions, concentrations, and so forth must be carefully monitored. For example, although one diuron concentration has been utilized extensively to study PSII, greater diuron concentrations influence several other biochemical processes. Very rarely does an exogenous compound produce only one reaction regardless of the concentration. Examples of metabolic control by the "second messenger" Ca^{2+} have shown that cytosol Ca^{2+} concentration is very tightly regulated. Variation from the optimal Ca^{2+} concentration results in massively modified cellular metabolism.

Root absorption of mineral nutrients and the growth of plants in relation to the concentrations and ratios of various ions have been studied by agriculturally oriented plant scientists for decades. These studies have benefited mankind tremendously in the production of food and fiber for a constantly increasing world population. But understanding how plant nutrients are absorbed into the root has been a puzzle. Recently, studies of human heart arrhythmia have led to the development of chemicals that control the transfer of Ca^{2+} through heart cell plasma membranes. An entire scientific discipline has developed that is concerned with the biochemistry and biophysics of the ion pumps and voltage-gated ion pores that control Ca^{2+} transport through the plant plasma membrane. These studies have been extended to plant roots and an explanation of how ions are transported through root plasma membranes may soon be more completely established. Additionally, these processes have been shown to vary between roots of cultivars of a single species, between organelles within a cell, and between (or at the interface of) specific tissues (i.e., phloem sieve tube elements or xylem elements). When confirmed and extended to different species, genomes, and so forth, these findings may help explain many correlations of plant growth and development that currently are totally inexplicable. These enzymes that control transport through membranes are also found in plant pathogens. And alteration of plant epicuticular chemistry has been shown to have a profound influence on the growth and infestation of some plant pathogens.

Natural PGRs (i.e., hormones) are present at different concentrations during the development of the plant. Factors determining concentration–response have been shown to include (a) species, (b) tissue, (c) age, (d) relative concentration of other PGRs, and (e) other *stresses* that develop in the tissue/organ/organism.

Thus, interactions and correlations of plant growth extend through a complete ecological array. One environmental parameter produces one primary response at specific growth stages, etc. However, side reactions also occur. And occasionally those side reactions have striking results. Because these various environmental stresses alter plant growth to differing degrees depending on time of stress and the particular plant response being measured, computer modeling of these factors offers hope of developing an integrated understanding of the entire process. But first the influence of individual stresses and other factors must be ascertained. This text is a compilation of a few of the correlations and interactions that are currently known. We make no claim for discussing all the known interactions, and more correlations will be discovered with additional research. Thus, we present some data for the perusal of students of plant growth and development. Extension of these concepts lies in the province of individual researchers. And, since 90–98% of the researchers who have ever worked throughout recorded history are alive and working today, we feel confident that much more will be learned about these interactions and correlations in the future.

This book constitutes a preliminary introduction to a possible study of a large and difficult subject. I hope that readers learn as much as I have learned while editing these chapters.

Robert E. Wilkinson

Contents

Preface   iii

Contributors   ix

Introduction   xi

1. Genetic Manipulation   1
    R. R. Duncan

2. The Role of Plant Hormones   39
    Frank B. Salisbury

3. Light and Plant Development   83
    Michael J. Kasperbauer

4. Acid Soil Stress and Plant Growth   125
    Robert E. Wilkinson

5. Mineral Nutrition   149
    Larry M. Shuman

6. Isoprenoid Biosynthesis   183
    Robert E. Wilkinson

7.   Whole-Plant Response to Salinity   199
     Michael C. Shannon, Catherine M. Grieve, and Leland E. Francois

8.   Symbiotic Nitrogen Fixation   245
     John G. Streeter

9.   Plant Response Mechanisms to Soil Compaction   263
     M. J. Vepraskas

10. Plant Response to Flooding   289
     S. R. Pezeshki

11. Stomata   323
     Thomas M. Hinckley and Jeffrey H. Braatne

12. Plant Response to Air Pollution   357
     James A. Weber, David T. Tingey, and Christian P. Andersen

13. Environmental Effects of Cold on Plants   391
     Maynard C. Bowers

14. Photosynthetic Response Mechanisms to Environmental Change
     in C3 Plants   413
     Rowan F. Sage and Chantal D. Reid

15. Plant Respiratory Responses to the Environment and Their Effects
     on the Carbon Balance   501
     Jeffrey S. Amthor

16. Barriers in the Wheat Leaf Rust Preinfection Phase   555
     Robert E. Wilkinson and John J. Roberts

Index   585

Contributors

Jeffrey S. Amthor, M.S., Ph.D. Assistant Scientist, Woods Hole Research Center, Woods Hole, Massachusetts

Christian P. Andersen, Ph.D. Research Plant Physiologist, Environmental Research Laboratory, U.S. Environmental Protection Agency, Corvallis, Oregon

Maynard C. Bowers, Ph.D. Professor, Biology Department, Northern Michigan University, Marquette, Michigan

Jeffrey H. Braatne, Ph.D. Research Assistant Professor, College of Forest Resources, University of Washington, Seattle, Washington

R. R. Duncan, Ph.D. Professor of Breeding and Stress Physiology, Department of Crop and Soil Sciences, Georgia Experiment Station, University of Georgia, Griffin, Georgia

Leland E. Francois, B.S., M.S. Research Agronomist, Department of Plant Science, U.S. Salinity Laboratory, ARS, USDA, Riverside, California

Catherine M. Grieve, Ph.D. Plant Physiologist, Department of Plant Science, U.S. Salinity Laboratory, ARS, USDA, Riverside, California

Thomas M. Hinckley, Ph.D.   Bloedel Professor of Forestry, College of Forest Resources, University of Washington, Seattle, Washington

Michael J. Kasperbauer, Ph.D.   Research Plant Physiologist, Coastal Plains Soil, Water, and Plant Research Center, ARS, USDA, Florence, South Carolina

S. R. Pezeshki, M.S., Ph.D.   Associate Professor, Wetland Biogeochemistry Institute, Center for Coastal, Energy, and Environmental Resources, Louisiana State University, Baton Rouge, Louisiana

Chantal D. Reid   Research Associate, Department of Botany, University of Georgia, Athens, Georgia

John J. Roberts, Ph.D.   Research Plant Pathologist, Georgia Experiment Station, ARS, USDA, Griffin, Georgia

Rowan F. Sage, Ph.D.   Assistant Professor, Department of Botany, University of Georgia, Athens, Georgia

Frank B. Salisbury, M.A., Ph.D.   Professor of Plant Physiology, Plants, Soils, and Biometeorology Department, Utah State University, Logan, Utah

Michael C. Shannon, M.S., Ph.D.   Plant/Research Geneticist, Department of Plant Science, U.S. Salinity Laboratory, ARS, USDA, Riverside, California

Larry M. Shuman, M. S., Ph.D.   Professor of Soil Chemistry, Department of Crop and Soil Sciences, Georgia Experiment Station, University of Georgia, Griffin, Georgia

John G. Streeter, Ph.D.   Professor, Department of Agronomy, The Ohio State University, Wooster, Ohio

David T. Tingey, M.A., Ph.D.   Program Leader, Global Processes and Effects Program, U.S. Environmental Protection Agency, Corvallis, Oregon

M. J. Vepraskas, Ph.D.   Professor, Department of Soil Science, North Carolina State University, Raleigh, North Carolina

James A. Weber, Ph.D.   Research Plant Physiologist, Environmental Research Laboratory, U.S. Environmental Protection Agency, Corvallis, Oregon

Robert E. Wilkinson, Ph.D.   Professor, Department of Crop and Soil Sciences, Georgia Experiment Station, University of Georgia, Griffin, Georgia

Introduction

Terrestrial eukaryophytes are an intermediate phase in the soil–atmosphere continuum. The plant is basically a highly responsive quivering hunk of protoplasm that must overcome every stress, environmental signal, change in mineral nutrition, water quantity and quality alteration, temperature change, exogenous chemical (natural and synthetic), and so forth, in an effort to produce the next generation of seed. Different plant growth stages have different responses to each stress. Different genomes have permitted plants to survive in different environmental situations. This is not a factor of "adaptation to" but a measure of "survival in" the different stress situations. Therefore, the important factor is what mechanisms increase survivability. This compilation of chapters details *some* of the mechanisms by which plants respond to environment. Choice of a different group of stresses would result in a similar but different text. This text is not intended to be an encyclopedia but is designed to offer some insight into basic plant responses.

Obviously, plant response to one environmental stress intergrades into a response to a second stress. Or, plant response to environmental stress is an integrated composite of responses to several factors. As plant scientists, we must recognize and integrate these responses. Yet each of us spends a lifetime studying a particular stress response by plants, and the success of this syndrome is best illustrated by the remarkable bulk of knowledge that has accumulated in one lifetime. However, to understand the plant, we must learn to mentally integrate the responses of plants to all the environmental stresses.

1

Genetic Manipulation

R. R. Duncan

University of Georgia
Griffin, Georgia

PLANT ADAPTATIONS TO EDAPHIC ENVIRONMENTS

Constraints to crop production evolve from environmental stresses such as erratic rainfall distribution (drought stress), inopportune temperature stresses, and pest and pathogen encroachment. When production inputs are relatively inexpensive and readily available, soil environments are traditionally altered to fit plant requirements. Luxurious use of fertilizers, pesticides, and lime dominates production practices.

Sustainable agriculture practices are dictating a shift away from manipulating the edaphic environment to fit plant needs to genetic alteration of the plant to tolerate environments with reduced inputs. Mineral stress resistance, herbicide resistance, heat stress tolerance, pathogen resistance, saline/sodic soil tolerance, pest and pathogen resistance, and water and fertilizer uptake/utilization efficiency are potentially useful tactics to reduce production inputs, minimize environmental pollution, and maintain a sound ecological system. Most of these attributes are genetically controlled. However, plant breeders have had limited success in improving these predominantly quantitatively inherited traits. Most of the mechanisms governing tolerance/resistance reactions are complex and not completely understood.

Plants constantly confront erratic environmental stresses. Survivability is governed by how well plants adapt to short- or long-term stresses.

Different plant growth stages or physiological processes result in variable sensitivities to stress (3). This chapter will discuss the genetic control of mechanisms governing tolerance or sensitivity reactions to various environmental stresses. Complicating the problem is plant response (persistence) to interacting and simultaneous multiple stresses, i.e., heat and drought, cold and wet, salinity and drought. These interrelations may involve interactions between two adaptations to the same stress, such as stress avoidance interacting with and controlling the development of another adaptation (stress tolerance) to the same stress (1,2). Avoidance of drought-induced water stress has been found essential for development of tolerance (drought-induced acclimation to water stress). While a plant may react immediately to current environmental constraints, the survival of the genome results from adaptation over time.

EDAPHIC CONSTRAINTS AND PLANT STRESS

Nutrient availability, mineral toxicities and deficiencies, and plant species interact with climatic factors such as moisture and temperature to govern plant productivity. Soil properties determining nutrient effects on plants include pH, nutrient content and availability, base saturation, permeability, cation exchange capacity, organic matter content, sesquioxide content, and moisture retention capacity (55). In addition, soil physical properties (high soil strength, low soil oxygen levels, and high soil temperatures) can indirectly influence root growth, root plasticity, and nutrient uptake. Several of these properties will be discussed in relation to plant responses and mechanisms governing their responses.

Two metabolic mechanisms function to attract nutrient ions from soil solution and govern their subsequent uptake at specific rates and ratios during ontogeny (227). One active metabolic mechanism is an auxin-dependent H-proton pump located in the leaves. The other mechanism is located in the root and involves excretion of negatively charged ions (OH^-, HCO_3^-) to ensure (a) uptake of phosphate and NO_3^-, (b) functional NO_3^- reduction in cells, and (c) later transport of negative charges via carboxylates from the leaves (227). Nutrient uptake and utilization involves closely synchronized events: (a) ion movement from the soil solution to the root surface, (b) ion transport through root membranes, (c) centripetal ion translocation to and into xylem vessels, and (d) acropetal translocation and distribution via the xylem. Root selectivity has been shown to promote preferentially positive absorption of K^+ and NO_3^--N, negative selectivity for Ca^{2+}, and no selectivity for Mg^{2+}, $H_2PO_4^-$, or Cl^- (228).

Nitrogen

This essential nutrient and major productivity-enhancing amendment has garnered considerable research attention for non–N_2-fixing plants. Actual breeding efforts for improving N efficiency traits has resulted in minimal advancements because of the complexity of N metabolism and its multifaceted genetic control (36,59–62). Major variables involved in N use efficiency include (a) NO_3^- and NH_4^+ uptake, (b) translocation of nitrogenous compounds, (c) reduction of NO_3^-, and (d) assimilation of reduced N into organic compounds. Assessment of N use efficiency can be accomplished by measuring efficiency ratios (unit dry matter produced per unit N accumulated), N harvest index (proportion of N in harvested components), N remobilization efficiency (fraction of N translocated from vegetative to reproductive or harvested components after peak accumulation in the vegetative parts), and total N accumulation. Nitrogen uptake differences may account for half of the genetic variation in N use efficiency contribution to yield and three fourths of the contribution to protein production in wheat (*Triticum aestivum* L. em Thell) (64).

A summary of physiological and morphological factors that are associated with N use efficiency is presented in Table 1. A thorough discussion can be found in Barker (19). These factors include root proliferation, uptake rate per unit root mass or area, ion concentration (balance among influx, efflux, and growth), translocation, and utilization. Each stage of N assimilation is under genetic control. Nitrate is transported to the cell and reduced to ammonia by two soluble enzymes: nitrate reductase (a pyridine nucleotide-dependent enzyme found in the cytoplasm) cofactored with molybdenum-pterin and nitrite reductase (a ferredoxin-dependent enzyme located in the plastids) cofactored with Fe-containing hydrochlorin (siroheme) (224). Two genes encode these biochemically distinct enzymes.

Nitrate induces nitrate and nitrite reductase activities by altering gene expression, mainly by enhancing transcription of the respective genes (354). Nitrate also signals developmental and physiological changes in the plant. Primary responses [rapid activation (within minutes), selectivity (limited number of genes), directness (no protein synthesis)] when exposed to the stimulus include (a) gene induction for nitrate and nitrite reductase, (b) nitrate uptake and translocation systems, and (c) DNA regulatory proteins required for expression of secondary response gene systems. The secondary responses (requiring new protein synthesis for their expression) include more complex responses such as (a) proliferation of the root system, (b) enhancement of respiration, and (c) physiological changes in the plant (354).

Table 1 Efficiency of N Use Associated With Physiological Parameters

Efficiency factor	Basic efficiency response	Ref.
Root Proliferation		
Absorption efficiency	Specific uptake rates/unit mass or area; high affinity for NO_3^- or NH_4^+	4–6
Selective ion absorption	Discrimination between NO_3^- and NH_4^+ or capacity to absorb NO_3 in presence of NH_4^+	7–11
Tolerance to NH_4^+	Age vs. root physiology, lateral to primary root proportions, root/shoot ratios	10–17
Nitrate Uptake Efficiency		
Uptake system development	Rapid induction of ion-absorbing mechanism	10,20,21
Uptake stimulation/inhibition	Capacity to absorb NO_3^- in presence of NH_4^+, additional ion stimulation of absorption	5,22–28
Preanthesis N absorption	Shift in competition between vegetative and reproductive growth	11,22
Nitrogen Translocation Efficiency		
Site of NO_3^- reduction	Response to external N supplies; nitrate reductase regulation; reduce N partitioning; root development; root/shoot ratios; NO_3^- accumulation	10,16,17, 29–35
N harvest index/remobilization efficiency	N translocation/transport to harvested organs; protein concentration in harvested organs	5,25,27, 36,37,65
Nitrate Reductase Activity Efficiency		
Enzyme species	Heritability of variations in nitrate reductase/NADH/NADPH specificity, inducible or constitutive enzymes	38,39,66
Enzyme activity	Rate limitation in N assimilation, increased yields vs. protein production, enhanced N accumulation, genetic capacity to assimilate N	5,39–44
Enzyme distribution	Root morphology, root/shoot ratio, responses to external supplies, NO_3^- accumulation	5,19,29, 30–32,34,45
Ammonium Nutrition Tolerance		
NH_4^+ root assimilation	NH_4^+ incorporation into root organic compounds = detoxification	46
Enzyme metabolism of NH_4^+	Isozymes catalyzing NH_4^+ assimilation	47–51
Potassium accumulation	K^+ association with NH_4^+ toxicity resistance (tomato)	52,53

The physiological efficiency index of absorbed nitrogen (PEN = ratio kg grain production to kg N absorbed in above-ground dry matter at maturity) is closely related to genetics of cultivars and is highly correlated ($R = 0.98$) to grain yield (65). High PEN cultivars at low fertility also have high PEN at high fertility. A recent treatise summarizes the current knowledge of nutrient uptake and use efficiency (56).

Phosphorus

As with nitrogen, phosphorus efficiency includes absorption efficiency, translocation, and internal utilization efficiency, particularly under suboptimal availability conditions. Intraspecific genotypic differences may result from differential physical or metabolic compartmentation or functional use after translocation to specific tissues. P partitioning between organic and inorganic forms may be a major determinate of efficiency in maize (*Zea mays* L.) (67) but was determined not to be a factor in sorghum [*Sorghum bicolor* (L.) Moench] (69). Since a high vacuolar affinity for inorganic P might be detrimental to P remobilization, a specific P fraction could be cycled in different genotypes (68,69).

Physiological efficiency of P absorption is the relative ability to maximize P concentration with minimal fixed carbon investment in root growth and metabolism (70). Efficient P utilization is expressed as dry matter production per increment of P accumulated (67). Genotypes having a high ability to absorb or utilize P under deficiency conditions may not necessarily be responsive to high P supplies (71,72).

Since P has a central role in energy transfer and protein metabolism in plants, accumulation of insoluble P compounds in young leaves under P deficiency may indicate that P in nucleic acids, proteins, and lipids represents an important fraction for cell growth and functioning (69,73). Maintenance or increased accumulation of total root P during deficiency stress may be related to an enhanced change of assimilate distribution of the insoluble fraction favoring the roots (74). In a P-efficient sorghum cultivar that produced more dry matter per unit of P absorbed, inherent "growth-promoting factors" contributed to the intraspecific P efficiency by stimulation of P redistribution intensity and a subsequent compensation for lower root P absorption capacity (69). Pigeon pea [*Cajanus cajan* (L.) Millsp.] was extremely efficient in utilizing Fe-bound phosphates under P deficiency stress conditions (75). This increased utilization efficiency was attributed to piscidic acid (and its *p-O*-methyl derivative) in the root exudates, which releases P from the Fe-P complex by chelating Fe^{3+}.

In a P-efficient soybean (*Glycine max* L.) cultivar, sucrose levels were higher during phosphate stress, while hexose (glucose, fructose) levels were equally high compared to the inefficient cultivar (74). The shoot of an N-efficient cultivar supplied its root with an increased amount of energy via carbohydrates that enhanced P uptake efficiency by supplying structural material to maintain membrane systems or an energy source to activate proton pump deposition required for the membrane potential establishment involved in phosphate uptake.

Grain yield at low-to-moderate P fertility and low levels of P incorporation in the grain as phytic phosphate have been used as specific traits to select for improved P use efficiency in wheat (76,77). Grasses are generally more tolerant to low soil P conditions than legumes, and this difference has been attributed to more extensive root systems coupled with the internal ability to regulate P between roots and shoots.

More detailed summaries on P efficiency improvements can be found in Clark and Duncan (57) and Baligar and Duncan (56).

Potassium

Root growth and morphology, ion uptake efficiency, ion efflux, ion translocation, and ion use efficiency should receive consideration when addressing improved K nutrition traits in breeding programs (78). Root regulation of ion uptake and synchronization of plant growth rate to ion supply is important in ecological adaptation (81). Gene-controlled ion transport carrier synthesis is responsible for the variation in maximum K uptake efficiencies among plant species (80). Uptake mechanisms and synthesis of ion transport carriers are genetically controlled in plants (82).

Differential K uptake efficiency is associated with a membrane mechanism dependent on ploidy level in wheat, sugarbeet (*Beta vulgaris* L.), and tomato (*Lycopersicon esculentum* Mill.). Potassium utilization rather than K uptake, translocation, or accumulation is the major factor associated with efficiency (83–85). Sodium substitution capacity for K in metabolic functions can be used to enhance K efficiency breeding efforts (83,86,87) in tomato, but not in snapbeans (*Phaseolus vulgaris* L.) (84,85). Potassium sufficiency is also involved in the enhancement of the Fe stress response mechanism in soybean and tomato (88,90) via a specific role in root reduction of Fe^{3+} as well as H^+ and reductant release. In muskmelon (*Cucumis melo* L.) roots, adequate K levels were essential only to maximize the rate of Fe^{3+} reduction (89).

Iron Deficiency Chlorosis

Iron deficiency occurs on neutral and alkaline soils because of ferric iron insolubility. "Iron-efficient" plants have the ability to respond to Fe deficiency stress by inducing biochemical reactions that release Fe in an available and usable form (110). Three different Fe acquisition strategies (strategies I, II, and III) have been identified (92–94,113). The strategy I response involves an inducible Fe deficiency stress mechanism (91,95) characterized by (a) enhanced reduction of Fe^{3+} to Fe^{2+} at root plasmalemma (96), (b) enhanced H^+ release that lowers rhizosphere pH and escalates ferric solubilization or reduction of Fe^{3+} to Fe^{2+} (97), (c) enhanced reductant (such as organic acids, particularly citric acid) released into the xylem, which chelated and transported Fe to plant tops (91,98), and (d) released reducing compounds such as phenolics by roots that maintained Fe^{2+}.

Dicotyledonous plants with strategy I Fe nutrition induce a cell surface reduction system in response to Fe deficiency stress. This response may be accompanied by a release of protons and reductants that alter chemical conditions in the rhizosphere and increase Fe solubility (94). Extracellular chelate reduction involves an inducible transmembrane reductase that utilizes a cytosolic electron donor to release Fe from chelates on the root surface (113). This indirect shuttle mechanism (112) involves an initial reduction of Fe^{3+} to Fe^{2+}, a reoxidation back to Fe^{3+}, and a precipitation in the cell wall as an extracellular Fe pool (115). Iron transport may occur via diffusion to an ion channel (pore) or Fe carrier protein (pump). Synthetic chelates such as ethylenediaminetetraacetate (EDTA) are involved in the efficient operation of this mechanism in nutrient solution studies (112). Potassium has a specific role in controlling the release of H^+ and reductants via enzyme stimulation that reduces Fe^{3+} to Fe^{2+} (136,137). Iron uptake efficiency has also been linked to a greater capacity to acidify the internal medium (by negativization of the transmembrane electrical potential through the activation of H^+ extrusion mechanisms at the plasmalemma), to a higher (hyperpolarization) transmembrane electrical potential difference (-145 mV vs. -105 mV), and to a higher H^+-ATPase activity ($+30\%$) (135). This mechanism is controlled by the concentration of Fe^{2+} in the cytoplasm.

The strategy II response involves release of Fe-solubilizing substances (biosynthetic chelates or phytosiderophores, which are composed of low molecular weight nonproteinaceous amino acids, particularly avenic and mugineic acid) by grass roots that enhance Fe solubility and availability (99–107). Iron phytosiderophores might be transported across the plasma membrane without involvement of the Fe^{3+} to Fe^{2+} reduction step by a plasma membrane–bound reductase (93). The efficacy of phytosidero-phores depends on their concentrations produced in various soils, their chelation properties with competing metals (copper, zinc, manganese), their resistance to microbial degradation, and their resistance to interaction from other chelators or Fe-absorbing microbes (112). Biosynthetic chelates are related to nicotinamine, a compound functioning in Fe metabolism and internal Fe transport (360). These phytosiderophores are secreted into the rhizosphere and are eventually transported by a specific uptake mechanism on the root surface (94). The Fe-efficient plant may have the ability to continuously reduce Fe^{3+} in its roots and release phytosiderophores to solubilize Fe to make it more readily available for uptake by the roots (134).

Gramineae species have been categorized by their innate ability to produce varying quantities of phytosiderophores: barley > wheat > oat > rye >> maize >> sorghum > rice (107). Except for rice, the C_3 (barley, wheat, oat, and rye) and C_4 (maize, sorghum, and millet) grasses have evolved in oxidized soil environments (99). The ability to produce phytosiderophores could potentially relate to their C_3 or C_4 status, since C_3 species produce more under Fe deficiency stress than C_4 species (99). However, maize has an ability to utilize ferrated microbial siderophores (108, 112) and nonspecific Fe^{3+} phytosiderophores (109) as Fe sources, indicating a variety of Fe availability avenues (99, 111). Iron efficient species perhaps have several active factors (high release capability, high uptake affinity, highly effective phytosiderophore structures in sequestering Fe, and an affinity for ferrated microbial siderophores), whereas less efficient species lack one or more factors (99).

Strategy III (113) involves plant uptake, transport, and use of Fe sources from microbial siderophores for Fe-efficient species involving both previously discussed strategy groups. The mechanisms by which plants acquire Fe from siderophores in soils is not understood. Indirect processes such as passive diffusion, extracellular reduction, and chelate degradation may contribute to root uptake of Fe mobilized by microbial siderophores (112). The microbial siderophore transport system in oat is induced by Fe deficiency stress, is dependent on metabolic energy, is saturable, and functions within physiological concentrations produced and used by microbes (112). Ferrioxamine B (FOB) and rhodotorulic acid (RA) are better utilized by plants than ferrichrome, ferrichrome A, or coprogen, and all siderophores are apparently utilized by a common transport system (112). FOB utilization involves a receptor-mediated transport system that is dependent on metabolic energy provided by respiration but is not coupled to the electronegative potential generated by the plasmalemma ATPase (112). Plant nitrate reductase catalyzes reductive Fe release from microbial siderophores using $NADH^+$ as an electron source. This enzyme is thought to function in reduction of Fe^{3+} sideophores taken up by diffusion or by a siderophore transporter system (112,114). At least one or more reductases can utilize siderophores as an electron acceptor and plants can remove Fe either at the cell surface or from the cytosol (112).

Until the transport mechanisms are completely identified and the genes encoding for Fe deficiency stress-induced cell membrane transport systems (receptors, membrane transporters and reductases, siderophore and phytosiderophore biosynthesis and degradation enzymes, and regulatory enzymes) are better understood, little progress will be made in genetic and plant breeding programs. Breeding for improved Fe utilization in plants has advanced methodically but slowly (116), and only interdisciplinary efforts will lead to the understanding necessary to escalate the program (57).

Toxicities

Mechanisms governing toxic metal tolerance can be grouped into two major categories based on the site of metal detoxification and immobilization, or on the root location for adaptation to toxic nutrient stress: apoplasm (external) or symplasm (internal) (117). The negative response of plants to an excessively available nutrient (tolerance) may be due to both external and internal mechanisms. The plasma membrane and cell wall are key sites for exclusionary responses via selective permeability or polymerization. Formation of pH barriers (chelating ligands or mucilage) at the root–soil interface are involved. Internal chelation by organic acids (carboxylic acid, citrate) or metal-binding proteins, vacuolar compartmentation, and formation of metal toxicity–tolerant enzymes can detoxify toxic metals and allow growth and development to proceed. Tolerance generally involves more than one mechanism.

Aluminum toxicity tolerance in plants has been associated with both physiological characteristics [rhizosphere pH changes, organic acid content, root CEC, nonmetabolic site compartmentation, root phosphatase activity, drought tolerance (118), and herbicide tolerance (119)] as well as with nutritional characteristics (P and Fe use efficiency, Si concentration, NH_4^+, NO_3^-, Ca^{2+}, and Mg^{2+} uptake and transport) (118, 126,352). Aluminum-tolerant plants have mechanisms that encompass active exclusion (127), lower the CEC of cell wall material (128,129), which alters membrane selectivity and nutrient absorption (117), and increase ability to maintain normal ion fluxes and membrane potentials across the plasmalemma of root cells in the presence of Al (130), retain a higher affinity for Mg^{2+} by transport proteins in presence of Al (131), and emit high amounts of citric acid (132,133,140,353) or transaconitic and malic acids (199), which could chelate positively charged Al externally in the rhizosphere.

Plant roots may respond to environmental stress signals by way of a common regulatory system (355). Stimulus–response Al tolerance mechanisms may involve (a) signal perception by the root cap where it is translated into growth or regulatory responses involving the transduction of signals between interacting cell populations, (b) Ca^{2+} as a secondary messenger system via coupling of the primary signal/response system with a number of physiological reactions, (c) linkage between Ca^{2+} cap cell differentiation and mucilaginous cap secretions, (d) role of cap secretions and peripheral cap Golgi apparatus in root physiology, (e) impact of cap secretions (as influenced by Al, microgravity of space, and depletion of apoplastic Ca^{2+} resources) and linkage between the peripheral cap cells and the polarity of central cap cells, and (f) growth responses that reflect the action of the endogenous cap inhibitor (355).

Manganese toxicity tolerance has been linked to mechanisms involved in the oxidizing power of roots (120), silicon alleviation (121,357), Mn absorption and translocation rates, nonmetabolic site entrapment (118), and high internal tolerance (125). Manganese and Fe uptake are positively interrelated and both ions are mobilized by similar root processes in the rhizosphere (138). Dicotyledonous plants primarily utilize a chemical reduction mechanism, whereas monocotyledonous plants utilize $Mn^{2+}-Fe^{3+}-$organic complexation to mobilize soil Mn and Fe. Boron toxicity tolerance functions via an exclusion mechanism (122) or is governed by the amount of B accumulation rather than internal distribution (139). Copper toxicity tolerance has been traced to metallothionein protein binding (123). Zinc toxicity tolerance was due to metal inactivation at the cell wall (124).

Salinities

Plants can be grouped into two categories based on their response to salinity environments. Halophytes are salt-tolerant plants native to saline habitats. Dicotyledonous halophytes generally have stimulated growth as salinity increases up to approximately 15,000 mg/L (250 mM) sodium chloride, whereas monocotyledonous halophytes grow poorly at salt concentrations exceeding 10,000 mg/L (170 mM) sodium chloride (141). Some halophytes take up excessive levels of Na and Cl at high rates, accumulate these elements in their leaves, and use these salts for osmotic adjustment to the low soil water potential (141,159). Both low soil osmotic potentials (due to dissolved salts) and low soil matrix potential (reduced soil water content) cause lower water potentials in plants (164). Ions are normally compartmentalized in the leaf cell vacuoles, which maintain low salt concentrations in the cytoplasm and in organelles where enzymatic and metabolic functions are not compromised. Restrictive transport of Na^+ to the shoot may be part of the tolerance mechanism (159,160). Osmotic adjustment in the cytoplasm is accomplished with enzymatically and metabolically compatible organic solutes such as glycinebetaine and proline (nitrogenous compounds) or sorbitol (sugar alcohol) (141). Potassium is maintained in the cytoplasm at about 4000 mg/L (100 mM). Calcium utilization efficiency may also be involved in salt tolerance (165). In addition, the tonoplast (membrane separating the vacuole from the cytoplasm) aids osmotic adjustment by regulating transport mechanisms for handling steep solute gradients and for maintaining turgor pressures necessary for sustained growth under salt stress.

Glycophytes respond to salinity in the range from very sensitive to moderately tolerant. Tolerant plant species include bermudagrass (*Cynodon dactylon* L. Pers.), sugarbeet (*Beta vulgaris* L.), wheatgrass (*Agropyron* spp.), cotton (*Gossypium hirsutum* L.), and barley (*Hordeum vulgare* L.) (143). Adaptive mechanism strategies for these tolerant species may include avoidance (unlikely), exclusion (low permeability of external toxic salts at the root plasmalemma, active efflux pumps, compartmentalization), or physiological tolerance involving simultaneous maintenance of favorable water relations via regulation of osmotic adjustment by organic and inorganic solutes, tolerance of cells to toxic ions, inorganic ion regulation and maintenance of essential nutrients, and efficient use of metabolic energy (bioenergetics) (141,144).

Salinity stress is first sensed in the roots, but osmotic adjustment, ion toxicity, and growth inhibition occur in the shoot. Salt in the root zone decreases soil solution osmotic potential. The plant reaction is turgor loss, followed by osmotic adjustment (reduction in the internal osmotic potential to compensate for the lower external osmotic potential) (143). Turgor reduction detrimentally affects cell division and elongation, as well as stomatal closure in salt-sensitive plants. Gas exchange (photosynthesis and transpiration) can be reduced and growth inhibition can result (143,161).

Nutritional effects of salinity include (a) direct toxicity due to accumulation of Na and Cl, and (b) nutrient imbalances (and interactions) due to excess element accumulation (143). Sodium is the predominate cation in sorghum roots (146), while chloride is preferentially accumulated in leaf sheaths (147). Partitioning of Cl is ion-specific and results from a combination of the sheath tissue's ability to sequester Cl at high concentrations and the blade tissue's ability to regulate Cl concentrations at moderate levels. Quaternary ammonium compounds such as glycinebetaine are preferentially partitioned into leaf blade tissues of salt-stressed sorghum leaves (148). The reciprocal Cl sheath and glycinebetaine leaf blade partitioning may allow leaf osmotic adjustment to lower water potentials associated with high external salt levels. The Cl partitioning into leaf sheaths and midribs may provide a relatively stable ionic environment in the blade tissue by sequestering potentially toxic Cl levels away from primary sites of leaf carbon assimilation (147,163). Moderate-to-severe salt stress (-0.4 MPa or higher) is required to induce significant glycinebetaine concentration increases (148).

An Na^+ exclusion mechanism operates in sorghum (145) and functions in the root rhizodermis cells (163). Maize cultivar differences in Na^+ exclusion at the root surface results from a lower passive Na^+ permeability of the plasmalemma in epidermal and cortical cells (166). Sodium may be excluded by efflux pumps (extrusion from the roots) (149), removal from the xylem before the cation reaches the leaves (150), or retranslocation from the leaves back to the root (151). Preference for K over high Na concentrations provides salt tolerance (152). The ability to maintain high intracellular leaf K levels under stress is involved in salinity tolerance (144,145). Sorghum primarily uses inorganic rather than organic solutes for osmotic adjustments (146). Sucrose was the solute used for osmotic adjustment under mild saline stress, whereas proline was involved in adjustment under severe stress conditions. Proline concentrations did not increase until the monovalent cation cell concentration crossed a threshold value of 200 μmol (per g fresh wt) while sucrose concentrations began increasing when total monovalent cations exceeded 100 μmol (per g fresh wt) (146). A sorbitol increase or pinitol accumulation has been found in plants subjected to NaCl stress and could be involved in osmoregulation (162). Cytokinin production may form the innate trigger mechanism for growth regulation and salt resistance reactions (153). Organ-specific changes occur in the enzymatic (arginine decarboxylase and ornithine decarboxylase) activities involved in polyamine biosynthesis as a result of NaCl stress in mungbean [*Vigna radiata* (L.) Wilczek] (154). Precursors are incorporated into polyamines in roots and putrescine accumulates in the leaves of salt-stressed plants. Polyamines may stabilize membrane function and structure associated with plant response to salt stress (154).

Plant adaptation to salt may be associated with osmotin (cationic protein) accumulation in vacuolar bodies, cytoplasm, plasma membrane, and tonoplast vesicles (155). Osmotin synthesis is induced by abscisic acid (ABA) and the protein may be associated with resumption of growth following initial exposure to salt stress (156). This stable alteration in gene expression of osmotin accumulation probably affects salt tolerance (155). The adaptive response to high salinity may result from a modulation of genome expression during extended exposure to nonlethal NaCl concentrations (157,158). Concomitantly with this ability to grow in high salinity, adaptation also comprises a developed capacity to regulate internal Na and Cl concentrations. ABA enhances growth during the period of initial exposure to nonlethal salinity levels and escalates the adaptation to high salinity (193,195).

The ionic components of salt stress interact with cell membrane functioning. K^+-Na^+ selectivity in the root is affected by Ca^{2+}-salinity interactions (141,165). Anion effects (chloride and sulfate) are also important in plant responsiveness to salinity. Sorghum is more sensitive to sulfate than chloride salinity (141).

Basic tolerance mechanisms to salt stress may involve (159,163) (a) efficient Na^+ exclusion or compartmentation and retranslocation, (b) high K^+ concentrations, (c) tolerance to tissue dehydration, (d) compatible solute synthesis (betaine, proline), and (e) preferential Cl^- partitioning.

Drought Stress

Uneven distribution of water during critical growth stages is a serious deterrent to crop production. Plant survival under water stress is achieved by cryptobiosis through such survival devices as leaf area reduction (senescence) and depression of metabolic activity (332). Drought resistance encompasses two principal components: the ability of the roots to exploit available soil water and the ability of shoots to restrict water loss through control of transpiration during high atmospheric evaporative demand (314,315). The most drought-resistance genotypes generally have greater root mass, root volume, and higher root/shoot ratios than drought-sensitive types (321). Roots act as sensors of soil water deficits. Plant growth regulators transfer signals from the root to the shoot indicating soil water deficits (322).

Drought tolerance in plants can be grouped into several major categories (54,259,323,331,333):

1. *Dehydration avoidance*: Physiological factors that maintain a relatively high leaf water potential at a given level of soil moisture stress, including osmoregulation, leaf senescence (green leaf retention), canopy temperature, epicuticular leaf wax load, organ pubescence, leaf orientation, and stomatal density/conductance.
2. *Drought escape*: Mechanisms that involve rapid root biomass distribution patterns and plasticity, small leaf size and leaf area per plant, short growth duration, and traditionally low-yielding genotypes. Plants in this category generally complete their life cycle via rapid phenological development and encumber developmental plasticity prior to serious water deficits (346).
3. *Dehydration tolerance*: Tissue and cellular factors that support cellular life functions in conjunction with reduced tissue water potential

under stress, such as cell membrane stability and stem reserve mobilization.

4. *Drought adaptation*: Factors such as organ growth rate, recovery after rehydration, and functioning of photosynthetic system components and enzyme systems.

No single drought tolerance trait is predictive of plant response to stress. Multiple physiological selection criteria are required because of the complex nature of the tolerance mechanisms (Table 2). Plant response to stress certainly involves more than one attribute.

Osmoregulation and leaf senescence are apparently more important factors than stomatal closure (325). Phaseic acid and total ABA metabolic concentrations are negatively correlated to percentage reduction in grain yield under stress, implying a higher ABA conversion efficiency in drought-resistant than sensitive genotypes (318). Drought-avoiding genotypes generally have early maturity, maintain high turgor, and have low leaf ABA concentrations (326). Drought-tolerance genotypes generally have late maturity and high leaf ABA concentrations for turgor pressure regulation (318). Root cytokinin (CK) production is markedly reduced, which contributes to increased senescence associated with water stress (332). Water stress may cause a general decline in mitochondrial oxidative properties, diminishing phosphorylative efficiency and resulting in improper adjustment of cell osmolarity (327). Cell osmolarity is a primary factor controlling plant water relations and is often coupled with other plant adaptations: high root/shoot ratios, small cell size, increased succulence, altered stomatal behavior, and thick cuticles (339). Poor development of osmotically stressed plants could be explained by a high level of mitochondrial cyanide and rotenone resistance leading to excessive use of carbohydrates for maintenance respiration and diminished growth (327). The effect of water stress could be a solubilization of cytochrome c from the inner mitochondrial membrane and a permeabilization of the membranes that contribute to a loss of proton gradients, decreased mitochondrial integrity, and subsequent impairment of phosphorylative efficiency (327).

ABA-induced stomatal closure is one integrated response to short water supply in plants. Farnesol, as a regulator of transpiration, is the agent responsible for altering the permeability of chloroplast envelop membranes, allowing the release of ABA into the cytoplasm (340,341). During stomatal opening, guard cells exchange K^+ from the surrounding epidermis for H^+ within its plasmalemma. ABA blocks the H^+ efflux from the guard cells (340). ABA also increases the hydraulic conductivity of the root to water and increases the rate of xylem ion influx. ABA inhibits the active proton flux associated with cotransport of sucrose in phloem loading. The decreased sugar transport out of the leaf enables mesophyll cells to retain solutes for turgor maintenance (340,341). Genotypic variation of stress-induced ethylene production is not related to drought sensitivity or resistance (342).

Plant water loss can be divided into three distinct stages (306,328): stage I occurs when soil water is freely available and plant stomatal conductance of water vapor loss is at a maximum; stage II begins when the water uptake rate cannot match potential transpiration and stomatal conductance decreases to maintain the plant water balance; stage III is triggered by continued soil drying to the point where the stomata cannot retard the leaf water loss rate sufficiently to offset the low soil water uptake rate. Stomata reach minimum aperture and the epidermal transpiration rate is regulated by the leaf epidermis vapor conductance in conjunction with the differential vapor pressure gradient between the leaves and the atmosphere. As water is lost, leaves dehydrate and senesce. Therefore, crop survival is dependent on the relative water content at leaf senescence and the epidermal transpiration rate. A combination of these two plasticity traits may be useful in selecting genotypes with an enhanced ability to survive severe water stress (303,306).

When stomata attain minimum aperture, peristomal transpiration (from teichodes, or holes in the external cell walls of guard cells and subsidiary cells) (329) continues and can be a major component of epidermal transpiration as well as an important environmental response mechanism in leaves (306). Genetic variation for epidermal conductance is related to stomatal density (306). When epidermal conductance becomes less than 200 mmol/m^2/sec, water evaporates from epidermal cell walls after diffusing through a waxy cuticle. Consequently, epidermal conductance is closely linked to epicuticular wax load (307).

Transpiration ratios under stress are lower for resistance than for sensitive genotypes (259). This capability for greater osmotic adjustment is also coupled with a lower carbon exchange rate per unit leaf area (259,311). Other factors such as drought escape, phenotypic plasticity, root characteristics, and heat tolerance may be mechanistically operative in some plants (330).

A rapidly penetrating and extensive network of primary and secondary roots is essential for stable yields of plants grown in water-stressed environments (331,333–335,337–338). The primary root must be capable of rapid proliferation in both depth and surface area, since frequent and severe drying of the seedbed may restrict development of lateral and crown roots (334). Deep soil penetration associated with a greater partition of dry matter to roots enhances drought tolerance (334). Water flow resistance and conservative water behavior can be accomplished by reducing the number of seminal roots during the vegetative stage,

Table 2 Possible Selection Criteria for Various Plant Efficiency or Tolerance/Resistance Traits

Trait	Selection criteria	Correlation	Source
N uptake efficiency	Nitrate reductase activity	Positive association with grain yield, grain N; paternal contribution; heterotic effect	58
N uptake efficiency	High lateral/primary root ratio	Inheritance of branded roots contributed to hybrid vigor	63
N metabolism	Glutamate dehydrogenase	Oxidizes glutamate; ensures C skeletons for tricarboxylic acid cycle; regulatory function	225
NO_3^- uptake/transport	Arginine "residue"	Essential for uptake by NO_3^- carrier in plant roots	226
Salinity tolerance	Cl^- sheath partitioning	Leaf osmoticum adjustment; intracellular compartmentation	147
	Glycinebetaine	Organic cytoplasmic osmoticum: balance sugar accumulation in the vacuole, increase cell permeability	148
	Sucrose, proline	Organic solutes for osmotic adjustment	146
	Cytokinin concentration (benzyladenine)	Decreased levels is first response to salt stress	153
	Putrescine	Accumulation stabilizes membrane function	154
	Proline, asparagine, pinitol	Involved in osmoregulatory and tolerance mechanisms	356
Drought-salinity tolerance	Abscisic acid accumulation	Escalates adaptation and enhances growth	193,195
Phosphorus efficiency	Sucrose, glucose, fructose	Increased carbohydrate source for enhanced uptake	74
Phosphorus use efficiency	Excreted phosphate starvation–inducible acid phosphatase enzyme	Pleiotropic effects suggests modification of regulatory apparatus controlling coordinated changes in expression of a multigene system	351
Potassium deficiency	Pyruvate kinase	Detects low-K status	205
Iron efficiency	Ferrioxamine B, rhodotorubic acid	Siderophore production to release Fe^{3+} for plant utilization	112
	Peroxidase	Assesses FE status in diverse plants	205
	O-Phenanthroline	Reactive with Fe^{2+} and Fe^{3+}	206
Aluminum toxicity	Citric acid; *trans*-aconitic and malic acid	Chelates Al^{3+}	132,133,140, 199

Trait	Measurement	Criterion / Association	References
	tion rate		
Copper toxicity	Metallothionein	Protein binding	123
Copper deficiency	Ascorbic acid oxidase	Cu deficiency causes reduced enzyme activity	205
Zinc deficiency	Carbonic acid anhydrase	Enzyme's activity detects Zn deficiency	205
Drought tolerance			
1. Dehydration avoidance	Leaf rolling, wilting	Ability to recover turgor under stress vs. injury	230,231,242
	Leaf firing; green leaf retention	Leaf senescence negatively correlated with yield	232–235,277, 303,304
	Canopy temperature (infrared thermometer)	Negative correlation between temperature and yield under stress	230,234, 236–242,305
	Leaf water retention	Greater retention, higher the yield	243–251
	Osmoregulation	Turgor maintenance positively associated with yield	258–263, 265,344
	Proline	Increased deposition maintains root elongation at low water potentials	357
	Stomatal conductance (leaf diffusive resistance)	High conductance positively associated with yield	264,266–268, 278,279
	Stomatal density	Low epidermal transpiration indicative of survivability	306,308
Epicuticular wax load		Association between water retention and cuticular transpiration	245,249
	Glaucousness	Positive association with yield under stress	269–274,300
	Leaf erectness	Paraheliotropic leaf movement (angle or orientation) minimizes solar radiation reception, provides better leaf water status under stress	280,281 282–284
2. Drought escape	Root distribution pattern	Faster and deep root penetration and more extensive development is associated with stability	
	Deep sand, topographical gradient		252
	Herbicide zone		253
	Rainout shelters		253
	Short growth duration	Early maturity capitalizes on available water	254–257

Table 2 Continued

Trait	Selection criteria	Correlation	Source
3. Dehydration tolerance			
	Membrane integrity	Leakage from dehydrated tissue is related to susceptibility	54
	Desiccation	Water retention, leaf senescence, and survivability are associated with postanthesis stress	285,292,323
	Stem reserve mobilization	Positive relationship between stem dry matter loss after anthesis and grain production capacity	286–291
4. Drought adaptation			
	Survival technique: exposure to successive drought cycles	Effective screening method	317
	Carbon assimilation rate	Correlated with whole-plant water use efficiency	311,312
	Awnness	Water use–efficient organ, higher transpiration ratio than flag leaf or glumes, selective advantage	54,293,294
	Leaf-stem-silk growth rate	Rapid growth under stress is positive trait	302
	Leaf hydration	Positively correlated with osmotic potential and adjustment	310
5. Field Selection			
	Line source sprinkler irrigation	Accessions subjected to water stress gradient that increases with distance from water source	295–297
	Saddle effect	Border effect/plant density vs. moisture competition	233
	Charcoal pit technique	Combines interactive effects of high soil temperature and moisture stress	319,320
	Aerial infrared photography	Nonavoiding genotypes are detected	324
	Selection index: infrared thermometry, growth under stress, leaf firing, synchro-	Greatest selection differential and best association with yield under stress was canopy temperature	237

Selection index: infrared thermometry, chemical desiccation, yield differential, awnness	Better osmotic adjustment, large amount of awns, wax, glaucous leaves, early maturity	34,355
Selection index: canopy temperature, grain yield, foliar senescence rate, male–female flowering interval, leaf-stem extension rate	Improved drought adaptive traits, no change in maturity	301
Selection index: canopy temperature, vapor pressure deficit	Warmer genotypes (increased leaf temperature is related to decreased transpirational cooling via stomatal closure) and those less sensitive to VPD (vapor pressure deficit) changes produce panicles under stress	313
Selection index: growth rate, percentage moisture loss, base temperature, grain yield	Components of grain yield stability	314,315
Selection index: thick leaves, high electrolyte levels, high chlorophyll content, few stomata, vigorous root growth	Promotes greater water uptake	316
Selection index: low free ABA, low free and conjugated IAA concentrations in drought-stressed plants, seed number per panicle, grain yield	Low drought-induced hormonal concentrations improves drought resistance qualities	318
Selection index: stomatal conductance and CO_2 assimilation	Useful in vegetative stage for preanthesis drought resistance assessement	358

thereby enhancing water availability during heading and conserving water for later growth stages (335). If forage rather than grain production is the objective, this approach would not be practical. The plant should have a good distribution of roots in the top 60 cm of the soil profile (tolerance attribute) coupled with capability for deeper rooting (90+ cm) (avoidance attribute). Root plasticity is the key component in a plant's adaptivity to variable water-stressed environments.

Plant response to water stress can be categorized at two major growth stages: preanthesis (vegetative stage) and postanthesis (reproductive stage) (54,233,343–345). These response phases coincide with major physiological and biochemical changes in the plant.

Preanthesis drought stress symptoms may include leaf rolling, uncharacteristic leaf erectness, leaf bleaching, leaf tip and margin burn, delayed flowering, "saddle effect," poor panicle/head exsertion, floret abortion or sterility, and smaller reproductive organs for harvest (233). Osmotic adjustment during preanthesis stress minimizes the reduction in grain yield by maintaining grain number (343,346). Higher levels (15–40%) of osmotic adjustment contributed to this higher seed number and subsequently higher grain yields (range 15–34%; $\bar{x} = 23\%$) via greater distribution indices and larger harvest indices (343). Plants had better panicle exsertion, higher dry matter production, and more efficient water use during stress (344). Water stress prior to anthesis reduced yield more than postanthesis stress of the same intensity (344). The panicle differentiation process was more sensitive to environmental stress than was the leaf expansion process or the ability to accumulate dry matter (346). Postanthesis drought stress symptoms may include premature senescence, stalk lodging, predisposition to stalk rots or other diseases, and reduction in size of harvestable product (233). Higher (range 14–39%; $\bar{x} = 21\%$) relative yields of postanthesis stressed plants was not due to more effective osmotic adjustment after anthesis but rather to grain yield being less sensitive to the postanthesis stress. Osmotic adjustment was equally effective in both phenological stages. When stress occurred after anthesis, higher harvest indices and grain yields were associated with better retranslocation of dry matter produced before anthesis (344). Grain yield under severe moisture stress has also been associated with maintenance of leaf, stem, and silk extension, canopy temperature (transpiration) at anthesis, and maintenance of green leaf area (nonsenescence) during grain filling (301).

The extent to which osmotic adjustment minimizes the reduction in grain yield caused by pre- or postanthesis water stress depends on (344):

1. Degree of osmotic adjustment induced by a particular level of stress

2. Sensitivity of each phenological stage to water stress when that stress occurred
3. Mode of action for osmotic adjustment at the particular phenological stage

Irrespective of osmotic adjustment level, preanthesis water-stressed plant yields were consistently lower than postanthesis stressed yields. Genotypes selected for high osmotic adjustment produced 21–23% higher yields when water stress reduced yields in both developmental stages than low osmotic adjusting types (344).

Photosynthetic rate and leaf conductance have been primarily responsible for dry matter accumulation and water loss via transpiration, even though neither process was highly correlated with any specific growth or developmental process (346). High photosynthetic rates were not directly related to high leaf conductance. Consequently, plants capable of producing more dry matter per unit water transpired can be developed (346). Water use efficiency can be improved under surplus or adequate conditions and available water supply could be extended under deficit conditions.

If a plant cannot maintain a high tissue water status to avoid desiccation, it must tolerate drought at low leaf water potentials via turgor pressure maintenance or cellular integrity maintenance (346). Turgor maintenance can be accomplished through a lowering of solute potential (osmotic adjustment). Regression of tissue solute potential against relative water content (change in slope with increasing dehydration) is a good indicator of osmotic adjustment capability (346). The greatest degree of osmotic adjustment occurred in nonstressed leaves. During grain filling, nonstressed plants have greater degrees of osmotic adjustment than stressed plants. A drought-tolerant genotype grown under stress conditions is capable of lowering its solute potential to a greater extent at all relative water contents than its nonstressed, sensitive counterpart. Concurrently, tolerant lines can maintain higher than average photosynthetic rates under stress during grain filling. Since photosynthate has been used in the osmotic adjustment process, photosynthetic stability under stress conditions may be critically important in the response mechanism (346).

Senescing genotypes depended more on preanthesis assimilates than nonsenescing genotypes. Stressed plants relied more heavily on preanthesis assimilates than did nonstressed plants. Nonsenescing hybrids produced more daily assimilate due to greater leaf area duration (303) and greater photosynthetic rate per unit leaf area (346). Consequently, breeding programs should incorporate water conservation mechanisms such as photosynthesis per unit conductance with tolerance mechanisms such as root pattern plasticity and osmotic adjustment to engineer improvements in drought stress tolerance (233,301,331,335,337,346). Several selection indices are revealed in Table 2.

Several components of drought tolerance and their gene control are presented in Table 3. Because of the complexity of the physiologically oriented control mechanisms, the lack of knowledge concerning functions of the mechanisms, and the lack of definitive marker-assisted evaluation techniques that are conducive to breeding efforts, little progress has been made in the improvement of drought-tolerant plants. This fact is also apparent in summarizing some of the breeding techniques that have been used in developmental programs (Table 4).

Table 3 Genetic Control of Plant Stress Response Traits

Trait	Gene action	Source
Nitrate reductase activity	Single, dominance	39,42
	Polygenic	39,41,
	Two genes	224
Phosphate uptake	Multigenic	351
Phosphorus efficiency	Multigenetic: additive, dominance, epistasis	167,17
	Multigenetic, 3 chromosomes	173
	Multigenetic, additive	176
Phosphorus assimilation	Single, partially dominance gene	348
Potassium efficiency	Predominantly additive, but dominance and epistasis important in certain crosses	167
	Single gene, recessive	167,17
Iron efficiency	Single locus, dominant	168,21 219
	Major gene governs uptake, additional genes with additive gene action	169,17
	Major + minor genes	171
	Polygenic, additive	179
	Two complementary dominant genes	180
	Quantitative	170,21
Boron efficiency	Single gene	172
Cooper efficiency	Single gene, chromosome 5 locus	173
Aluminum toxicity	Single gene, dominant	181
	2–3 major dominant genes plus several modifiers, chromosome 5D	182
	2 dominance genes, additive	183
	1 dominant, 1 recessive gene, nonadditive	183
	Multiple alleles, 1 locus	184
	Quantitative inheritance	185
	1–3 genes, additive with varying degrees of dominance	188,18

ble 3   Continued

ait	Gene action	Source
anganese toxicity	Multigenetic, maternal effects	174
	Additive, no maternal effects	175
	Additive, little or no dominance	186
	1–4 genes	187
	Multigenic, additive	56
lt/sodic toxicity	Additive gene effects	190,192
	Additive and dominance effects	198
	Overdominance, transgressive segregation for resistance	191
lt tolerance	Paternal inheritance, several nonallelic genes, dominant or overdominant expression	347
⁻ exclusion	Single dominance gene	196
	Quantitative inheritance	197
scisic acid level	Small number major genes	193
ycinebetaine level	Predominantly additive	216
rought tolerance		
Leaf diffusive resistance (stomatal conductance)	Additive and dominance effects	268
Epicuticular wax load	Single dominant gene governs waxy bloom	269
	Two different loci govern bloomlessness	270
	Homozygous recessive alleles at 3 loci govern sparse bloom	270
Leaf glossiness (epicuticular wax plus wetting characteristics)	Simple recessive	271
	Series of 10 genes	271
	56 gene loci	272
Glaucousness (dense pubescence)	Series of dominant or recessive allelic genes on either of 2 genomes	273
Cuticular transpiration (leaf water retention)	Simple genetic control, dominant genes	245–249, 274
Seedling recovery after drought stress	1–3 genes, partial to complete dominance	275
Flower/pod sustainability under high heat and drought stress	Single dominant gene	276
	2 genes with epistatic effects	276
Carbon assimilation rate	Maternal effects	312
Root-pulling resistance	Transgressive segregation, dominant and additive gene action	298
Desiccation tolerance	Dominance	299

Table 4 Breeding Techniques for Stress Tolerance Traits

Characteristic	Breeding method	Sourc
Total residual root N concentration	Phenotypic recurrent selection	200
P efficiency	Inbred backcross line method, nutrient solution evaluation before field testing	201
	Suspension cell cultures, somaclonal variation	207
	In vitro selection–phosphate starvation	351
Fe deficiency chlorosis	Population improvement, recurrent selection (S1 testing)	170,208-
	Mass, pedigree (single- or three-way cross), backcross	116,2(
	Recurrent phenotypic selection	116
Al toxicity	Divergent recurrent selection	202
Acid soil tolerance complex (50% Al saturation)	Pedigree backcross selection	203
	Genetic male-sterile–facilitated random mating and recurrent selection	188,2(
	Tissue culture, somaclonal variation, mature embryo explants	204
Salt tolerance	Cell suspension, somaclonal variation, cotyledon explants	211
	Fl anther culture	212
	Agar and suspension culture, somaclonal variation, ovule explants	213
	Recurrent selection	214
Water/salt stress (using glycinebetaine level)	Isopopulation approach	216
Heat tolerance	Dual-step selection procedure; exposure of seedlings to high temperature followed by a growth period in cool temperatures	350
Drought tolerance		
Root-pulling resistance	Line "mean" selection	298,29
Index: rate of leaf/ stem extension, male–female flowering interval, foliar senescence rate, grain yield, canopy temperature	Recurrent full sib progeny selection	301
Osmotic adjustment	Divergent selection	309
Index: early maturity, disease resistance, hardiness, stomatal behavior (transpiration)	Composite cross, male-sterile facilitated populations	332

GENE CONTROL OF STRESS TOLERANCE

Improvement in a plant's ability to cope with edaphic stress environments depends on an understanding of the biochemical and physiological aspects of gene action (epigenetics) (194). Because of the complexity of the plant genome, the ecological and physiological processes related to survival and production under stress conditions, and the quantitative inheritance of many of these processes, little breeding progress has been made. This complex phenomenon is reflected by the data in Table 3. In general, a comparison of each trait with documentation on number of genes and subsequent gene action emphasizes the confusion encumbered in breeding programs. Appropriate knowledge about mechanisms, specific physiological/biochemical markers for various traits, and appropriate screening/evaluation technology to select for variability (assuming it will become available) is needed to escalate advances in stress tolerance.

SELECTION CRITERIA FOR STRESS TOLERANCE TRAITS

Breeders are constantly confronted by the dilemma of what selection criteria to use in breeding programs. They must have genetic variability for the targeted trait; diagnostic criteria must be available for easy identification of large segregating populations. Since stress response varies with stage of development, the problem is compounded. Ion and water balances, membrane integrity, and hormonal balances all interact to influence growth and development of plants grown under stress. Multiple selection criteria at various growth stages are ultimately utilized and must parallel selection of characteristics governing productivity. Several possible selection criteria are listed in Table 2. The need for standardized tolerant and susceptible checks for specific traits is critical to the documentation of advancements (56).

BREEDING TECHNIQUES USED IN STRESS TOLERANCE SELECTION PROGRAMS

Many stress tolerance traits are polygenic, being governed by levels of organization from subcellular to organismic (215). Imposition of typical environmental interactions on genetic differences has led to slow progress in actual breeding programs. The mechanisms are complicated and not well understood. Highly correlated physiological or biochemical markers are critically needed if substantial progress is to be made. Problems in breeding for stress tolerance have been addressed by Shannon (220) and Blum (54).

Data in Table 4 summarize some of the breeding methods attempted in efforts to improve stress tolerance traits. Recurrent selection or cell culture–induced somaclonal variation are predominant methods in many breeding programs. Inherent in any breeding effort to improve stress tolerance per se is a minimization of environmental effects. Use of appropriate standardized tolerant and sensitive checks (56), selection under controlled laboratory conditions and subsequently under differentiating field conditions, and delaying of selection until later generations (such as F5 and beyond) by bulking within families to maximize the number of recombinants will help to increase the correspondence between genotype and phenotype (220) under stress conditions. Recurrent selection programs within families should include replicated evaluations at several locations to minimize interactions between stress and environment. Reconstitution of the best stress-tolerant, advancing-generation parents will eventually lead to quantitative improvement over time (220). If the field stress environment is quite discerning, selection indices of 1–5% may be the rule rather than the exception. Inclusion of good agronomic characteristics (particularly those directly associated with yield components) in the selection program will help to prevent erosion of this critically important attribute.

MULTIPLE ENVIRONMENTAL STRESS INTERACTIONS

Plant responses to multiple field stresses are even more complicated. Physiological adaptations, most of which are genetically controlled, are central to plant stress adaptations. Stressed-induced root growth cessation, nutrient deficiencies or toxicities, drought, high evaporative demand, high root and shoot temperatures, high levels of radiation, low relative humidity, and large leaf canopies may result in shoot growth termination. High ABA levels in conjunction with low indoleacetic acid (IA), gibberellic acid (GA), and CK concentrations may occur under these conditions (222). Drought or salt stress during any growth period causes increased ABA and decreased IAA and CK, slowing growth and possibly detrimentally affecting productivity. The single stress–single process concept of studying these plant adaptation responses may help to narrow the factors associated with understanding these processes, but plants must normally adjust to multiple environmental stresses, such as heat and drought or avoidance and tolerance (221,223). A good treatise on some of the interactive limitations in crop production can be found in Wallace (229). Interdisciplinary research into basic physiological, biochemical, and genetic processes is critically needed to elicit further plant developmental progress in stress tolerance.

REFERENCES

1. J. Levitt (1985). *Plant Cell Environ.*, 8:287.
2. J. Levitt (1990). *Hort. Sci.*, 25:1363.
3. T.C. Hsiao (1973). *Annu. Rev. Plant Physiol.*, 24:519.
4. A.V. Barker and H.A. Mills (1980). *Hort. Rev.*, 2:395.
5. P.B. Cregan and P. Van Berkum (1984). *Theor. Appl. Genet.*, 67:97.
6. A.D.M. Glass (1989). *Hort. Sci.*, 24:559.
7. J.D. Dubois and R.H. Burris (1986). *Plant Soil.*, 93:79.
8. P.H. Harvey (1939). *Genetics*, 24:437.
9. P.L. Minotti, D.C. Williams, and W.A. Jackson (1969). *Crop Sci.*, 9:9.
10. W.L. Pan, W.A. Jackson, and R.H. Moll (1985). *J. Exp. Bot.*, 36:1341.
11. W.L. Pan, W.A. Jackson, and R.H. Moll (1985). *Plant. Physiol.*, 77:560.
12. S.A. Barber (1984). *Soil Nutrient Bioavailability.* John Wiley and Sons, New York.
13. M.B. Hatlitligil, R.A. Olson, and W.A. Compton (1984). *Fertil. Res.*, 5:321.
14. A.D. MacKay and S.A. Barber (1986). *Agron. J.*, 78:699.
15. D.B. Mengel and S.A. Barber (1974). *Agron. J.*, 66:399.
16. A. Oaks (1985). *Ann. Rev. Plant Physiol.*, 36:345.
17. A. Oaks, A. Aslam, and L.L. Boesel (1979). *Plant Physiol.*, 59:391.
18. D.N. Maynard and A.V. Barker (1969). *J. Am. Soc. Hort. Sci.*, 94:235.
19. A.V. Barker (1989). *Hort. Sci.*, 24:584.
20. W.A. Jackson (1978). In: *Nitrogen in the Environment, Vol. 2, Soil-Plant-Nitrogen Relationships* (D.R. Nielsen and J.G. MacDonald, eds.), Academic Press, New York, p. 45.
21. M.A. Morgan, W.A. Jackson, and R.J. Volk (1985). *J. Exp. Bot.*, 36:859.
22. E.G. Beauchamp, L.W. Kannenberg, and R.B. Hunter (1976). *Agron. J.*, 68:418.
23. D.C. Cosgrove, J.B. Jones, Jr., and H.A. Mills (1985). *Hort. Sci.*, 20:427.
24. M.C. Cox, C.O. Qualset, and D.W. Rains (1986). *Crop Sci.*, 26:737.
25. E.J. Kamprath, R.H. Moll, and N. Rodriguez (1982). *Agron. J.*, 72:955.
26. W.S. McElhannon and H.A. Mills (1978). *Agron. J.*, 70:1027.
27. H.A. Mills and W.S. McElhannon (1982). *Hort. Sci.*, 17:743.
28. R.H. Moll, E.J. Kamprath, and W.A. Jackson (1982). *Agron. J.*, 72:562.
29. H. Lorenz (1976). *Plant Soil*, 45:169.
30. F.C. Olday, A.V. Barker, and D.N. Maynard (1976). *J. Am. Soc. Hort. Sci.*, 101:219.
31. F.C. Olday, A.V. Barker, and D.N. Maynard (1976). *J. Am. Soc. Hort. Sci.*, 101: 217.
32. J.S. Pate (1983). In: *Nitrogen As An Ecological Factor* (J.A. Lee, S. McNeill, and I.H. Rorison, eds.), Blackwell, Oxford, p. 225.
33. J.I. Sprent and R.J. Thomas (1984). *Plant Cell Environ.*, 7:637.
34. R.J. Thomas, V. Feller, and K.J. Erismann (1979). *New Phytol.*, 82:657.
35. P.B. Vose (1963). *Herb. Abstr.*, 33:1.
36. R.B. Clark (1983). In: *Genetic Aspects of Plant Nutrition* (M.R. Saric and B.C. Loughman, eds.), Martinus Nijhoff, Boston, p. 49.

37. M.M.A. Valera and B.R. Murty (1985). *Theor. Appl. Genet.*, *69*:353.
38. R.H. Hageman (1979). In: *Nitrogen Assimilation of Plants* (E.J. Hewitt and C.V. Cutting, eds.), Academic Press, New York, p. 591.
39. G. Sorger, D.O. Gooden, E.D. Earle, and J. McKinnon (1968). *Plant Physiol.*, *82*:473.
40. D.J. Cove (1979). In: *Nitrogen Assimilation in Plants* (E.J. Hewitt and C.V. Cutting, eds.), Academic Press, New York, p. 289.
41. S. Fortini, G. Galterio, B.M. Mariani, and D. Sgrulletta (1975). *Maydica*, *20*:133.
42. L.W. Gallagher, K.M. Soliman, C.O. Qualset, R.C. Huffaker, and D.W. Rains (1980). *Crop Sci.*, *20*:717.
43. K.P. Rao, D.W. Rains, C.O. Qualset, and R.C. Huffaker (1977). *Crop Sci.*, *17*:283.
44. R.L. Warner, R.H. Hageman, J.W. Dudley, and R.J. Lambert (1969). *Proc. Natl. Acad. Sci. USA*, *62*:785.
45. W. Wallace (1986). *Physiol. Plant*, *66*:630.
46. D.N. Maynard and A.V. Barker (1969). *J. Am. Soc. Hort. Sci.*, *94*:235.
47. M.W. Fowler and R.J. Barker (1979). In: *Nitrogen Assimilation in Plants* (E.J. Hewitt and C.V. Cutting, eds.), Academic Press, New York, p. 489.
48. G.V. Givan (1979). *Phytochemistry*, *18*:375.
49. R.J. Haynes (1986). In: *Mineral Nitrogen in the Plant–Soil System* (R.J. Haynes, ed.), Academic Press, New York, p. 303.
50. P.J. Lea and B.J. Mifflin (1979). In: *Nitrogen Assimilation in Plants* (E.J. Hewitt and C.V. Cutting, eds.), Academic Press, New York, p. 475.
51. F.M. Robert and P.P. Wong (1986). *Plant Physiol.*, *81*:142.
52. A.V. Barker and W.H. Lachman (1986). *J. Plant Nutr.*, *9*:1.
53. D.N. Maynard, A.V. Barker, and W.H. Lachman (1968). *Proc. Am. Soc. Hort. Sci.*, *92*:537.
54. A. Blum (1988). *Plant Breeding for Stress Environments*. CRC Press, Boca Raton, FL.
55. R. Dudal (1976). In: *Plant Adaptation to Mineral Stress in Problem Soils* (M.J. Wright, ed.), Cornell Univ. Press, Ithaca, NY, p. 3.
56. V.C. Baligar and R.R. Duncan (1990). *Crops as Enhancers of Nutrient Use*. Academic Press, San Diego.
57. R.B. Clark and R.R. Duncan (1991). *Field Crops Res.*, *26*:219–240.
58. S. Ramani and S. Kannan (1986). *Plant Breed.*, *97*:334.
59. R.B. Clark (1983). In: *Genetic Aspects of Plant Nutrition* (M.R. Saric and B.C. Loughman, eds.), Martinus Nijhoff, Boston, p. 49.
60. G.C. Gerloff and W.H. Gabelman (1983). In: *Inorganic Plant Nutrition* (A. Lauchli and R.L. Bieleski, eds.), Springer-Verlag, Berlin, p. 453.
61. P.J. Goodman (1979). In: *Nitrogen Assimilation in Plants* (E.J. Hewitt and C.V. Cutting, eds.), Academic Press, New York, p. 165.
62. P.B. Vos (1984). In: *Crop Breeding: A Contemporary Basis* (P.B. Vose and S.G. Blixt, eds.), Pergamon Press, Elmsford, NY, p. 67.
63. S.N. Smith (1934). *J. Am. Soc. Agron.*, *26*:785.
64. D.A. Van Sanford and C.T. MacKown (1986). *Theor. Appl. Genet.*, *72*:158.

65. D. Isfan (1990). *J. Plant Nutr.*, *13*:907.
66. N.M. Crawford and W.H. Campbell (1990). *The Plant Cell*, *2*:829.
67. G.C. Elliot and A. Lauchli (1985). *Agron. J.*, *77*:399.
68. O. Biddulph (1959). *Plant Physiology* (F.C. Steward, ed.), Academic Press, New York, p. 553.
69. J. Wieneke (1990). *Plant Soil*, *123*:139.
70. J.L. Brewster and P.B.H. Tinker (1972). *Soil Fertil.*, *35*:355.
71. G.C. Gerloff (1976). *Plant Adaptation to Mineral Stress in Problem Soils* (M.J. Wright, ed.), Cornell Univ., Ithaca, NY, p. 161.
72. A.S. Lyness (1936). *Plant Physiol.*, *11*:665.
73. C.E. Barr and A. Ulrich (1963). *J. Agric. Food Chem.*, *11*:313.
74. P. Burauel, J. Wieneke, and F. Fuhr (1989). *Plant Soil*, *123*:169.
75. N. Ae, J. Arihara, K. Okada, T. Yoshihara, and C. Johansen (1990). *Science*, *248*:477.
76. G.D. Batten (1986). *Ann. Bot.*, *58*:49.
77. G.D. Batten (1986). *Cereal Chem.*, *63*:384.
78. S. Pettersson and P. Jensen (1983). *Plant Soil*, *72*:231.
79. J.R. Caradus (1980). *J. Agric. Res.*, *23*:75.
80. P. Jansen and S. Pettersson (1978). *Physiol. Plant*, *42*:207.
81. G. Cacco, G. Ferrari, and G.C. Lucci (1976). *J. Agric. Sci. Camb.*, *87*:585.
82. E. Epstein and R.C. Jeffries (1964). *Annu. Rev. Plant Physiol.*, *15*:169.
83. A. Makmur, G.C. Gerloff, and W.H. Gabelman (1978). *J. Am. Soc. Hort. Sci.*, *103*:545.
84. P.F. Shea, W.H. Gabelman, and G.C. Gerloff (1967). *Proc. Am. Soc. Hort. Sci.*, *91*:286.
85. P.F. Shea, G.C. Gerloff, and W.H. Gabelman (1968). *Plant Soil*, *28*:337.
86. S.S. Figdore, W.H. Gabelman, and G.C. Gerloff (1987). *Plant Soil*, *99*:85.
87. S.S. Figdore, W.H. Gabelman, and G.C. Gerloff (1989). *J. Am. Soc. Hort. Sci.*, *114*:322.
88. V.D. Jolley, J.C. Brown, M.J. Blaylock, and S.D. Camp (1988). *J. Plant Nutr.*, *11*:1159.
89. D.C. Hughes, V.D. Jolley, and J.C. Brown (1990). *J. Plant Nutr.*, *13*:1405.
90. V.D. Jolley and J.C. Brown (1985). *J. Plant Nutr.*, *8*:527.
91. R.A. Olsen and J.C. Brown (1980). *J. Plant Nutr.*, *2*:629.
92. H. Marschner, V. Romheld, and M. Kissel (1986). *J. Plant Nutr.*, *9*:695.
93. V. Romheld and H. Marschner (1986). *Plant Physiol.*, *80*:175.
94. V. Romheld (1987). *Physiol. Plant*, *70*:231.
95. J.C. Brown (1978). *Plant Cell Environ.*, *1*:249.
96. R.L. Chaney, J.C. Brown, and L.O. Tiffin (1972). *Plant Physiol.*, *50*:208.
97. R.A. Olsen and J.C. Brown (1980). *J. Plant Nutr.*, *2*:647.
98. H.F. Bienfait, R.J. Bino, A.M. vander Bliek, J.F. Duvenvoorden, and J.M. Fontain (1983). *Physiol. Plant*, *59*:196.
99. C.M. Lytle and V.C. Jolley (1991). *J. Plant Nutr.*, *14*:341.
100. J.C. Brown, V.D. Jolley, and C.M. Lytle (1991). In: *Iron Nutrition and*

Interactions in Plants (Y. Chen and Y. Hardar, eds.), Kluwer Academic, The Netherlands, p. 189.

101. S. Takagi (1976). *Soil Sci. Plant Nutr.*, *22*:423.
102. S. Takagi, K. Nomoto, and T. Takemoto (1984). *J. Plant Nutr.*, *7*:469.
103. Y. Sigiura and H. Tanaka (1981). *J. Am. Chem. Soc.*, *103*:6969.
104. W. Shi, M. Chino, R. Youssef, S. Mori, and S. Takagi (1988). *Soil Sci. Plant Nutr.*, *34*:585.
105. S. Mori, N. Nishizawa, S. Kawai, Y. Sato, and S. Takagi (1987). *J. Plant Nutr.*, *10*:1003.
106. Y. Mino, T. Ishida, N. Ota, M. Inoue, K. Nomoto, T. Takemoto, H. Tanaka, and Y. Suguira (1983). *J. Am. Chem. Soc.*, *105*:4671.
107. S. Kawai, S. Takagi, and Y. Sato (1988). *J. Plant Nutr.*, *11*:633.
108. D.E. Crowley, C.P.P. Reid, and P.J. Szaniszlo (1987). In: *Iron Transport in Microbes, Plants and Animals* (G. Winkelmann, D. van der Helm, and J.B. Neilands, eds.), VCH, Weinheim, p. 375.
109. J.C. Brown, V.D. Jolley, and C.M. Lytle (1990). *Plant Soil*, *130*:157.
110. J.C. Brown and V.D. Jolley (1989). *Bioscience*, *39*:546.
111. C.M. Lytle, C.M. Jolley, and J.C. Brown (1991). In: *Iron Nutrition and Interactions in Plants* (Y. Chen and Y. Hadar, eds.), Kluwer Academic, The Netherlands, p. 197.
112. D.E. Crowley, Y.C. Wang, C.P.P. Reid, and P.J. Szaniszlo (1991). In: *Iron Nutrition and Interactions in Plants* (Y. Chen and Y. Hadar, eds.), Kluwer Academic, The Netherlands, p. 213.
113. H.F. Bienfait (1989). *Acta Bot. Neerl.*, *38*:105.
114. J. Smarelli and D. Castignetti (1986). *Biochim. Biophys. Acta*, *882*:337.
115. H.F. Bienfait, W. van den Briel, and N.T. Mesland-Mul (1985). *Plant Physiol.*, *78*:596.
116. S.R.R. de Cianzo (1991). *Plant Soil*, *130*:63.
117. G.J. Taylor (1988). *Commun. Soil Sci. Plant Anal.*, *19*:1179.
118. C.D. Foy (1984). In: *Soil Acidity and Liming* (F. Adams, ed.), Am. Soc. Agron., Madison, WI, p. 57.
119. R.E. Wilkinson, E.L. Ramseur, R.R. Duncan, and L.M. Shuman (1990). In: *Genetic Aspects of Plant Mineral Nutrition* (N. El Bassam, ed.), Kluwer Academic, The Netherlands, p. 263.
120. T. Horiguchi (1987). *Soil Sci. Plant Nutr.*, *33*:595.
121. T. Horiguchi (1988). *Soil Sci. Plant Nutr.*, *34*:65.
122. R.O. Nable (1988). *Plant Soil*, *112*:45.
123. W.E. Rauser and N.R. Curvetto (1980). *Nature*, *287*:563.
124. R.G. Turner and C. Marshall (1972). *New Phytol.*, *71*:671.
125. Y. Kohno and C.D. Foy (1983). *J. Plant Nutr.*, *6*:877.
126. G.J. Taylor and C.D. Foy (1985). *Can. J. Bot.*, *63*:2181.
127. G. Zhang and G.J. Taylor (1989). *Plant Physiol.*, *91*:1094.
128. C.W. Kennedy, W.C. Smith, Jr., and M.T. Ba (1986). *J. Plant Nutr.*, *9*:1123.
129. I.M. Mugwira and S.M. Elgawhary (1979). *Soil Sci. Soc. Amer. J.*, *43*:736.

130. S.C. Miyasaka, L.V. Kochian, J.E. Shaff, and C.D. Foy (1989). *Plant Physiol.*, *91*:1188.
131. Z. Rengel and D.L. Robinson (1989). *Plant Physiol.*, *91*:1407.
132. K. Ojima and K. Ohira (1988). *Commun. Soil Sci. Plant Anal.*, *19*:1229.
133. H. Koyama, R. Okawara, K. Ojima, and T. Yamaya (1988). *Physiol. Plant.*, *74*:683.
134. V.D. Jolley and J.C. Brown (1989). *J. Plant Nutr.*, *12*:923.
135. G. Zocci and S. Cocucci (1990). *Plant Physiol.*, *92*:908.
136. V.D. Jolley, J.C. Brown, M.J. Blaylock, and S.D. Camp (1988). *J. Plant Nutr.*, *11*:1159.
137. D.F. Hughes, V.D. Jolley, and J.C. Brown (1990). *J. Plant Nutr.*, *13*:1405.
138. B.T. Warden and H.M. Reisenauer (1991). *J. Plant Nutr.*, *14*:7.
139. R.O. Nable (1991). *J. Plant Nutr.*, *14*:453.
140. S.C. Miyasaka, J.G. Buta, R.K. Howell, and C.D. Foy (1991). *Plant Physiol.*, *96*:737.
141. A. Lauchli and E. Epstein (1984). *Calif. Agric.*, *38*:18.
142. E.V. Maas (1984). *Calif. Agric.*, *38*:20.
143. D. Pasternak (1987). *Annu. Rev. Phytopathol.*, *25*:271.
144. S.J. Stavarek and D.W. Rains (1983). *Iowa St. J. Res.*, *57*:457.
145. Y.W. Yang, R.J. Newton, and F.R. Miller (1990). *Crop. Sci.*, *30*:775.
146. R. Weimberg, H.R. Lerner, and A. Poljakoff-Mayber (1984). *Physiol. Plant*, *62*:472.
147. P. Boursier, J. Lynch, A. Lauchli, and E. Epstein (1987). *Aust. J. Plant Physiol.*, *14*:463.
148. C.M. Grieve and E.V. Maas (1984). *Physiol. Plant*, *61*:167.
149. H. Nassery and D.A. Baker (1974). *Ann. Bot. (London)*, *38*:141.
150. A.R. Yeo, D. Kramer, A. Lauchli, and J. Gullasch (1977). *J. Exp. Bot.*, *28*:17.
151. E. Winter (1982). *Aust. J. Plant Physiol.*, *9*:227.
152. T.J. Flowers, P.F. Troke, and A.R. Yeo (1977). *Annu. Rev. Plant Physiol.*, *28*:89.
153. D. Kuiper, J. Schuit, and P.J.C. Kuiper (1990). *Plant Soil*, *123*:243.
154. Friedman, R., A. Altman, and N. Levin (1989). *Physiol. Plant*, *76*:295.
155. P.C. LaRosa, N.K. Singh, P.M. Hasegawa, and R.A. Bressan (1989). *Plant Physiol.*, *91*:855.
156. N.K. Singh, P.C. LaRosa, A.K. Handa, P.M. Hasegawa, and R.A. Bressan (1987). *Proc. Natl. Acad. Sci. USA*, *84*:739.
157. G.N. Amzallag, H.R. Lerner, and A. Poljakoff-Mayber (1990). *J. Exp. Bot.*, *41*:29.
158. A.A. Watad, R.H. Lerner, and L. Reinhold (1985). *Physiol. Vegetale*, *23*:887.
159. T. Matoh, N. Matsushita, and E. Takahashi (1988). *Physiol. Plant*, *72*:8.
160. S. Ramani and S. Kannan (1986). *J. Plant Nutr.*, *9*:1553.
161. R.W. Kingsbury, E. Epstein, and P.W. Pearcy (1984). *Plant Physiol.*, *74*:417.

162. J. Gorham, D. Hughes, and R.G. Wyn Jones (1981). *Physiol. Plant,* *53*:27.
163. P. Boursier and A. Lauchli (1989). *Physiol. Plant,* *77*:537.
164. S.G. Richardson and K.J. McCree (1985). *Plant Physiol.* *79*:1015.
165. A.H. Khan and M.Y. Ashraf (1988). *Acta Physiol. Plant,* *10*:257.
166. S. Schubert and A. Lauchli (1990). *Plant Soil, 123*:205.
167. W.H. Gabelman and G.C. Gerloff (1983). *Plant Soil, 72*:335.
168. M.G. Weiss (1943). *Genetics, 28*:253.
169. W.R. Fehr (1982). *J. Plant Nutr.,* *5*:611.
170. W.R. Fehr (1983). *Iowa St. J. Res.,* *57*:393.
171. J.C. Brown and E.V. Wann (1982). *J. Plant Nutr.,* *5*:623.
172. D.T. Pope and H.M. Munger (1953). *Proc. Am. Soc. Hort. Sci., 61*:481.
173. R.D. Graham (1984). *Adv. Plant Nutr.,* *1*:57.
174. J.C. Brown and T.E. Devine (1980). *Agron. J.,* *72*:898.
175. T.E. Devine (1982). In: *Breeding Plants for Less Favorable Environments* (N. Christiansen and C.F. Lewis, eds.), John Wiley and Sons, New York, p. 143.
176. A.M.C. Furlani, R.B. Clark, W.M. Ross, and J.M. Maranville (1984). In: *Genetic Aspects of Plant Mineral Nutrition* (W.H. Gableman and B.C. Loughman, eds.), Martinus Nijhoff, Boston, p. 287.
177. I. Fawole, W.H. Gabelman, G.C. Gerloff, and E.V. Nordheim (1982). *J. Am. Soc. Hort. Sci., 107*:94.
178. P.F. Shea, W.H. Gabelman, and G.C. Gerloff (1967). *Proc. Am. Soc. Hort. Sci., 91*:286.
179. S.R. DeCianzio and W.R. Fehr (1982). *Crop Sci., 22*:433.
180. H.Z. Zaiter, D.P. Coyne, and R.B. Clark (1987). *J. Am. Soc. Hort. Sci., 112*:1019.
181. D.A. Reid (1971). In: *Proc. 2nd Intl. Barley Genetics Symp* (R.A. Nilan, ed.), Washington St. Univ. Press, Pullman, WA, p. 409.
182. L.G. Campbell and H.N. Lafever (1981). *Cereal Res. Commun., 9*:281.
183. J. Mesdag and A. Balkema-Boomstra (1984). *Fertil. Res.,* *5*:213.
184. R.D. Rhue (1979). In: *Stress Physiology in Crop Plants* (H. Mussell and R.C. Staples, eds.), John Wiley and Sons, New York, p. 61.
185. R.A. Culvenor, R.N. Oram, and D.J. David (1986). *Aust. J. Agric. Res.,* *37*:409.
186. L. Dessureaux (1959). *Euphytica, 8*:260.
187. A.H. Eenink and G. Garretsen (1977). *Euphytica, 26*:47.
188. R.A. Borgonovi, R.E. Schaffert, and G.V.E. Pitta (1987). In: *Sorghum for Acid Soils* (L.M. Gourley and J.G. Salinas, eds.), CIAT, Cali, Colombia, p. 271.
189. S.R. Boye-Goni and V. Marcarian (1985). *Crop Sci., 25*:749.
190. M. Ashraf, T. McNeilly, and A.D. Bradshaw (1986). *Euphytica, 35*:935.
191. S. Moeljopawiro and H. Ikehashi (1981). *Euphytica, 30*:291.
192. F.M. Azhar and T. McNeilly (1989). *Euphytica, 43*:69.
193. S.A. Quarrie (1981). *Plant Cell Environ.,* *4*:147.
194. M. Tal (1985). *Plant Soil, 89*:199.

195. G.N. Amzallag, H.R. Lerner, and A. Poljakoff-Mayber (1990). *J. Exp. Bot.*, *41*:1529.
196. G.H. Abel and A.J. Mackenzie (1964). *Crop. Sci.*, *4*:157.
197. W.J.S. Downton (1984). *CRC Crit. Rev. Plant Sci.*, *1*:183.
198. F.M. Azhar and T. McNeilly (1988). *Plant Breed.*, *101*:114.
199. J. Cambraia, F.R. Galvani, M.M. Estevao, and R. Sant'Anna (1983). *J. Plant Nutr.*, *6*:313.
200. G.H. Heichel, D.K. Barnes, C.P. Vance, and C.C. Sheaffer (1989). *J. Prod. Agric.*, *2*:24.
201. T.M. Schettini, W.H. Gabelman, and G.C. Gerloff (1987). *Plant Soil*, *99*:175.
202. T.E. Devine (1977). In: *Plant Adaptation to Mineral Stress in Problem Soils* (M.J. Wright, ed.), Cornell Univ. Press, Ithaca, NY, p. 65.
203. R.R. Duncan (1988). *Commun. Soil Sci. Plant Anal.*, *19*:1295.
204. R.R. Duncan, R.M. Waskom, D.R. Miller, R.L. Voigt, G.E. Hanning, D.A. Timm, and M.V. Nabors (1991). *Crop Sci.*, *31*:1396–1397.
205. D. Bouma (1983). In: *Encyclopedia of Plant Physiology, Vol. 15A. Inorganic Plant Nutrition* (A. Lauchli and R.L. Bieleski, eds.), Springer-Verlag, Berlin, p. 121.
206. B.P. Singh, R.A. Singh, M.K. Sinha, and B.N. Singh (1985). *J. Agric. Sci. Camb.*, *105*:193.
207. P.C. Bagley and N.L. Taylor (1987). *Iowa St. J. Res.*, *61*:459.
208. R.W. Hintz, W.R. Fehr, and S.R. de Cianzio (1987). *Crop Sci.*, *27*:707.
209. H.J. Jessen, M.B. Dragonuk, R.W. Hintz, and W.R. Fehr (1988). *J. Plant Nutr.*, *11*:717.
210. E.R. Williams, W.M. Ross, R.B. Clark, G.M. Herron, and M.D. Witt (1986). *J. Plant Nutr.*, *9*:423.
211. R.K. Jain, S. Jain, and J.B. Chowdhury (1991). *Ann. Bot.*, *67*:517.
212. J.M. Ye, K.N. Kao, B.L. Harvey, and B.G. Rossnagel (1987). *Theor. Appl. Genet.*, *74*:426.
213. P. Spiegel-Roy and G. Ben-Hayyim (1985). *Plant Soil*, *89*:243.
214. F.M. Azhar and T. McNeilly (1987). *J. Agron. Crop Sci.*, *159*:269.
215. E. Epstein and D.W. Rains (1987). In: *Genetic Aspects of Plant Mineral Nutrition* (W.H. Gabelman and B.C. Loughman, eds.), p. 113.
216. R. Grumet, T.G. Isleib, and A. D. Hanson (1985). *Crop Sci.*, *25*:618.
217. M.E. McDaniel and J.C. Brown (1982). *J. Plant Nutr.*, *5*:545.
218. E.P. Williams, R.B. Clark, W.M. Ross, G.M. Herron, and M.D. Witt (1987). *Plant Soil*, *99*:127.
219. C. Shifriss and E. Eidelman (1983). *J. Plant Nutr.*, *6*:699.
220. M.C. Shannon (1985). *Plant Soil*, *89*:227.
221. A.N. Lakso (1990). *Hort. Sci.*, *25*:1365.
222. S. Seeley (1990). *Hort. Sci.*, *25*:1369.
223. J. Levitt (1990). *Hort. Sci.*, *25*:1363.
224. N.M. Crawford and W.H. Campbell (1990). *Plant Cell*, *2*:829.
225. S.A. Robinson, A.P. Slade, G.G. Fox, R. Phillips, R.G. Ratcliffe, and G.R. Stewart (1991). *Plant Physiol.*, *95*:509.

226. M. Ni and L. Beevers (1990). *Plant Physiol.*, *94*:745.
227. N.I. Mitreva (1989). *Plant Soil*, *115*:29.
228. P. Morard, A. Bernadac, and V. Valles (1990). *J. Plant Nutr.*, *13*:249.
229. A. Wallace (1990). *J. Plant Nutr.*, *13*:309.
230. T.T. Chang and G.C. Loresto (1985). *Proc. Natl. Sem. Breeding for Stress Resistance in Plants*, Haryana Agric. Univ., India.
231. J.C. O'Toole and R.T. Cruz (1979). *Plant Physiol.*, *64*:428.
232. N. Seetharama, B.V. Subba Reddy, J.K. Peacock, and F.E. Bidinger (1982). *In Drought Resistance in Crop Plants With Emphasis on Rice*, IRRI, Los Banos, Philippines, p. 317.
233. D.T. Rosenow, J.E. Quisenberry, C.W. Wendt, and L.E. Clark (1983). *Agric. Water Manage.*, *7*:207.
234. K.S. Fisher, E.C. Johnson, and G.O. Edmeades (1976). *Breeding and Selection for Drought Resistance in Tropical Maize*, CIMMYT, Mexico City, p. 28.
235. R.M. Castleberry (1983). In: *Crop Reaction to Water and Temperature Stresses in Humid, Temperate Climates* (C.D. Raper, Jr. and P.J. Kramer, eds.), Westview Press, Boulder, CO, p. 277.
236. W.C. Hofmann, M.K. O'Neill, and A.K. Dobrenz (1984). *Agron. J.*, *76*:223.
237. A. Blum, J. Mayer, and G. Gozlan (1982). *Field Crops Res.*, *5*:137.
238. H.W. Gausman, J.E. Quisenberry, J.J. Burke, and C.W. Wendt (1984). *Field Crops Res.*, *9*:373.
239. D.S. Harris, W.T. Schapaugh, and E.T. Kanemasu (1984). *Crop Sci.*, *24*:839.
240. S.B. Idso, R.J. Reginato, R.D. Jackson, and P.J. Pinter, Jr. (1981). *Irrig. Sci.*, *2*:205.
241. S.B. Idso, R.J. Reginato, D.C. Reicosky, and J.L. Hatfield (1981). *Agron. J.*, *73*:286.
242. N.C. Turner, J.C. O'Toole, R.T. Cruz, O.S. Manuco, and S. Ahmad (1986). *Field Crops Res.*, *13*:257.
243. J.M. Clarke and T.N. McCaig (1982). *Can. J. Plant Sci.*, *62*:571.
244. J.M. Clarke and T.N. McCaig (1982). *Crop Sci.*, *22*:503.
245. W. Dedio (1975). *Can. J. Plant Sci.*, *55*:369.
246. H.G. Nass and J.D.E. Sterling (1981). *Can. J. Plant Sci.*, *61*:283.
247. M.H. Salim, G.W. Todd, and C.A. Stutte (1969). *Agron. J.*, *61*:182.
248. A. Jaradat and C.F. Konzak (1983). *Cereal Res. Commun.*, *11*:179.
249. J.C. Clarke and T.F. Townley-Smith (1986). *Crop Sci.*, *26*:289.
250. D.G. Stout, T. Kannangara, and G.M. Simpson (1978). *Can. J. Plant Sci.*, *58*:225.
251. S.E. Simmelgaard (1976). *Physiol. Plant.*, *37*:167.
252. B. Mambani and R. Lal (1983). *Plant Soil*, *73*:73.
253. B.M. Robertson, A.E. Hall, and K.W. Foster (1985). *Crop Sci.*, *25*:1084.
254. L.P. Reitz (1974). *Agric. Meteorol.*, *14*:3.
255. W.R. Jordan, W.A. Dougas, Jr., and P.J. Shouse (1983). *Agric. Water Manage.*, *7*:281.

256. M. Saeed and C.A. Francis (1983). *Crop Sci.*, *23*:683.
257. A. Blum (1970). *Crop Sci.*, *62*:333.
258. J.M. Morgan (1984). *Annu. Rev. Plant Physiol.*, *35*:299.
259. A. Blum and C.W. Sullivan (1986). *Ann. Bot. (London)*, *57*:835.
260. A. Blum, Mayer, Jr., and G. Gozlan (1983). *Plant, Cell Environ.*, *6*:219.
261. M. McGowan, P. Blanch, P.J. Gregory, and D. Haycock (1984). *J. Agric. Sci. (Camb.)*, *102*:415.
262. R.C. Johnson, H.T. Nguyen, and L.I. Croy (1984). *Crop Sci.*, *24*:957.
263. J.M. Cutler, K.W. Shanon, and P.L. Steponkus (1980). *Crop Sci.*, *20*:314.
264. A. Kumar, P. Singh, D.P. Singh, H. Singh, and H.C. Sharma (1984). *Ann. Bot. (London)*, *54*:537.
265. J.M. Morgan (1983). *Aust. J. Agric. Res.*, *34*:607.
266. R.G. Henzell, K.J. McCree, C.H.M. van Bauel, and K.F. Schertz (1975). *Crop Sci.*, *15*:516.
267. A.P. Gay (1986). *Ann. Bot. (London)*, *57*:361.
268. B. Roarke and J.E. Quisenberry (1977). *Plant Physiol.*, *59*:354.
269. G.N.R. Ayyangar and B.W.X. Ponnaiya (1941). *Curr. Sci.*, *10*:408.
270. G.C. Peterson, K. Suksayretrup, and D.W. Weibel (1982). *Crop Sci.*, *22*:63.
271. I. Tarumoto (1980). *Jap. J. Breed.*, *30*:237.
272. P. von Wettstein-Knowles (1972). *Planta*, *106*:113.
273. R.A. McIntosh (1983). *Proc. 6th Intl Wheat Genet. Symp.*, p. 1197.
274. R.M. Karamayshev and N.N. Kozhushko (1981). *Tr. Prikl. Bot. Genet. Sel.*, p. 71.
275. T.V. Williams, R.S. Snell, and C.E. Cress (1969). *Crop Sci.*, *9*:19.
276. J.C. Bouwkamp and W.L. Summers (1982). *J. Hered.*, *73*:385.
277. A.D. Hanson, C.E. Nelson, A.R. Pedersen, and E.H. Everson (1979). *Crop Sci.*, *19*:489.
278. D. Shimshi and J. Ephrat (1975). *Agron. J.*, *67*:326.
279. J.A. Bunce (1981). *Can. J. Bot.*, *59*:769.
280. D.A. Johnson, R.A. Richards, and N.C. Turner (1983). *Crop Sci.*, *23*:318.
281. R.A. Fischer and J.T. Wood (1979). *Aust. J. Agric. Res.*, *30*:1001.
282. M.M. Ludlow and O. Bjorkman (1984). *Planta*, *161*:505.
283. V.R. Babu, P.S.S. Murty, G.H.S. Reddy, and T.Y. Reddy (1983). *Environ. Exp. Bot.*, *203*:183.
284. K.R. Stevenson and R.H. Shaw (1971). *Agron. J.*, *63*:327.
285. J.D. Bewley (1979). *Annu. Rev. Plant Physiol.*, *30*:195.
286. T.C. Hsiao (1973). *Annu. Rev. Plant Physiol.*, *24*:519.
287. J.S. Boyer (1976). *Philos. Trans. R. Soc. London, Ser. B.*, *273*:501.
288. I.F. Wardlaw (1967). *Aust. J. Biol. Sci.*, *20*:25.
289. H.M. Rawson, A.K. Bagga, and P.M. Bremner (1977). *Aust. J. Plant Physiol.*, *4*:389.
290. P.K. Aggarwal and S.K. Sinha (1984). *Ann. Bot. (London)*, *53*:329.
291. R.B. Austin, R.B. Morgan, M.A. Ford, and R.D. Blackwell (1980). *Ann. Bot. (London)*, *45*:309.

292. A. Blum, H. Poiarkova, G. Golan, and J. Mayer (1983). *Field Crops Res.*, 6:51.
293. A. Blum (1985). *J. Exp. Bot.*, 36:432.
294. K.R. Keim, F.R. Miller, and D.T. Rosenow (1983). *Agron. Abstr.*, p. 69.
295. J.R. Hanks, J. Keller, V.P. Rasmussen, and G.D. Wilson (1976). *Soil Sci. Soc. Am. Proc.*, 40:426.
296. M.K. O'Neill, W. Hofmann, A.K. Dobrenz, and V. Marcarian (1983). *Agron. J.*, 75:102.
297. C.J. Gerards and W.D. Worrall (1986). *Agron. J.*, 78:348.
298. L.H. Penny (1981). *Crop Sci.*, 21:237.
299. H. Morishima, I. Hirata, and H.I. Oka (1962). *Ind. J. Genet. Plant Breed.*, 22:1.
300. W.R. Jordan, R.L. Monk, F.R. Miller, D.T. Rosenow, L.E. Clark, and P.J. Shouse (1983). *Crop Sci.*, 23:552.
301. K.S. Fischer, G.O. Edmeades, and E.C. Johnson (1989). *Field Crops Res.*, 22:227.
302. J.S. Boyer and H.G. McPherson (1975). *Adv. Agron.*, 27:1.
303. R.R. Duncan, A.J. Bockholt, and F.R. Miller (1981). *Agron. J.*, 73:849.
304. M. Santamaria, M.M. Ludlow, and S. Fukai (1986). In: *Proc. 1st Aust. Sorghum Conf.* (Ma. Foale and R.G. Henzell, eds.), Gatton, QLD, Australia, p. 4.127.
305. J.C. O'Toole, N.C. Turner, O.P. Namuco, M. Dingkuhn, and K.A. Gomez (1984). *Crop Sci.*, 24:1121.
306. R.C. Muchow and T.R. Sinclair (1989). *Plant, Cell Environ.*, 12:425.
307. W.R. Jordan, P.J. Shouse, A. Blum, F.R. Miller, and R.L. Monk (1984). *Crop Sci.*, 24:1168.
308. G.H. Liang, A.D. Dayton, C.C. Chu, and A.J. Casady (1975). *Crop Sci.*, 15:567.
309. H. Thomas and C. Evans (1989). *Ann. Bot.*, 64:581.
310. H. Thomas (1987). *J. Exp. Bot.*, 38:115.
311. S.P. Kidambi, D.R. Krieg, and D.T. Rosenow (1990). *Plant Physiol*, 92:1211.
312. S.P. Kidambi, D.R. Krieg, and H.T. Nguyen (1990). *Euphytica*, 50:139.
313. U.N. Chaudhuri, M.L. Deaton, E.T. Kanemasu, G.W. Wall, V. Marcarian, and A.K. Dobrenz (1986). *Agron. J.*, 78:490.
314. W.G. Wenzel (1989). *Angew. Botanik.*, 63:7.
315. W.G. Wenzel (1978). *Angew. Botanik.*, 61:529.
316. C.G. Kuo, B.J. Shen, H.M. Chen, H.C. Chen, and R.T. Opena (1988). *Euphytica*, 39:65.
317. A.E.M. Nour, D.E. Weibel, and G.W. Todd (1978). *Agron. J.*, 70:509.
318. R.C. Durley, T. Kannangara, N. Seetharama, and G.M. Simpson (1983). *Can. J. Plant Sci.*, 63:131.
319. M.K. O'Neill and M. Diaby (1987). *J. Agron. Corp Sci.*, 159:192.
320. G.L. Wilson, P.S. Raju, and J.M. Peacock (1982). *Ind. J. Agric. Sci.*, 52:848.
321. A.E.M. Nour and D.E. Weibel (1978). *Agron. J.*, 70:217.

322. N.C. Turner (1986). *Aust. J. Plant Physiol.*, *13*:175.
323. A. Blum (1983). *Agric. Water Manage.*, 7:195.
324. A. Blum, K.F. Schertz, R.W. Toler, R.I. Welch, D.T. Rosenow, J.W. Johnson, and L.E. Clark (1978). *Agron. J.*, *70*:472.
325. D.G. Stout and G.M. Simpson (1978). *Can. J. Plant Sci.*, *58*:213.
326. T. Kannangara, R.C. Durley, G.M. Simpson, and D.G. Stout (1982). *Can. J. Plant Sci.*, *62*:317.
327. N. Schmitt and P. Dizengremel (1989). *Plant Physiol. Biochem.*, 27:17.
328. T.R. Sinclair and M.M. Ludlow (1986). *Aust. J. Plant Physiol.*, *13*:329.
329. O.L. Lange, R. Losch, E.D. Schulze, and L. Kappen (1971). *Planta (Berlin)*, *100*:76.
330. F.R. Bidinger, V. Mahalakshmi, B.S. Talukdar, and G. Alagarswamy (1982). In: *Drought Resistance in Crops, with Special Emphasis on Rice*, IRRI, Los Banos, Philippines, p. 357.
331. K.H. Asay and D.A. Johnson (1983). *Iowa St. J. Res.*, 57:441.
332. D. Atsmon (1973). In: *Agricultural Genetics* (R. Moav, ed.), John Wiley and Sons, New York, p. 157.
333. C.Y. Sullivan (1983). *Iowa St. J. Res.*, 57:423.
334. W.R. Jordan and F.R. Miller (1980). In: *Adaptation of Plants to Water and High Temperature Stress* (N.C. Turner and P.J. Kramer, eds.), John Wiley and Sons, New York, p. 383.
335. D.A. Johnson, M.D. Rumbaugh, and K.H. Asay (1981). *Plant Soil*, 58:279.
336. J.B. Passioura (1972). *Aust. J. Agric. Res.*, *23*:745.
337. E.A. Hurd (1969). *Euphytica*, *18*:217.
338. R.O. Slatyer (1960). *Bot. Rev.*, *24*:331.
339. M.A.L. Smith, L.A. Spomer, and E.S. Skiles (1989). *J. Plant Nutr.*, *12*:233.
340. T.A. Mansfield, A.R. Wellburn, and T.J.S. Moreira (1978). *Phil. Trans. R. Soc. London B*, *284*:471.
341. R. Fenton, T.A. Mansfield, and A.R. Welburn (1976). *J. Exp. Bot.*, 27:1206.
342. J. Zhang and M.B. Kirkham (1990). *Euphytica*, *46*:109.
343. J.M. Santamaria, M.M. Ludlow, and S. Fukai (1990). *Aust. J. Agric. Res.*, *41*:51.
344. M.M. Ludlow, J. M. Santamaria, and S. Fukai (1990). *Aust. J. Agric. Res.*, *41*:67.
345. R.A. Fischer (1981). *Plant Soil*, 58:249.
346. D.R. Krieg and R.B. Hutmacher (1982). In: *Proc. 37th Ann. Corn & Sorghum Industry Res. Conf.* (H.T. Loden and D. Wilkinson, eds.), Chicago, IL. Am. Seed Trade Assoc., Washington, DC, p. 37.
347. N.S. Spivakou (1990). *Sou. Agric. Sci.*, 4;25.
348. N.S. Spivakou (1990). *Dok. Vsesoy. Ord. Len. Ord. Trud. Krasn. Znam. Akad. Sel'skokh. Nauk*, *11*:2.
349. J.S. Moroni, K.G. Briggs, and G.J. Taylor (1991). *Plant Soil*, *136*:1.
350. D. Levy, E. Kastenbaum, and Y. Itzhak (1991). *Theor. Appl. Genet.*, *82*:130.

351. A.H. Goldstein (1991). *Theor. Appl. Genet.*, *82*:191.
352. H. Marschner (1991). *Plant Soil*, *134*:1.
353. S.C. Miyasaka, J.G. Buta, R.K. Howell, and C.D. Foy (1991). *Plant Physiol.*, *96*:737.
354. M.G. Redinbaugh and W.H. Campbell (1991). *Physiol. Plant.*, *82*:640.
355. R.J. Bennet and C.M. Breen (1991). *Plant Soil*, *134*:153.
356. F. Fougere, D. Le Rudulier, and J.G. Streeter (1991). *Plant Physiol.*, *96*:1228.
357. G.S. Voetberg and R.E. Sharp (1991). *Plant Physiol.*, *96*:1125.
358. S.H. Al-Hamdani, J.M. Murphy, and G.W. Todd (1991). *Can. J. Plant Sci.*, *71*:689.
359. L. Galves, R.B. Clark, L.M. Gourley, and J.W. Maranville (1989). *J. Plant Nutr.*, *12*:547.
360. I. Benes, K. Schreiber, R. Ripperger, and A. Kirscheiss (1983). *Experientia*, *39*:261.

2

The Role of Plant Hormones

Frank B. Salisbury

Utah State University
Logan, Utah

INTRODUCTION

During the last decade or two, our ideas about hormones have been changing. The term *hormone* has itself become controversial; some authors say we should not use it in relation to plants. It is necessary, then, before beginning the discussions of how hormones may help transduce environmental stimuli into plant responses to consider the concept of the plant hormone as it has been and as it might be used by plant physiologists.

There was little controversy in 1969 when Cleon Ross and I tried to summarize accepted definitions of plant hormones and growth regulators as follows (1, p. 437):

> **Plant Hormone:** An organic substance synthesized in one part of an organism and *translocated* to another part, where very small amounts have a controlling or regulatory effect: it causes a physiological response.

"Small amounts" excluded such products of primary metabolism as sugars, amino acids, and fatty acids. "Synthesized in . . . an organism" excluded mineral ions and synthetic substances that do not occur naturally in organisms. The concept of translocation in the definition emphasized that hormones act as *messengers*. Already there were doubts, however, so we added this statement: "To merit the term hormone, the substance must be translocated, but sometimes a compound normally a hormone in this sense is active in the same cells in which it is synthesized." We also described plant growth regulators (1, p. 438):

> Compounds exist that influence plant growth and development in a manner similar or identical to that of the hormones. These agents, which need not be natural plant components and are not necessarily translocated within the plant, are called *plant growth regulators*. This broad term includes the true plant hormones. It also includes synthetic growth regulators that act much like hormones. . . . Synthetic growth regulators . . . are often referred to as being "hormone-like."

The agricultural-chemical industry has essentially usurped the term *plant growth regulator* by using it almost exclusively with synthetic, "hormone-like" substances. The term *phytohormone* has appropriately been used as a synonym for plant hormone, whereas *plant growth substance* was more or less synonymous with plant growth regulator but more likely to be associated with natural substances. The term *substance*, however, is rather ambiguous, and hormones influence more than growth (especially development).

Our early ideas about plant hormones were based on definitions of animal hormones developed early in this century. The key concept was *translocation*—the "messenger" idea. But it is clear now (and was in 1969) that several of the substances that we would intuitively like to call plant hormones are not always translocated but can act in the same cells in which they are synthesized (2,3). Hence, some authors have suggested that the term hormone not be used. Nevertheless, Davies (4) notes that the original Greek origin of the term hormone (*hormon*, present participle of *horman*, to stir up, set in motion, stimulate; from *horme*, impulse, assault) says nothing about translocation. Furthermore, Fitting (5,6), who in 1909 and 1910 was the first person to introduce the word hormone into plant physiology, considered hormones merely as stimulative substances, without stressing the importance of their transport. (He found that a substance or mixture of substances extracted from orchid pollen could cause swelling of the ovary; this substance was subsequently identified as auxin.) Thus the term hormone is still appropriate for use in plant physiology—although, for many readers, it still carries the connotation of translocation.[1] After some discussion of our modern concepts

---

[1] One could say that hormones are translocated from their point of origin in a cell to their point of action in the same cell, as Ross and I did in the 1991 edition of our text (7), but that would apply to all other endogenous compounds as well.

and an attempt at a definition, I will use the term hormone without further apology.

Just how *can* we define a plant hormone? What factors need to be considered? In addition to translocation, three important factors come to mind:

Concentration: Traditionally this is an important aspect of a hormone definition (most act in the micromolar range), but many other substances also act in the micromolar range (see below).

Mode of action: We might imagine that this could be the best foundation for a definition if only we knew how hormones act, but current evidence suggests various modes of action for various hormones, weakening this criterion.

Role in the plant: This probably comes closest to our intuitive feelings about what a hormone should be.

With these ideas in mind, and to gain some perspective, let us try a rough classification of molecules based primarily on their role in the plant. Of course there is significant overlap; it is difficult to force nature into highly distinct categories:

Energy nutrients (active and storage): Carbohydrates, lipids, and proteins used for respiration and thus ATP production. These compounds can strongly influence growth and development but in less subtle ways and usually at relatively high concentrations (1–50 mM).

Mineral nutrients: Usually simple ionic substances that act as enzyme activators and often as osmotic solutes (especially K^+ along with various anions). Some are effective at very low concentrations (range: $\sim 4\,\mu M$ MoO_4^{2-} to $0.5\,M\,K^+$ in guard cells of open stomates). An important example is Ca^{2+}, which works in the micromolar range.

Structural materials: Cell wall carbohydrates, cytoskeleton materials (some proteins, etc.), amino acids, membrane materials, etc. Some cytoskeleton components occur at low concentrations.

Machine molecules: The *enzymes* that control metabolism (machines that build and dismantle the molecules of life). They are generally effective at low concentrations (millimolar to micromolar range).

Activating molecules: The coenzymes and allosteric activators (but also many mineral nutrients) that are typically essential for enzyme function, including many molecules that are recognized as animal vitamins. These molecules are effective at low concentrations (millimolar to micromolar). Some hormones might fit into this class, but there is little or no evidence that they do.

Information and translation molecules: DNA and RNA in all their forms including the RNA-containing ribosomes. They are active at relatively low concentrations.

Hormones or growth substances: What characteristics diffrentiate these from the above groups?

It is clear that we cannot specify that translocation is a necessary feature of hormone action. Nevertheless, while some substances thought of as hormones are not always translocated, translocation is a unique feature of each of the principle hormones acting in one or more capacities. That is, it is possible to find at least one example of translocation for each of the principle hormones.

Because some substances act in concentrations as low as those of hormones, activity in the micromolar range is not a sufficient criterion to define a plant hormone. Nor can we define plant hormones in terms of their mode of action because, considering what little is known, it seems clear that hormones can act in more than one way: activate genes or second messengers or enzymes (little evidence for this so far, but could they act as coenzymes? allosteric activators?), etc. Probably the most critical point in our intuitive feelings about plant hormones is that they regulate or are essential for various steps in plant growth and development (hence, the term plant growth regulators).

With these ideas in mind, it is time to try a definition of a plant hormone (phytohormone):

An organic compound that is synthesized in a plant and that, at micromolar concentrations, regulates or is essential for various steps (growth and development) in a plant's life cycle, including various responses to environmental changes. In some but not all cases, hormones act as chemical messengers (i.e., they are transported from a site of synthesis to a site of action, as is the rule in animals). The term should not be applied to mineral nutrients, primary metabolites (e.g., sucrose), nor to compounds not synthesized in plants but with similar effects in plants (i.e., many plant growth regulators).

There is evidence that hormones (perhaps all of them in one capacity or another) are produced at certain steps in the life cycle of a plant even under constant environmental conditions, but in response to the genetically controlled developmental program (8). We need to know much more about the role of hormones in the normal, genetically controlled growth and development of a plant. Hormones can also act as mediators in plant response to environment, which is the point to be emphasized in this chapter. As some environmental factor changes, one or more hormone response systems is changed, and this leads to changes in growth and development. That is, hormones act as part of the transduction system in a sequence of events that includes *perception, transduction,* and *response.* The first two of these merit some discussion here. We'll refer to all three in the remainder of the chapter.

Perception: I know of no example in which the hormone is thought to be part of the perception mechanism although it is tempting to classify phytochrome as a hormone. Two kinds of perception mechanisms have been extensively studied: *pigments* (chlorophyll, phytochrome, cryptochrome, probably others) and *statoliths* (in-plant responses to gravity: amyloplasts, $BaSO_4$ crystals, others?). Changing temperatures change enzyme activities and membrane characteristics (permeabilities, at least). Drought might influence membrane characteristics (i.e., the turgor applied to the membrane might influence its characteristics) or (less likely) enzyme activities. The very term perception implies that something changes in the plant in direct response to changing environment.

Transduction: The changed perception mechanism often changes something else that in its turn leads to a changed response. The something else that is changed constitutes transduction. (This step can be bypassed; the changed perception mechanism can directly affect the response system, as is the case when enzyme activity is changed by changing temperature, for example.) The exciting thing about the discovery and study of hormones was that they might well be involved in at least some transduction mechanisms; evidence has accumulated to support this notion. In this chapter, I will ignore the cases where they might not be involved and describe situations in which hormones are directly or indirectly involved in transduction.

A question has governed our thinking about this for many decades: When a hormone is part of the transduction mechanism between environmental change and plant response, how does the hormone regulate or control? The wording of the question suggests that a changed perception mechanism leads to a changed concentration of hormone, which then leads to changing response. This is integral to the messenger concept and even to the transduction concept itself. The hormone concentration could change in three ways: First, as in the Cholodny–Went–postulated mechanism for photo- and gravitropism (discussed below), the hormone might be transported from one place to another, increasing in concentration at the target site and decreasing somewhere else. This would be a true messenger mechanism. Second, the synthesis or destruction machinery of the hormone might be influenced by the changed perception mechanism. There is evidence that this happens in some cases. Third, there could be a combination of these two effects (possibly acting as backup systems).

If we restate the question, we begin to see other possibilities: If a hormone is involved in a plant response to environment, what role does it play? It is quite possible that the hormone is an essential link in the response chain but that its concentration does not change: that the environment changes some other step in the process. That is, the hormone might be essential but not controlling, in which case we might decide not to think of it as part of the transduction mechanism proper.

Most plant physiologists are just now beginning to recognize the importance of this second possibility as emphasized by Trewavas in 1981 (3). It is possible that the changed perception mechanism might in some way act on the receptor sites in the target cells while hormone concentration remains constant. This could lead to changing sensitivity or responsiveness of the cells to the hormone. The changed perception mechanism might even act at some point beyond the hormone receptor sites, but this could be difficult to determine.

There are two possible effects of genes or environment on receptor sites: First, the number of sites might be affected. By analogy with Michaelis–Menten enzyme kinetics, this would be an effect on V_{max} sensitivity. Second, the tenacity of binding between hormone and binding site might be affected. In Michaelis–Menten kinetics, this is an effect on K_m sensitivity. The second effect is somewhat unexpected, but there is good evidence that it occurs, as discussed below in relation to gravitropism in the section "Transduction: Possible Role of Changing Sensitivity to Auxin."

Roles for changing hormone concentration and changing sensitivity to hormones are not mutually exclusive. Some combination of changing concentration and changing sensitivity might also occur.

The hormone field is huge; several books have been written on the topic (e.g., 9–13). Table 1 summarizes the plant hormones known to me and most of their known effects in plants, organized according to steps in the life cycle. There are six well-accepted groups of hormones plus another set of proposed hormones. The importance of the environment is suggested in Table 1 although some cases of environmental control are much more clear-cut than others. The few cases in which changing sensitivity is known or strongly suspected to be a part of the transduction process are also indicated.

One could organize a discussion of the role of hormones in plant response to environment in at least three different ways: based on the different hormones, steps in the life cycle (14), or responses to different environmental factors. It would also be possible to emphasize the nature of stress and the importance of the biological clock (14), but these points will not be discussed here. The organization here is based on different environmental factors. Because the field is so large, only selected examples can be discussed.

Table 2 outlines the examples of hormones involved in transduction processes that are discussed in this chapter. The examples discussed in the chapter follow the table approximately but not exactly. The column on receptors emphasizes that they must always be involved in hormone action. Almost by definition, if a hormone is involved in growth or development, there must be a hormone receptor to complete the step and bring about the response (although there could be more steps after the receptor has been activated and before the response occurs). It is this universality of receptors that emphasizes the potential importance of changing sensitivity to hormones. The environment could be influencing hormone concentrations or hormone receptors or both.

LIGHT

Light Quality and Seed Germination

Germination of seeds from many species is promoted by red light, which produces the P_{fr} form of phytochrome. Phytochrome is the perception mechanism in these cases. What happens after P_{fr} forms? This problem has been studied intensively, and much has been learned. For example, many studies have shown that applied gibberellins (GAs) and sometimes cytokinins will overcome the light requirement for germination; furthermore, abscisic acid (ABA) has often been implicated in causing dormancy in seeds (see general texts; e.g., 7,15). Does this mean that P_{fr} normally causes synthesis of GAs and cytokinins or destruction of ABA? The most obvious first step is to look for positive (GAs and cytokinins) and negative (ABA) correlations in concentrations of these hormones with the breaking of seed dormancy. The expected correlations appear sometimes but not always. Two possible reasons for lack of correlation come to mind.

First, entire seeds have been analyzed, but it could be just the concentration in the embryo or other parts of the seed that affect germination. For example, when a seed germinates, the radicle almost always elongates first, and only a few cells in it could be responsible for this elongation, as in normal root growth. Hormones could even be transported from one part of the seed to another; i.e., P_{fr} could be affecting either synthesis or transport. Second, P_{fr} could also be affecting sensitivity to the hormones that take part in some phase of the germination mechanism, in which case hormone concentrations might not change. Indeed, the genetically controlled developmental program (rather than environment) might affect hormone concentration or sensitivity to hormones or both.

Table 1 Summary of Known Hormone Effects in Plants
(Mostly from Application Experiments)

Plant Responses	AX	GAs	CKs	C_2H_4	ABA	TC	BRS	SA	BT	JA	TG	L
Basic Cellular Processes:												
cell enlargement	+	+	+	+/-								
cell division	+	+	+									
The Seed (formation):												
induces transport of photosynthate towards developing seeds					+							
induces storage protein synthesis in seeds					+							
The Seed (dormancy and germination):												
release from dormancy (induce germination)		+S	+	+	–							
enzyme (α-amylase) production after germination (barley)		+			–							
germination of liverwort gemmae												
The Vegetative Plant (whole plant):												
vascular tissue differentiation	+											
tropistic responses	+S											
assimilate partitioning	+											
shoot and root growth and differentiation				+								
stimulates growth and yield when sprayed on whole plants at low concentration		+				+						
increase sensitivity to auxins							+					
resistance to certain pathogens including TMV								+				
growth of liverwort thallus												+
The Vegetative Plant (roots):												
adventitious root initiation	+			+								
The Vegetative Plant (stems):												
apical dominance	+											
stem growth		+S										
shoot and bud initiation (plants, tissue culture)			+									
growth of lateral buds	–		+									
potato tuber formation		–	+							+		

Table 1 Continued

Plant Responses	AX	GAs	CKs	C_2H_4	ABA	TC	BRS	SA	BT	JA	TG	LA
Vegetative Plants (leaves):												
Leaf expansion			+									
Stomatal opening			+	−								
Chloroplast development			+									
Closing of *Mimosa* and other leaflets											+	
The Reproductive Plant (flowering):												
Bolting in long-day plants		+										
Flower induction	−/+	+/− S	+/− S	−/+	−							
Flowering of bromeliads	+			+S								
Flower opening				+								
Growth of flower parts	+											
Promotes femaleness in dioecious flowers	+			+								
Promotes maleness in dioecious flowers		+										
Flower senescence				+								
Heat production in *Arum* lilies								+S				
Reproductive Plant (fruit maturation):												
Fruit setting and growth	+	+	+									
Fruit ripening	+			+								
Fruit drop	−											
Senescence and Dormancy:												
Leaf senescence	−	−	−	+	+					+		
Leaf and fruit abscission	−/+	−	−	+	+							
Release from dormancy		+	+	+	−							
Dormancy in bulbils of yam (*Dioscorea*)										+		

Key to Hormone Abbreviations

AX = auxins
GAs = gibberellins
CKs = cytokinins
C_2H_4 = ethylene
ABA = abscisic acid
TC = triacontanol
BRS = brassinosteriods
SA = salicylic acid
BT = batasins
JA = jasmonic acid & related
TG = turgorins
LA = lunularic acid

Key to Shading and Symbols

 Strongly involved in response to environment

 Suspected to be involved in response to environment

 Indirectly or not known to be involved in response to environment (genetically programmed?)

+ promotes the process (or no effect in some cases)

− inhibits the process (or no effect in some cases)

S changing sensitivity to the hormone is known to be involved.

Table 2 Some Examples of Hormones Involved in Transduction

Environmental change	Perception mechanism	Transduction		Rec[a]		Response
Light						
Quality (may detect irradiance level)	Phytochrome	→ GAs, cytokinins, ABA (ethylene?)	→	Rec	→	Seed germination
	Phytochrome	→ GA₁	→	Rec	→	Stem elongation, leaf unrolling, etc.
	Cryptochrome	→ Flavoprotein	→	Rec	→	Stem elongation decreased
Duration	Phytochrome	→ Clock→ GAs, etc.	→	Rec	→	Stem elongation, tuber formation, autumn syndrome, etc.
	Phytochrome	→ Clock→ Florigen	→	Rec	→	Flowering
	Phytochrome?	→ Clock→ salicylic acid	→	Rec	→	Heat generation in *Sauromatum spadex*
Direction (phototropism)	Cryptochrome	→ Synthesis of auxin/inhibitor	→	Rec	→	Stem cell elongation = stem bending toward the light
	Cryptochrome	→ Transport of auxin/inhibitor Auxin?	→	Rec	→	Solar tracking
Temperature						
Autumn cooling (often combined with SDs)	Enzymes?	→ Various hormones?	→	Rec	→	Induction of dormancy (hardiness, etc.)
Winter cold (sometimes combined with LDs)	Enzymes?	→ Various hormones?	→	Rec	→	Breaking of dormancy (growth, loss of hardiness, etc.)
Drought	Cell turgor in leaves and roots	→ ABA (synthesis/transport?)	→	Rec	→	Stomatal closure
		→ ABA (synthesis?/transport)	→	Rec	→	Stomatal closure
Gravity	Statoliths	→ Auxin synthesis?/transport?	→	Rec	→	Stem cell or root cell elongation leading to gravitropic bending
	Statolith	→ Bypass	→	Rec	→	Stem cell or root cell elongation leading to gravitropic bending
Touch	Cell turgor?	→ Turgorins	→	Rec	→	Leaf closing, possibly other responses

Studies with gibberellin mutants have provided some evidence about the relative roles of concentration and sensitivity (8,16–19). The dwarf mutants have blocks in gibberellin synthesis or response. In most cases, their seeds germinate normally, suggesting that GAs might not be involved—but they were selected for dwarfism, not germination ability. Most are cultivated species with no seed dormancy. Netherlands researchers (20–22) found GA-deficient, dwarf mutants of tomato and *Arabidopsis thaliana*, the seeds of which do require applied GAs to germinate. Tomato seeds showed no photodormancy in either wild-type or mutant; hence, there is no evidence that P_{fr} induces GA synthesis in that case. Wild-type *Arabidopsis*, on the other hand, requires P_{fr} or GAs. A mixture of GA_4 and GA_7 is especially effective (1000 times less required than GA_3). The mutant required either *light* plus 1 μM GA_{4+7} or, in the dark, 100 μM GA_{4+7}. Thus, applied GAs overcome both the genetic block and the light requirement. Does this mean that light normally causes GA synthesis? Perhaps. But the concentration difference suggests that *light* (P_{fr}) *increases the sensitivity to GAs*. It appears that 100 times less GA_{4+7} is required when P_{fr} is present. The sensitivity possibility needs study in these systems.

In highly dormant celery cultivars, the requirement for light is not effectively replaced by either gibberellic acid (GA_3) or a cytokinin, but application of both hormones does break dormancy in darkness (23). This could again suggest that sensitivity to GA_3 or cytokinin is very low except in the presence of P_{fr}—or that P_{fr} causes synthesis of both GAs and cytokinins.

Ethylene cannot break photodormancy. But ethylene can partially overcome the dormancy caused by high temperatures in lettuce seeds and can overcome certain photodormancy problems in cocklebur seeds (24).

Light Quality and Stem Elongation

Much study over many years (see basic texts) has examined the effects of light on stem elongation, and it is well known that GAs (ultimately GA_1, in some if not all species) can strongly promote stem elongation. Does this act through P_{fr}? In many seedlings, blue and far-red (less P_{fr}) or only far-red light inhibits stem elongation, but in mature plants that normally grow in the sun, far-red light (leaf shade) promotes stem elongation. Applied GAs will overcome the inhibitory effects in some cases (brief review in 25).

As long ago as 1963, Kende and Lang (26) measured the levels of chromatographic fractions that were probably identical with GA_1 and GA_5 in pea stems grown in the light or in the dark. Irradiation with weak red light (i.e., production of P_{fr}) inhibited elongation of dark-grown stems but did not change the levels of either GA fraction. Application of GA_1 or GA_5 in the dark strongly promoted stem elongation, but after irradiation GA_1 promoted elongation but GA_5 did not. Kende and Lang concluded that light (P_{fr}) did not change tissue levels of the GAs but reduced the sensitivity of the stems to endogenous GA_5 but not GA_1. If, as seems likely, GA_5 must be converted to GA_1 to promote stem elongation, light (P_{fr}) might someway inhibit the conversion system.

Recent work, much of it relating to light control of stem elongation and the early stages of seedling development, has shown that P_{fr} can control gene expression, especially including the expression of the gene that controls the production of phytochrome itself. For example, Cotton et al. (27) studied a 4.2-kilobyte messenger RNA species that codes for phytochrome in 5-day-old, dark-grown cucumber seedlings. Two hours after irradiation with red light (i.e., production of P_{fr}), this phytochrome mRNA was reduced to 40% of its initial abundance; 10 hr after treatment, the mRNA had returned to the time-zero, dark-control level. Benzyladenine, a cytokinin, also reduced the mRNA to 45% of its dark abundance, but the mRNA level did not recover after benzyladenine treatment (perhaps because the benzyladenine was still present in the cells?). Thus production of P_{fr} by red light downregulates the gene that produces phytochrome, and cytokinins can have the same effect. Does this mean that P_{fr} acts to downregulate itself by leading to a production of cytokinin? Or does P_{fr} directly influence gene expression? Future research will be required to find out.

Light Quality and Unrolling of Grass Leaves

Unrolling of such grass leaves as wheat, barley, and maize is controlled by a typical phytochrome reaction: Low fluences of red light cause the cells on the concave surface to grow more than those on the convex surface, so that the rolled leaf (as it emerges from the seedling) flattens. Subsequent exposure to far-red light reverses the effect of the red light (27a). Applied GAs and, in some species, cytokinins replace the need for light (28). These results suggest that P_{fr} causes rolled leaves to form GAs or cytokinins that then cause unrolling, and red light does promote GA synthesis and release from young plastids in rolled wheat and barley leaves before the leaves unroll (29). It is also possible that P_{fr} changes sensitivity to the GAs that are already present in the cells of the concave surface. There is no evidence as yet relating to P_{fr} effects on cytokinins.

Light Duration: Photoperiodism

Photoperiodism has been widely studied in many plants for many years. Several of these studies have involved hormones, and GAs have been strongly implicated. There are many responses to changing photoperiod but not many valid generalizations about these responses. There is one good criterion for a true photoperiod effect: The long-day response must occur when the dark period is interrupted with a brief interval of light. This is the *night break phenomenon*. (Often more light is required to promote a long-day response than to inhibit a short-day response.)

With many species, long days produce stem elongation and short days produce the autumn syndrome: shorter stems, reduced chlorophyll production and increases in other pigments, leaf senescence and abscission, development of winter hardiness, and bud dormancy. Many specific effects have been documented (reviews: 7,30–34). Effects may be produced by short days (SDs), long days (LDs), or various combinations of day lengths. Examples include flowering of many plants (absolute control by photoperiod in some, quantitative promotion in others, and virtually no effect in day-neutral species); seed germination of a very few seeds; tillering of grasses (formation at the crown of separate flowering stems, typically an SD response, but an LD response in rice); development of storage organs (tubers and roots); vegetative reproduction (e.g., foliar plantlets on *Bryophyllum* under LDs); and breaking of winter dormancy in the buds of a few woody plants under LDs.

What is the role of phytochrome and of hormones? Phytochrome has often been implicated (as in effects on stem elongation independent of photoperiod, which can be a complication in studies of photoperiod effects on stem growth), but the plants must also measure the length of day and/or night (good evidence that both are important, although the dark period seems to be especially critical because it is easier to inhibit the process by interrupting the dark period with light than the light period with dark). Thus, the biological clock must play a role. The perception mechanism must involve phytochrome (P_{fr}, but P_r could also be important) and some timing mechanism. It is possible that the timing mechanism, interacting with phytochrome, might be part of the transduction system. Interaction of phytochrome and timing in turn might affect the synthesis of hormones (GAs, especially), which in turn control stem elongation (and flowering?). It is also possible that this phytochrome/clock system might influence the sensitivity to hormones— or both responses might occur together. This is another vast field; we'll consider just a few examples.

Stem Elongation in Silene armeria

Both flowering and stem elongation in *Silene armeria* are promoted by 3–6 LDs, and stem growth continues under subsequent SDs although at a reduced rate. Stem elongation is induced by application of GA_3, but flowering is not. Furthermore, stem elongation is inhibited by a GA inhibitor (tetcyclacis) but flowering is not affected; thus, flowering is apparently not dependent on GAs in this plant.

Talon and Zeevaart (35) identified and measured GAs in this plant under SDs, LDs, and subsequent SDs. All the GAs were members of the early 13-hydroxylation pathway leading to GA_1 and thus stem elongation. All were present under both SDs and LDs, but GA_{53} was highest under SDs and decreased under LDs, whereas GA_{19}, GA_{20}, and GA_1 increased in plants when they were transferred to LDs but then declined. When transferred back to LDs, GA_{53} increased, while GA_{19}, GA_{20}, and GA_1 decreased back to their levels under SDs. But elongation continued! The authors concluded that photoperiod modulates GA_1 in shoot tips upon transfer to LDs, which in turn leads to stem elongation. This is an example of control by changing concentration. The continuing stem elongation upon return to SDs, however, suggests that *sensitivity* to GAs has also been increased by the LD treatment. This is an excellent example of transduction that involves influences of environment (the perception mechanism) on both hormone concentration and sensitivity to the hormone. Note, however, that increased sensitivity could be caused either by the LDs or by the increased GAs activating their own receptors, as auxin might do in other systems (e.g., indirect indication in 36).

The Induction of Flowering

Excellent evidence for a translocated flowering hormone, called *florigen*, has accumulated since the 1930s (see above reviews of photoperiodism):

1.   The leaf responds, but the bud becomes the flower. For example, with an SD plant such as cocklebur, one can enclose a single sensitive leaf (the most rapidly growing leaf) in an envelope made of black paper for the necessary 9 hr or more, and the plant will flower. Some stimulus is clearly transported from the leaf to the bud.

2.   In numerous experiments, induced plants have been grafted to plants held under noninducing conditions (either LDs or SDs as the case may be), and the flowering stimulus has been transmitted through the graft union, causing the noninduced plants to flower. In one of the most impressive of these experiments, Chailakhyan (the original formulator of the florigen concept) and Frolova in Moscow, cooperating with Lang of Michigan State University (37), grafted plants of both SD and LD cultivars of tobacco onto a day-neutral (DN) cultivar and then grew them under various light conditions. When exposed to SDs, the SD graft partners caused the DN plants (DNPs) to flower earlier, and under LDs, LD partners also promoted earlier flowering of DN partners. Clearly flowering promoters were present. When the SD graft partners were maintained under LDs, flowering in the DN species was slightly or not at all retarded, but under SDs, the LD tobacco completely prevented

flowering of the DN partner. That is, noninductive conditions resulted in production of a flowering inhibitor—an *antiflorigen*.

3. Cocklebur (*Xanthium strumarium*) can be defoliated to a single, highly responsive leaf, and this leaf will respond to a single dark period longer than the critical night by inducing the buds to flower. When a 16-hr dark period is given and leaves are removed immediately after the dark period, the plant fails to flower—but plants flower as well as nondefoliated controls when the induced leaf is allowed to remain on the plant for a dozen or more hours (depending on conditions) after the end of the dark period. It is as though early defoliation removes the florigen before it has had a chance to move out of the leaf; some hours later, the florigen has left the leaf, which is therefore no longer needed for induction (see discussion in 7). The experiment has also been done with other species that require only a single inductive cycle. The LD grass *Lolium temulentum* (darnel), for example, produces a flowering inhibitor during the dark period of SDs. Removal of the leaves at various times during this dark period (while other leaves are in the light) gives a translocation curve for the inhibitor much like that for the *Xanthium* promoter (38).

In spite of numerous attempts to extract the hypothetical florigen, so strongly supported by these three indirect evidences, or to find compounds that will act like florigen, successes have been limited and difficult to reproduce (summaries in 32,39). There have also been numerous attempts to influence flowering with applied hormones, growth regulators, and antimetabolites. Applications have been at various times in relation to the inductive cycles (best with plants that respond to a single cycle). Many successful experiments of this type have been reported: both inhibitions and promotions (30,34).

There have also been many measurements of hormones as a function of time during the inductive process. The story is highly complex, and it is seldom possible to say that the hormones are really part of transduction between stimulus perception (typically phytochrome and a clock mechanism) and the flowering response. Nevertheless, here are a few examples.

CYTOKININS Applied cytokinins promote flowering in several species and inhibit flowering in a few. Often their effects are strongly fortified by the presence of other hormones, especially GAs but also indoleacetic acid (IAA). In some cases, it can be shown that sensitivity of the apex to cytokinins is important and is even influenced by environment. *Arabidopsis* plants must be grown in noninductive SD for at least 3 months so that the central zone of their meristems is in the "intermediate" configuration; i.e., ideally sensitive (40). With *Pharbitis nil*, there is evidence that the inductive long night sensitizes the apical meristem to exogenous cytokinin (41).

Some dramatic changes in cytokinin levels related to flowering have been reported. In Douglas fir, for example, the largest measured difference between vegetative and reproductive buds was an increase in cytokinins; GAs, IAA, and ABA were also measured (42). *Xanthium*, on the other hand, exhibits a sharp decrease in cytokinin activity in leaves, buds, and root exudate following a single inductive long night, and this (as well as flowering) is nullified by a night interruption with light (43). But cytokinin in the phloem sap exported from induced leaves does increase. It was postulated that a signal of unknown nature moves quickly in the phloem from induced leaves of *Xanthium* and *Sinapis* to the roots, where it alters the course of cytokinin production and/or release.

GIBBERELLINS   GAs can elicit a flowering response in many LD- and cold-requiring rosette plants grown in noninductive conditions (reviewed in 44), but their role seems to be strongly species-dependent. GAs cause both bolting and flowering in many of these plants grown on SDs, and in some cases both effects are overcome by growth retardants (anti-GAs). In other cases (as the *Silene* mentioned above), bolting depends on GAs but flowering does not (is not inhibited by anti-GAs). Flowering of caulescent species is seldom promoted by GAs, and GAs are clearly inhibitory to flowering of some species (e.g., the SD strawberry and *Ribes*, the LD *Fuchsia*, the short-long-day *Poa* and *Bromus*, and the DN tomato, apple, and *Citrus*).

There are three important complications in the studies of GA effects on flowering. First, the particular GA that is used can make a large difference. Pinaceae react to GA_{4+7}, not to GA_3; in apple, GA_3 is inhibitory and GA_4 promotive; there are many other examples. Second, the GA effect may depend on an association with some adjunct treatment such as water stress or root pruning in Pinaceae (45). Third, the timing of GA application is often critical, as best illustrated with plants that require only a single inductive cycle. In dwarf *Pharbitis*, for example, GA_3 is promotive when applied just before the inductive long night and inhibitory when applied just after (46), but this sequence is reversed in some other species (e.g., in *Fuchsia* and tomato; see 12). Clearly, the genetic background is critical to how plants respond to GAs (and to other hormones as well; examples in 30).

AUXINS   Auxin can both promote and inhibit flowering in both LDPs and SDPs, but inhibition is by far the more common response (30). Ethylene inhibits flowering of many species (but see discussion of bromeliads below), and ethylene production in response to auxin is most likely one complication. The applied concentration and timing are also important (as with all hormones); changes in sensitivity to auxin could well be involved. Auxin levels in SDPs are generally lower just before flower inhibition, suggesting that auxin levels could be part of transduction in a few species.

ETHYLENE   Exogenous ethylene inhibits flowering in many SDPs, but in such bromeliads as pineapple and mature *Guzmania*, flowering is produced by applied ethylene or 1-amino-cyclopropane-1-carboxylic acid (ACC, ethylene precursor), and by shaking (which mechanical stress causes ethylene production in many species); flowering is inhibited by the ethylene inhibitor aminoethoxyvinylglycine (AVG), and this is overcome by application of ACC (47). Apparently, ethylene is the only factor controlling flowering of mature *Guzmania* and possibly (probably?) other bromeliads. Ethylene does not cause flowering of juvenile *Guzmania*, suggesting that the plants must develop sensitivity to the ethylene before they can respond (i.e., ripeness to respond may well be a function of developing sensitivity to ethylene).

ABSCISIC ACID   Although applied ABA inhibits flowering of several SDPs and LDPs in inductive conditions, endogenous ABA levels do not bear consistent relationships with day length (30). Hence, ABA does not appear to be part of the transduction mechanism in flowering controlled by photoperiod.

Salicylic Acid and the Voodoo Lily

Sauromatum guttatum, the voodoo lily, is related to such other species in the Araceae as skunk cabbage and jack-in-the-pulpit (collectively called *Arum* "lilies"). The inflorescence, which develops from a corm and can be 80 cm tall, consists of a phallus-shaped organ called the *spadix* (with the upper part called an *appendix*) surrounded by a large bract, the *spathe*. Soon after the spathe unfolds to reveal the spadix, the female flowers are receptive to pollen, and the appendix generates heat, which volatilizes amines and indoles that have a putrescent odor and attract insects. The heat generated raises the temperature of the appendix as much as 14°C above ambient, but the temperature returns to normal in the evening. A second episode of heating occurs later in the night in the lower portion of the spadix hidden inside the pollination chamber. During the rapid, cyanide-resistant respiration that generates the heat, oxygen consumption is as high as that of a hummingbird in flight. Flies attracted by the stench are imprisoned by hairs (rudimentary male flowers) at the mouth of the spathe until the male flowers below them mature and the hairs wither, allowing the flies dusted with pollen to escape to another plant with mature female flowers. Similar although often less spectacular heating also occurs in the inflorescences of plants in a few other families (e.g., cycad, water lily, and palm families; see 48).

In 1937, van Herk in The Netherlands (see 49,50) suggested that heating is initiated by "calorigen," a water-soluble substance produced in the staminate flower primordia just below the appendix. For many years Bastiaan Meeuse at the University of Washington in Seattle and his colleagues studied the *Arum* lilies and attempted to isolate calorigen. Their efforts were finally successful (50), and calorigen proved to be 2-hydroxybenzoic acid, commonly called salicylic acid. (Aspirin is acetylsalicylic acid.) A calorigen extract (initially that of van Herk), or salicylic acid after calorigen's identification, caused warming of the appendix tissue when applied on the day preceding the day of heating.

In this system, a compound synthesized in one part of the plant (the staminate flower primordia) is transported to another plant part (the appendix) where it exercises absolute control over a metabolic process (the generation of heat by respiration via the cyanide-resistant pathway). In this sense, it fits the classical definition of a hormone. The temperature of the appendix is proportional to the amount of salicyclic acid that is applied, so concentration plays a controlling role. Only 200 μL of 15 μM salicylic acid applied to a 3-g slice of appendix caused the heating, so the operative concentration is in the range of effectiveness of other plant hormones.

In spite of the clear-cut role of changing salicylic acid concentration, sensitivity to salicylic acid also plays an important role. The maximum heat release always occurred after 4.5 hr of light exposure of the appendix, and this timing was independent of salicylic acid concentration. Furthermore, the time of application of salicylic acid the previous day strongly influenced the amount of heat that was produced—again without influencing the timing of heat production. Clearly, the appendix tissue is undergoing changes in sensitivity to salicylic acid, and these changes are controlled by the biological clock as it is exhibited in photoperiodism. (Possibly phytochrome is interacting with the biological clock as in other cases of photoperiodism; this has not yet been studied in suitable detail.)

Tuber Formation in Potatoes

Potato tubers form at the ends of so-called *stolons* (actually, underground rhizomes), typically in response to short days (but always with considerable temperature interaction and depending on cultivar; reviews in 7, 51, and 52).

SDs result in lowering of GAs and (probably consequent) reduction in stolon elongation; application of GAs will halt tuber formation, but this is reversed by simultaneous application of the GA inhibitor, CCC or Cycocel. Ethylene also causes stolen elongation to slow or stop, but ethylene may also stop tuber formation (53). Clearly, cytokinins must be present for tuber formation to occur, and they do increase during tuber initiation, but the increase is smaller than might be expected, and they decline about 4 days after tuber initiation has begun. Applied cytokinins do not induce tubers.

Tuber initiation is more than a response to lowered GAs or increased cytokinins; there is good evidence for a *tuber-inducing substance* that forms in the leaves of most cultivars in response to SDs. An induced leaf can be grafted to a noninduced plant, causing it to form tubers. There is a critical night and expected effects of a night break (54). In some cultivars, the photoperiod control is lost when night temperatures are low (e.g., below 20°C; optimal at 12°C). In sensitive cultivars, the shoots must be exposed to low temperatures as well as SDs.

Recently, a tuber-inducing substance was isolated by Yasunori Koda and coworkers (55) in Japan. It proved to be a relative of jasmonic acid (identified, according to 52, as 3-oxo-2(5-?-D-*p*-glucopyranosoloxy-2-*cis*-pentyl)cyclopentane-1-acetic acid). It will induce tubers in vitro (single-node stem segments) at concentrations of 3×10^{-8} M (0.03 μM), in the range of effective concentrations for most hormones. This appears to be another excellent example of a hormone that is a translocated, chemical messenger.

Light Direction: Phototropism

Phototropism, in which plants bend toward a unilateral light source, is an extremely complex phenomenon (reviews: 7, 56, 57). As the Darwins (58) discovered over a century ago, canary grass (*Phalaris canariensis*) coleoptiles bend little or not at all unless their tips are irradiated, but oat (*Avena sativa*) coleoptiles will react phototropically even when tips are removed although the base response is 1000 times less sensitive than the tip response. The localization of the response is less clear in dicot hypocotyls, but the response to different light levels is closely similar. There is a so-called first positive curvature (increased bending toward the light with increasing light energy, called *fluence*) that obeys the law of reciprocity (irradiance times duration equals a constant) and a second positive curvature that does not but that occurs at the same time after beginning of irradiation, more or less independently of fluence rate (59).

Blue light is most effective in causing the response, and there is evidence that a flavoprotein is the photoreceptor pigment (called *cryptochrome* until it is positively identified). Carotenes, which also absorb blue light, might also be involved, if only as screening pigments that increase the steepness of the light gradient across the coleoptile. Phytochrome influences the extent of the response but does not account for it; red light (P_{fr}) desensitizes the tissue to blue light so that 10 times as much blue light is required to produce a given degree of first positive curvature when plants have first been irradiated with red light. Action spectra for alfalfa and *avena* are given by Baskin and Iino (60), and they are closely similar for the two species.

There are many related responses such as the movement of leaves that produces leaf mosaics or the solar tracking that is observed in many species, often with cup-shaped leaves. Relatively little is known about the hormone relations (transduction mechanisms) in these cases, so they will not be discussed here.

The transduction for stem phototropism has been discussed since early in this century: Blaauw in 1909 and 1918 (61,62) documented that blue light was most effective in phototropism and suggested that it *inhibited growth on the irradiated side*. In 1926, Cholodny (63) and Went (64) theorized and Went demonstrated that auxin apparently migrates from the irradiated to the shaded side of a unilaterally irradiated coleoptile tip. Indeed, this was part of the initial discovery of auxin. This Cholodny–Went hypothesis has guided research for nearly 70 years.

During recent years, however, the picture has become more complex rather than simpler. The evidence for Cholodny–Went is based on Went's (64) classical experiment. Briggs (65) repeated this with maize coleoptile tips placed on agar blocks in four ways: (a) in the dark (controls), (b) with unilateral irradiation, (c) with such irradiation and a mica sheet dividing the light from the shaded sides of the tip almost to the top of the tip, or (d) dividing the tip all the way to the top. The agar blocks were then put on prepared coleoptiles in the classical *Avena* curvature test for auxin. There was about the same total "auxin" (actually, curvature) in all four cases. This led to the conclusion that there was no destruction of auxin as predicted by Blaauw (61,62). There was nearly twice as much auxin in the block under the shaded side as under the light side when the tip was not completely divided, but there was about the same amount in both blocks when the tip was completely divided. This was considered strong evidence for lateral transport; the results provided primary data to support the definition of a plant hormone as a compound that must be transported.

Briggs and Baskin (66) summarized the criteria [originally formulated by Firn and Digby (67)] that must be met if the Cholodny–Went hypothesis is to be validated—and they concluded that all the criteria could be met. I have summarized their summary as follows:

1. An auxin gradient must form in response to unilateral irradiation. Several studies (e.g., 68, 69) seemed to show that auxin transport can be influenced by unilateral light—although there were negative studies and several complications as well (see below).

2. It must be shown that auxin is indeed a factor limiting growth in the responsive organ and that the auxin gradient is sufficient to account

for the observed differences in growth rate. Baskin et al. (70) applied auxin to coleoptile tips and reported that auxin was indeed limiting for growth and that there was an approximately linear relation between applied auxin concentration and growth over a range that spanned the growth rates occurring in phototropic bending. Furthermore, an auxin gradient created at the coleoptile tip was sustained during its basipetal transport.

3. In a phototropically curving organ, acceleration of growth on the shaded side should accompany retardation of growth on the irradiated side. Franssen et al. (71) reported that when high light was used in the unilateral irradiation, growth on the irradiated side stopped almost instantaneously upon the beginning of irradiation, whereas growth of the shaded side continued, typically at about the same rate as before the beginning of unilateral irradiation. This finding clearly supports Blaauw, but if coleoptiles are irradiated with red light before the unilateral irradiation with blue light, then growth is compensatory as predicted by Cholodny–Went (e.g., 72–74).

Although Briggs and Baskin (66) concluded that all the criteria had been met and that Cholodny–Went was thus strongly supported, there are other recent complications. The contrary evidence has become so persuasive that an unmodified Cholodny–Went mechanism now seems almost untenable. Franssen and Bruinsma in 1981 (75) failed to find an auxin gradient across phototropically stimulated sunflower hypocotyls, for example, but they did demonstrate a gradient of an inhibitor, xanthoxin, with 60–70% of the inhibitor on the lighted side.

Furthermore, in 1989 Hasegawa et al. (76) isolated and characterized three inhibitors that formed a similar gradient across phototropically stimulated radish hypocotyls. They further studied the classic *Avena* coleoptiles, duplicating the original experiments of Went (and of Briggs) with the agar blocks. They obtained the same results: more curvature of test coleoptiles with blocks that were under the shaded side of the partially divided tips. But when they analyzed the blocks with a physicochemical assay, they found equal amounts of IAA in both blocks, and there was about 2.5–7 times as much as was indicated by the curvature test. They suspected an inhibitor in both blocks but especially in those under the lighted side—and they found two as yet unidentified inhibitors in the blocks. Such a finding, if confirmed by future research, calls into question all the previous experiments with agar blocks and the *Avena* curvature test; all could be compromised by the presence of inhibitors. Of course, such inhibitors might well qualify as hormones.

Thus, in spite of the careful analysis of Briggs and Baskin (66), it appears that the Cholodny–Went mechanism of transduction as originally formulated for phototropism is strongly suspect. It could survive in a reverse sense if the inhibitor is transported from the dark to the light side of the unilaterally irradiated coleoptile. Or Blaauw could prove to be right in a reverse sense if the inhibitors are synthesized on the lighted side rather than a promoter (auxin) being destroyed on the lighted side. Talk of an inhibitor brings to mind the classical experiments of Boysen-Jensen (77) in which a sheet of mica was inserted in the tissue just below the tip of a unilaterally irradiated coleoptile. It had little or no effect when placed on the light side but inhibited phototropism when placed on the dark side. This was taken as evidence for a promoter rather than an inhibitor. (Actually, Boysen-Jensen was thinking of a kind of nervous stimulus rather than a chemical hormone.)

It is not impossible to imagine that auxin is essential for growth (i.e., of the dark side) whether an inhibitor is important on the light side or not. And the mica on the dark side would prevent movement of both the inhibitor and the essential auxin. Note that the inhibitor concept was considered by several of the early workers in the field of auxinology, as reviewed by Went and Thimann in 1937 (78), but it was rejected in favor of the idea that auxin could account for nearly all of the phenomena. Also note that virtually none of these studies has even considered effects of light on sensitivity to auxin or other hormones. [In 1992, a so-called forum edited by Anthony J. Trewavas (78a) was published; it contains 17 brief statements on the topic: What Remains of the Cholodny–Went Theory? Most of the contemporary people who have contributed to this field of research are represented in this forum.]

TEMPERATURE

As noted, many plants of temperate zones undergo a number of changes in response to the shortening days of late summer and autumn, accentuated by the decreasing temperatures of that season. This can be called the *autumn syndrome,* and its most obvious manifestations are dormancy in seeds and perennial buds and senescence in annuals (which we won't discuss). There are many other plant responses to changing temperature, at least some of which might involve hormones in their transduction mechanisms, but we will confine our discussion to the autumn syndrome and the release from dormancy in the spring.

The perception mechanisms in temperature responses have seldom been studied. Perhaps there is a tacit assumption that enzymes are being influenced, but it is conceivable that genes or gene regulators are more or less directly influenced. Enzymes could certainly be involved in synthesis or destruction of hormones—or even in changing sensitivity to hormones.

Consider some of the phenomena of the autumn syndrome: Leaves and buds of evergreen perennials (e.g., conifers) exhibit a reduced metabolic activity as autumn progresses, and deciduous perennials lose their leaves and form special frost-resistant, inactive buds, often covered with special bud scales. Seeds of most species in cold regions are dormant or quiescent during winter; in such a condition, they can withstand subfreezing temperatures. Growth in the spring often depends on prolonged exposure of dormant buds and seeds to the cold of winter—and sometimes to the long days of spring. The buds or seeds accumulate or sum the periods of exposure to cold and thereby measure the length of winter and anticipate spring. (Discussions of dormancy and quiescence and the terminology that has been applied to these concepts are in 7, 79–82.)

What roles, if any, do hormones play in the induction of dormancy or quiescence and the breaking of dormancy in spring? It is commonly assumed that dormancy is hormonally controlled, and there is evidence to support this assumption, but the basic biochemical mechanisms of dormancy induction and termination remain unknown. Most attention has been paid to ABA, GAs, and cytokinins. There is little evidence that auxins or ethylene play any controlling roles in dormancy.

ABSCISIC ACID ABA was originally discovered by a group working on dormancy, and it was thought that ABA would be strongly implicated in the induction of dormancy. Decreases in ABA were thought to be the basis for the breaking of dormancy during the cold of winter. Some correlations became evident (see review in 81), but often they failed to appear. When ABA was measured in buds or seeds that were exposed to both cold and warm storage conditions, it declined in both; thus, its decline was not correlated with the cold that broke dormancy. In other cases, it declined well before there was any diminution in dormancy.

There has been much controversy over the role of an inhibitor such as ABA or a promoter such as GAs or cytokinins. Rudnick in 1969 (83) found that much of the ABA had disappeared after 3 weeks of stratification of apple seeds although limited germination was possible at that time. Germination continued to improve during further stratification, and Rudnick found that it required increasing concentrations of applied ABA to inhibit it. Rudnick's interpretation was that a promoter was being generated during stratification and that it took increasing amounts of ABA to overcome the effects of the increasing amounts of promoter. But the actual results simply show that the seeds were becoming less and less sensitive to applied ABA as stratification progressed. This could be caused by changed ABA receptors as well as by increasing amounts of some promoting substances that acts opposite to ABA.

GIBBERELLINS   Several studies have shown an increase in GAs either during the breaking of dormancy or shortly after return to warm temperatures, but again definitive evidence is lacking that this is part of the dormancy-breaking process. It may be more closely associated with normal growth and development of seedlings than release from dormancy. Physiological dwarfs often appear when embryos are removed from mature unchilled seeds that have a chilling requirement; these seedlings have a rosette of leaves and greatly shortened internodes. Repeated applications of GAs or treatment with cold will often result in stem elongation and thus more normal seedlings. Applied GAs will also overcome at least part of the chilling requirement in buds of peach or apricot (84). Arias et al. (85) measured GAs in the embryonic axis and in the food storage cotyledon cells of Hazel seeds (*Corylus avellana*), a species in which GAs fully overcome the prechilling requirement for germination. Although GAs did not accumulate to any great extent during stratification, the embryonic axis synthesized 300 times as much GAs after return to 20°C than did the cotyledons. In this case, it seems clear that stratification increased the potential of the seedling axis to synthesize GAs rather than causing an increase in GA synthesis during the cold. Such treatment could also influence the sensitivity to GAs. In any case, it is important to analyze individual parts of seeds and buds rather than the entire seed or bud.

CYTOKININS   Application of cytokinins sometimes promotes bud growth or seed germination, and measured endogenous levels may correlate well with these same events. In many studies, dramatic effects of applied cytokinins on bud growth occurred with buds that were apically dominated rather than in winter dormancy. In some cases, cytokinin application did overcome seed dormancy; in other cases it did not (reviewed in 81). The negative cases are difficult to evaluate because they could have been caused by the wrong concentrations, the wrong cytokinins, measurement of the wrong tissues, and so forth. As with the other hormones, there are many reports of the expected changes in cytokinin concentrations in dormant buds or seeds during release from dormancy, but there are also reports of studies that failed to confirm such changes in hormone levels. Sensitivity changes could be involved but have seldom if ever been studied.

DROUGHT

Much study has shown that stomates open in response to light or CO_2-free air as guard cells take up water, and this occurs as K^+ ions move into the guard cells. Such anions as Cl^- may enter with the K^+, or organic acids (primarily malic) may be synthesized from starch in guard cells, with H^+ ions moving out in exchange for incoming K^+. Stomates close in response to water stress, and ABA at micromolar concentrations is strongly implicated as the messenger hormone that controls this closure (textbook reviews in 7, 15).

When leaves are subjected to water stress, ABA builds up in them. Pierce and Raschke (86) reported that the perception mechanism was cell turgor; changes in solute potential or water potential had little effect. For several years, one piece of evidence cast doubt on the role of ABA: When leaves were removed from the plant and allowed to dry on the table, stomates closed *before* ABA built up to measurable concentrations in the leaves. But two factors can help us to understand what was happening: First, under such severe stress, water can evaporate directly from the guard cells (hydropassive closure), causing them to lose turgor and close. Second, ABA is redistributed within the leaf from various pools to the guard cells. ABA is synthesized slowly in leaf mesophyll cells, where it is stored in chloroplasts. The pH of the chloroplast stroma is normally higher than that of the cytosol, and this leads to an accumulation of ABA, which is a weak acid. Dehydration lowers the pH of the chloroplasts, which releases some ABA; the apoplast increases in pH, resulting in a net transfer of ABA from chloroplasts to apoplast and eventually to the guard cells (87,88). Thus guard-cell ABA could easily change although total leaf ABA remained constant.

Harris et al. (89) showed that ABA in guard cells represents only 0.15% of total ABA in the leaf, so that changes in guard-cell ABA could easily go undetected if only total leaf ABA is measured (as was often done in early studies), even if newly synthesized ABA were transferred to guard cells. Harris and coworkers used an enzyme-amplified immunoassay for ABA that was about 100 times more sensitive than previously used methods. They could measure ABA in individual guard cells. When the fresh mass of detached leaves decreased 10% by transpiration, guard-cell ABA increased approximately 20-fold (to 7–13 μM). ABA in all the leaf cells increased in response to the water stress. The authors showed that the ABA was probably not restricted to the apoplast because it didn't wash out in 30 min of immersion in water—but did wash out after 4 hr of immersion, suggesting that such a loss could account for the lower concentrations reported by some other workers. It appears that ABA can act as a messenger from drying mesophyll cells in the leaf to guard cells, causing them to close.

Apparently, ABA can also act as a messenger from drying roots, moving to the leaves in the transpiration stream and causing stomates to close. This has been studied by dividing the root system of a plant into two pots, keeping one watered, and letting the other dry. Stomates close even though leaves remain completely hydrated (90,91). Here are two examples: Zhang et al. (92) measured a large increase in ABA concentration in roots in the dry pot, and ABA content of epidermal strips was closely related to the degree of stomatal closure. Zhang and Davies (93) found a close correlation between soil drying, decreased stomatal conductance, and ABA in xylem sap; leaf ABA was not as closely correlated with stomatal conductance.

If the interpretation of the results presented here does not change as future data accumulate, this will provide another excellent example of a hormone that acts as a messenger, being synthesized in one place and translocated to another at micromolar concentrations, where a response is produced. Furthermore, there seem to be two sites of perception: leaves and roots. If upper soil layers dry while deeper layers still contain moisture, this moisture will be conserved as the upper roots synthesize or release the ABA that is transferred to leaves in the transpiration stream, causing stomates to close. If atmospheric conditions are such that transpiration exceeds uptake of water from the soil, water-stressed leaf meosphyll cells will release ABA from chloroplasts and ultimately synthesize more ABA, which reaches the stomates, again causing them to close and conserve water.

GRAVITY

Along with phototropism, studies on gravitopism were important in early formation of the hormone concept (work of Dolk, especially; see below). Roots and stems and some other organs (e.g., nodes of grasses) respond differently, so that it is necessary to discuss them separately, although stem and root systems do have much in common. I will emphasize the transduction phase of gravitopism although it will first be necessary to outline the phenomenon in general terms for each system. (Reviews are in 94–97a; much of this section was modified and updated from Salisbury and Ross, 7.)

Gravitropism in Roots

Primary roots are *positively gravitropic*, bending down toward the earth's center; secondary roots tend to be *diagravitropic*, growing more or less horizontally; and tertiary roots may be virtually *agravitropic*, growing in any direction. Roots of some cultivars of maize are diagravitropic until they are irradiated with red light, which induces them to become positively *orthogravitropic* (i.e., they grow vertically downward as do most primary roots). Phytochrome (P_{fr}) has been implicated in this response. Thus, when the root grows close enough to the soil surface to be irradiated with light, it becomes positively gravitropic and turns downward. This response has also been studied in radish, and no doubt it occurs in many species besides the maize cultivar that has been most extensively studied. In any case, these gravitropic responses allow the root system to explore the soil most efficiently.

The Perception Mechanism
It was clear from the work of Ciesielski in 1872 (98) in Poland and that of the Darwins in 1880 (58; the Darwins cited the work of Ciesielski) that the site of gravity perception was the root cap, while actual bending took place at some distance above the cap. If the root cap is carefully removed so as not to inhibit growth, there is no gravitropic bending until the same or another cap has been replaced or a new cap has regenerated (96,99). Both Ciesielski and the Darwins suggested that some message—later to be called a hormone—must move from the tip to the elongating cells.

At the turn of the century, Haberlandt (100,100a) in Austria and Nemec (101) in Czechoslovakia suggested that it was the *amyloplasts*, each containing two or more starch grains, that settle in *statocytes* (cells that perceive gravity) and thus act as the primary perception mechanism. Amyloplasts acting in this capacity would be called *statoliths*. There has been controversy about this hypothesis, but much supporting evidence has accumulated, and it is now widely accepted although in somewhat modified form (reviews: 95,96,97a,102–104). Correlative evidence is as follows:

1. Virtually all organs that respond to gravity also have sedimentable, starch-filled amyloplasts.

2. The presentation time (minimum time required to elicit a gravitropic response) is closely correlated with the rate of amyloplast settling.

3. If roots (or coleoptiles) are treated with gibberellins and kinetin at elevated temperatures, all starch in the amyloplasts disappears and so does the response to gravity in some plants, at least (105,106).

4. Gravitropic sensitivity reappears at the same time that starch grains reappear in amyloplasts and the amyloplasts can begin to settle. (They usually remain suspended in the cytoplasm for a few hours after they appear; 107.)

One recent evidence seemed to contradict the role of starch-filled amyloplasts as statoliths: A mutant of *Arabidopsis* was found that could not synthesize starch but that could respond to gravity (108). Kiss et al. (109) reported, however, that the gravitropic response was significantly slower in starchless plants and that the starchless amyloplasts did settle in root cap statocytes of those plants but more slowly than did the starch-filled amyloplasts of the wild type. Thus is appears that starch may not be essential to a gravitropic response, but sedimentable amyloplasts might well be necessary.

Some other organelles can also sediment in cells, but their rate of settling is too slow for them to be serious candidates for the role of statolith (110).

After this chapter was first submitted, I became aware of recent work in the laboratory of Randy Wayne that has produced evidence for a perception mechanism that is an alternative to the starch-statolith hypothesis (110a–110d; see summary in 97a). It is based on the idea that the lower membrane of each cell that responds to gravity detects the slight weight of the cell. Wayne and his coworkers studied protoplasmic streaming in *Nitellopsis obtusa* and *Chara corallina*, algae in the Characeae, a family characterized by large cells often several centimeters long. Downward streaming is about 10% faster than upward streaming, which is the response to gravity that was studied. There are no amyloplasts in these cells, so a starch-statolith mechanism cannot be invoked. If gravity is perceived as the weight on the protoplast against the lower wall, then immersion in a fluid with a density equal to or greater than that of the protoplast should obviate the gravity response. In solutions of ethylene glycol, sucrose, and Optiray (an iodated compound used for X-ray enhancement), this proved to be the case; indeed, in solutions more dense than the protoplast, the rates of streaming were reversed. The group also tested the concept with rice plants, finding that gravitropic bending could be stopped in dense solutions.

However, such solutions also have a negative osmotic potential that reduces and eventually stops water uptake and consequently growth, which is needed for gravitropic curvature. Actually, there was some response to gravity after a time interval even in a 20% sucrose solution; bending was stopped at first while some growth continued, but bending began after elongation had almost ceased (see discussion in 97a). On theoretical grounds, the model is difficult to accept. The weight of a fluid is detected as an increase in pressure of the fluid. I calculated that the pressure on the bottom of a typical cell caused by its weight would be about 6.47×10^{-17} MPa, but in a typically turgid cell, there is already a pressure on all sides of the cell of about 0.5 MPa—some 18 orders of magnitude greater than that caused by the weight of the cell! Can a cell membrane detect such a slight difference in pressure against such a high-pressure background? Wayne and his coworkers think so, but much research might be required to settle the question once and for all.

Transduction: Role of an Inhibitor

Several experiments have convinced workers in the field that gravitropism in roots occurs as an inhibitor from the root cap builds up in the bottom tissues of a primary root turned to the horizontal. For example, if a portion of the root cap is removed, bending occurs toward the remaining portion of the cap, or if a tiny sheet of mica is inserted in one side of the root just above the cap, bending occurs toward the other side (111,112). Measurements have shown that growth virtually stops on the concave side of a bending root but continues on the convex side.

Near the end of the 1970s, it seemed clear that the inhibitor from root caps was ABA, but this now seems highly unlikely, and most workers accept the idea that IAA (auxin) is the inhibitor (113). For one thing, 100–1000 times higher concentrations of ABA than IAA are required to inhibit root growth, and such ABA concentrations apparently do not occur naturally in roots. Maize mutants that do not synthesize ABA nevertheless respond to gravity, as do roots grown in the presence of a compound that inhibits synthesis of ABA. Roots will also respond to gravity by bending down even when they are immersed in relatively high concentrations of ABA that would most likely eliminate any ABA concentration gradient within the root.

If the root cap is removed, and a small block of agar containing IAA is added to one side, the root bends to that side. Radioactive IAA applied uniformly to a horizontal root moves to the lower side, and auxin-transport inhibitors overcome the root's ability to bend in response to gravity (but the transport inhibitors also inhibit growth of the root, weakening this evidence). Thus IAA may be the effective inhibitor in root gravitropism, although other substances cannot be completely ruled out (114–116).

If IAA is indeed the effective inhibitor in root gravitropism, then the question that must be answered to make transduction understandable concerns how IAA moves to the bottom of the root cap and root in response to settling statoliths. (Note that small molecules in solution cannot respond directly to gravity because of their kinetic activity.)

Calcium ions (Ca^{2+}) may provide part of the answer. Radioactive Ca^{2+} (used as a tracer) moves to the bottom of a gravistimulated root along with the IAA, and strong ligands of Ca^{2+}, such as EDTA, applied to roots will completely prevent gravitropic bending. Excess Ca^{2+} will overcome this effect without inhibiting root growth, suggesting that the ligand is indeed having its effect by chelating Ca^{2+} (117). A block of agar containing Ca^{2+} added to one side of a root will prevent growth on that side so that the root curves, sometimes 360° or more into a loop.

Changes in electric currents around the root tip have been measured, and it has been suggested that these result from the counterflow of H^+ ions moving in a direction opposite to moving Ca^{2+} ions (118–121). (For almost a century, various authors have attempted to implicate electric currents in the gravitropic transduction mechanism.)

White and Sack (122) observed that amyloplasts of maize and barley are closely associated with microfilament bundles. They suggested that these bundles could link settling amyloplasts with plasma membranes and other cellular organelles, especially the endoplasmic reticulum, which has often been suggested to play a role in transduction (123). Evans et al. (113) noted that the ER is known to be rich in Ca^{2+} and that some Ca^{2+} might be released when the ER is contacted or pulled on by amyloplasts. The released Ca^{2+} might in turn activate *calmodulin*, a small protein known to be a powerful activator of many enzymes important to cellular function. The calmodulin could then activate both calcium and auxin pumps in cell membranes, accounting for the observed movements of Ca^{2+} and auxin in gravistimulated roots.

Evans et al. (113) outlined a model in which auxin normally flows through the stele of the root toward the root cap where it is redistributed to the cortex. If the root is vertical, the redistribution is symmetrical with equal amounts of auxin going to cortex tissues all around the root. If the root is gravistimulated, some mechanism such as that involving Ca^{2+} might control movement of more IAA into bottom cortex tissues. Accumulating evidence seems to be supporting this model of transduction.

Gravitropism in Stems and Coleoptiles

When a stem is turned to the horizontal, part of the top surface of the most actively growing zone stops growing almost immediately and may even shrink slightly as the bottom surface continues to grow (sometimes at about the same rate, sometimes somewhat faster or slower). This differential growth causes upward bending of the stem tip. It has long been known that the stem epidermis and cortical cells play a critical role in stem growth (124–126). These outer layers are under tension relative to the more central layers (pith in many stems), so that the outer layers tend to resist the growth of the inner layers. If a stem is split longitudinally, it will usually bend outward because of this tension/pressure differential. (Children split dandelion peduncles and watch them curl in response to this differential.)

The gravitropic response is so strong that a stem can be tied down in a horizontal position for several hours so it cannot bend upward, and when it is released it will curve upward (as shown by Bateson and Darwin in 1888; 127) in 1–10 sec (128). Cells on the bottom of such a restrained stem swell in diameter whereas cells on top are stretched and narrow. Upon release, bottom cells quickly narrow and elongate while top cells shorten and thicken, accounting for the rapid bend (129).

Perception

In stems, perception is not limited to the tip (as it is in the root cap) but occurs all along the elongating region where upward bending occurs. A portion of the stem can be removed, and the remaining stem will respond to gravity providing only that it includes rapidly growing cells. The Darwins (58; pp. 511–512 in 1896 edition) removed coleoptile tips from *Phalaris* and noted that the organs "bowed themselves upwards as effectually as the unmutilated specimens in the same pots, showing that sensitiveness to gravitation is not confined to their tips." (This has been overlooked by many textbook authors who have stated, perhaps by analogy with roots, that only stem tips are sensitive to gravity.)

As with roots, amyloplasts are thought to be the statoliths of stems. The evidence is mostly correlative and identical to that described above for roots, and the controversies that surround the topic (e.g., the recent work of Randy Wayne's group; 110a–110d) apply to stems as well as roots. In many angiosperms, the amyloplasts are confined to one or two layers of cells, called the *starch sheath*, just outside the vascular bundles or just inside in coleoptiles.

Transduction

For almost 70 years research on stem gravitropism has been guided by the Cholodny–Went hypothesis. Early support came from the work of Dolk (130; reviewed in detail by Went and Thimann, 78). Dolk performed experiments closely comparable to those performed by Went to understand phototropism. In his experiments, agar blocks, separated by a sheet of mica or a razor blade, were used to collect auxin diffusing from upper and lower halves of coleoptile tips that had been turned to the horizontal. Approximately 60% of the total auxin (actually, curvature in the *Avena* test) was found in the lower block. This work is still widely cited.

During recent years, the Cholodny–Went hypothesis as it applies to stem gravitropism has been questioned by several workers (e.g., 67, 131; and especially the recent forum on the Cholodny–Went hypothesis: 78a). For one thing, it is somewhat difficult to account for the rapid cessation of growth of top tissues when a stem is laid on its side. Rates of growth of top and bottom surfaces can differ by a factor of 10 or more—especially when top surfaces actually shrink. Could top tissues be depleted of auxin rapidly enough to account for this halting of growth? Furthermore, many workers have detected only minimal auxin gradients across a horizontal stem—or no gradients at all (132). Phillips and Hartung (133) could not detect auxin transport across horizontal sunflower hypocotyls. Could the gradients that have been observed be the result of gravitropic bending rather than its cause? Could these gradients be a backup mechanism rather than the primary gravitropic response?

Measurement of gradients is admittedly difficult, and most studies have simply measured top and bottom halves (e.g., Bandurski et al., 134, who reported 56–57% of the IAA in maize mesocotyl lower halves and 43–44% in upper halves). MacDonald and Hart (135) proposed that dicot stem tissues differ in their sensitivity to auxin, the epidermis being much more auxin-responsive than subepidermal tissue. Auxin transport during gravistimulation might be limited in distance, with auxin moving out of responsive epidermal cells on top of the stem and into cortical tissue and moving from cortical tissue into lower epidermal cells. Such short movements might be difficult to measure and would not appear when entire stem halves are analyzed.

Brauner and Hager (136) removed the cotyledons and tips from sunflower hypocotyls (presumably the source of auxin), waited 4 days for the auxins to be depleted (they assumed), and then gravistimulated the decapitated hypocotyls. No bending occurred. When the hypocotyls were returned to the vertical, however, and an IAA solution was applied to the cut stumps (in small glass tubes), bending occurred in the expected direction. This seems to demonstrate that auxin is indeed essential to the gravitropic response. When mature stems of castor bean, cocklebur, or tomato are decapitated and gravistimulated, bending is increased somewhat by adding auxin in lanolin to the entire cut stumps. Such an application increases the auxin apparently required for growth of the stems, and it probably does so in an indiscriminate way that might swamp out any existing lateral auxin gradient (129).

A few other hormones have been implicated in stem gravitropism although their relative importance remains to be determined. Gibberellins, for example, build up in bottom tissues of gravistimulated stems although this may not occur until after bending is well underway (reviewed in 97). In the leaf sheath false pulvini of grasses, however, the GA gradients did form during bending (137).

Some experiments have demonstrated a positive role for ethylene in stem gravitropism (138,139). Four inhibitors of ethylene action or synthesis (Ag^+, Co_2, Co^{2+}, and AVG) reduced the rate of gravitropic bending in mature stems of cocklebur, tomato, and castor bean. Some of the original rate could be restored by surrounding the stems with low concentrations of ethylene. There is also a buildup of ethylene in bottom tissues of these and other species—which seems contrary to expectations. Ethylene is generally known to inhibit stem growth, so it was surprising to find it building up in the most rapidly growing bottom tissues. It has been reported that ethylene plays no role in the graviresponse of dandelion peduncles (140), cereal leaf-sheath false pulvini (141), or tomato hypocotyls (142).

Calcium ions apparently play a role in shoot as well as root gravitropism. For one thing, Ca^{2+} concentrations were found to be higher in upper stem tissues (143), and Ca^{2+} is known to inhibit cell elongation, perhaps by overcoming auxin effects.

Transduction: Possible Role of Changing Sensitivity to Auxin

If Cholodny–Went isn't the full explanation of stem gravitropism, what are the alternatives? Sensitivity to auxin might change in response to gravistimulation. If bottom tissues become more sensitive than top tissues to the auxin that is already present, the lower cells will grow more, causing upward bending, whether or not auxin concentrations change. Although most workers in this field did not consider this possibility until recently, Brauner and Hager (136) over three decades ago suggested it as one possible explanation for their "gravitropic memory" outlined above. If gravistimulation of decapitated, aged hypocotyls did not cause bending during gravistimulation, the gravity treatment might have changed the sensitivity to the auxin that was presented later, accounting for the delayed bending. Brauner and Hager also gravistimulated intact hypocotyls in the cold, but no bending occurred until the hypocotyls were returned to warm conditions. They suggested that the endogenous auxin could not function in the cold but that sensitivity to auxin could change. (They also suggested other explanations for their phenomenon such as a transported inhibitor.)

Brauner went on to examine the sensitivity hypothesis for over a decade until shortly before his death (144; reviewed in 145). For example, he and his coworkers applied auxin in agar to induce different auxin gradients across sunflower hypocotyls and then measured bending and tissue auxin concentrations. A given measured auxin gradient in vertical hypocotyls produced a given bending, but the same measured auxin gradient produced much more bending in horizontal hypocotyls, provided that the highest concentrations were in bottom tissues. This result is difficult to explain unless the tissue sensitivity to auxin changed in response to gravistimulation. (Some of Brauner's papers were cited in various reviews, but until the late 1980s such reviews never mentioned his ideas on changing sensitivity; e.g., 97, 145a.)

What is sensitivity? The dictionary defines it as the capacity of an organism or physical system such as a microphone or a photocell to respond to a stimulus. Sensitivity is thus synonymous with *responsiveness*. To measure it quantitatively, one varies the stimulus or dose and measures the response, producing a dose–response curve.

When stem or coleoptile segments are immersed in or floated on auxin solutions, the dose–response curve for elongation as a function of auxin concentration (log scale) is typically bell-shaped. In the lower ranges, increasing concentration results in increased growth rate up to a maximum rate at the optimum concentration, and then still higher concentrations cause an inhibition.

Except for the descending (inhibition) part of the curve, the auxin–response curve can be thought of much as the Michaelis–Menten S-shaped (saturation) curves obtained when reaction rates are plotted as a function of substrate concentration in an enzyme-controlled reaction. As noted in the introduction, in the Michaelis–Menten analysis, two features of such curves are especially noteworthy: the maximum reaction rate (called V_{max}) and the substrate concentration that causes half of V_{max}, called the Michaelis constant, K_m, an indication of the tenacity of binding between enzyme and substrate.

The maximum growth rate produced by the optimum concentration of auxin is analogous to V_{max} and the auxin concentration that causes half of this response is then analogous to K_m (see 145 for discussion of Michaelis–Menten kinetics as applied to the auxin dose–response curves). Such an approach to the auxin dose–response curve was taken by Cleland in 1972 (146) and by Foster et al. in 1952 (147). In an ezyme-controlled reaction, V_{max} is determined by the amount of enzyme present; when sufficient substrate is present to react with all the enzyme active sites, saturation is achieved. By analogy, V_{max} in the auxin dose–response curve would be controlled by the number of auxin binding sites. A greater number of binding sites would result in a higher V_{max} sensitivity. In the enzyme-controlled reaction, K_m is determined by the binding strength of the enzyme for its substrate; the more actively the enzyme forms bonds with its substrate, the lower the concentration of substrate required to produce half of V_{max}. Thus, smaller values of K_m translate to greater K_m sensitivity of the enzyme toward its substrate—or to greater K_m sensitivity of auxin binding sites for auxin.

In a series of experiments with sunflower and soybean hypocotyls, my colleagues and I (145,148) immersed hypocotyl sections in a wide range of auxin concentrations (0, 10^{-8}–10^{-2} M IAA). The sections were either turned to the horizontal or left in the vertical position. Photographs were taken at 30-min intervals, and the projected images (negatives) were analyzed with a digitizer-computer system. Bending was measured as were top and bottom lengths along the section surfaces. Bending could be plotted as a function of time or auxin concentration, as could growth of upper and lower surfaces.

As auxin concentrations increased beyond about 10^{-7} M IAA (with considerable variation from experiment to experiment but always the same general pattern of response), hypocotyl bending decreased until at about 10^{-4} M IAA no bending occurred. Sections typically bent down instead of up at concentrations around 10^{-3} M IAA (and appeared to be dead or damaged at 10^{-2} M IAA although some downward bending usually occurred).

The dose–response curve for growth of the lower surface usually had its highest point at the lowest or near-lowest auxin concentration (0, 10^{-8} M IAA), and increasing concentrations above that level resulted in less and less surface growth. Apparently, the auxin already in the tissue was capable or nearly capable of producing growth at the V_{max} rate, and any increase above that level acted in an inhibitory way. Clearly, bottom tissue was extremely sensitive to auxin with a high V_{max} and a very low K_m (unmeasurably low because of auxin already in the tissue).

The dose–response curve for the upper surface, however, was typically bell-shaped. Growth was low at the lowest auxin concentrations and increased with increasing auxin until, often at concentrations of about 10^{-4} M IAA, growth was at its maximum (V_{max}). This value was almost never as high as it was for lower surfaces, indicating that V_{max} sensitivity of top tissues was typically lower than that for bottom tissues. K_m values appeared to be around 10^{-7}–10^{-6} M IAA, much higher than the extremely low (but unknown) K_m values for the lower tissues. Thus K_m sensitivities of upper tissue were strikingly lower than those of lower tissues, perhaps by a factor of 1000 or more.

Vertical hypocotyl sections nearly always had lower V_{max} values than both surfaces of gravistimulated sections, and K_m values appeared to be intermediate between those of upper and lower surfaces. Actually, dose–response curves for the vertical sections were sometimes bimodal and thus difficult to interpret. They need to be repeated with smaller intervals between auxin concentrations and perhaps with more sections per auxin concentration (to improve the statistical treatment; there is considerable variability among sections in the same solution).

By the definition of sensitivity given above, it was clear that both V_{max} and especially K_m sensitivities of upper and lower tissues to applied auxin were strongly influenced by gravistimulation. If these sensitivities represent the status of auxin binding sites as suggested above, then gravistimulation apparently increases the number of effective binding sites (V_{max}) in lower compared with upper tissues—and in both tissues compared with vertical controls. The striking differences in K_m for the three kinds of tissues suggest that gravistimulation induces radical changes in the ability of binding sites to interact with auxin. It remains to be seen if these conclusions applied to hypocotyl sections immersed in auxin solutions will hold up when applied to intact plants in their natural setting. If sensitivity to auxin does change as suggested by these experiments, we should remember that this could be instead of *or* in addition to changes in auxin concentration (i.e., auxin transport as suggested by Cholodny–Went). If sensitivity changes are indeed involved, the next problem will be to understand the mechanism of this transduction response. Incidently, Brock and Kaufman (149) reported evidences for changes in tissue sensitivity to auxin in the false pulvini of oats. The experiments were basically similar to those described for sunflower and soybean.

PHYSICAL CONTACT

Thigmonasty, plant movement resulting from touch, is widespread, especially in the subfamily Mimosoideae in the family Fabaceae (Leguminosae). If leaflets of *Mimosa* pudica, the sensitive plant, are touched, shaken, heated, rapidly cooled, or treated with an electrical stimulus, the leaflets and leaves rapidly fold together. The first movements at the point of contact are evident almost instantly, and the folding continues back along the compound leaf until all the leaflets and the leaf itself have folded or collapsed, usually within a minute or two. The movements are caused by water moving out of motor cells in pulvini, and this occurs as K^+ and water move out of the cells.

Transmission of signals in *Mimosa* has been studied for many years (reviews: 150–153). There is evidence for both an electrical and a chemical stimulus. The electrical stimulus is exhibited as an action potential (a change in voltage between two points that forms a characteristic peak when plotted as a function of time; 154), and action potentials exhibited by *Mimosa* are similar to those measured along animal nerve cells except that they are much slower (about 2 cm/sec compared with m/sec in nerve cells).

The chemical response was first studied by Ubaldo Ricca in 1916 (155,156), who cut through a stem and then connected the cut ends with a narrow, water-filled tube. The chemical stimulus would pass through the tube; for many years it was called *Ricca's factor*. During the 1980s, Hermann Schildknecht (157,158) and his group in Germany isolated and identified from several species a number of compounds that, like Ricca's factor, activate pulvini in the leaves of such plants as *Mimosa* and *Acacia karroo* (which is not sensitive to touch but folds its leaves at night much as *Mimosa* does). For a bioassay, a *Mimosa* leaf is placed in a solution with the suspected active substance, which then moves in the transpiration stream to the pulvini, causing the leaf pinnules to fold up if the substance is active.

Schildknecht's group identified over half a dozen active substances, and they suggest that they form a new group of plant hormones called *turgorins*. Chemically, most are derivatives of the β-glucosides of gallic acid. The most active compounds were the β-D-glucoside-6-sulfate and β-D-glucoside-3,6-disulfate of gallic acid, which the researchers called periodic leaf movement factors 1 and 2 (PLMF 1 and PLMF 2). They were active at concentrations of about 2×10^{-7} M, in the same range as other hormones, although some other turgorins were somewhat less active (e.g., 10^{-5} M). Extracts from *Mimosa* contained PLMF 1 plus one other related compound. Another active substance proved to be a derivative of protocatechuic acid rather than gallic acid, but glucose-6-sulfate was again part of the molecule. It will be interesting to see if sensitivity to the turgorins plays a role in their activity. A start has been made by Kallas et al. (159) who demonstrated the presence of a PLMF-1-specific receptor (presumably a protein) in the outer side of the plasma membranes of *Mimosa*.

SOME CONCLUSIONS

Many regulatory compounds are known in plants, and most if not all fit the definition of plant hormones if that definition does not require translocation from a synthesis site to a receptor site. Actually, we've seen that many of these compounds are translocated and thus act as chemical messengers.

During most of this century, we have assumed that plant hormones control plant functions as their concentrations change, and in many cases this seems to be true, but we are beginning to appreciate that transduction processes between environmental stimuli and plant response might act on hormone receptors (binding sites) rather than or in addition to hormone concentrations; that is, that *sensitivity* to the hormone might be influenced.

It is likely that changing hormone concentrations or changing receptors (quantity or activity) are often under control of the genetic developmental program, but many such changes are mediated by changes in the environment and are part of the transduction step between stimulus perception and response to the stimulus.

There is much to learn, but much progress has been made since Fitting spoke of plant hormones over eight decades ago (5).

ACKNOWLEDGMENTS

Preparation of this chapter was supported by the Utah Agricultural Experiment Station, Utah State University, Logan, Utah 84322-4810, and by grants from the NASA Space Biology Program and the NASA

Controlled Ecological Life-Support System (CELSS) Program. Approved as UAES journal paper no. 4463. I thank the following people who reviewed the manuscript and made many valuable suggestions: Robert E. Cleland (University of Washington), Cleon Ross (Colorado State University), and Schuyler Seeley (Utah State University).

REFERENCES

1. F.B. Salisbury and C. W. Ross (1969). *Plant Physiology*, 1st ed. Wadsworth, Belmont, CA.
2. P. J. Davies (1987). *Plant Hormones and Their Role in Plant Growth and Development* (P. J. Davies, ed.), Martinus Nijhoff, Boston, pp. 12–23.
3. A. Trewavas (1981). *Plant Cell Environ.*, 4:203–228.
4. P. J. Davies (1987). *Plant Hormones and Their Role in Plant Growth and Development* (P. J. Davies, ed.), Martinus Nijhoff, Boston, pp. 1–11.
5. H. Fitting (1909). *Zeit. Botanik*, 1:1–86.
6. H. Fitting (1910). *Zeit. Botanik*, 2:255–267.
7. F. B. Salisbury and C. W. Ross (1991). *Plant Physiology*, 4th ed. Wadsworth, Belmont, CA.
8. J. B. Reid (1987). *Plant Hormones and Their Role in Plant Growth and Development* (P. J. Davies, ed.), Martinus Nijhoff, Boston, pp. 318–340.
9. C. M. Chadwick and D. R. Garrod (1986). *Hormones, Receptors and Cellular Interactions in Plants*, Cambridge University Press, Cambridge, UK.
10. P. J. Davies (ed.) (1987). *Plant Hormones and Their Role in Plant Growth and Development*, Martinus Nijhoff, Boston.
11. J. E. Fox and M. Jacobs (eds.) (1987). *Molecular Biology of Plant Growth Control*, Alan R. Liss, New York.
12. G. V. Hoad, M. B. Jackson, J. R. Lenton, and R. K. Atkin (eds.) (1987). *Hormone Action in Plant Development*, Butterworths, London.
13. T. C. Moore (1989). *Biochemistry and Physiology of Plant Hormones*, Springer-Verlag, Berlin.
14. F. B. Salisbury and N. G. Marinos (1985). *Hormonal Regulation of Development III, Role of Environmental Factors*, vol. 11 (R. P. Pharis and D. M. Reid, ed.), *Encyclopedia of Plant Physiology*, Springer-Verlag, Berlin, pp. 707–766.
15. L. Taiz and E. Zeiger (1991). *Plant Physiology*, Benjamin/Cummings, Redwood City, CA.
16. P. Hedden and J. R. Lenton (1988). *Beltsville Symposia in Agricultural Resources, Vol. 12, Biomechanisms Regulating Growth and Development*, Kluwer Academic, Boston, pp. 175–204.
17. J. B. Reid (1990). *J. Plant Growth Reg.*, 9:97–111.
18. I. M. Scott (1990). *Hormonal Regulation of Development, I. Molecular Aspects of Plant Hormones*, Vol. 9 (J. MacMillan, ed.), Springer-Verlag, Berlin, pp. 281–444.
19. I. M. Scott (1990). *Physiologia Plantarum*, 78:147–152.
20. S. P. C. Groot and C. M. Karssen (1987). *Planta*, 171:525–531.

21. S. P. C. Groot, B. Kieliszewska-Rokicka, E. Vermeer, and C. M. Karssen (1988). *Planta, 174*:500–504.
22. C. M. Karssen, S. Zagorski, J. Kepczynski, and S. P. C. Groot (1989). *Ann. Bot., 63*:71–80.
23. T. H. Thomas (1989). *J. Plant Growth Reg., 8*:255–261.
24. Y. Esashi, R. Kuraishi, N. Tanaka, and S. Satoh (1983). *Plant. Cell Environ., 6*:493–499.
25. J.-P. Métraux (1987). *Plant Hormones and Their Role in Plant Growth and Development* (P. J. Davies, ed.), Martinus Nijhoff, Boston, pp. 296–317.
26. H. Kende and A. Lang (1964). *Plant Physiol., 39*:435–440.
27. J. L. S. Cotton, C. W. Ross, D. H. Byrne, and J. T. Colbert (1990). *Plant Mol. Biol., 14*:707–714.
28. J. A. De Greef and H. Frédéricq (1983). *Encyclopedia of Plant Physiology, Vol. 16A, Photomorphogenesis*, W. Shropshire, Jr. and H. Mohr, eds., Springer-Verlag, Berlin, pp. 401–427.
29. H. I. Virgin (1989). *Physiologia Plantarum, 75*:295–298.
30. G. Bernier (1988). *Ann. Rev. Plant Physiol. Plant Mol. Biol., 39*:175–219.
31. G. Bernier, J. Kinet, and R. M. Sachs (1981). *The Physiology of Flowering, Vol. 1, The Initiation of Flowers, Vol. 2, Transition to Reproductive Growth*, CRC Press, Boca Raton, FL.
32. F. B. Salisbury (1982). *Hort. Rev., 4*:66–105.
33. F. B. Salisbury (1981). *Physiological Plant Ecology I, Vol. 12A* (O. L. Lang, P. S. Nobel, C. B. Osmond, and H. Ziegler, eds.), *Encyclopedia of Plant Physiology* (new series), Springer-Verlag, Berlin, pp. 135–167.
34. D. Vince-Prue (1975). *Photoperiodism in Plants*, McGraw-Hill, London.
35. M. Talon and J. A. D. Zeevaart (1990). *Plant Physiol., 92*:1094–1100.
36. M. J. Vesper and M. L. Evans (1978). *Plant Physiol., 61*:204–208.
37. A. M. Lang, M. Chailakhyan, and I. A. Frolova (1977). *Proc. Natl. Aca. Sci. USA, 74*:2412–2416.
38. L. T. Evans (1960). *Aust. J. Biol. Sci., 13*:429.
39. C. F. Cleland (1978). *BioScience, 28*:265–269.
40. C. Besnard-Wibaut (1981). *Physiologia Plantarum, 53*:205–212.
41. S. S. Abou-Haidar, E. Miginiac, and R. M. Sachs (1985). *Physiologia Plantarum, 64*:265–270.
42. E. Miginiac, G. Pilate, and M. Bonnet-Masimbert (1987). *Proc. of 14th International Botanical Congress*, West Berlin, p. 105.
43. P. F. Wareing, R. Horgan, I. E. Henson, and W. Davis (1977). *Plant Growth Regulation* (P.-E. Pilet, ed.), Springer-Verlag, Berlin, pp. 147–153.
44. R. P. Pharis and R. W. King (1985). *Ann. Rev. Plant Physiol., 36*:517–568.
45. R. P. Pharis and S. D. Ross (1986). *Handbook of Flowering, Vol. 5* (A. H. Halevy, ed.), CRC Press, Boca Raton, FL, pp. 269–286.
46. R. King, L. Evans, R. P. Pharis, and L. N. Mander (1987). *Plant Physiol., 84*:1126–1131.

47. M. De Proft, R. Van Dijck, L. Philippe, and J. A. De Greef (1985). *Proc. 12th International Conference on Plant Growth Substances*, Heidelberg, p. 93 (abs.).

48. H. Skubatz, P. S. Williamson, E. L. Schneider, and B. J. D. Meeuse (1990). *J. Exp. Bot.*, *41*:1335–1339.

49. B. J. D. Meeuse and I. Raskin (1988). *Sexual Plant Reprod.*, *1*:3–15.

50. I. Raskin, A. Ehmann, W. R. Melander, and B. J. D. Meeuse (1987). *Science*, *237*:1601–1602.

51. E. E. Ewing (1987). *Plant Hormones and Their Role in Plant Growth and Development* (P. J. Davies, ed.), Martinus Nijhoff, Boston, pp. 515–538.

52. D. Vreugdenhil and P. C. Struik (1989). *Physiologia Plantarum*, *75*:525–531.

53. A. M. Mingo-Castel, O. E. Smith, and J. Kumamoto (1976). *Plant Physiol.*, *57*:480–485.

54. H. W. Chapman (1958). *Physiologia Plantarum*, *11*:215–224.

55. Y. Koda, E.-S. A. Omer, T. Yoshihara, H. Shibata, S. Sakamura, and Y. Okazawa (1988). *Plant Cell Physiol.*, *29*:1047–1051.

56. D. S. Dennison (1979). *Physiology of Movements*, Vol. 7 (W. Haupt and M. E. Feinleib, eds.), *Encyclopedia of Plant Physiology* (new series), Springer-Verlag, Berlin, pp. 506–508.

57. D. S. Dennison (1984). *Advanced Plant Physiology* (M. B. Wilkins, ed.), Pitman, London, pp. 149–162.

58. C. Darwin (assisted by F. Darwin) (1880). *The Power of Movement in Plants*, Murray, London.

59. B. Steyer (1967). *Planta*, *77*:277–286.

60. T. I. Baskin and M. Iino (1987). *Photochemistry and Photobiology*, *46*(1):127–136.

61. A. H. Blaauw (1909). *Rec. Trav. Botica Neerlandica*, 5:209–272.

62. A. H. Blaauw (1918). *Mededelingen Landbouwhogeschool Wageningen*, *15*:89–204.

63. N. Cholodny (1926). *Jahrb. Wiss. Bot.*, *65*:447–459.

64. F. W. Went (1926). *Proc. K. Akad. Wet. Amsterdam*, *30*:10–19.

65. W. R. Briggs (1963). *Plant Physiol.*, *38*:237–247.

66. W. R. Briggs and T. I. Baskin (1988). *Botanica Acta*, *101*:133–139.

67. R. D. Firn and J. Digby (1980). *Ann. Rev. Plant Physiol.*, *31*:131–148.

68. B. G. Pickard and K. V. Thimann (1964). *Plant Physiol.*, *39*:341–350.

69. J. Shen-Miller, P. Cooper, and S. A. Gordon (1969). *Plant Physiol.*, *44*:491–496.

70. T. I. Baskin, W. R. Briggs, and M. Iino (1986). *Plant Physiol.*, *81*:306–309.

71. J. M. Franssen, R. D. Firn, and J. Digby (1982). *Planta*, *155*:281–286.

72. T. I. Baskin (1986). *Planta*, *169*:406–414.

73. M. Iino and W. R. Briggs (1984). *Plant Cell Environ.*, *7*:97–104.

74. T. C. G. Rich, G. C. Whitelam, and H. Smith (1987). *Plant. Cell Environm.*, *10*:303–311.

75. J. M. Franssen and J. Bruinsma (1981). *Planta*, *155*:281–286.

76. K. Hasegawa, M. Sakoda, and J. Bruinsma (1989). *Planta, 178*:540–544.
77. P. Boysen-Jensen (1913). *Berichte der Deutschen botanischen Gesellschaft, 31*:559–566.
78. F. W. Went, and K. W. Thimann (1937). *Phytohormones*, Macmillan, New York.
78a. A. J. Trewavas (ed.) (1992). *Plant Cell Environ., 15*:759–794.
79. G. A. Lang, J. D. Early, N. J. Arroyave, R. L. Darnell, G. C. Martin, and G. W. Stutte (1985). *Hort Sci., 20*:809–812.
80. G. A. Lang, J. D. Early, G. C. Martin, and R. L. Darnell (1987). *HortScience, 22*:371–377.
81. L. E. Powell (1987). *Plant Hormones and Their Role in Plant Growth and Development* (P. J. Davies, ed.), Martinus Nijhoff, Boston, pp. 539–552.
82. F. B. Salisbury (1986). *Hort. Sci., 21*:185–186.
83. R. Rudnick (1969). *Planta, 86*:63–68.
84. A. H. Hatch and D. R. Walker (1969). *J. Am. Soc. Hort. Sci., 94*:304–307.
85. I. Arias, P. M. Williams, and J. W. Bradbeer (1976). *Planta, 131*:135–139.
86. M. Pierce and K. Raschke (1980). *Planta, 148*:174–182.
87. K. Cornish and J. A. D. Zeevaart (1985). *Plant Physiol., 78*:623–626.
88. W. Hartung, J. W. Radin, and D. L. Hendrix (1988). *Plant Physiol., 70*:908–913.
89. M. J. Harris, W. H. Outlaw, R. Mertens, and E. W. Weiler (1988). *Proc. Natl. Acad. Sci. USA, 85*:2584–2588.
90. W. J. Davies, J. Metcalfe, T. A. Lodge, and A. R. da Costa (1986). *Aust. J. Plant Physiol., 13*:105–125.
91. E.-D. Schulze (1986). *Ann. Rev. Plant Physiol., 37*:247–274.
92. J. Zhang, U. Schurr, and W. J. Davies (1987). *J. Exp. Bot., 38*:1174–1181.
93. J. Zhang and W. J. Davies (1990). *Plant, Cell, and Environment, 13*:277–285.
94. P. B. Kaufman and I. Song (1987). *Plant Hormones and Their Role in Plant Growth and Development* (P. J. Davies, ed.), Martinus Nijhoff, Boston, pp. 375–392.
95. F. D. Sack (1991). *Int. Rev. Cytol., 127*:193–252.
96. D. Volkmann and A. Sievers (1979). *Pysiology of Movements*, Vol. 7 (W. Haupt and M. E. Feinleib, eds.), *Encyclopedia of Plant Physiology*, (new series), Springer-Verlag, Berlin, pp. 573–600.
97. M. B. Wilkins (1979). *Physiology of Movements*, Vol. 7 (W. Haupt and M. E. Feinleib, eds.), *Encyclopedia of Plant Physiology*, (new series), Springer-Verlag, Berlin, pp. 601–626.
97a. F. B. Salisbury (1993). *Hort. Rev., 15*:233–278.
98. T. Ciesielski (1872). *Beiträge Biol. Pflanzen, 1*:1–30.
99. B. E. Juniper, S. Groves, B. Landua-Schachar, and L. J. Audus (1966). *Nature, 209*:93–94.
100. G. Haberlandt (1901). *Ber. Dtsch. bot. Ges., 18*:261–272.
100a. G. Haberlandt (1902). *Ber. Dtsch. bot. Ges., 20*:189–195.

101. B. Nemec (1901). *Jahrb. Wiss. Bot.*, *36*:80–178.
102. L. J. Audus (1979). *J. Exp. Bot.*, *30*:1051–1073.
103. B. E. Juniper (1976). *Ann. Rev. Plant Physiol.*, 27:385–406.
104. M. B. Wilkins (1984). *Advanced Plant Physiology* (M.B. Wilkins, ed), Pitman, London, pp. 163–185.
105. T. H. Iversen (1969). *Physiologia Plantarum*, *22*:1251–1262.
106. T. H. Iversen (1974). *K. Norske Vidensk, Selsk, Mus. Miscellanea*, *15*:1–216.
107. S. K. Hillman and M. B. Wilkins (1982). *Planta*, *155*:267–271.
108. T. Caspar and B. G. Pickard (1989). *Planta*, *177*:185–197.
109. J. Z. Kiss, R. Hertel, and F. D. Sack (1989). *Planta*, *177*:198–206.
110. J. Shen-Miller and R. R. Hinchman (1974). *BioScience*, *24*:643–651.
110a. S. W. Lee, S. Y. Kim, A. Jeyabalasinkham, R. Wayne, K. Schneider, M. L. Vaughan, and T. J. Mulkey (1993). *Plant Physiol.*, *99*(Suppl.):63.
110b. M. P. Staves, R. Wayne, and A. C. Leopold (1991). *The Physiologist*, *34*(Suppl.):S-70–S-71.
110c. R. Wayne, M. Staves, and A. C. Leopold (1990). *Protoplasma*, *155*:43–57.
110d. R. Wayne, M. Staves, and A. C. Leopold (1992). *J. Cell Sci.*, *101*:611–623.
111. H. Wilkins and R. L. Wain (1974). *Planta*, *121*:1–8.
112. M. B. Wilkins (1975). *Current Advances in Plant Science*, *8*:317–328.
113. M. L. Evans, R. Moore, and K.-H. Hasenstein (1986). *Sci. Am.*, *255*(6):112–119.
114. L. J. Feldman (1981). *Planta*, *153*:471–475.
115. M. B. Jackson and P. W. Barlow (1981). *Plant Cell Environ.*, *4*:107–123.
116. T. Suzuki, N. Kondo, and T. Fujii (1979). *Planta*, *145*:323–329.
117. J. S. Lee, T. J. Mulkey, and M. L. Evans (1983). *Science*, *220*:1373–1376.
118. H. M. Behrens, D. Gradmann, and A. Sievers (1985). *Planta*, *163*:463–472.
119. H. M. Behrens, M. H. Weisenseel, and A. Sievers (1982). *Plant Physiol.*, *70*:1079–1083.
120. T. Björkman and A. C. Leopold (1987). *Plant Physiol.*, *84*:847–850.
121. T. Björkman and A. C. Leopold (1987). *Plant Physiol.*, *84*:841–846.
122. R. G. White and F. D. Sack (1990). *Am. J. Bot.*, *77*(1):17–26.
123. A. Sievers, H. M. Behrens, T. J. Buckhout, and D. Gradmann (1984). *Z. Pflanzenphysiol.*, *144*:195–200.
124. J. M. Diehl, C. J. Gorter, J. G. Van Iterson, and A. Kleinhoonte (1938). *Recueil travauxx botaniques néerlandais*, *36*:709–798.
125. K. V. Thimann, and C. L. Schneider (1938). *Am. J. Bot.*, *25*(8):627–641.
126. J. von Sachs (1882). *Textbook of Botany*, 2nd English ed., Clarendon, Oxford.
127. A. Bateson and F. Darwin (1888). *Ann. Bot.*, *2*:65–68.
128. W. J. Mueller, F. B. Salisbury, and P. T. Blotter (1984). *Plant Physiol.*, *76*:993–999.
129. J. E. Silwinski and F. B. Salisbury (1984). *Plant Physiol.*, *76*:1000–1008.

130. H. E. Dolk (1930). Geotropie en groestof. (Geotropism and growth substances.), dissertation, Utrecht.
131. J. Digby and R. D. Firn (1976). *Curr. Adv. Plant Sci.*, *8*:953–960.
132. R. Mertens and E. W. Weiler (1983). *Planta*, *158*:339–348.
133. I. D. J. Phillips and W. Hartung (1976). *New Phytologist*, *76*:1–9.
134. R. S. Bandurski, A. Schulze, P. Dayanandan, and P. B. Kaufman (1984). *Plant Physiol.*, *74*:284–288.
135. I. R. MacDonald and J. W. Hart (1987). *Plant Physiol.*, *84*:568–570.
136. L. Brauner and A. Hager (1958). *Planta*, *51*:115–147.
137. R. P. Pharis, R. L. Legge, M. Noma, P. B. Kaufman, N. S. Ghosheh, J. D. LaCroix, and K. Heller (1981). *Plant Physiol.*, *67*:892–897.
138. P. A. Balatti and J. G. Willemöes (1989). *Plant Physiol.*, *91*:1251–1254.
139. R. M. Wheeler and F. B. Salisbury (1986). *Plant Physiol.*, *82*:534–542.
140. P. E. Clifford, D. M. Reid, and R. P. Pharis (1983). *Plant Cell Environm.*, *6*:433–436.
141. P. B. Kaufman, R. P. Pharis, D. M. Reid, and F. D. Beall (1985). *Physiologia Plantarum*, *65*:237–244.
142. M. A. Harrison and B. G. Pickard (1986). *Plant Physiol.*, *80*:592–595.
143. R. D. Slocum and S. J. Roux (1983). *Planta*, *157*:481–492.
144. L. Brauner and R. Diemer (1971). *Planta*, *97*:337–353.
145. F. B. Salisbury, L. Gillespie, and P. Rorabaugh (1988). *Plant Physiol.*, *88*:1186–1194.
145a. B. G. Pickard (1985). *Hormonal Regulation of Development. III. Role of Environmental Factors*, Vol. 11 (R. P. Pharis & D. M. Reid, eds.), *Encyclopedia of Plant Physiology*, Springer-Verlag, Berlin, pp. 193–281.
146. R. E. Cleland (1972). *Planta*, *104*:1–19.
147. R. J. Foster, D. H. McRae, and J. Bonner (1952). *Proc. of the Natl. Acad. Sci. USA*, *38*:1014.
148. P. A. Rorabaugh and F. B. Salisbury (1989). *Plant Physiol.* *91*:1329–1338.
149. T. G. Brock and P. B. Kaufman (1988). *Plant Physiol.*, *87*:130–133.
150. G. Roblin (1982). *Zeit. Pflanzenphysiol.*, *106*:299–303.
151. M. Samejima and T. Sibaoka (1980). *Plant Cell Physiol.*, *21*:467–479.
152. P. J. Simons (1981). *New Phytologist*, *87*:11–37.
153. K. Umrath and G. Kastberger (1983). *Phyton*, *23*:65–78.
154. B. G. Pickard (1973). *Bot. Rev.*, *39*:172–201.
155. U. Ricca (1916). *Arch. Ital. Biol. (Pisa)*, *65*:219–232.
156. U. Ricca (1916). *Nuova Giorn. Bot. Ital. N. S.*, *23*:51–170.
157. H. Schildknecht (1984). *Endeavour, New Series*, *8*(4):113–117.
158. H. Schildknecht (1983). *Angewandte Chemie Int. Edition English*, *22*:695–710.
159. P. Kallas, W. Meier-Augenstein, and H. Schildknecht (1990). *J. Plant Physiol.*, *136*:225–230.

3

Light and Plant Development

Michael J. Kasperbauer

Coastal Plains Soil, Water, and Plant Research Center
ARS, USDA
Florence, South Carolina

INTRODUCTION

Each plant is the product of its genetics and the total environment in which it is grown. Some grow in harsh environments whereas others grow under milder conditions. Some plants complete their life cycle from seed germination through production of the next generation of seed within a single year (annuals). Others germinate and begin growth one year, survive over winter, and produce seed during the second year (biennials). Still others grow for many years and produce many crops of seed (perennials). Regardless of its growth habitat, the "strategy" of each plant is to survive in its environment long enough to produce the next generation. Therefore, the plant must be able to sense various aspects of its total environment and activate or repress genes that regulate adaptation of the plant to the environment in which it is growing.

The total growth environment is constantly changing during the lifetime of a plant. It includes available soil moisture, quantities and solubilities of mineral nutrients, acidity of the soil solution, diseases, insects, air and soil temperatures, and light. Because of the growth and developmental processes associated with adaptation to the many environmental variables, plants of the same genotype can differ significantly in size and chemical composition.

The most widely studied aspect of light–plant relationships is photosynthesis, in which the energy of light is absorbed by the growing plant

83

and used in combining carbon, hydrogen, and oxygen into simple sugars. However, an equally important aspect of plant survival and productivity is the partitioning and use of that photosynthate within the plant. This chapter will concentrate on aspects of plant growth and development called photomorphogenesis, as influenced by photoperiod (day length), light quantity, and light quality. Some of the light-regulated processes are modified by temperature. Many of the examples used in this chapter are from research conducted by the author and his colleagues from the late 1950s to the present time.

PHOTOMORPHOGENESIS

In addition to its very significant role in photosynthesis, light is involved in natural regulation of how and where photosynthate is used within a developing plant. Plants contain photoreceptor systems that sense, or measure, various aspects of the light environment and initiate physiological processes that regulate adaptation of the plant to increase its probability of survival and reproduction in that environment. Plants respond to day length as an initiator of seasonal events, such as flowering or development of fleshy roots and tubers. They also respond to light quality (spectral distribution) and light quantity (photon flux density). Under experimental conditions in controlled environment chambers, each of these factors can be studied in detail while the other environmental components are held constant. However, in the real world plants grow in constantly changing total environments, including natural day lengths and changes in light quantities associated with season, time of day, and competition from other plants. Light spectral distribution is also influenced by reflection from competing plants. Even reflection from different colored soils, plant residues, and mulches can affect morphological development. This chapter will concentrate on five main phases in the accumulation of knowledge on light regulation of plant development: (a) discovery of photoperiodism and its significance, (b) discovery of a photoreversible pigment system, phytochrome, (c) phytochrome-regulated developmental responses under controlled environments, (d) the importance of far-red light reflected from other plants under field conditions, and (e) the theory and use of colored soil covers (mulches) to regulate spectrum of upwardly reflected light and its affect on field-grown plants.

Photoperiodism

The research that led to the discovery of photoperiodism was begun in the early 1900s by W. W. Garner and H. A. Allard, who worked with the Maryland type of tobacco (*Nicotiana tabacum* L.). They conducted research on the old U.S. Department of Agriculture farm at Arlington, Virginia, close to where the Pentagon now stands. Research was less specialized at that time and the same scientist often studied a broad range of plant problems. Thus, the research of Garner and Allard involved various aspects of crop production including plant nutrition, virus and disease resistance, and cultivation of tobacco and some other agricultural plants. The program also involved cross-breeding and development of improved genetic lines and varieties.

Genetic materials with desired disease resistances and other characteristics were grown in field plots and evaluated for possible use in cross-breeding combinations. The tobacco being evaluated in the field included some "mammoth" plants that developed many leaves but flowered long after the other genetic lines had set seed. According to Garner and Allard (1), some mammoth (initially called "giant") plants were observed in the field as early as 1906. Since leaf production is important in tobacco, they became interested in possible incorporation of the leaf factor into some of their genetic combinations. The fact that flowering of the mammoth strain was not synchronized with flowering of other desirable strains was a problem. As was common practice among tobacco breeders, some potentially useful plants were transferred from the field to a greenhouse in autumn before the first killing frost. The intact mammoth plants flowered and set seed in the greenhouse, as did the regrowth from stumps of plants that were cut back before transfer to the greenhouse. The procedure was cumbersome, but it was possible to produce seed. One of the existing concepts was that the mammoth strain had to be older than other strains before it could flower. One year, some seeds of the mammoth strain were sown very early in the greenhouse so that the plants would be old enough to flower in the field at the same time as the other genetic lines. Contrary to the plant age hypothesis, the early-sown plants flowered in the greenhouse at a relatively small size. Clearly, something about the greenhouse conditions in winter resulted in altered time of flowering. Garner and Allard suspected that the number of hours of light per day had something to do with timing of flower development. Another possibility was lower temperature. They tested the day length concept by giving some plants extra hours of light each day. Other treatments consisted of moving some plants into darkness before sunset to shorten their day length. The scientists suspected that other plant species might also respond to day length, and the studies included soybean [*Glycine max* (L.) Merr.] and some other species in addition to the "Maryland Mammoth" strain of tobacco. The classic paper that described the discovery of photoperiodism was published by Garner and Allard in 1920 (1).

Some of the "tools" used in Garner and Allard's research are shown in Figure 1. The upper photograph shows a box of field-grown soybean plants being moved into a dark chamber during the day. In this way it was possible to determine whether a shorter-than-natural day length would alter time of flowering. As a historic note, the first treatment in the dark chamber began at 4 p.m. on July 10, 1918, when a box of the "Peking" cultivar of soybean and three pots of Maryland Mammoth tobacco were placed in the ventilated dark chamber (1). The plants were removed from the dark chamber at 9 a.m. the next morning and the sequence was repeated each day until the seeds of soybean and tobacco were mature. Plants that received the shortened day treatment in the dark chamber matured earlier than control plants left on natural days. Treated plants were moved into and out of the dark chamber by hand during that initial experiment in 1918.

A larger "dark house" was constructed the following spring (Fig. 1, lower photograph). It was designed for easier moving of plants into and out of the dark chambers (1). There were four steel tracks, each entering the building by a separate door. Low-platform trucks were mounted on the tracks to allow the boxes of plants to be moved into or out of the dark chambers. By utilizing different chambers within the building, it was possible to give several different light/dark (day length) combinations at the same time. For example, some treatments involved darkness from 4 p.m. to 9 a.m. whereas others were in darkness from 6 p.m. until 6 a.m., etc. Some were even moved into darkness at 10 a.m. and back to daylight at 2 p.m. to break the natural day into two shorter days.

Other experiments during the autumn of 1919 utilized greenhouses to compare flowering of Maryland Mammoth tobacco on natural winter day lengths with similar plants grown on natural winter day lengths that were extended from 4:30 p.m. until 12:30 a.m. with supplemental light from tungsten filament lamps.

The combination of experiments with tobacco and soybean compared plant responses to shortened vs. natural day lengths in summer and natural vs. extended day lengths in winter. Maryland Mammoth plants flowered earlier and at smaller size when grown on short days. Garner and Allard (1) suggested the term *photoperiod* to describe length of day and *photoperiodism* to describe the response of an organism to the relative length of day. Many subsequent experiments were done with many plant species.

Extending the natural day length with supplemental light resulted in delayed flowering in species such as tobacco, soybean, and cocklebur (*Xanthium pensylvanicum* Wallr.). The term "short-day plants" was coined to describe this group because their flowering time was hastened

Figure 1 "Tools" used by W. W. Garner and H. A. Allard in the discovery of photoperiodism: dark chamber used in 1918 (upper photograph) and the larger "dark house" used in 1919 (lower photograph).

by short days and delayed by long days. Time of flowering of some other plants, such as barley (*Hordeum vulgare* L.), was hastened if the plants received supplemental light to extend the natural days. This group was called "long-day plants." A third category appeared to be indifferent to day length and became known as "day-neutral" types.

The discovery of photoperiodism was highly significant for plant breeders who could then use supplemental light to hasten and synchronize flowering time of long-day plants. They could also use supplemental lighting to keep short-day plants vegetative for a while, then provide short days by moving them into dark rooms (or covering them with light-tight curtains) from about 5 p.m. until 8 or 9 a.m. to synchronize flowering time for short-day species.

A dramatic example of photoperiodic control of flowering of Maryland Mammoth tobacco is shown in Figure 2. Both of the plants grew from the same lot of seed supplied to this author by Dr. James E. McMurtrey, Jr. in 1962 when he was in charge of the USDA tobacco physiology research (about 40 years earlier he was a junior colleague of Garner and Allard). The small plant was grown on 8-hr days alternated with uninterrupted 16-hr nights. When photographed, it was about 0.6 meters tall, had 23 leaf nodes, and had already flowered and set seed. The large plant was grown in a greenhouse that received natural day lengths plus several hours of supplemental light so that it always received long days. Even though the large plant was started earlier than the small one, it had grown to a height of more than 4.5 m, had more than 190 leaves, and had not yet flowered when this photograph was taken. Shortly thereafter, the plant grew beyond the supplemental light fixture and flowered.

Following the classic discovery by Garner and Allard, many scientists throughout the world published papers showing that other species sensed photoperiod and used that environmental signal to initiate flowering. As the papers appeared, it became apparent that the photoperiod-sensing mechanism was sometimes highly modified by temperature. Also, after the term photoperiod (for day length) was firmly established in the scientific literature, it became apparent that the number of hours of uninterrupted darkness rather than the hours of light was the dominant factor involved in the timing mechanism (2). The next major step in the research was based on the fact that a short period of darkness during the day did not affect flowering time, whereas a short period of light near the middle of the night delayed flowering of short-day plants and hastened flowering of long-day plants.

Figure 2 Maryland Mammoth plants grown on 8-hr days (small plant with seed) or in a greenhouse with natural day lengths plus several hours of supplemental light to provide long-day treatment (tall plant).

Discovery of Phytochrome

In the mid-1930s, a new USDA research team was organized at Belts-ville, Maryland, to study the nature of photoperiodism and its significance to agriculture. The new team was headed by Harry A. Borthwick (a botanist) and Marion W. Parker (a plant physiologist). The approach was to discover the light-sensing mechanism involved in photoperiodic control of flowering and other aspects of plant development. Two "photoperiod houses" were constructed at Beltsville. They were similar to the earlier "dark house" at Arlington in that boxes of plants were mounted on carts and moved into and out of the buildings on steel rails. The new buildings were equipped with electricity, and light-tight curtains were used to separate treatment compartments within the buildings. This allowed use of natural outdoor daylight alternated with various timing and light combinations when the plants were inside the photoperiod houses.

Some of the research was done in greenhouses equipped with various supplemental light sources and adjacent dark rooms. The plants were grown on warehouse carts rather than on fixed benches in order to allow more orderly movement to adjacent rooms for various supplemental light and temperature combinations during the night.

Other aspects of the research required a more completely controlled light environment and resulted in construction of artificially illuminated growth rooms so that light intensity during the daily light period would not vary with season, as it did in the greenhouse and outdoors next to the photoperiod houses. In order to obtain adequate light for plant growth in those early rooms, the team used a carbon arc lighting system supplemented with white incandescent filament lamps arranged in a circle around the carbon arc (Fig. 3). The table used to support growing plants was also circular in shape and placed below the incandescent lamps (see Fig. 3). The carbon arc system was surrounded by glass in order to filter out the ultraviolet light before it reached the plants. Occasionally a window broke and an experiment was ruined. However, this lighting system was the best available at the time of its construction in 1937 (3), and it was used successfully until 1962 (4), when it was replaced by a combination of white fluorescent and incandescent filament lamps. The carbon arc growth room was instrumental in development of the 8-hr light period as the standard "short day." This came about quite naturally because the carbons would burn for about 8 hr and 15 min before needing replacement. Thus, many of the early growth room experiments with soybean and cocklebur (both short-day plants) involved 8 hr of the bright light, and other light combinations given in adjacent rooms where the

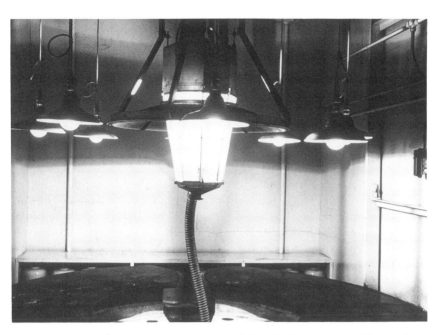

Figure 3 An early plant growth room at Beltsville illuminated with a carbon arc lamp supplemented with incandescent filament lamps.

plants were treated with various colors, durations, and intensities of light during the 16-hr night.

While some experiments were done in the artificially illuminated growth rooms, others were done with a combination of outdoor daylight and timing of supplemental light during the night in the photoperiod houses. Still other experiments were done in the greenhouse. Some of these experiments were designed to test which color of light was most effective as a night interruption. The rationale was that the effectiveness of different colors would indicate absorption characteristics of the pigment system involved in the photoperiodic responses of plants, and responsiveness to timing of the supplemental light would provide evidence toward a mechanism of action.

Also in the 1930s, scientists at the Smithsonian Institution in Washington, DC, constructed a number of small irradiation chambers that allowed testing the effects of narrow wavebands of light on seed germination. At that time, the Smithsonian research group worked within the "castle" building on the Mall and the USDA seed research was conducted about a block away in the USDA building. Flint, of the USDA, and McAllister, of the Smithsonian Institution, experimented with a selection of lettuce (*Lactuca sativa*) seed that had a low percentage of germination in uninterrupted darkness but a high percentage if exposed to white light (5,6). Using a series of fixed filter chambers, they found that germination of this selection of seed was increased by exposure to wavelengths at about 660 nm. However, germination was decreased below that of the dark controls after exposure to wavelengths at about 730 nm.

At Beltsville, Borthwick, Parker, and colleagues experimented with whole intact plants using white light filtered through broad-band colored glass filters at various times during the night. They found that the red component of white light was most effective in regulating time of flowering. After confirmation of the effectiveness of red light with a number of plant species, Borthwick and Parker decided to conduct quantitative studies on involvement of light color in regulation of the flowering process. It was decided that the approach would require action spectra (i.e., efficiency of different wavelengths of light) for control of flowering as a means of learning more about the light sensor within the plant. Sterling B. Hendricks (a physical chemist interested in botany) joined the group and built a double-prism spectrograph largely from surplus items and two prisms borrowed from the Smithsonian Institution. The theory, design, and construction of the spectrograph were evidence of the combined innovation, resourcefulness, and scientific genius of Sterling Hendricks, Harry Borthwick, and Marion Parker. The light source was a 12-kW carbon arc projector that was once used to provide the "spotlight" on the stage of a nearby theater. (Hendricks explained to this author that it was Parker who "rescued" the carbon arc projector as it was being discarded by a burlesque theater in Baltimore.) The prisms borrowed by Hendricks from the Smithsonian Institution were also historic. They had been used by Samuel Pierpont Langley (1834–1906), a noted astronomer, physicist, and aeronautics pioneer. A diagram of the "Beltsville spectrograph" is shown in Figure 4.

During operation of the spectrograph, the beam of light from the carbon arc first passed through a narrow vertical slit to a front-surfaced mirror, then through the prisms and through a door (which served as a manually operated shutter) to the treatment table. Large plants such as soybean and cocklebur were trimmed to a single leaf and that leaf was placed in a specific waveband, as diagrammed in Figure 4. The procedure for small plants such as *Chenopodium* required a modification. For these materials, the "rainbow" of colors was beamed at a front-surfaced mirror above the treatment table and reflected downward onto the seedlings. The treatment table was movable and could be placed

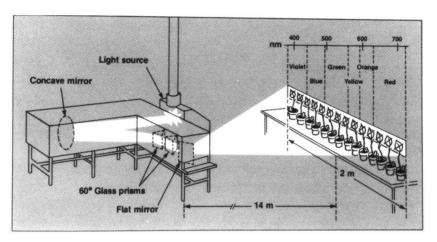

Figure 4 Diagram of the Beltsville spectrograph showing path of light from carbon arc through prisms to treatment table where soybean plants trimmed to a single leaf are being treated.

closer to or farther from the spectrograph. Increasing the distance allowed greater resolution of wavebands but decreased photon flux density at a given point on the treatment table. The spectrograph was used in development of many action spectra for control of flowering, seed germination, leaf movement, stem elongation, and many other morphological responses by scientists from around the world. It was a very important "research tool" from the time of its construction in the 1940s (7) until it was dismantled by Hendricks just before he retired in 1970.

The spectrograph separated white light into its component colors and allowed exposure of plants to a range of wavelengths for various durations, usually near the middle of the night following 8 hr of bright light in the nearby growth room. To obtain an action spectrum for control of flowering in soybean, for example, it was necessary to first determine how old the plants had to be before they were florally responsive to photoperiod, and how many short days were needed to cause flowering of "control" plants that received 16-hr uninterrupted nights alternated with 8-hr light periods (7). It was also necessary to determine whether plants such as soybean or cocklebur could be trimmed to a single leaf so that the treated part of the plant could be put in exactly the same position in the spectrum each day for the three or four consecutive days of treatment (see Fig. 4). In order to get different energy levels at each waveband, it was necessary to use different exposure durations. For example, one set of plants would be arranged as shown in Figure 4, irradiated in the "rainbow" of colors for one minute and then returned to darkness. Then another set would be put in the same positions for 2 min, followed by return to darkness. Other sets of identically pretreated plants received 4-, or 8-, or 16-, or 32-min treatments before returning to darkness. The same procedure with the same plants had to be done for several consecutive days. The early action spectra for control of flowering (7,8) were based on relatively few plants, and the red action peak seemed to be a broad band from about 640 to 660 nm. Germination of light-requiring seed offered the possibility of greater precision because the small size of seeds (relative to the size of intact plants) allowed use of more experimental units (seeds) on the spectrograph.

Borthwick et al. (9) tested germination of lettuce seeds and found a prominent peak for promotion of germination near 660 nm and a depression of germination near 730 nm. When seeds that had been pretreated with enough red light to promote germination were placed on the spectrograph, the scientists found an inhibition peak near 730 nm. They hypothesized that the effect of red light at 660 nm was reversed or negated by exposure to far-red (then called near-infrared) at 730 nm. To test the theory, they irradiated seeds repeatedly with red and far-red (Table 1). It was apparent that the effects of red could be reversed by far-red and vice versa. This observation became a very important step toward discovery of a photoreversible regulatory system. In subsequent spectrographic experiments with whole plants, the plants were often irradiated for a few minutes with red light before placement in the far-red

Table 1 Germination Responses of Grand Rapids Lettuce Seeds to Repeated 1-min Irradiations with Red (R) Alternated with 4-min Irradiations with Far-red (FR) Light

Irradiation	Germination (%)
None (dark control)	9
R	98
R, FR	54
R, FR, R	100
R, FR, R, FR	43
R, FR, R, FR, R	99
R, FR, R, FR, R, FR	54
R, FR, R, FR, R, FR, R	98

Source: Adapted from Ref. 9.

part of the spectrum in order to test action spectra for photoreversible control. The far-red action peak was near 730 nm for control of seed germination and floral induction in whole plants. However, an apparent discrepancy between red action peaks for seed germination and control of flowering was noted. The red action peak in whole plants appeared to be near 650 nm (7,8), whereas the peak for seed germination was near 660 nm (9).

The research on photoreversible control of seed germination (9), stem elongation (10,11) and hypocotyl hook opening (12) provided evidence of the presence of a photoreversible pigment system that responded to low energies of red and far-red light. It was apparent that one form of the pigment absorbed red light and became the far-red–absorbing form, which then absorbed far-red and became the red-absorbing form, etc.

With information drawn from the spectrographic studies, reversible responses to red and far-red, and the expertise of K. H. Norris (an engineer), W. L. Butler (a physicist), and H. W. Siegelman (a chemist), the team was able to build a dual-wavelength photometer and to photometrically determine the presence of a red/far-red photoreversible pigment in dark-grown seedlings. The paper was published by Butler et al. (13) in the *Proceedings of the National Academy of Science USA* in 1959. As other studies were published, the name *phytochrome* became firmly established.

After the discovery of phytochrome, there was an explosion of interest in laboratories around the world. Many followed the lead of Siegelman and Hendricks and concentrated on purification and identification of phytochrome from dark-grown etiolated seedlings. Preliminary studies showed that dark-grown seedlings contained more phytochrome and obviously less chlorophyll (which also absorbs red light) than light-grown seedlings. Some scientists questioned whether the "first phytochrome" in dark-grown seedlings would be the same as that in light-grown green plants. However, extraction from dark-grown seedlings offered more promise of success than extraction from green plants. The rationale was that if one knew the chemistry of phytochrome it would lead to better understanding of regulatory mechanisms within growing plants. Although many studies have been conducted, progress toward chemical characterization of phytochrome was much slower than originally envisioned by Hendricks.

Other scientists, including this author, concentrated on phytochrome regulation of various processes and endproducts in green plants. The objectives of that approach were to learn about the basic regulatory action of phytochrome in growing plants and to use that information to improve quantity and quality of plant products in a crop production system. The approach was to begin with precise spectrographic studies, follow these with observations of real crop production problems in the field, then use controlled environments and broad-band red and far-red light sources to study plant responses, and to relate these phytochrome-mediated responses to field growth under modified production practices. This author joined the Beltsville group in 1961 on a postdoctoral fellowship after completing doctoral research with Walter E. Loomis at Iowa State University on interactions of photoperiod and temperature on flowering and shoot/root partitioning in a biennial legume plant, sweetclover (*Melilotus alba* L.). It was apparent during the Iowa research that some photoperiodic responses such as flowering differed with temperature, and others such as shoot/root biomass ratios in first-year biennial plants were dominated by photoperiod (Table 2). After arriving at Beltsville, the first step was to adapt and use small-size seedlings to develop highly refined action spectra on the spectrograph. An Iowa strain of pigweed (*Amaranthus retroflexus* L.) and a Canadian strain of *Chenopodium rubrum* (L.) were compared for early responsiveness. Both could be induced to flower on short days (alternated with uninterrupted long nights) soon after emergence. However, the *Chenopodium* was responsive in the cotyledonary stage (Fig. 5). Thus, the series of experiments was done with this species. More than 100 *Chenopodium* seedlings could be placed in the same space needed for one soybean or cocklebur plant. Use of the miniplant system greatly improved precision of the action spectra and quickly became a team effort with Borthwick and Hendricks. A tray of *Chenopodium rubrum* seedlings old enough for treatment on the spectrograph and a seedling of the Beltsville strain of cocklebur that is also old enough for treatment are shown in Figure 6. For treatment, the *Chenopodium* seedlings were thinned to 10 per row and the rows were

Table 2   Shoot/Root Biomass Ratios in First-Year Biennial Sweetclover Plants Grown Under Three Photoperiods from July 15 until November 15 in a Greenhouse and Outdoors[a] at Ames, Iowa (42°N Latitude)

Photoperiods and locations				
24 hr		Natural Outdoors	9 hr	
Greenhouse	Outdoors		Greenhouse	Outdoors
(shoot/root biomass ratios)				
3.9	3.8	0.5	0.4	0.3

[a] Outdoor temperatures were lower than those in the greenhouse late in the season.
Source: Adapted from Ref. 14.

Figure 5 A *Chenopodium rubrum* seedling that was induced to develop a single floret between its cotyledons, without leaf formation. Note that the seed coat is still evident.

about 1.5 cm apart. A template was used to mark the rows so that row spacings in all trays were exactly the same. This allowed placement of successive trays (and rows within each tray) in exactly the same wavebands for treatment.

In a typical experiment, Hendricks calibrated the spectrum on the treatment table and determined the wavelength and energy to be received at each row position. Kasperbauer determined row positions and placed the plants, and Borthwick manually opened and closed the shutter (door) on the spectrograph to obtain the desired irradiation time. After

Figure 6 A tray of *Chenopodium rubrum* seedlings old enough for treatment on the spectrograph and a cocklebur plant also at an age suitable for treatment (units of measurement are in inches). The *Chenopodium* seedlings would be thinned to 10 per row, and the cocklebur trimmed of all but the most recently expanded leaf for treatment, as diagrammed for soybean in Figure 4.

five consecutive days of such treatment, the plants were grown in a noninductive greenhouse until the dark controls had visible florets. Then Kasperbauer placed all plants in a refrigerator to stop floral development, and examined and staged all plants. Floral development of each plant was assigned a numerical stage: 0.0 was completely vegetative and 9.0 indicated that florets were detectable without magnification (Fig. 7). Stages other than 9.0 were determined with the aid of dissecting needles and a magnifier. Floral stages (means for 10 plants) were plotted according to wavelength and energy received on the spectrograph to develop the action spectrum (efficiency of various wavelengths in control of floral development). The red action peak was at about 645 nm in these green seedlings (Fig. 8) and not at 660 nm as it is in light-requiring seed and in vitro. The shift from 660 to 645 nm was attributed to competitive absorption by chlorophyll at 660 nm in green plants (4). The treatment of red-irradiated plants in the far-red part of the spectrum showed that a few minutes at about 730 nm reversed the effect of red, but continued irradiation at that waveband caused a second reversal.

Treatment in the far-red part of the spectrum was done in two different ways: (a) following a saturation (7-min) exposure to red or (b) directly from darkness (Table 3). Exposure to far-red at about 735, 755, 775, and 795 nm took progressively longer to obtain a far-red effect (i.e., reversal of the inhibitory effect of a brief exposure to red light that was

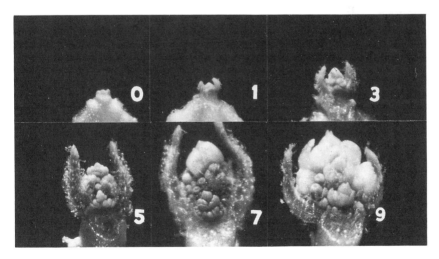

Figure 7 Floral stages of *Chenopodium rubrum* used to develop action spectra on the Beltsville spectrograph. (From Ref. 4.)

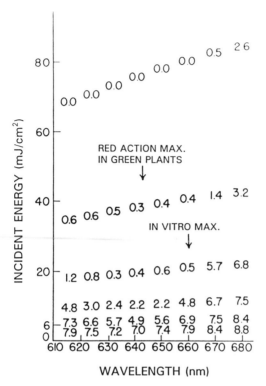

WAVELENGTH (nm)

Figure 8 Red action spectrum for inhibition of flowering in *Chenopodium rubrum* seedlings irradiated at different wavelengths on the Beltsville spectrograph in the middle of the 16-hr night. Seedlings kept in darkness during these treatments (controls) attained stage 9.0. Each value is the average floral stage for 10 plants. (Source: Adapted from Ref. 4.)

applied just before placement of the plants in the spectrum). However, prolonged exposure to those wavelengths also produced progressively less red effect. The red effect of prolonged exposure to far-red was interpreted to be due to the overlapping of the Pr absorption curve into the far-red part of the spectrum and maintenance of a small but effective amount of phytochrome in the biologically active form long enough to inhibit floral induction (4). The different responses to far-red at 735 and 755 or even 775 nm (Table 3) later became a significant factor in interpreting plant responses to light reflected from other plants and from different colored soils, plant residues, and mulches (see a later section of this chapter). Similarly, the shifted red action peak from 660 nm (the in vitro absorption peak) to 645 nm in green plants (see Fig. 8) is also very important in the interpretation of plant responses under field conditions.

Table 3 Effects of Prolonged Irradiation at Selected Wavebands Immediately After (A) a Saturation Exposure to Red Light (Sufficient to Completely Inhibit Floral Development, i.e., Stage 0.0) or (B) Directly from Darkness (Dark Controls = Stage 9.0)

Waveband on spectrograph (nm)	Duration of irradiation (min)							
	0.5	1	2	4	8	16	32	64
A. After saturation exposure to red (controls = 0.0)								
645	0.0	0.0	0.0	0.0	0.0	0.0	0.0	0.0
735	2.8	6.8	7.3	7.5	7.7	7.2	5.9	1.1
755	0.7	6.3	7.1	7.4	8.0	8.2	6.3	5.2
775	0.1	1.7	5.7	7.2	8.1	8.1	6.7	6.3
795	0.0	0.0	0.4	3.1	6.8	7.2	6.8	6.3
B. Directly from darkness (controls = 9.0)								
645	2.6	0.3	0.1	0.0	0.0	0.0	0.0	0.0
735	9.0	9.0	9.0	8.9	8.7	8.0	6.6	0.6
755	9.0	9.0	9.0	9.0	8.8	8.4	8.0	7.1
775	9.0	9.0	9.0	9.0	9.0	8.7	8.3	7.6
795	9.0	9.0	9.0	9.0	9.0	8.9	8.7	8.2

Source: Adapted from Ref. 4.

Evidence obtained on the spectrograph and with narrow-band fixed filters suggested that immediately after the treatment with red light, the amount of the far-red–absorbing form of phytochrome (Pfr) would be high but would then decrease in amount while the amount of Pr would increase in darkness over a several hour period. Inhibition of flowering of short-day test plants was accomplished if the amount of Pfr was above a "critical level" for an "adequate duration." However, the critical level and adequate duration seemed to vary with plant species and stage of growth (15,16). Conversely, maintaining Pfr above a critical level for an adequate period resulted in promotion of flowering in long-day plants.

Since a brief irradiation with red light converted most of the phytochrome to Pfr and the proportion of phytochrome in the Pfr form diminished in darkness at a temperature-dependent rate, a series of short exposures to red light alternated with short dark periods could be almost as effective as continuous light for some plant species. Flowering of some short-day plants like *Chenopodium* would be inhibited by one brief exposure to red near the middle of the night, and repeated brief exposures to red did not further enhance that floral inhibition. However, chrysanthemum (*Chrysanthemum morifolium*) required longer exposures to light in order to completely inhibit flowering (17). Thus, chrysanthemum responded equally well to several hours of continuous light or to short light–dark cycles during the same several hour period near the middle of the night (18,19). Flowering of long-day sweetclover plants also responded very little to a single brief exposure to light in the middle of the night. However, sweetclover (like chyrsanthemum) responded about the same to several hours of continuous light or to much less light applied for 10% of the time in short on–off cycles repeated over the same duration (20).

The spectrographic studies on photoconversion and timing of dark reversion (21) contributed to greenhouse studies with H. M. Cathey and Borthwick on cyclic lighting to regulate time of flowering. Cathey and Borthwick investigated short-day chrysanthemum (19) and Kasperbauer et al. the long-day sweetclover (20). Responses of long-day sweetclover plants to different cycle lengths are shown in Figure 9. Cycle length could be less frequent with light from cool-white fluorescent lamps than from incandescent filament lamps because these sources differed in the far-red/red ratio in the light that they emitted. The fluorescent lamps emitted very little far-red. Consequently, light from fluorescent lamps put a greater fraction of phytochrome in the Pfr form and more time could elapse before the Pfr level dropped below the "critical" level for control of flowering. While fluorescent lamps were more efficient in cycle length, incandescent filament flood lamps were more convenient and equally effective if the cycle length was shorter. Cyclic lighting effectively controlled time of flowering with only 10–20% of the electrical energy that is needed for continuous lighting. During recent years, cyclic lighting has been used to make the mountains of the tropics a desirable place to produce cut flowers commercially all year. The natural day lengths are about the same all year near the equator, and temperatures differ with elevation. With these natural background conditions, cyclic lighting can be used to delay floral induction of short-day plants until they attain suitable size, after which they can be brought to flower under the natural length days. On the other hand, long-day plants such as carnation can be kept vegetative on the natural days until they are ready to be placed under cyclic lighting to induce flowering. It is obvious that basic studies on the photoconversion of phytochrome and timing of dark reversion were important aspects of the real world use of this information.

Controlled Environment Studies

The use of prolonged exposures (up to 90 min) to far-red light, and the energy requirements for photoconversion and the timing of dark reversion of phytochrome (4,21) in regulation of a physiological process, flowering, provided the foundation for many controlled-environment and

Figure 9 Sweetclover (long-day) plants grown under 8 hr of sunlight and supplemental incandescent filament light applied during the 16-hr night as follows (left to right): 16 hr continuous, 1.5 min every 15 min, 6 min every 60 min, 24 min every 4 hr, 96 min centered at midnight, and no supplemental light. Plants were 3 months old when treatments began. Photographs were taken after 6 weeks of the daily treatment. (From Ref. 20.)

field studies. There are numerous excellent examples of phytochrome regulation of plant physiological processes under controlled environments in many laboratories around the world. However, because of space limitations this section will be confined to some selected experiments with intact green plants that led to a better understanding of phytochrome action in field growth and development of crop plants. Many of the examples are with tobacco, the same species that launched the scientific curiosity of Garner and Allard (1) and led to their discovery of photoperiodism.

The controlled environment experiments discussed herein were started with tobacco in the early 1960s. They involved determination of phytochrome regulation of leaf shape and thickness, internode length, chlorophyll concentration, chlorophyll a/b ratios, chloroplast structure, photosynthetic efficiency of leaves, and accumulation of compounds such as sugars, starch, epicuticular alkanes, fatty acids, amino acids, organic acids, alkaloids, and polyphenolics. The objective of the controlled-environment experiments was to learn how and why plants responded as they did to light variables.

For most of these studies, plants within a given experiment were grown under identical conditions in the same controlled environment for about 23 hr and 50 min each day. At the end of the bright light period each day, plants were moved to adjacent rooms and irradiated with either 5 min of red, 5 min of far-red, or 5 min of far-red followed immediately by 5 min of red light to test for photoreversible control. This approach allowed study of phytochrome regulation of developmental processes when temperature and photosynthetic light were kept constant among all plants. The red light put most of the phytochrome in the Pfr form whereas the far-red put most of the phytochrome in the Pr form at the beginning of the night. This allowed the phytochrome form to initiate physiological events during the night that regulated how the plants invested (partitioned) the photosynthate that had accumulated at the end of the photosynthetic period. The working hypothesis was that responsiveness to red and far-red was related to adaptation of plants to various environments in the real world. Some of the controlled-environment responses that became highly relevant in the interpretation of field plant responses and their management are summarized below.

Tobacco plants that received a brief exposure to far-red at the end of each day developed stem and leaf characteristics that were dramatically different from those that received a brief exposure to red at that time. Also, plants that received 5 min of far-red followed immediately by 5 min of red responded to the kind of light received last. This photoreversible regulation of developmental responses was evidence that phytochrome was involved in initiating physiological events in plant development. Since phytochrome action was shown to be dependent on the photoequilibrium between the two forms of phytochrome (21), it was reasonable to think of end-of-day far-red as either a low red/far-red ratio or as a high far-red/red ratio. Because responses to far-red were very dramatic in growing seedlings and there is much competitive absorption of red at 660 nm (the phytochrome absorption peak) in green plants, far-red was projected as the more important variable in nature (22). Thus, this author uses the far-red/red ratio. This concept is consistent with a study by Vogelmann and Bjorn (23) who inserted fiber optic probes into fleshy leaves to compare the amount of far-red light that reached different depths within the leaf tissue relative to the amount received at the exterior surface. They detected higher amounts of far-red (at 750 nm) inside the fleshy leaves, which they attributed to photon scattering within tissue that had relatively little competitive absorption of far-red light.

Leaves that developed on plants that received far-red (a high far-red to red ratio) at the end of the daily photosynthetic period grew longer and narrower than those that received a low ratio at the end of each day (Fig. 10). The petioles were longer as were the stem internodes (Fig. 11).

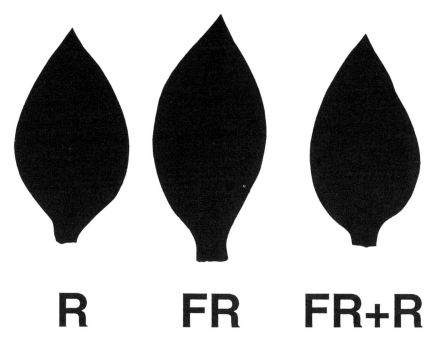

Figure 10   Tobacco leaves from plants that received 5 min of red (R), 5 min of far-red (FR), and 5 min of far-red followed immediately by 5 min of red (FR + R) at the end of each day during development. (Adapted from Ref. 24.)

Figure 11 Tobacco plants that received (from left to right, respectively) 5 min of red, 5 min of far-red, or 5 min of far-red followed immediately by 5 min of red at the end of each day for 21 days.

Other plant species such as soybean responded similarly. As with tobacco, biomass partitioning among leaves, stem, and roots in soybean seedlings was highly influenced by the far-red/red light ratio received just before darkness (see Table 4). In addition to the differences in leaf shape (Fig. 10), leaves that received the higher far-red/red ratios were thinner, had a higher chlorophyll a/b ratio (24), and a higher concentration of light-harvesting chlorophyll protein (LHC-II) (26). Chloroplasts in leaves that developed with the higher far-red/red ratio (far-red treatment) had more but smaller grana and smaller starch grains (27). Far-red–treated

Table 4 Effects of End-of-Day Red (R) or Far-Red (Low and High FR/R Ratio, Respectively) on Percentages of Dry Biomass Partitioned to Leaves, Stems, and Roots of Soybean Seedlings Under Controlled Environments

End-of-day		Dry biomass % in:			
Light[a]	FR/R ratio	Leaf blades	Stems and petioles	Roots	Shoot/root ratio
R	Low	43.9	23.6	32.5	2.1
FR	High	43.6	33.2	23.2	3.3
FR, R	High, low	43.4	22.8	33.8	2.0

[a] R and FR treatments were for 5 min at the end of each day for 20 consecutive days. The FR, R treatment received 5 min FR followed immediately by 5 min R each day.
Source: Adapted from Ref. 25.

leaves had higher concentrations of sugar and organic acids, and lower concentrations of amino acids (Table 5). In addition to being thinner with higher chlorophyll a/b ratios, the leaves that developed on plants that received the higher far-red/red ratio were more efficient photosynthetically (29). That is, they fixed more CO_2 per mass of leaf tissue (Fig. 12) even though they did not differ on a leaf area basis. Those combined observations in controlled environments suggested that the amount of far-red and the far-red/red ratio played a major role in development of plant characteristics that could favor survival while competing with other plants. It was apparent that the amount of phytochrome in the Pfr form relative to the total amount of phytochrome (P), particularly at the beginning of a dark period, plays a critical role in signaling photosynthate distribution and developmental patterns.

It was not clear, however, whether a low Pfr/P ratio (the consequence of irradiation with far-red) triggers a chain of metabolic events leading to "competition-adapted" development or whether the events occur because the Pfr/P ratio is too low to signal a chain of events leading

Table 5 Concentrations of Free Sugars, Organic Acids, and Amino Acids in Tobacco Plants that Received 5-min Far-Red or 5-min Red (High or Low FR/R Ratio, Respectively) at the End of Each Day During Development

| | End-of-day radiation and plant part | | | | | |
| | FR (high FR/R) | | | R (low FR/R) | | |
Component	Leaf blade	Mid-rib	Stem	Leaf blade	Mid-rib	Stem
Free sugars (mg/g dry matter)						
Sucrose	7.5	8.3	12.5	6.3	7.5	17.5
Glucose	3.0	19.2	45.0	1.5	2.7	10.0
Fructose	2.8	10.3	40.0	2.0	1.3	10.1
(Total)	(13.3)	(37.8)	(97.5)	(9.8)	(11.5)	(37.6)
Organic acids (mg/g dry matter)						
Malic	16.3	50.1	13.0	12.5	50.1	12.5
Citric	2.7	1.5	<0.5	2.5	1.4	<0.5
Succinic	3.0	3.0	4.0	2.5	3.0	3.8
Fumaric	1.3	2.5	<0.5	0.5	2.5	<0.5
Ascorbic	1.8	6.3	3.0	1.5	6.3	2.8
(Total)	(25.1)	(63.4)	(21.0)	(19.5)	(63.3)	(20.1)
Free amino acid (µM/g dry matter)						
(Total)	(44.2)	—	—	(66.4)	—	—

Source: Adapted from Ref. 28.

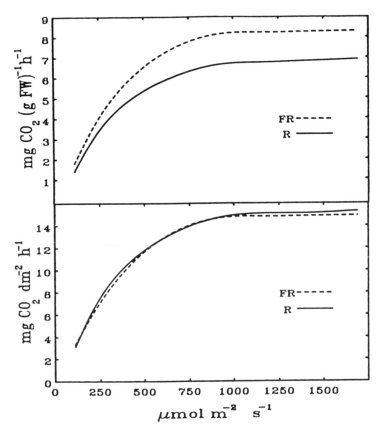

Figure 12 Net CO_2 assimilation rates of tobacco leaves that received 5 min of red (R, a low FR/R ratio) or far-red (FR, a high FR/R ratio) at the end of each day during development. Data are expressed on the basis of fresh weight of leaf lamina, and leaf area. (Curves are drawn from Ref. 29.)

to "sun-adapted" development. Whichever the case, some effects of far-red are similar to those of exogenous gibberellic acid (GA). The thin leaves, light color, and somewhat elongated internodes of plants treated with GA (30–32) and end-of-day far-red (29) suggested that both treatments may involve the same metabolic pathway. Both GA- and far-red-treated plants have decreased total chlorophyll, and the reduction of chlorophyll b is greater than the reduction of chlorophyll a resulting in an altered chlorophyll a/b ratio. It was suggested that end-of-day far-red, through its influence via the phytochrome system during the dark period, may initiate shifts in the balance of naturally occurring growth regulators such that the imbalance tips in favor of gibberellins much the same as when exogenous GA is added to a plant. Free amino acid concentrations in tobacco leaves also differed with photoperiod as well as with the far-red/red light ratio received at the end of each day (Table 6). Wilkinson hypothesized that photoperiod and plant competition for light are sufficient to alter endogenous GA and enzyme systems that regulate metabolic pathways and endproducts in a field environment (34). Epicuticular alkanes differed significantly between two ages of leaves from the same field-grown tobacco plants (35). Since the field-grown leaves developed at different times under changing natural environments, plants of a genetically uniform line of tobacco were grown under different photoperiods in controlled environments. Epicuticular alkanes, fatty acids, and fatty alcohols were all highly influenced by growth environment (Table 7). Other investigators (37) have shown that exogenous [^{14}C]-fatty acids could be incorporated into intracellular lipids. Also, when $^{14}CO_2$ was applied to leaves with or without GA_3 treatment, there was a greater translocation of ^{14}C from the GA-treated leaves (38) even though the GA treatment did not result in differences in total ^{14}C fixed in the leaves. And synthesis of some enzymes has been induced by GA_3 (39) whose synthesis is photoperiodically controlled (40). Modified epicuticular content and composition associated with light parameters are examples of metabolic alterations in response to environmental changes that may increase the probability of survival.

Table 6   Free Amino Acid Content of Tobacco Leaves that Developed at 25°C Under 8- or 16-hr Daily Light Periods that Ended With 5 min Red (R, Low FR/R Ratio) or 5 min Far-red (FR, High FR/R Ratio) Each Day

Photoperiod	End-of-day Light	FR/R ratio	Amino acid group Oxalacetate	α-Ketoglutarate	Pyruvate
			μM/g dry wt [a]		
Short	R	Low	24 a	15 a	13 a
	FR	High	14 c	13 b	10 b
Long	R	Low	14 c	10 c	8 c
	FR	High	17 b	8 d	6 d

[a] Values are means for 15 plants. Within each column, values followed by the same letter do not differ significantly at the 5% level.
Source: Adapted from Ref. 33.

Table 7 Epicuticular Fatty Acid, Fatty Alcohol, and Alkane Concentrations in Tobacco Leaves Grown in Controlled Environments Under Short or Long Photoperiods at 28°C

Photoperiod	Fatty acids	Fatty alcohols	Alkanes (n)			
			C_{27}	C_{29}	C_{31}	C_{33}
	ng/cm^2		$\mu g/cm^2$			
Short	1306	1524	12	40	266	221
Long	2549	840	45	50	586	710
Signif.	*	*	*	*	*	*

*Indicates that differences are statistically significant at the 5% level of confidence.
Source: Adapted from Refs. 35 and 36.

The controlled-environment studies suggested that the ratio of far-red relative to red photons received by developing leaves of a field-grown plant could influence adaptive morphological development of the plant and photosynthetic efficiency of the leaves. An interesting analogy would be to think of the phytochrome system within the growing plant as a variable sensor that is constantly monitoring the far-red/red ratio as an indicator of competition from other plants and as an initiator of physiological events that favor survival of the plant among that perceived competition. For example, a higher far-red/red ratio leads to a longer stem with fewer branches and a more efficient photosynthetic system. This adaptive response would favor survival by increasing the probability of keeping some leaves in sunlight above competing plants and perhaps by having leaves that are more efficient in utilizing light within the plant canopy. It was evident from the controlled-environment studies that the phytochrome system can initiate physiological events leading to adaptive morphological development such that the plant is better suited to compete with other plants in its growth environment, and that genetically identical plants can differ significantly in quantities of various chemical constituents, depending on growth environment.

Field Plant Response to Far-Red Reflected from Other Plants

This author's field plant population density studies were started with tobacco in the mid-1960s and extended to other species in the early 1980s. The initial field observations of end vs. mid-row tobacco plants showed that the mid-row plants were taller, had slightly longer internodes and slightly thinner leaves. These characteristics were in the direction of controlled-environment plants that received the higher far-red/red ratios.

Growth of pretransplant tobacco seedlings followed the same pattern. That is, close-spaced seedlings had longer and thinner stems, narrower leaves, and a smaller root system relative to seedlings that were wider apart. The characteristics began to appear even before mutual shading occurred among the close-spaced seedlings. These outdoor observations combined with controlled-environment observations strongly suggested that the far-red/red ratio and phytochrome might be involved in tobacco response to nearness of other plants. The "tools" available to study field population effects on light spectra were somewhat primitive in the 1960s; however, critical measurements were made. The available portable spectroradiometer could measure only 16 fixed wavebands. Of these, 11 were between 390 and 700 nm, one was at 725 nm, and another at 791 nm. The light detector head contained four rows of four fixed filters, each 2.5 cm × 2.5 cm, and there was a manual switch to change wavebands. Thus, it was cumbersome by today's standards. However, the detector head was on a 3-m cable and it was possible to measure transmission through a single tobacco leaf and at various places within the grown canopy (22). The data showed that green leaves absorbed most of the blue and red, but transmitted some of the green and much of the far-red. These measurements also revealed that a higher percentage of light was transmitted at 791 than at 725 nm. The measurements documented differences in spectral balance of light within and below a plant canopy (Table 8). However, the differences in far-red/red ratios within and below the tobacco canopies could not explain why the upper leaves (those in sunlight above competing leaves) were longer, narrower, and thinner when plants were grown close together. These observations suggested (see Figs. 10 and 11) that the upper leaves in close-spaced tobacco received more far-red or at least a higher far-red/red ratio than upper leaves on an isolated plant. As part of the measurements and comparisons, light at each measured waveband in the canopy was expressed as a percentage of the incoming sunlight at the same waveband. An observation that began as a puzzle in the late 1960s became the critical factor in relating field plant population density to phytochrome-regulated adaptive morphological development.

When light measurements at various points within and below the tobacco canopy were compared with incoming sunlight, some dramatic differences were observed at the 791-nm waveband depending on whether the incoming light was measured on a road away from other plants or in a small patch of sunlight on the ground within the tobacco fields (see footnote, Table 8). Subsequent comparison of the spectra of light on the roadway with that on the soil surface in the tobacco field showed that the light close to plants had more far-red than that on the

Table 8 Percentages of Incoming Sunlight Received Within and Below a
Canopy of 190-cm-tall Tobacco at About 1 p.m. on September 1, 1967 near Lexington, KY

Peak wavelength (nm)	Percentage of incoming sunlight[a] detected:		
	Below a single leaf	Within canopy	Below canopy
391	1.7	0.9	0.5
432	0.5	0.7	0.3
448	0.7	0.7	0.3
483	0.9	0.6	0.4
511	3.3	0.8	0.6
543	22.7	11.0	6.5
576	14.7	5.0	3.4
601	10.8	2.6	2.1
629	7.9	1.7	1.4
658	6.1	2.3	1.7
686	6.6	2.2	1.9
725	27.5	11.6	8.8
791	49.5	36.3	20.3

[a] The incoming sunlight was measured on a road, away from tall plants.
Note: Light at 791 nm was about 15% higher in sunflecks on the ground near tobacco plants than it was above the road, away from tall plants.
Source: Adapted from Ref. 22.

roadway. The only logical explanation was that the extra far-red in the tobacco field was reflected from the nearby plants. Far-red reflection from other plants also helped explain why the upper (unshaded) leaves of close-spaced tobacco had a different shape than upper leaves on isolated plants. Similarly, the stem length and leaf shape differences of close-spaced and isolated tobacco pretransplant seedlings could be explained when far-red reflected from other seedlings was considered as a contributor to the far-red/red ratio received by the growing plants.

Beginning in the early 1970s tobacco plants were routinely set at three population densities to study plant morphological development and leaf chemistry. These studies were highly relevant as a background for possible alternate production procedures, and also as a test of plant spacing, the far-red/red ratio, and the resultant physical and chemical characteristics of plants as related to closeness of other plants. Close-spaced plants were usually 30 × 30 cm, normal spacing was 45 cm apart in rows that were 100 cm apart, and wide-spaced plants were 120 × 120 cm apart. Although precise spectral measurements were difficult with the available portable spectroradiometer, higher far-red levels were detected at a midpoint between the close-spaced plants than at the mid-point between wide-spaced plants. That is, the detected level of far-red was higher when the spectroradiometer light collector was closer to a growing plant.

The physical and chemical effects of spacing on the plants were more easily documented. Close-spaced plants began growing taller than wide-spaced plants even before mutual shading began. Representative plants taken from the field 6 weeks after transplanting are shown in Figure 13. Notice that close-spaced plants were taller (longer internodes), with thinner stems and narrower leaves. All of these features were consistent with those associated with a higher far-red/red ratio in the controlled-environment studies (see Figs. 10 and 11). In addition to being taller with narrower leaves, the close-spaced plants had thinner leaves, higher chlorophyll a/b ratios, lower alkaloid concentrations, and higher chlorogenic acid concentrations (Table 9). These plant responses followed the same trends as plants that received higher far-red/red ratios in the controlled-environment studies (Table 10).

Figure 13 Tobacco plants after 6 weeks of growth in (from left to right, respectively) close, normal, and wide spacings in a field.

Table 9 Effects of Field Plant Spacing on Concentrations of Alkaloids and Phenolics (Chlorogenic Acids) in Mature Tobacco Leaves

| | Plant spacing during growth | | |
Component	Close	Normal	Wide
	(mg/g)		
Total alkaloids	13 c*	32 b	54 a
Chlorogenic acid	0.57 a	0.32 b	<0.05 c

*Values in the same line that are followed by the same letter do not differ significantly at the 5% level.
Note: The close-spaced plants received higher FR/R ratios during growth.
Source: Adapted from Ref. 41.

In 1983, we extended the studies to other plant species and obtained a much improved portable spectroradiometer that was capable of measuring radiation from 300 to 1100 nm at 2-nm intervals. With a remote light collector on a 1.5-m fiber optic probe, we measured the light spectra received at the upper surface of soybean canopies in north–south vs. east–west rows and in various other spacing combinations. Light measurements were taken near the tops of growing plants because Parker and Borthwick (43) had shown that the most recent fully expanded leaves were very efficient in sensing morphogenic light signals that regulate developmental responses in the growing parts of the plant.

Table 10 Concentrations of Alkaloids and Phenolics in Leaves of Tobacco Seedlings that Received 5 min of Far-Red (FR, High FR/R Ratio) or 5 Min Red (R, Low FR/R Ratio) at the End of Each Day in Controlled Environments for 18 Days

| | End-of-day radiation and plant part | | | |
| | FR (high FR/R ratio) | | R (low FR/R ratio) | |
Component	Blades	Mid-ribs	Blades	Mid-ribs
Alkaloids		(mg/g dry matter)		
Total	3.9	0.5	7.0	0.8
Phenolics				
Chlorogenic acid	20.0	10.0	18.8	6.3
Total	24.0	20.3	22.1	13.1

Source: Adapted from Ref. 42.

The absorption, reflection, and transmission spectra were also determined from individual leaves of a number of species, including soybean.
Representative spectra from soybean leaves are shown in Figure 14. It
was clearly evident that each green leaf reflected almost half of the far-
red that reached it and that the reflection "plateau" begins at about 750
nm. Thus, it was reasonable to expect that the number, nearness, and
size of competing plants would influence the amount of reflected far-red
and the far-red/red ratio received by a nearby growing plant. Also, it
was considered possible that row direction might further influence the
far-red/red ratio received because of heliotropic movement of leaves,
causing them to be directional far-red reflectors. Further, it is important
to note that the far-red/red ratio in incoming sunlight also increases near
sundown.

Clearly, the far-red/red ratio at the surface of the upper leaves was
largely affected by the amount of far-red reflected from nearby plants
(44). The ratio was also influenced by heliotropic movement of leaves of
the broad-leaf, long-petiole soybean plants, which had the effect of directional far-red reflectors, especially near the end of the day. Thus, it was

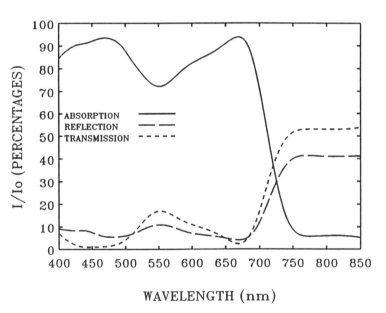

WAVELENGTH (nm)

Figure 14 Light absorption, reflection, and transmission spectra from a soybean leaf. The absorption percentages were calculated by subtracting the combined values for reflection plus transmission from 100 at each measured wavelength. (Redrawn from Ref. 25.)

hypothesized that incoming sunlight with a higher far-red/red ratio that is reflected off of the heliotropic leaves could contribute to an important end-of-day "signal" of the relative amount of competition from other plants, and this might be influenced by row direction in broad-leaf, long-petiole plants.

Incoming light from all four directions was measured near the tops of bean plants growing in north–south vs. east–west rows at different stages of growth. When numerous readings taken throughout the day were averaged, plants in north–south rows received higher far-red/red ratios. An example is shown in Table 11. As predicted, this pattern of higher far-red/red ratio in north–south rows was most evident near the end of the day when the heliotropic movement of the leaves had the effect of reflecting light back to the adjacent row.

As projected from the earlier controlled-environment experiments, a higher amount of reflected far-red and the associated higher far-red/red ratio resulted in slightly taller shoots and less branching of soybean seedlings in north–south vs. east–west rows (Table 12). There were several examples in which the larger shoots produced more seed or fruit, when there was no moisture or soil nutrient stress (25,45).

In a wheat (*Triticum aestivum* L.) experiment with Karlen (46), we measured light spectra at the soil surface within different populations of field-grown seedlings in early spring (Fig. 15). As expected, close-spaced plants received more reflected far-red and higher far-red/red ratios. The

Table 11 Photosynthetic and Photomorphogenic Light Received at the Shoot Apex of Bush Bean Plants Grown in North–South (N-S) vs. East–West (E-W) Rows, and Plant Productivity

	Row orientation	
Characteristic	N-S	E-W
Light (means of 24 readings) [a]		
Photosynthetic (μmol/m^{2}s)	389 ± 62	393 ± 59
Photomorphogenic (FR/R photon ratio)	1.85 ± 0.23	1.48 ± 0.13
Plant productivity		
Green beans (g fresh wt/plant)	59.0 ± 8.3	43.0 ± 4.1

[a] Light coming to the shoot of two representative plants from each row orientation was measured from the north, south, east, and west at 11:00 a.m., 1:30 p.m., and 3:30 p.m. on a cloudless day near Frankfort, KY. Each light value in the table is the mean ± SE for the two plants, four directions, and three times during the day (i.e., means are for 24 separate readings).

Source: Adapted from Ref. 45.

Table 12 Row Orientation Effects on Characteristics of Soybean Plants Grown in North–South (N-S) vs. East–West (E-W) rows in Irrigated Loamy Sand in Field Plots near Florence, SC

	Row orientation	
Characteristic	N-S	E-W
At 6 weeks (means/plant)		
Stem length (mm)	348 ± 6	324 ± 5
Nodes/stem (no.)	8.1 ± 0.1	8.2 ± 0.1
Branches/plant (no.)	1.8 ± 0.4	3.0 ± 0.2
At harvest (dry matter/1-m row)		
Seed (g)	158.2 ± 7.8	142.8 ± 7.5
Pods (g)	58.2 ± 3.0	53.0 ± 4.0
Stem (g)	43.8 ± 1.9	40.8 ± 3.2
Seed/straw (ratio)	1.55	1.52

Source: Adapted from Ref. 25.

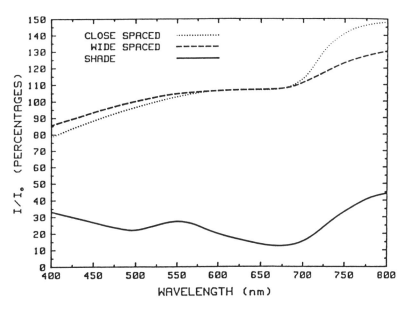

Figure 15 Spectral distribution of light in shade (solid line) and in sunflecks (dashed lines) at soil level in a field of close-spaced and wide-spaced wheat seedlings (about 8–10 cm tall) near Florence, SC in early afternoon in mid-March 1984. Values are expressed as percentages of incoming sunlight at each measured wavelength. (From Ref. 46.)

plant response was to develop fewer tillers (analogous to fewer branches on soybean seedlings) and longer leaves, again showing a phytochrome response to reflected far-red under field conditions. This was an example of how the phytochrome system within the wheat seedlings could sense (measure) the nearness of competing plants and then initiate physiological events that regulated the amount of tillering. That is, when soil moisture and nutrients were not limiting, the phytochrome system measured the far-red/red ratio, sensed the amount of competition, and regulated the amount of tiller development. Representative wheat seedlings from wide- and close-spaced plantings are shown in Figure 16.

There were a number of other population and spacing studies, some of which involved the use of plants in containers imbedded in soil among the various populations. This allowed determination of effects of spacing and the far-red/red ratio on shoot-root biomass ratios as a result of phytochrome regulation of partitioning within a growing plant, when the container-grown plants in all population densities had the same volume of the same soil mixture. Again, as projected from controlled-environment studies (see Table 4), plants that were closer together and received the higher far-red/red ratios had the higher shoot/root biomass ratios. Plants that received the higher far-red/red ratios in the field as well as those in controlled environments partitioned higher percentages of the new photosynthate to growing stems, and less to branches (or tillers) or new root growth. This is a reasonable adaptive response because the far-red/red ratio would be sensed as an indicator of competition, and plants with longer stems would have a greater probability of keeping some leaves in sunlight above competitors and surviving long enough to produce the next generation.

Upwardly Reflected Light from Soil Surfaces

When it was clear that plants respond to spectral composition of light reflected from other living plants, P. G. Hunt (a soil scientist) and I decided to find out whether plants would respond morphologically to spectral differences in light reflected from different colored soils, plant residues, or other soil covers (mulches). In the initial studies during 1984 and 1985, the spectra of upwardly reflected light were measured 10-cm above five different colors of soil (47). This height above soil was selected because it is in the seedling establishment zone, and seedlings are extremely responsive to spectral composition of light (4,21). Measurements were made over dry or wet soil and over the soils when they were about 80% covered with plant residue from a previous crop. The different colored soils and plant residues reflected different far-red/red ratios.

Figure 16 Field-grown wheat seedlings (about 8 to 10 cm tall) from wide-spaced (left) and close-spaced (right) population densities in mid-March 1984. (From Ref. 46.)

The next step was to determine whether soil surface color could influence reflected light sufficiently to modify seedling growth. Soybean seedlings were started in pots of soil and placed on greenhouse benches about 60 cm apart in groups of four. Each group of four soybean seedlings was covered with a 122 × 122 × 2 cm insulation panel that had four 2.5-cm holes 60 cm apart so that the seedlings could grow through the insulation panels. The panels were covered with about 5 mm of the different colored soils, or soil that was about 80% covered with straw. In this manner, root temperature differences were minimized below the different soil surface colors. Plants over the brick-red soil and over the straw residue received higher reflected far-red/red light ratios than those grown over the white soil (47,48); and they grew taller, had less root growth, and developed higher shoot/root biomass ratios (48). These initial studies were very significant because soils are of many colors as are plant residues left on the soil surface in many no-tillage or other conservation tillage procedures.

Other experiments were done in which the insulation panel surfaces were painted instead of being covered with different colored soils or plant residues. Plants responded the same to either painted or soil-covered surfaces if they reflected the same spectrum of light. That is, plants grown over the red painted surfaces received higher reflected far-red/red ratios than plants grown over white, and they grew taller and had higher shoot/root biomass ratios. Subsequently, painted surfaces were used for outdoor experiments because soils and plant residues were affected by wind and rain. Clearly, soil surface color could affect the reflected far-red/red light ratio sufficiently to influence photosynthate partitioning and biomass distribution within growing seedlings.

In late 1985, D. R. Decoteau (a horticulturist) observed the soybean seedling responses to light reflected from different colors of painted or soil-covered surfaces (described above). He proposed that the concept be extended to irrigated field-grown tomatoes (*Lycopersicon esculentum*) for the 1986 season. Since black or white plastic mulches were widely used for soil and water conservation as well as to control weeds in the production of tomato and other high-value food crops, we then explored the possibility that an altered surface color on the mulch could maintain those benefits and have an added favorable affect on plant productivity. The working hypothesis (based on many previous experiments that involved controlled environments, reflection from other plants, and reflection from colored soils, plant residues, and colored panels) was that an upwardly reflected far-red/red ratio higher than that in incoming sunlight would signal the plant to partition more of the new photosynthate to shoots, whereas a lower ratio would favor partitioning to roots. Irrigated tomatoes grown in sunlight over mulches with red surfaces produced significantly increased fruit yield relative to those grown with conventional black or white mulches (49). This response was consistent with the hypothesis, i.e., the red paint used to change the surface color of the plastic mulch reflected a higher far-red/red ratio, and the tomato plants partitioned more photoassimilate to shoots, including fruit. Subsequent experiments with a wide range of colored mulches and a number of mulching materials and plant species have confirmed that the spectrum (particularly the far-red/red ratio and the quantity of blue) of upwardly reflected light over colored soil surface covers (mulches) can regulate fruit number and size, leaf shape and thickness, concentrations of chlorophyll and light-harvesting chlorophyll protein, root size of turnip, and even the yield of cotton (26,50).

SUMMARY

Photomorphogenesis plays a very important role in utilization of photosynthate within the growing plant. It is important to realize that the strategy of each plant is to survive long enough in its existing environment to produce the next generation. Thus, the plant must be able to sense the total environment, integrate the information, and adapt to the constantly changing environmental conditions. Examples presented in this chapter involved the light environment primarily as it is affected by season and competition from other plants. The phytochrome system within the growing plant functions as a constant sensor of photoperiod and competition from other plants, and then regulates initiation of metabolic events that result in adaptive responses such as stem length, leaf shape and thickness, leaf waxes, amount of branching (or tillering), relative root size, and flowering.

REFERENCES

1. W.W. Garner and H.A. Allard (1920). *J. Agric. Res.*, 18:553.
2. K.C. Hamner and J. Bonner (1938). *Bot. Gaz.*, 100:388.
3. M.W. Parker, and H.A. Borthwick, (1949). *Plant Physiol.*, 24:345.
4. M.J. Kasperbauer, H.A. Borthwick, and S.B. Hendricks (1963). *Bot. Gaz.*, 124:444.
5. L.H. Flint and E.D. McAllister (1935). *Smithson. Misc. Collec.*, 94(5):1.
6. L.H. Flint and E.D. McAllister (1937). *Smithson. Misc. Collec.*, 96(2):1.
7. M.W. Parker, S.B. Hendricks, H.A. Borthwick, and N.J. Scully (1946). *Bot. Gaz.*, 108:1.
8. H.A. Borthwick, S.B. Hendricks, and M.W. Parker (1948). *Bot. Gaz.*, 110:103.

9. H.A. Borthwick, S.B. Hendricks, M.W. Parker, E.H. Toole, and V.K. Toole (1952). *Proc. Natl. Acad. Sci. USA*, 38:662.
10. R.J. Downs (1955). *Plant Physiol.*, 30:468.
11. R.J. Downs, S.B. Hendricks, and H.A. Borthwick (1957). *Bot. Gaz.*, 118:199.
12. R.B. Withrow, W.H. Klein, and V. Elstad (1957). *Plant Physiol.*, 32:453.
13. W.L. Butler, K.H. Norris, H.W. Siegelman, and S.B. Hendricks (1959). *Proc. Natl. Acad. Sci. USA*, 45:1703.
14. M.J. Kasperbauer, F.P. Gardner, and I.J. Johnson (1963). *Crop Sci.*, 3:4.
15. H.A. Borthwick and S.B. Hendricks (1960). *Science*, 132:1223.
16. S.B. Hendricks (1960). *Cold Spring Harbor Symp. Quant. Biol.*, 25:245.
17. N.W. Stuart (1943). *Proc. Am. Soc. Hort. Sci.*, 42:605.
18. H.M. Cathey, W.A. Bailey, and H.A. Borthwick (1961). *Flor. Rev.*, 129(3330):21.
19. H.M. Cathey and H.A. Borthwick (1961). *Proc. Am. Soc. Hort. Sci.*, 78:545.
20. M.J. Kasperbauer, H.A. Borthwick, and H.M. Cathey (1963). *Crop Sci.*, 3:230.
21. M.J. Kasperbauer, H.A. Borthwick, and S.B. Hendricks (1964). *Bot. Gaz.*, 125:75.
22. M.J. Kasperbauer (1971). *Plant Physiol.*, 47:775.
23. T.C. Vogelmann and L.O. Bjorn (1984). *Physiol. Plant.*, 60:361.
24. M.J. Kasperbauer and A.J. Hiatt (1966). *Tob. Sci.*, 10:29.
25. M.J. Kasperbauer (1987). *Plant Physiol.*, 85:350.
26. J.A. Bradburne, M.J. Kasperbauer, and J.N. Mathis (1989). *Plant Physiol.*, 91:800.
27. M.J. Kasperbauer and J.L. Hamilton (1984). *Plant Physiol.*, 74:967.
28. M.J. Kasperbauer, T.C. Tso, and T.P. Sorokin (1970). *Phytochemistry*, 9:2091.
29. M.J. Kasperbauer and D.E. Peaslee (1973). *Plant Physiol.*, 52:440.
30. H. Kende and A. Lang (1964). *Plant Physiol.*, 39:435.
31. J.A. Lockhart (1956). *Proc. Natl. Acad. Sci. USA*, 42:841.
32. I. Szalai (1969). *Physiol. Plant.*, 22:587.
33. R.E. Wilkinson, M.J. Kasperbauer, and C.T. Young (1981). *J. Agric. Food. Chem.*, 29:658.
34. R.E. Wilkinson (1972). *Phytochemistry*, 11:1273.
35. R.E. Wilkinson and M.J. Kasperbauer (1972). *Phytochemistry*, 11:2439.
36. R.E. Wilkinson and M.J. Kasperbauer (1980). *Phytochemistry*, 19:1379.
37. P.G. Roughan, G.A. Thompson, Jr., and S.H. Chao (1987). *Arch. Biochem. Biophys.*, 259:481.
38. B. Aboni, J. Daie, and R.E. Wyse (1986). *Plant Physiol.*, 82:962.
39. M.J. Chrispeels, A. J. Tenner, and K.J. Johnson (1973). *Planta*, 113:35.
40. S.J. Gilmour, J.A.D. Zeevaart, L. Sehwenen, and J.E. Graebe (1986). *Plant Physiol.*, 82:190.
41. W.S. Schlotzhauer, M.J. Kasperbauer, and R.F. Severson (1989). *Tob. Sci.*, 33:47.

42. T.C. Tso, M.J. Kasperbauer, and T.P. Sorokin (1970). *Plant Physiol.*, 45:330.
43. M.W. Parker and H.A. Borthwick (1940). *Bot. Gaz.*, 101:906.
44. M.J. Kasperbauer, P.G. Hunt, and R.E. Sojka (1984). *Physiol. Plant.*, 61:549.
45. K. Kaul and M.J. Kasperbauer (1988). *Physiol. Plant.*, 74:415.
46. M.J. Kasperbauer and D.L. Karlen (1986). *Physiol Plant.*, 66:159.
47. M.J. Kasperbauer and P.G. Hunt (1987). *Plant and Soil*, 97:295.
48. P.G. Hunt, M.J. Kasperbauer, and T.A. Matheny (1989). *Crop. Sci.*, 29:130.
49. D.R. Decoteau, M.J. Kasperbauer, and P.G. Hunt (1989). *J. Am. Soc. Hort. Sci.*, 114:216.
50. M.J. Kasperbauer and P.G. Hunt (1991). *Plant Physiol. Suppl.*, 96(1):118.

4

Acid Soil Stress and Plant Growth

Robert E. Wilkinson

University of Georgia
Griffin, Georgia

The influence of mineral nutrition on plant growth has been studied voluminously since the first plant physiologists began to try to discern how plants grow (1–3). Over a period of several decades, mineral nutrient requirements were measured and correlated with plant growth. These requirements have, in general terms, shown that the root environment is most favorable when the rooting medium has a pH of 5.6–7.0. At this pH range, requisite mineral nutrients are most readily available and root growth is close to optimum if other environmental conditions are favorable.

Based on these "optimum" conditions, soil scientists have developed sophisticated soil amendments (i.e., fertilizer) recommendations that favor plant growth. But acid soils (i.e., pH <4.9) are not optimum for plant growth. What, then, are the conditions presented to plants grown on acid soils and how do the plants respond to these conditions?

Acid soils, pandemic in the humid tropic and subtropic areas of the world, are characterized as presenting excesses of H^+, Mn^{2+}, and Al^{3+} and deficiencies of Ca^{2+}, Mg^{2+}, and PO_4^{3-} to the plant (1,4). Additionally, while air pollution may induce specific foliage responses, when that air pollution is due to SO_2 or other acidifying substances that reach the soil through various pathways, acid soil stress develops. And plants not growing in humid tropic or subtropic regions are then exposed to acid soil stress adaptation problems. Thus, plant response to the individual factors

125

in acid soil stress must be correlated (5–8). However, extensive field observation has indicated that acid soil stress may influence plant ontogeny differently during various growth and development stages (9); also, differences between species are extant (5,10). Thus, while individual plant responses are considered herein, the reader must recognize that this discussion can only be a preliminary introduction to a massively complicated research area which includes the entire ecological plant soil atmosphere continuum. And, while individual escape mechanisms, nutrient absorption variations, or enzymatic biosynthetic processes may explain individual species or cultivar responses to the various acid soil stress parameters, plant response to acid soil stress is always a composite of all of these individual stresses and growth patterns. *Or*, simplistic explanations do not correlate with the facts.

EXCESS H^+

Sorghum [*Sorghum bicolor* (L.) Moench.] cultivars vary in response to acid soil stress from sensitive to tolerant (9). "Tolerance" may be a relative term. In sorghum it means "less sensitive, but not immune." Relative tolerance/sensitivity of several cultivars is shown in Table 1.

When several cultivars of a wide range of acid soil stress sensitivity/ tolerance were grown in white quartz flintshot sand watered with NaAc (0.01 M) buffer at pH 6.0, 5.5, 5.0, 4.5, or 4.0, shoot lengths after 7 days showed different growth rates at pH 6 but relatively uniform rates when grown at pH 4.0 (Fig. 1) (11). These curves resemble bioassay data obtained during the measurement of 2,4-D (2,4-dichlorophenoxyacetic acid) formulations in water (12,13); and, in those bioassays (12), the data were converted to a percentage of the untreated control. Using growth in

Table 1 *Sorghum bicolor* (L.) Moench. Cultivar Relative Tolerance to Acid Soil Stress

Cultivar	Relative acid soil tolerance (%)
SC283	90
SC574	90
SC599	60
GP140	50
RTx430	5
TAM428	5

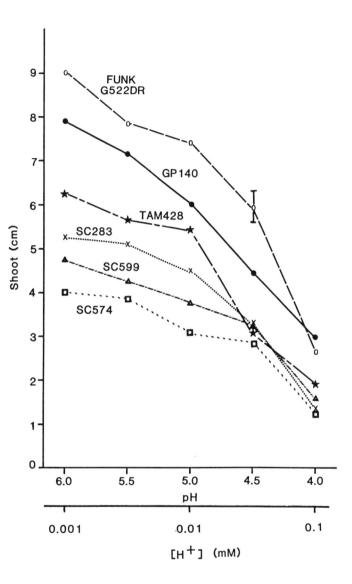

Figure 1 Shoot length (cm) of *Sorghum bicolor* (L.) Moench cultivars grown for 7 days in sand at known pH. (Reprinted from Ref. 11.)

length of each sorghum cultivar at pH 6 as the standard for the respective cultivar, the data showed a relatively uniform response among cultivars. Thus, a single mechanism was involved that influenced all cultivars equally (Fig. 2) (11).

Germinating grass seeds are known to produce gibberellic acid (GA) in the scutellum followed by diffusion of the GA to the aleurone layer where GA induces the synthesis of α-amylase (14). [This paradigm has been accepted for decades. Recently, Grosselindemann et al. (67) showed evidence that the complete process may not be as straight forward as has been previously believed.] α-Amylase is required for the metabolism of endosperm-stored starch that serves as an energy supply for the seedling until it attains a viable self-sustaining size (14–16). Sorghum is also known to have a double layer of dead waxy cells between the scutulum

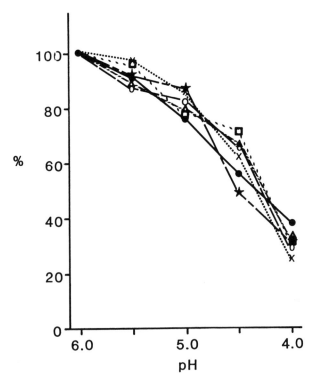

Figure 2   Normalization of sorghum cultivar shoot growth where the growth at pH 6.0 was used as the base. Symbols are the same as in Figure 1. (Reprinted from Ref. 11.)

and the aleurone (17). Therefore, if the diffusion of GA from the scutellum to the aleurone were pH-influenced, the responses shown in Figure 2 would be explicable. Water–ethyl acetate partitioning of GA_3 as influenced by pH and pH vs. growth of Funk G522DR at pH 4.0–6.0 showed a very high correlation (Fig. 3) (11). Thus, presumably, excess H^+ influenced GA movement from the scutellum to the aleurone by a physical mechanism of sequestering the GA in the scutellum membranes so that GA could not move across the low-pH water barrier in sufficient quantity to induce a highly active α-amylase activity in the aleurone. Proof of the pH influence on α-amylase activity is shown in Figure 4 (18).

Since plants maintain a very strong pH buffering capacity between the external soil solution and the cytoplasm (19), emerged sorghum shoots escape from an immediate influence of excess H^+ on their growth; and, acid–soil stress influence on vegetative or reproductive structure develop-

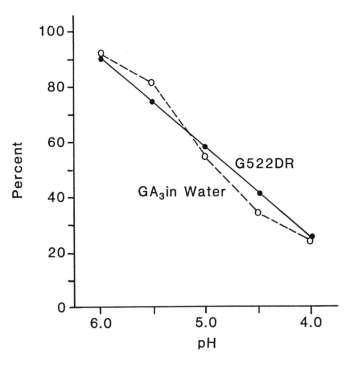

Figure 3 GA_3 partitioning from ethyl acetate into water as a function of solution pH and growth of Funk G522DR shoots in sand watered at known pH. (Reprinted from Ref. 11.)

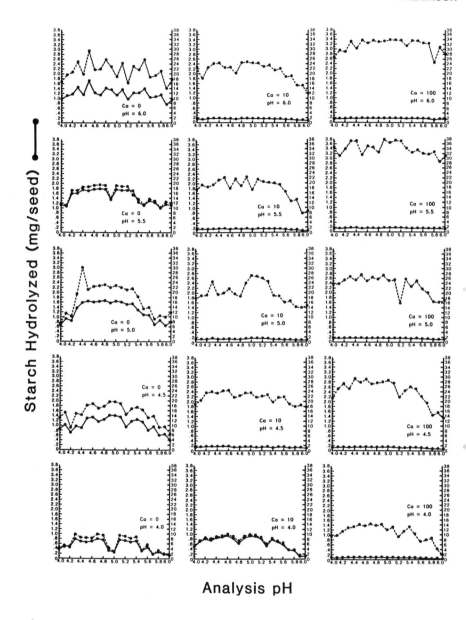

Figure 4 α-Amylase activity of Funk G522DR seed (4-day-old) when the seeds were germinated at known pH ± Ca^{2+} (0, 10, 100 ppmw). (Reprinted from Ref. 11.)

ment must be due to another factor in the acid soil stress ecological continuum (1–4). But roots remain constantly exposed to the excess H^+ in the soil solution.

Sorghum root growth was influenced by pH (Fig. 5) (20), and similar bioassay logic showed that the sorghum cultivars responded equivalently

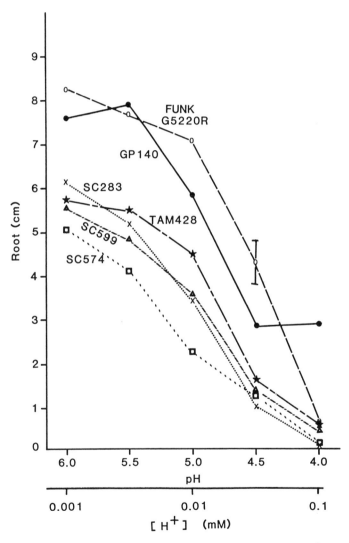

Figure 5   Root length (cm) of sorghum cultivars grown for 7 days in sand at known pH. (Reprinted from Ref. 11.)

(Fig. 6) (20). Roots are not highly responsive to GA but do respond readily to indoleacetic acid (IAA) (20). The influence of pH on the water solubility of IAA and the growth of SC283 at pH 4.0–6.0 showed that these parameters are equivalent (Fig. 7) (20). Since IAA is moved from the interior root tip cells to the exterior cells (21–24), root growth is IAA concentration–dependent (Fig. 8) (20); and IAA water solubility is controlled by the pH-dependent ionization constant. The spaces between root tip plasmalemmas are highly water-soluble, and plasmodesmata are relatively deficient in the root tip (25); therefore these data demonstrate a sequestering of IAA at the site of synthesis (i.e., root cap area) and a lack of ability of IAA translocation to other cells. Root elongation ceases or is greatly decreased at pH <4.5. When endodermis development shuts off the major water absorption areas in the sorghum primary roots, the vegetative plant is highly susceptible to water deficiency.

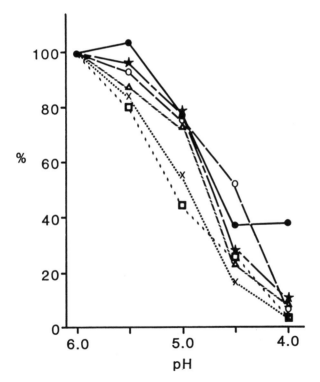

Figure 6 Normalization of sorghum cultivar root growth where the growth at pH 6.0 was used as the base. Symbols are shown in Figure 5. (Reprinted from Ref. 11.)

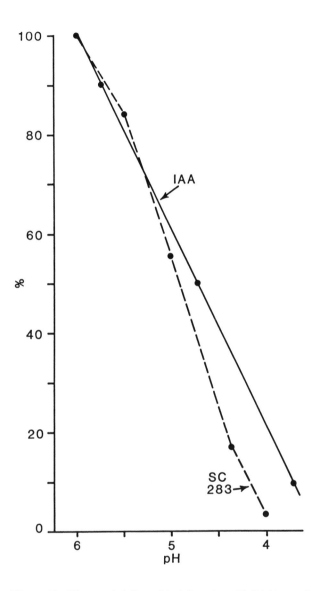

Figure 7 Water solubility of indoleactic acid (IAA) as a function of pH and the root growth of SC283 in sand for 14 days at different pHs. (Reprinted from Ref. 11.)

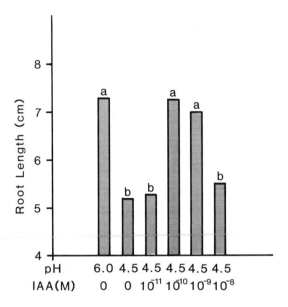

Figure 8 Influence of pH and exogenous IAA concentrations on SC283 root length at 7 days after initial watering. (Reprinted from Ref. 11.)

Thus, in plants exposed to excess H^+, the shoots escape the primary response, but the roots continue to be stressed. In ecological systems with sufficient water (i.e., irrigation or high precipitation), the roots can continue to function (i.e., water absorption), at least partially, so that a secondary acid soil stress parameter (i.e., excess Mn^{2+} or aluminum, or Ca^{2+} and PO_4^{3-} deficiencies) becomes the dominant controlling factor in vegetative growth. Thus, sorghum cultivars show differential acid soil stress sensitivity/tolerance (Table 1) but respond equivalently to excess H^+ (Figs. 1–7). Therefore, excess H^+ per se cannot be the factor in acid soils that explains sorghum cultivar differential acid soil stress sensitivity/ tolerance unless there are additional metabolic processes influenced by excess H^+ concentration which induce a different stress on plant growth.

EXCESS Mn^{2+}

Manganese is a required plant micronutrient that influences many enzymatic reactions involving electron transfer (1). Excellent monographs have been published on this complex subject (60). Mn^{2+} deficiencies have been shown to severely influence PSII (see Chapter 14) (26) and some steps in isoprenoid biosynthesis (see Chapter 6). Excess Mn^{2+} influences these same reactions but possibly by slightly different means. Sorghum cultivars vary in response to acid soil stress (Table 1). Yet the Mn^{2+} content of vegetative and reproductive tissues from sorghum grown under acid soil stress did not vary significantly (Table 2) (27). Thus, differential Mn^{2+} uptake and translocation could not entirely explain the differences in sorghum cultivar responses to acid soil stress (9).

Excess Mn^{2+} induces a decrease in overall isoprenoid synthesis and GA biosynthesis (see Chapter 6). And excess Mn^{2+} induces a decreased sorghum seedling shoot growth in addition to the decreased growth of seedling shoots caused by excess H^+ (11,20) that was explicable as (a) decreased GA synthesis in shoots and (b) increased IAA degradation in roots (11,20). In these tissues, addition of exogenous supplies of the respective growth regulator induces a partial reversal of the growth inhibitions (11,20). In roots, the reversal was IAA concentration–dependent (Fig. 8) (20). Thus, Mn^{2+} concentration at the individual sites of activity is a major influence. Vegetative Mn^{2+} concentrations must, axiomatically, be influenced by root ion uptake and translocation to the leaves from the roots. While sorghum cultivars grown under acid soil stress in the field did not show significant differences in vegetative Mn^{2+} contents (Table 2), sorghum and wheat (*Triticum aestivum* L. em Thell) exposed to different Mn^{2+} levels had significantly different Mn^{2+} concentrations (28–30). Upward translocation from the roots to the leaves was strictly correlated with transpiration (31) so that sequestration, complex-

le 2 Manganese Content of Reproductive Bud and Leaf Tissue of *Sorghum bicolor* Moench Cultivars (μM Mn/g DW)

tivar	Bud Soil pH >6.0	Bud Soil pH 4.2	Leaf Soil pH >6.0	Leaf Soil pH 4.2
74	0.502 ± 0.164[a]	1.068 ± 0.129[b,c]	0.434 ± 0.092[a]	0.998 ± 0.166
99	0.796 ± 0.022[a]	1.490 ± 0.191[b,c]	0.482 ± 0.050[a]	0.807 ± 0.197
430	1.074 ± 0.067[a]	0.783 ± 0.087	0.783 ± 0.136	0.858 ± 0.146
M428	0.439 ± 0.114[a]	1.296 ± 0.151	0.374 ± 0.083[a]	1.001 ± 0.206

Significant difference between pH >6.0 and pH 4.2 within a tissue.
Significant difference between bud and leaf tissue at pH >6.0.
Significant difference between bud and leaf tissue at pH 4.2.
ce: Reprinted from Ref. 27.

ation, chelation, and precipitation become mechanisms of escape from the influence of excess Mn^{2+} (1). Biological activity in leaves is a function of the Mn^{2+} concentration at the site of activity. This parameter has not been successfully quantitated (1). Isocitric lyase (IL) is a requisite enzyme of the glyoxylate cycle (61) that ultimately converts fats to carbohydrate. Thus, IL is requisite in the germination of oil-rich seeds and the establishment of seedlings from seeds whose major energy storage products are triglycerides. The reactive IL complex contains Mg^{2+} (61–63) and the enzyme reaction is dependent on the Mg^{2+} enzyme complex (61–63). Mn^{2+} can function in the reactive complex, but the Mn^{2+} dissociation from the enzyme is so much slower than Mg^{2+} dissociation (63) that Mn^{2+} is an effective IL inhibitor (63). Acid soils present excess Mn^{2+} to plants in the soil solution (1,4) that amounts to a 35-fold increase when soil pH decreased from pH 5.64 to pH 4.41 (64). An increase in plant tissue Mn^{2+} content when the plants are grown on acid soils (64) would result in a massive decrease of IL activity (63). Thus, seed germination and seedling establishment of triglyceride-rich (and dependent) crops would be very difficult under acid soil stress conditions. Genetic control of IL activity in sorghum was demonstrated (18) where acid soil stress sensitive Funk G522DR seed had two to three times the triglyceride content of acid soil stress tolerant cultivars (SC574 and SC283) and over five times the IL activity present in the acid soil stress cultivars (18).

In cell-free enzyme systems, mevalonic kinase (32), phytoene synthetase (33), and ent-kaurene synthase (34) have decreased activity under excess Mn^{2+} concentration conditions. Thus, synthesis of all the isoprenoid compounds would be greatly influenced by excess Mn^{2+}. And GA, chlorophyll, carotene, xanthophyll, quinone, and sterol biosyntheses would be decreased. Loss of any of these cellular components would decrease growth. Thus, the response of plants to excess Mn^{2+} is a multifaceted problem that will vary among species, cultivars, and specific growing conditions.

EXCESS ALUMINUM

Al^{3+} is accumulated in plants (35,38). In solutions more acidic than pH 5, Al^{3+} exists as the octahedral hexahydrate, $Al(H_2O)_6^{3+}$, abbreviated as Al^{3+} (39,40). While some plants accumulate aluminum (37), Al^{3+} is present at very low concentrations (i.e., $< 10^{-12}$ M) at pH 7, but shifts to μM concentrations at pH 4 (Fig. 9) (41). Al^{3+} substitutes for Mg^{2+} in critical enzyme and regulatory sites, but the dissociation of the Al^{3+}–enzyme complexes is 10^7 slower than for Mg^{2+} (41). Therefore, 1 nm

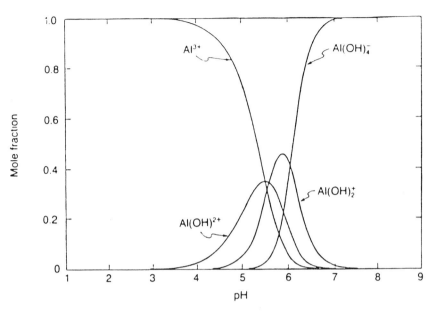

Figure 9 Distribution of soluble, mononuclear aluminum ion species in aqueous solution. Ordinate:mole fraction of aluminum ion occurring as each designated species. At any pH the individual mole fractions sum to unity. (Reprinted from Ref. 55.)

Al^{3+} competes very effectively with 1 mM Mg^{2+} in enzyme reactions utilizing Mg^{2+}; energy transfer and many biosyntheses are greatly decreased. In addition, Al^{3+} is associated with oxygen donor ligands (39,40) of which PO_4^{3-} is a major physiological example. Since PO_4^{3-} is utilized by so many requisite cellular components and biosynthetic processes, altered respiration, DNA and RNA synthesis and activity, ion competition, growth, ATP metabolism, all isoprenoid biosynthesis, and carbohydrate metabolism develop (42–44). Thus, Al^{3+} is a general toxicant inside the plasma membrane. Al^{3+} lethality depends on concentration, pH, organic acid chelator concentration, and other factors.

Citric acid is a major ligand for Al^{3+} (39,40). But at low pH, this ligand is ionically neutral and penetrates through membranes with great facility. Once inside the cytoplasm, physiological pH returns the Al^{3+} ligand to the same toxicant conditions as were present without the ligand. Therefore, major factors in the determination of Al^{3+} tolerance are plant-derived mechanisms that preclude Al^{3+} absorption through the plasma membrane.

ROOT MUCIGEL

Traditionally, plant scientists have visualized roots as in Figure 10A. Recently, the presence of a root mucilage has been shown (Fig. 10B). The mucilage is water-soluble, sloughed off by abrasion, and replaced rapidly (45). There appear to be two types of mucilage (45–49). Al^{3+} is complexed by the root tip mucigel to an extremely high extent so that Al^{3+} does not readily enter the root tip (45). This mucigel is relatively soluble in EDTA and is not present in plants grown in solution culture (45–49). Since scandium (Sc^{3+}) is not normally present in the environment, it can be utilized as a substitute for Al^{3+}, and Sc^{3+} can be measured by neutron activation. Exposure to roots to Sc^{3+} can be a good measure of Al^{3+} entry into (or absorption onto) roots.

Sorghum cultivars were grown in sand, exposed to 10 mM Sc^{3+}, washed in 10 mM EDTA, and Sc^{3+} contents were measured by neutron activation. Sc^{3+} contents in the EDTA wash were greatest in the most acid soil stress–tolerant sorghum cultivar (SC283) and least in the acid soil stress–sensitive cultivar (50) (Fig. 11). Thus, the quantity of mucigel present on the roots of sorghum cultivars influenced Sc^{3+} absorption. Additionally, in the acid soil stress–sensitive cultivar (Funk G522DR) the quantity of Sc^{3+} that was not EDTA-extractable increased as the pH decreased (Fig. 12). Concurrently, two possible explanations are obvious.

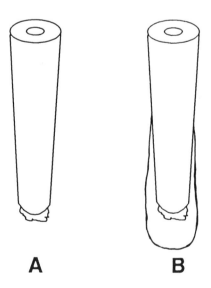

A   **B**

Figure 10 Presentation of rotos with root cap (A) and with mucigel (B).

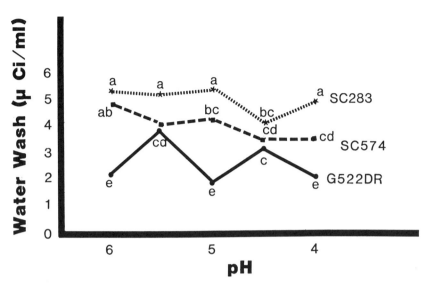

Figure 11 Extractable Sc^{3+} ($\times$ Ci/ml) washed off 5-mm root tips (10) of sorghum cultivars in deionized water (1 min) after the roots were grown at different pHs and exposed to 10 mM $ScCl_2$ (10 min). Points followed by the same letter are not significantly different at the 5% level. (Reprinted from Ref. 7.)

Either there was a major change in the cation exchange capacity (CEC) of the Funk G522DR roots as the pH decreased or there was a major increase in Sc^{3+} absorption through the plasma membrane as the pH decreased. Increased CEC would require an increased carboxyl group availability as the pH decreased (47,48). This would be most likely due to a change in the activity of pectin methylesterase (PE) (46). This concept is substantiated by known pH-influenced activities of pectin methylesterase (51).

This phase of ion exclusion from toxic ion absorption may be of major importance to the explanation of plant escape from the lethal affects of Al^{3+}. Much research remains to be done.

ION UPTAKE

Previously, plant root ion uptake was shown to be biphasic (Fig. 13). Ions absorbed in the initial phase were exchangeable whereas ions absorbed in the secondary phase are accumulated against a concentration gradient. Viable membranes and energy expenditure are required to maintain this gradient (40).

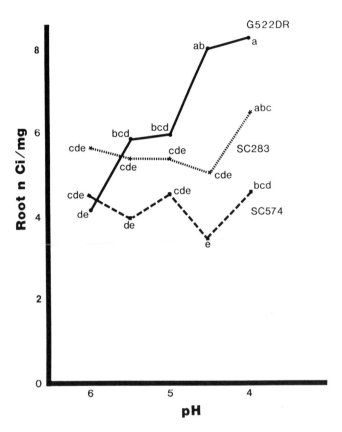

Figure 12 Nonextractable Sc^{3+} (nCi/mg) retained by roots of sorghum cultivars grown at different pHs. Points followed by the same letter are not significantly different at the 5% level. (Reprinted from Ref. 7.)

At first glance, this situation would appear to be a combination of (a) ion pores in the plasma membrane that allow equilibration of cytosol and external medium very quickly, and (b) ion H^+-ATPase pumps in the tonoplast that induce a slower accumulation of the specific ion in question. While this simplistic explanation may seen intuitively obvious, the real situation is much more complex. Tonoplasts are reported to have ion pores and H^+-ATPase activated ion pumps (52–54). Additionally, there appear to be more than one type of ion pore (52–54). Thus, the accumulation of specific ions in the vacuole is a function of the ΔpH between cytosol and vacuole solution plus the tonoplast voltage and specific enzymes (52,54). Similar controlling processes occur in the movement of ions through the plasma membrane (52,54). By current thinking, once

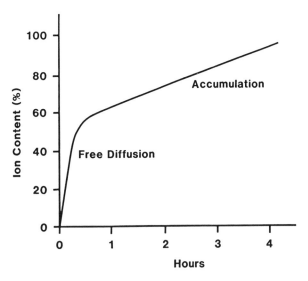

Figure 13 Biphasic ion accumulation.

ions have entered the cytosol at the root epidermis, transport through plasmodesmata is sufficient to account for ion movement to the stele (52–54). H^+-ATPase have been shown at the xylem and phloem (55). Thus, ion accumulation into the xylem is under metabolic control (55). Therefore, two membranes are of major import: epidermis and xylem parenchyma plasma membranes.

The plasma membrane constitutes a permeability barrier that is charged so that ions do not readily diffuse through the phospholipid layer. However, there are ion pumps and ion pores (channels) contained within the membranes (Fig. 14). Ion pumps are H^+-ATPase enzyme–mediated. The H^+-ATPase have multiple SH groups and at least one such SH group is located on the exterior surface of the plasma membrane. Therefore, an SH inactivator that does not penetrate through the plasma membrane would prevent ion absorption by the H^+-ATPase enzyme-mediated ion pumps. Similarly, ion pores (channels) are known to be voltage-gated openings with chemical receptor sites that can be inhibited.

When Ca^{2+} is applied in agar to one side of corn (*Zea mays* L.) roots in the zone of elongation, the root does not grow contiguous to the Ca^{2+} application and the root curves toward the side where the Ca^{2+} + agar was applied (49). Similar data have been reported in sorghum (50). Aluminum causes a similar reaction (Table 3) in at least two sorghum cultivars. Ca^{2+} must enter the cytoplasm to induce root curvature (50);

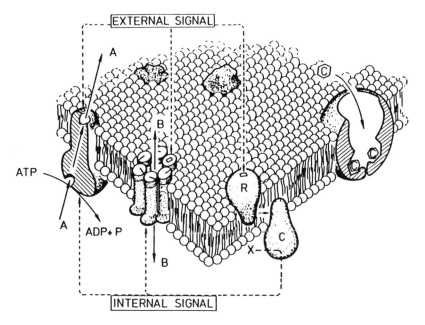

Figure 14 Mechanisms of ion transport and signal transduction in biological membranes. Three-dimensional diagram of the fluid mosaic model, showing the integral transport proteins embedded in the lipid layer. From left to right: (A) ion pump, (B) ion channel (pore), (R,C) coupling proteins for signal perception and transduction, (C) carrier. (Reprinted from Ref. 56.)

Table 3 Influence of the Sulfhydryl Inactivator p-chloromercuribenzenesulfonic acid (PCMBS) (10 mM) on the Curvature (degrees) of Sorghum Primary Roots Induced by Ca^{2+} (10 mM) or Al^{3+} (10 mM) in Agar Placed Unilaterally on the Root

| Cultivar | PCMBS | Check | Agar | | |
			0	Ca^{2+}	Al^{2+}
Funk G522DR	−	3.5d*	6.5c	10.0b	10.0b
	+	10.3b	12.4b	19.5a	24.5a
SC283	−	10.2c	10.6c	14.8b	28.2a
	+	4.3d	7.2c	10.4c	11.0c
SC574	−	11.5c	11.7c	18.8ab	22.2a
	+	11.5c	11.0c	17.8b	10.5c

*Values within a cultivar followed by the same letter are not significantly different at the 5% level.

and when an ion pump inhibitor (PCMBS) or an ion pore inhibitor (nifedipine, verapamil, or diltiazem) was used as pretreatment prior to unilateral application of Ca^{2+} + agar, Ca^{2+} entry into the roots was influenced by all of these inhibitors in Funk G522DR (sensitive genotype), nifedipine and diltiazem were active in SC574 (tolerance), and PCMBS and diltiazem influenced Ca^{2+} uptake in SC283 (tolerant) (Table 4). Thus, transport of these ions through the root plasma membrane varied on a genetic basis and was controlled by membrane-bound ion pumps and ion pores.

When sorghum roots were exposed to pH 6.0 → 4.0, $^{45}Ca^{2+}$ absorption decreased as the pH decreased (Table 5). However, variation between cultivars existed (Table 5). Ion channels and ion pumps are influenced by the plasma membrane (PM) potential (65). Excess cations neutralize the polyanionic charge of specific phospholipids (65). That neutralization results in a decreased PM potential and the PM ion chan-

Table 4 Curvature (degrees) of Sorghum Primary Roots After Unilateral Application Agar Containing Ca^{2+} (10 mM) or Al^{3+} (10 mM) ± Ion Pore Inhibitors

Compound	Cultivar	Inhibitor	Agar		
			0	Ca^{2+}	Al^{3+}
Nifedipine	Funk G522DR	−	13c*	18a	19a
		+		14b	20a
	SC574	−	5c	36a	30a
		+		16bc	19ab
	SC283	−	11c	18ab	20a
		+		17ab	14b
Verapamil	Funk G522DR	−	7c	19b	22ab
		+		12c	28a
	SC574	−	8c	23b	34a
		+		27b	21b
	SC283	−	8b	13a	10ab
		+		13a	8b
Diltiazem	Funk G522DR	−	3c	10a	6b
		−		3c	5b
	SC574	−	4c	9a	9a
		+		6b	4c
	SC283	−	7c	10b	11a
		+		6cd	5d

*Values within a cultivar in one inhibitor followed by the same letter are not significantly different at the 5% level.

Table 5 Influence of H^+ Concentration on $^{45}Ca^{2+}$ Absorption by *Sorghum bicolor* (L.) Moench Roots (%)*

Cultivar	pH				
	6	5.5	5	4.5	4.0
Funk G522DR	100a**	94b	87c	59d	37e
SC574	100a	53b	63b	55b	37c
SC283	100a	109a	96a	73b	63b

*Exposure for 0.5 hr in 0.01 M sodium acetate. Each value is the average of five applications.
**Values on a line followed by the same letter are not significantly different at the 5% level.

nels and ion pumps remain closed for longer periods of time (66) resulting in decreased cation uptake (Table 5). Therefore, excess H^+ does induce an acid soil stress. Concomitantly, the excess Al^{3+} and Mn^{2+} associated with acid soil stress can also induce similar results. These factors may eventually help explain much of the variability reported on aluminum toxicity in different species and cultivars. Certainly these data demonstrate that biological variability as a source of differences in response to acid soil stress has not been examined sufficiently.

Ca^{2+} AND PO_4^{3-} DEFICIENCIES

After GA induces α-amylase synthesis in grass seed aleurones, release of the amylase into the endosperm is Ca^{2+}-mediated (51,56). In sorghum (cv Funk G522DR) seed, α-amylase activity is both pH and Ca^{2+} concentration–mediated (Fig. 4) (18). And PO_4^{3-} is precipitated by aluminum and Ca^{2+} available in the soil solution (3,4). Since PO_4^{3-} is utilized in many plant enzyme reactions, deficiencies have a general debilitating influence. Association of the PO_4^{3-} with Al^{3+} (39–41) decreases the activity of PO_4^{3-}-bearing molecules (39–41). Thus, interactions between Al^{3+} and PO_4^{3-} in the growth of plants (53,54) is most explicable as a simple PO_4^{3-} availability.

ESCAPE MECHANISMS

Ion uptake and movement in the transpiration stream are relatively standard processes within each genome. The presence of ion pumps and/or ion pores plus the root endodermis guarantees that ion absorption and transport to the leaves will be under metabolic control. But variations occur.

Root cap mucilage has been shown to occur in two forms (57). The quantity of each form produced by a particular cultivar or species appears to be genetically determined (58,59). Aluminum is accumulated in the root tip mucilage and thereby root absorption of aluminum is decreased (45).

Dicarboxylic organic acids serve as ion chelators (41). Some plants contain sufficient chelators to detoxify aluminum inside the cell. Also, some species excrete chelators from the roots in sufficient quantity to prevent aluminum from being available for root absorption (41). Thus, the chelators function as an escape mechanism. However, when ligand-bound Al^{3+} becomes nonionized at low pH, these complexes may penetrate freely through membranes and increase the total Al^{3+} content in the plant (39–41).

When excess Mn^{2+} was absorbed by some species, X-ray analysis showed spots of accumulation that appeared to be outside the cell (1). This phenomenon was explained as Mn^{2+} excretion into extracellular spaces. Also detoxification of various substances by accumulation in the vacuole has been known for some time (1).

These few examples of escape mechanisms present in some plant species appear to b genetically controlled (1). The plants appear to be highly susceptible to the activity of the toxicant when that substance is present in the cytoplasm. When differences in ion accumulation into the transpiration stream are added to these considerations, the possibility of multiple escape mechanisms becomes important.

SUMMARY

Acid soil stress has been studied extensively and the literature is voluminous. A compendium of documented responses would be far beyond the scope of this volume. A few of the interactions have been referenced herein. However, just as differences in acid soil solution contents of Mn^{2+}, aluminum, Ca^{2+}, and PO_4^{3-} occur, differences in response interactions to these stresses by species at dispersed locations also result. Thus, although the general categories influencing all of these factors may be lumped into "acid soil stress," evaluation of plant morphology, physiology, genetics, and biochemistry must be integrated into a complete whole before any preliminary understanding can be attained. Plants do have common physiological and functional components among stresses. However, examination of supposedly uniform root membranes has shown major differences in ion absorption capability, and species or cultivar

differences in aluminum accumulation by root tip mucilages have not yet
been researched. Additionally, aluminum chelation makes the toxicant
inactive at some pH values, but evaluation of activity at low pH has not
been reported.

Species and environment have long been known to be determinative
factors in plant growth. Thus, in reality, acid soil stress presents a composite ecological problem to each plant. The importance of each factor is
a product of the total plant–soil ecosystem. As yet, we are still finding
pieces of the adaptations puzzle.

REFERENCES

1. R. D. Graham, R. J. Hannam, and N. C. Uren, eds. (1988). *Manganese in Soils and Plants*. Kluwer Academic, Dordrecht.
2. R. B. Clark (1982). In *Breeding Plants for Less Favorable Environments* (M. N. Christiansen and C. F. Lewis, eds.), John Wiley and Sons, New York, p. 71.
3. C. D. Foy (1983). *Iowa State J. Res.*, 57:355.
4. C. D. Foy (1984). In Adams, F., ed., *Soil Acidity and Liming*, Am. Soc. Agron. Monogr. No. 12, 2nd ed., Madison, WI.
5. G. W. Bates and P. M. Ray (1981). *Plant Physiol.*, 68:158.
6. T. B. Kinraide, and D. R. Parker (1987). *Plant Physiol.*, 83:546.
7. G. Fiskesjö (1989). *Hereditas*, 111:149.
8. A. D. Noble and T. Harding (1989). *S. Afr. J. Plant Soil*, 6:245.
9. R. R. Duncan and R. E. Wilkinson (1990). In *Plant Nutrition: Physiology and Application* (M. L. van Beusichem, ed.), Kluwer Academic, Dordrecht.
10. S. C. Miyasaka, L. V. Kochian, and C. D. Foy (1989). *Plant Physiol.*, 91:1188.
11. R. E. Wilkinson and R. R. Duncan (1989). *J. Plant Nutr.*, 12:1395.
12. R. E. Wilkinson (1964). *Weeds*, 12:69.
13. L. H. Yopp, L. H. Aung, and G. L. Steffens (eds.) (1986). *Bioassays and Other Specia Techniques for Plant Hormones and Plant Growth Regulators*. Plant Growth Regulators Soc. Am.
14. H. N. Krishnamoorthy (ed.) (1975). *Gibberellins and Plant Growth*. John Wiley and Sons, New York.
15. J. V. Jacobsen and T. J. V. Higgins (1982). *Plant Physiol.*, 70:1647.
16. R. L. Jones and J. V. Jacobsen (1983). *Planta*, 158:1.
17. H. Baggett (1988). *Sorghum*. John Wiley and Sons, New York.
18. R. E. Wilkinson and R. R. Duncan (1989). *J. Plant Nutr.*, 12:1483.
19. H. Felle (1988). *Physiol. Plant.*, 74:583.
20. R. E. Wilkinson and R. R. Duncan (1989). *J. Plant Nutr.*, 12:1379.
21. L. J. Feldman (1980). *Planta*, 49:145.
22. L. J. Feldman (1979). *Ann. Bot.*, 43:1.
23. J. J. Pernet and P. E. Pilet (1976). *Planta*, 128:183.

24. P. J. Davies, J. A. Doro, and A. W. Tarbox (1976). *Physiol. Plant.*, 36:333.
25. J. A. Raven (1975). *New Phytol.*, 74:163.
26. G. Sandmann and P. Böger (1983). In *Encyclopedia of Plant Physiology*, VOl. 15B (A. Lauchli and R. L. Bieleski, eds.), Springer-Verlag, New York.
27. R. E. Wilkinson and R. R. Duncan (1991) Z. *Naturforsch.* 46C:950.
28. R. E. Wilkinson and K. Ohki (1988). *Plant Physiol.*, 87:841.
29. K. Ohki (1985). *Crop Sci.*, 25:187.
30. K. Ohki (1984). *Agron. J.*, 76:213.
31. R. E. Wilkinson and K. Ohki (1991). *J. Plant Nutr.*, 14:93.
32. T. T. Tchen (1958). *J. Biol. Chem.*, 233:1100.
33. B. Maudinas, M. L. Bucholtz, C. Papastepkanou, S. S. Katiyar, A. V. Briedis, and J. W. Porter (1977). *Arch. Biochem. Biophys.*, 180:354.
34. J. Knotz, R. C. Coolbaugh, and C. A. West (1977). *Plant Physiol.*, 60:81.
35. Z. Rengel and D. L. Robinson (1989). *Agron. J.*, 81:208.
36. G. Zhang and G. J. Taylor (1989). *Plant Physiol.*, 91:1094.
37. R. C. Allen (1943). *Contrib. Boyce Thompson Inst.*, 13:221.
38. B. S. Meyer and D. B. Anderson (1952). *Plant Physiology*, Van Norstrand, New York.
39. T. L. MacDonald and R. B. Martin (1988). *Trends in Biochem. Sci.*, 13:15.
40. R. B. Martin (1986). *Clin. Chem.*, 32:1797.
41. N. V. Hue, G. R. Craddock, and F. Adams (1986). *Soil Sci. Soc. Am. J.*, 50:28.
42. A. Haug (1984). *Crit. Rev. Plant Sci.*, 1:345.
43. T. L. MacDonald and R. B. Martin (1988). *Trends in Biochem. Sci.*, 13:15.
44. R. B. Martin (1986). *Clin. Chem.*, 32:1797.
45. W. J. Horst, A. Wagner, and H. Marschner (1982). *Z. Pflanzenphysiol.*, 105:435.
46. F. P. C. Blamey, D. C. Edmeades, and D. M. Wheeler (1990). *J. Plant Nutr.*, 13:729.
47. R. J. Haynes (1980). *Bot. Rev.*, 46:75.
48. A. H. Knight, W. M. Crooke, and R. H. E. Inkson (1961). *Nature*, 192:142.
49. J. S. Lee, T. J. Mulkey, and M. L. Evans (1983). *Science*, 220:1375.
50. R. E. Wilkinson and R. R. Duncan (1992). *J. Plant Nutr.* 15:2559.
51. M. J. Chrispeels and J. E. Varner (1967). *Plant Physiol.*, 42:398.
52. R. Hedrich and J. I. Schroeder (1989). *Annu. Rev. Plant Physiol.*, 40:539.
53. R. Serrano (1989). *Annu. Rev. Plant Physiol.*, 40:61.
54. M. G. Pitman (1977). *Annu. Rev. Plant Physiol.*, 28:71.
55. A. Sarets-Soler, J. M. Pardo, and R. Serrano (1990). *Plant Physiol.*, 93:1654.
56. D. S. Bush, M.-J. Cornejo, C.-N. Huang, and R. L. Jones (1986). *Plant Physiol.*, 82:566.
57. B. M. McDougall and A. D. Rovira (1970). *New Phytol.*, 69:999.
58. R. E. Paull and R. L. Jones (1975). *Plant Physiol.*, 56:307.
59. S. F. Moody, A. E. Clarke, and A. Bacic (1988). *Phytochemistry*, 27:2857.
60. R. D. Graham, R. J. Hannan, and N. C. Uren (eds.) (1988). *Manganese in Soils and Plants.* Kluwer Academic, Dordrecht.

61. P. Vanni, E. Giachetti, G. Pinzauti, and B. McFadden (1990). *Comp. Biochem. Physiol.*, 95B:431.
62. E. Giachetti, G. Penzauti, R. Bonaccorsi, and P. Vanni (1988). *Eur. J. Biochem.*, 172:85.
63. E. Giachetti and P. Vanni (1991). *Biochem. J.*, 276:223.
64. R. E. Wilkinson, E. L. Ramseur, R. R. Duncan, and L. M. Shuman (1990). In *Genetic Aspects of Plant Mineral Nutrition* (N. El Basham, ed.), Kluwer Academic, Dordrecht, pp. 263–268.
65. L. Messiaen, H. Debmedt, G. Droogmans, F. Wuytack, L. Raeymaekers, and R. Castells (1990). *Biochem. Biophys. Acta*, 1023:449.
66. B. Hille (1984). *Ionic Channels and Excitable Membranes.* Sinauer Assoc., Sunderland, MA.
67. E. Grosselindemann, J.E. Graebe, D. Stockl, and P. Hedden. (1991). *Plant Physiol.*, 96:1099.

5

Mineral Nutrition

Larry M. Shuman

University of Georgia
Griffin, Georgia

INTRODUCTION

One of the major environmental factors to which plants respond is mineral nutrition. Plant nutrition has been defined as the supply and absorption of chemical elements necessary for growth and metabolism. The reactions that occur in the living cell that are necessary to maintain life and growth constitute metabolism (1). This chapter will discuss the mechanisms of plant response to the supply of inorganic nutrients whether they be in a deficient, sufficient, or toxic range. The positively identified essential plant elements include C, H, O, P, K, N, S, Ca, Fe, Mg, Mn, Cu, Zn, B, Mo, Na, and Cl. The mineral elements considered in this chapter are P, K, Ca, Mg, S, Mn, Cu, Fe, Zn, B, and Mo in that order. Nitrogen is covered in detail in another chapter by that title, and Na and Cl are found in the chapter on salinity.

Mengel and Kirkby (1) list the essential elements in four groups. The first includes C, H, O, N, and S. The only one of those considered here in S, which is taken up as the sulfate ion, is involved in enzymatic processes, and is assimilated by oxidation and reduction processes. Group 2 consists of P and B. Phosphorus is taken up as the phosphate ion and B as borate. These are esterified with native alcohol groups in the plant and P is a component of energy-rich compounds such as ATP. The third group consists of K, Na, Mg, Ca, and Cl. These are taken up as ions from the soil solution and are important in establishing osmotic potential. These ions also active enzymes including processes to optimize protein conformation, bridge reaction partners, balance anions, and control membrane permeability and electropotentials. The last group is that of Mn, Cu, Fe, Zn, and Mo. Taken up as ions from the soil solution, they are found mainly in plants incorporated in prosthetic groups. They are involved in electron transfer by oxidation–reduction reactions (1).

The objective of this chapter is to give an overview of plant mineral nutrition and plant responses to deficiencies and toxicities. The approach is to initially present general information relying heavily on two excellent reviews (1,2). Then more recent literature is reviewed selected from databases concentrating on the years 1986–1990. Each element is discussed in turn with subheadings that include uptake and assimilation, enzyme reactions, carbohydrate and protein synthesis, translocation, interactions, and plant morphology.

PHOSPHORUS

Uptake

Phosphorus is an essential plant macronutrient required at levels generally between 0.3% and 0.5% of the plant dry weight to be in the sufficiency range (2). Phosphorus is in the oxidized form in the plant unlike S and N which are in reduced forms. Phosphorus is absorbed rapidly into plant roots and is involved in many metabolic processes. The xylem sap P is 100–1000 times the concentration of that outside the root indicating that there is a very steep concentration gradient between the outside and inside of the root membranes. Thus, uptake is active and is driven by respiratory carbohydrate metabolism (1). The rate of phosphorus uptake is very pH-dependent with uptake at pH 4 being four times that at pH 8.7. The root can alter the rhizosphere pH through exudates, such as amino and organic acids, which make P more available (3). Plant species with more root hairs resulting in additional root surface have an advantage in being able to take up more P (4). The P efficiency (amount of growth per unit P taken up) is related to the root/shoot ratio and influx (5). Low efficiency is associated with low influx and a low root/shoot ratio whereas high efficiency is related to high influx or a high root/shoot ratio. Phosphorus uptake can also be influenced by other plant nutrients. For example, an increased N/P ratio increases P uptake at low N levels (6). (See also the discussion on interactions below.)

An approach to studying plant response to P is to develop mechanistic models. Barber (7) describes one such model to predict P uptake in terms of root growth and morphology, kinetics of nutrient uptake, and sites of supply of nutrients to the root by diffusion and mass flow in the soil. Plant response to P availability in terms of uptake is to activate various mechanisms including root exudates, more efficient uptake through development of root morphology, and increasing efficiency through higher influx rates.

Enzymes

Phosphorus in plants is found mainly as inorganic orthophosphate and as pyrophosphate (1,2). Phosphorous organic compounds are essential to metabolic processes and are especially active in energy transfers. In plants, as in all living things, energy transfers are linked by adenosine triphosphate (ATP), which is a "high-energy" compound. When it undergoes hydrolytic reactions it exhibits a large decrease in free energy. The reaction hydrolyzing ATP to ADP (adenosine disphosphate) releases 8000 cal at pH 7. Thus, many metabolic processes rely on P in the form of ATP to provide the energy to activate the reactions. Starch synthesis and carbon fixation are regulated by P as well as carbohydrate metabolism and sucrose translocation.

The lipophilic compounds that are important in biomembranes also include P. Another organic P compound is phytin, which is mainly found in seeds. Genetic materials such as RNA and DNA depend on P as a structural element. Low P in the plant increases the starch/sucrose ratio, which is correlated with changes in the total activity of the enzymes of starch and sucrose metabolism (8). The P status of plants has a direct effect on nitrate reductase concentrations, which are high at high P levels and low at low P levels (9). It has been found that acid phosphatase activity is inversely related to phosphate concentration in plants and the level of phosphatase has been suggested as a rapid test indicator of P stress (10). However, Caradus and Snaydon (11) reported that although root extracellular acid phosphatase did increase with decreasing P supply, it was not a sensitive screening agent for the tolerance of plants to low P conditions. The response of plant enzymes to P status is tied in with many of the plant's metabolic processes. Thus, plants react to low P status with a dysfunction in their starch and carbohydrate metabolism as well as in their genetic processes.

Carbohydrate and Protein Synthesis

Phosphorus deficiency decreases energy transport from the chloroplast to other parts of the plant. It also affects metabolic processes including protein synthesis and synthesis of nucleic acids (1). Low inorganic P causes an accumulation of starch and sucrose, and a reduction in sugar phosphates and adenylates (8,12), whereas an increase in P causes a reversal and decreases the starch in the leaves (13). Low P has less effect on photosyntheses than on leaf expansion. Phosphorus stress increases carbon concentration by decreasing carbon export from the leaves (14). At high P levels the soluble protein is also at high levels (9). Thus, as P deficiency alters the carbohydrate partitioning in leaves, it also can prevent senescence when in a sufficient range. Phosphorus deficiency effects on sources and sinks causes a reduction of the activity and number of sinks in shoot (grains and tillers) and reduced carbohydrate accumulation (15). Low P strengthens the source–sink interaction, which provides a regulatory mechanism for reducing carbohydrate accumulation. Phosphorus-deficient plants are less sensitive to UV-B light than non–phosphorus-deficient plants (16). Phosphorus-deficient plants accumulate flavenoids and have thicker leaves, and thus are less sensitive to this type of radiation. Thus, P deficiency has profound effects on plant photosynthetic processes including starch and sugars as well as on proteins.

Translocation

Phosphorus moves up and down readily in phloem and moves as inorganic phosphorus and as phosphorylcholine (1). Phosphorus deficiency decreases the hydraulic conductivity of the roots (17). Translocation of P, then, is not a significant factor in the response of plants to P deficiency.

Interactions

Phosphorus interacts with several other plant nutrients. High N and high P give a synergistic effect in that together they increase yield more than either alone (9,18). Phosphorus has an antagonistic effect on both Zn and Cu, causing the metals to be deficient at high levels. Increased Mg supply favors the uptake of P, which may be a result of magnesium's role in activation of almost all the reactions involving P transfer within the plant chlorophyll and dry matter and stimulating Fe uptake under Fe-deficient conditions (19). Plant stress from other limitations such as low water, high salinity, or high Cu can enhance P deficiency (20).

Morphology

Phosphorus deficiency causes reduced growth and often a reddish or purple coloration due to an increase in anthocyanin production by the leaves (2). The retarded growth is exacerbated by a decrease in leaf expansion. There is a relationship between P deficiency and a decrease in the number of flowers and a delay in the initiation of flowering due to changes in the phytochrome balance. Thus, the influence of P on reproductive processes leads to poor formation of fruits and seeds, low yield, and poor quality (1). Increased P utilization efficiency is associated with increases in leaf area (leaf expansion), decreases in the starch in the leaves and increases in dry matter (13). Phosphorus supply has a greater effect on root elongation than on root production rate (21). Phosphorus deficiency caused differences in the proportion of green leaf, dead leaf, stolon, and root of clover varieties (22). Finally, P supply is important in delaying senescence and improving pod retention in soybean (23).

POTASSIUM

Uptake

Potassium, the most abundant mineral element in plants, is absorbed in large quantities (24). Of the total cationic species in the phloem, potassium makes up about 80% (1). Potassium moves in the soil solution by mass flow and simple diffusion (24), and its uptake is regulated by diffusion from the bulk soil to the root surface (25). Potassium is taken up in the largest quantities during the vegetative stage of growth (1), after which there is a decrease in shoot demand due to the reduction of relative growth rate (26). In some plants such as tomato, the rate of K uptake is greatest during fruiting (27). Potassium maintains enzyme structure, is important in protein synthesis through its role in binding mRNA to ribosomes, and assists with the maintenance of electrostatic balance and turgor within the cell together with other cations (28).

The first phase in the absorption of K by the root is transport through the plasma membrane of the cortical cells. The rate of absorption is influenced by the tissue transport rate, the competing ion concentration, the availability of metabolic energy, and the magnitude of the membrane electrical potential differences across the plasma membrane. The K content of the tissue is regulated by the rate of K absorption. As the K concentration is increased, the rate of K absorption increases proportionately to a certain point. The rate of K absorption is directly related to the ATP content of the root cells. Energy for K transport into cells is available in the form of ATP, which supplies the energy for the potential differences between the sides of the membrane (24).

Enzymes

Potassium carries an electrical charge, is mobile, binds both oxygen and anions, and forms weak complexes (29). Potassium activates two types of enzyme reactions: phosphoryl transfer reactions and a heterogeneous group of elimination, hydrolytic reactions (29,30). Examples of the first

involve acetate kinase, acetyl thiokinase, carbamoyl phosphate synthetase, pyruvate kinase, orthophosphate dikinase, and adenosuccinate.
Examples of the second involve tryptophanase, propanediol dehydrase,
and type II fructose 1,6-biphosphate aldolase with a Zn cofactor (31).
The most widely studied is pyruvate kinase which also requires Mg.
Monovalent cations activate at or near the catalytic site by their effect on
an allosteric site. Important membrane phenomena, as mentioned above
under uptake, are generated by Na,K-ATPase. Potassium is necessary for
protein synthesis in the binding of tRNA to the ribosomes.

Potassium promotes the synthesis of ribulose biphosphate carboxylase, which in turn affects the rate of CO_2 assimilation. Potassium is
important in the transport of newly formed and stored photosynthates
(1). It is essential for full activity of the enzyme ATPase, which is crucial
for the exchange of nutrients and metabolites between the apoplast and
the symplasm (29). Potassium's role in enzyme activity involves the accumulation of toxic amines such as putrescine and agmatine in cases of
severe deficiency (1). Chloroplasts of K-deficient leaves contain reduced
activity of certain Calvin cycle enzymes (32).

There is an excellent correlations between H^+ pumping from root
cells and K uptake into root cells. ATPase acts as a pump which produces
the free energy gradient of H^+ required to accomplish the transport of K
and other ions. The plasma membrane is able to distinguish K from
chemically similar but nutritionally unessential ions through enzyme-like
carrier proteins (24). Luttage and Clarkson (28) give models for possible
pathways for one K ion entering and one H^+ ion exiting at the membrane interface (Fig. 1). There are channels for K movement through the
membrane. Tetraethylammonium is used as a specific K-channel inhibitor to study K movement through channels. Calcium, Mg, and Na are
involved in regulating the channel "gating" and thus in regulating K conductance (28).

Carbohydrate and Protein Synthesis

Potassium is essential for protein synthesis (29,33). The role of K is
difficult to prove experimentally because of its extreme solubility and
mobility making it difficult to localize (33). Potassium deficiency reduces
the rates of net photosynthesis and translocation and increases the rates of
dark respiration (32). In K-deficient leaves the activity and capacity of
the photosynthetic apparatus are reduced. Deficient leaves accumulate
sucrose and hexoses in response to reduced leaf water potentials. This
response is distinct from translocation effects and is due to enzyme
activity reductions (32). High K levels protect against water stress by

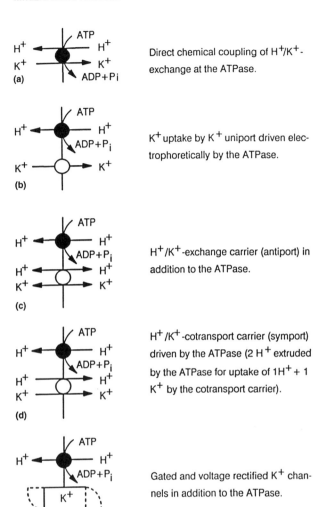

Figure 1 Possible mechanisms for one K^+ taken up and one H^+ released. (From Ref. 28.)

exchange of K in cytoplasm for stroma H, thus altering the stroma pH and restoring photosynthesis. This protective role has been observed in attached whole-leaf photosynthesis as well as in leaf slices (34). Potassium affects growth factors by changing the hydroxycinnamic acid levels and amine levels. Potassium deficiency also results in the collapse of the chloroplasts and mitochondria (1).

Translocation

Potassium is the only essential mineral cation species that is transported against an electrical gradient into plant cells (1,35). The uptake of water in plant tissues is frequently tied directly to K uptake. The lower water loss of plants well supplied with K is due to a reduction in transpiration (1). Potassium is needed for stomatal opening, and stomata open less under K deficiency, and intracellular CO_2 concentration remains unchanged or increases (35). Potassium contributes significantly to osmotic adjustment in the cytoplasm and turgor generation in the vacuole, when in the sufficiency range. Not much detail is known about the role of K in osmotic adjustment of plants under water stress (35), although it is known that K is required in the majority of plant cells for turgor buildup (30). Under K deficiency plants show a decrease in turgor and become flaccid easily under water stress (1). Also, lignification of the vascular bundles is impaired leading to lodging.

Potassium indirectly affects water uptake from the soil, water retention in plant tissue, meristematic growth, and long-distance transport through its effect on physiological processes (30). Processes that are affected by metabolism directly affect membrane permeability to K (36). Potassium transport is regulated by feedback control. In K-deficient plants the transport mechanism is responsive to changes in external and internal K concentration and is tightly coupled to metabolic energy. In high-K plants the transport is predominately passive. In low-K plants hormones interfere with transport. In high-K plants, hormones remain active but transport is inhibited only by feedback control (37).

Interactions

Potassium interacts with almost all of the essential elements. Increasing K levels reduce the uptake of B, Fe, and Mo, whereas the utilization of Cu, Mn, and Zn is increased (38). Increased K depresses Mg in the shoot but not in the root of wheat, although K did not affect the rate of Mg influx (25). Magnesium concentration did not affect the shoot or root concentration of K or the influx of K. Nitrate reductase activity is reduced in K-deficient wheat plants; thus K is important in N nutrition (39). In saline stress, K is replaced by Na. The process is reversed when K is supplied with the Na being replaced in the rhizodermal cells (40).

Morphology

Potassium affects plant growth mainly by its effect on cell extension (35,41). Adequate K supply results in greater synthesis of cell wall material creating walls that are thicker, thus helping crops to resist lodging, pests, and diseases. Morphogenic effects of K include an increased number of storage cells, affecting the sink capacity of the grain (41). Growth and bud formation are inhibited under both high and low K levels (42).

CALCIUM

Uptake

The Ca concentration required for plant growth is from 0.1% to about 5.0% of the dry weight. Calcium is the only mineral element, except perhaps B, that functions outside the cytoplasm in the apoplast where it stabilizes the cell wall (2). The high amount of Ca in plants is a result of high levels of Ca in the soil solution rather than efficiency of uptake. Calcium uptake is regarded as passive (1). Calcium uptake is less than K even though Ca concentration is ten times that of K in soil solution. Calcium is taken up only by root tips and can be competitively inhibited by monovalent cations. The Ca content of monocots is inhibited by monovalent cations. The Ca content of monocots is generally lower than that of dicots. In plant tissues, Ca occurs as free Ca, as Ca absorbed to undiffusible ions such as carboxylic and phenolic hydroxyl groups, and as oxalates, carbonates, and phosphates. These compounds are often found as deposits in cell vacuoles (1). In peanut, which has specific Ca requirements from the soil for the pod, the Ca is taken up by the peanut hull directly, but more is found in the outer tissue than in the inner. Both the lignified endocarp and the seed are lower in Ca than the other pod layers (43). Calcium uptake by rice is correlated with Ca activity in solution, but a more pronounced relationship exists between the uptake and the divalent charge fraction in the soil solution attributable to Ca and Mg (44).

Enzymes

Calcium has been found to activate enzymes, particularly those associated with membranes (1). However, Ca influences the activity of only a few enzymes compared to Mg (2). There are specific Ca-binding proteins in plants, particularly calcicoles (28). Both Ca and B are required for the auxin transport system (45). Calcium promotes β-glucan synthase activity in conjunction with calmodulin. It is also involved with reactions of phosphorylation and dephosphorylation in the regulation of enzyme activity. Calmodulin in the presence of Ca and ATP stimulates glucan synthase activity (46). Calcium activates calmodulin, an intermediate involved in the regulation of various biochemical processes in plants.

These include cell wall rigidity, membrane permeability, and enzyme activation. Figure 2 shows Ca activating calmodulin to enable it to bind to a protein that activates the protein enzyme. Calmodulin is involved in Ca pumps through Ca-ATPase and other metabolic reactions such as those involving NAD kinase and isofloridoside synthase (47). Calcium is a secondary messenger in plants conveying information from the plant environment and from primary messengers like hormones to the final response. Calmodulin, a Ca-dependent regulator protein, plays an important role in controlling plant membrane functions and enzyme activities. For plants, three processes are regulated by calmodulin: NAD kinase reactions, microsomal Ca transport, and microsomal Ca-ATPase reactions. Extracellular Ca affects several plant processes including growth hormone activity, cation transport into cells, leaf movement, abscission, senescence, auxin-mediated H^+ secretion, membrane damage and leakage. The intracellular functions of Ca are activating pyruvate kinase, alkaline lipase, phospholipid acylhydrolase, glutamine dehydrogenase, NAD kinase, and microsomal Ca transport and Ca-ATPase as mentioned above (48).

Calcium deficiency severely inhibits the ATPase activity of the plasma membrane. Thus, a decrease in the functional association of ATPase with phospholipids might be one of the physiological injuries in root cell membranes caused by Ca deficiency (49).

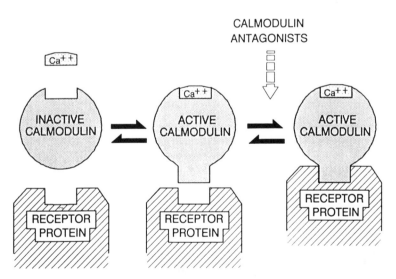

Figure 2 Mechanism for the mediation of calmodulin in the action of Ca. Site of calmodulin antagonist action indicated. (From Ref. 47.)

Calcium regulates membrane permeability and related processes, so that it plays a major role in plant biochemistry. The breakdown of cell walls and the collapse of affected tissues (petioles and stems) are Ca deficiency symptoms in plants. In Ca-deficient plants low molecular weight solutes and ions such as K leak across the cell membrane (1,2). Cell membranes are stabilized by Ca through its bridging of phosphate and carboxylate groups of phospholipids. Calcium plays an important role in mitochondria structure and function. In vacuoles Ca contributes to the cation–anion balance (2). In Ca deficiency growth is reduced and root tips turn brown and die, since Ca is needed for cell elongation and cell function. The symptoms of senescence are similar to Ca deficiency symptoms and can be retarded by adding Ca (1). Calcium deficiency in peanut leads to lower lipid content and an increased sugar content in the seed (50).

Translocation

The rate of flow of the transpiration stream determines the Ca translocation rate. Calcium movement in the root is restricted to the apoplastic or free space, which is only available to young roots. Both K and P are transported by the symplastic pathway that runs the whole length of the root. The downward movement of Ca is very low because of the low Ca concentration in the phloem. Plants deficient in Ca at the growing tip cannot use Ca from older leaves (1). An example of this low mobility is in sorghum where lower leaves were shown to have higher Ca concentration than upper leaves (51). Plant cell plasma membranes contain an active extrusion pump which results in a low Ca concentration in the cytoplasm. The mitochondria and microsomes also take part in the regulation of the cytoplasmic concentration of free Ca (48).

Interactions

Calcium interacts with several elements including N, K, and B. High N levels exacerbate Ca deficiency in sorghum (51). However, increased NO_3 supply in tomato increased the Ca uptake (52). Applied Ca increased N uptake on acid soils and decreased N uptake on calcareous soils for corn (53). This effect could be a soil effect and not a physiological effect. Potassium uptake by excised rice roots was increased by increasing Ca under both anaerobic and aerobic conditions (54). Increases in Ca supply caused less K requirement in perennial ryegrass. Calcium possibly substitutes for K in plant shoot tissue (55). Both Ca and B (essential, immobile elements) are required for the auxin transport process (45). Boron deficiency induces abnormal changes in the Ca metabolism of the cell wall. However, in tomato B deficiency slightly increased Ca uptake and inhibited Ca translocation to the upper leaves (56).

MAGNESIUM

Uptake

Magnesium is taken up in lower amounts than either Ca or K. The Mg content of plant tissue is about 0.5% (1,2), and the critical deficiency level for a legume (subterranean clover) is about 0.1% (57). Cation competition effects are important for Mg since these often lead to Mg deficiency in the field. Some cations that compete with Mg are K and NH_4 (1). The Mg influx is related to the Mg concentration at the root surface for low-pH soils. At higher pH levels the ratios of the concentration of cations (Mg/Ca, Mg/K, and Mg/Al) rather than Mg concentration alone are related to influx, which substantiates the complementary ion effect (58).

Enzymes

There is a long list of enzymes that require or are promoted by Mg. These include the phosphorous ATP reactions which have an absolute requirement for Mg (2). One of the major roles of Mg is as a cofactor in nearly every enzyme of the phosphorylation process. The Mg ion forms a bridge between the pyrophosphate structure of ATP or ADP and the enzyme (Fig. 3). Dehydrogenases and enolase are activated by Mg as well. In Mg-deficient plants protein N decreases whereas nonprotein N increases.

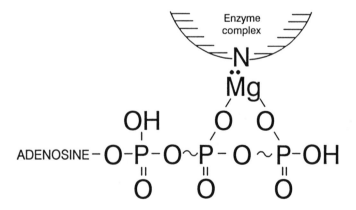

Figure 3 Adenosine triphosphate bridged to an enzyme with Mg. (From Ref. 1.)

The cause is the dissociation of the ribosomes into their subunits in the absence of Mg, which stabilizes the ribosomes into the configuration necessary for protein synthesis. Magnesium activates ribulose biphosphate carboxylase (1). The ADP and ATP complexes of Mg are fairly labile, unlike the tightly bound cations found in the metalloenzymes (29).

Carbohydrate and Protein Synthesis

Magnesium is at the center of the chlorophyll molecule, which is one of the best known functions of Mg. Along with K, Mg is also a component of ribosomes and chromosomes. Much of the Mg in plants acts in the regulation of pH in cells and in the balance of cations and anions. Along with K, Mg is a counterion for the organic acid anions and inorganic anions stored in vacuoles. Magnesium-deficient leaves are chlorotic since Mg is involved with the chlorophyll molecule. In Mg-deficient leaves, protein N is reduced whereas nonprotein N increases. Starch accumulates in the Mg-deficient leaves and there is a decrease in the starch content of storage tissues (2). Magnesium deficiency inhibits protein synthesis (1), as explained above, through the influence of ribosomes.

Translocation

Magnesium, like K, is mobile in the phloem. It can be readily transported from old leaves to younger leaves or to the apex (1,57). The phloem supplies mineral cations to fruit and storage tissue which are higher in K and Mg than Ca (1).

SULFUR

Uptake

The S concentration in the range of sufficiency is between 0.2% and 0.5% of the dry weight of plants (2). The plant takes up S mainly as SO_4^{2-}, and unlike NO_3^-, its uptake is not very sensitive to pH. There is relatively little interference from other ions. Plants can even use atmospheric SO_2 as a source (1). The uptake of S in higher plants is largely controlled by the intracellular levels of SO_4^{2-} (59). The uptake of S is active in that it requires an energy source (60).

Enzymes

The most important S-containing compounds are the amino acids cysteine and methionine, which are building blocks for proteins and also occur as free acids. The main function of S is the formation of disulfide bonds stabilizing polypeptide structure (Figure 4). Disulfide bonds

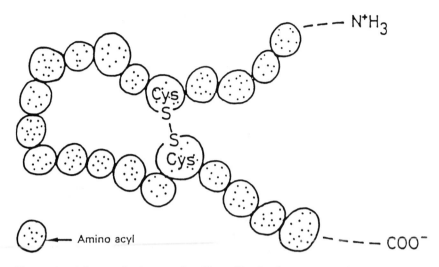

Figure 4   Polypeptide chain with sulfur-sulfur bridge. (From Ref. 1.)

between cysteine residues connect differing parts of a single or differing polypeptides resulting in crosslinking of the protein leading to secondary structure. Another important function is participation in enzyme reactions (1). Sulfur deficiency impairs the synthesis of various S-containing compounds including ferredoxin, biotin, and thiamine pyrophosphate (vitamin B_1) (2). Cysteine S becomes a component of thiamine, CoA, lipoic acid, and biotin which operates in CO_2 transfer. Methionine can serve as a source of non-S compounds such as methyl groups, ethylene, and polyamines through an S-adenyosylmethionine intermediate (60). Sulfur and Zn deficiency decrease the activity of malate dehydrogenase and glucose-6-phosphate dehydrogenase. Sulfur deficiency results in a decrease in phospholipids, free fatty acids, and triglycerides in mature peanut kernels (61). Under S deficiency conditions in wheat, ATP sulfurylase activity is significantly higher, whereas the activity is lower when S is applied (62).

Carbohydrate and Protein Synthesis

Sulfur as a constituent of the amino acids cysteine and methionine, and thus of proteins, acts as a functional group or as a structural constituent. Sulfur deficiency inhibits protein synthesis and the chlorophyll content of the leaves declines while starch may build up. Sulfur is a component of sulfolipids and therefore serves as a structural component of biological membranes. The inhibition of protein synthesis causes an accumulation of soluble N and nitrate under S-deficient conditions, and leads to the chlorosis of plant tissues (2). Storage forms such as SO_4^{2-} are important in plants in that they are the principal mobile forms, supply S when uptake is inadequate, and are important quality factors (59). Sulfate in the chloroplast is activated, reduced to a -2 valence state, and incorporated into cysteine. A large part of the cysteine is used to form methionine, and are incorporated into proteins. Sulfur is important in the regulation of the process of nitrate reduction (60).

Caryopses of barley are able to use sucrose and glutamine to make starch and convert the starch to protein. However, they cannot convert external cysteine into protein (63). Sulfur and Zn deficiency decrease the protein and starch content of seeds and increase the soluble sugar in peanut (61). Sulfur deficiency causes changes in the synthesis and accumulation of both salt-soluble and salt-insoluble polypeptides in wheat. In S-deficient wheat plants, storage protein synthesis in developing seeds has been detected earlier than in plants with adequate S (64). Sulfur deficiency has been shown to cause gliadins to reach a constant proportion of the dry weight of the endosperm earlier than in normal wheat (65).

Translocation

Sulfate transport is mostly in an upward direction in the plant with low downward transport (1). After absorption, S is transported to the endodermis where it moves into the xylem and is transported to the leaf in the transpiration stream (60).

Morphology

Sulfur gross deficiency symptoms are reduced growth and chlorosis (59), which are very general for most deficiencies. Various responses are observed from supplying S to deficient plants including increased shoot dry weight, total S content of the leaves, nodule number per plant, and pod yield in peanut (66). Plant proteins that are lower in S content are produced when S is deficient, which influences grain quality. The decreased number of S proteins in the wheat flour reduces disulfide bridging and lowers the baking quality (2).

MANGANESE

Uptake

Manganese uptake rates are lower than other divalent species such as Ca and Mg. Manganese uptake is considered to be active and can be inhibited by other cations. Manganese shows properties of both alkali earth cations (Ca and Mg) and transition metals (Zn and Fe); thus its uptake and translocation can be affected by these (1). Manganese is absorbed as the divalent ion and forms the weakest bonds with complexes of any of the essential transition metals (Fe, Zn, Cu, Mn, Mo). Because of the lability of Mn with respect to binding, Mn can substitute for Mg in many reactions. Because of its role in photosynthesis Mn deficiency lowers the level of soluble carbohydrates, particularly in the roots. Manganese toxicity often shows up as brown spots on leaves because of precipitation of MnO_2. Manganese critical toxicity levels vary widely depending on plant species in contrast to critical deficiency levels, which are in a narrower range (2).

Critical Mn deficiency and toxicity levels for physiological measures such as chlorophyll, carbonic anhydrase, and nitrate reductase are similar to those for plant growth (67). Excess Mn in plants is deposited in the roots as MnO_2, which constitutes up to half the amount in the plant, but Mn is also deposited in the leaves (68). Manganese from a chelated source is taken up by leaves in foliar applications on tomato and pea to a lesser extent than inorganic Mn (69). Root exudates can affect Mn uptake. Chemical, microbial, morphological, physical, and physiological gradients along the length of the root are important in exudates. The function of root exudates in making nutrients more plant-available involves interactions among the plant, the soil, and the associated microbial population (70).

Enzymes

Manganese and Mg are similar in their biochemical functions. They both bridge ATP with the enzyme complex as in phosphokinases and phosphotransferase, Manganese activates decarbosylases and dehydrogenases of the TCA (tricarboxylic acid) cycle as well as IAA (indolacetic acid) oxidases (1). The essentiality of Mn stems from its being a structural element, an active binding site, and a component of Mn-containing enzymes. These enzymes include superoxide dismutase and acid phosphatase. Manganese substitutes for Mg in the NADPH-specific decarboxylating malate dehydrogenase, the malic enzyme, and the isocitrate dehydrogenase or in redox reactions in metalloproteins (2). Manganese deficiency alters enzymes like superoxide dismutase isoenzymes (71).

Carbohydrate and Protein Synthesis

Manganese is necessary for photosynthetic O_2 evolution where it participates in the Hill reaction whereby water is split to give O_2 in photosynthesis. Thus, in Mn-deficient plants photosynthesis capability decreases as chlorophyll content is reduced. Membrane constituents such as glycolipids and polyunsaturated fatty acids are reduced under Mn deficiency (1), and there are also changes in lipid composition and content (72). Manganese deficiency causes a decline in cell elongation more than cell division. The critical deficiency levels of Mn range from 10 to 20 mg/g dry weight (73). Chloroplasts contain one atom of Mn per 14–600 chloroplasts, whereas one Mn per 50–100 chlorophyll molecules is required in the system that produces O_2 (73).

Manganese is involved in oxidation–reduction in the photosynthetic electron transport system. The influence of Mn on photosynthesis causes an indirect effect on nitrate reduction. In Mn-deficient plants, nitrate sometimes accumulates because of reduced photosynthesis. Chloroplasts are the most sensitive of the cell organelles to Mn deficiency (1). Chlorophylls a and b and photosystem II are altered by Mn deficiency (71). Chloroplast pigment, thylakoid Mn, total phytoene, *ent*-kaurene, and giberellic acid are altered with both Mn deficiency and toxicity (75).

Translocation

Manganese is relatively immobile in the plant being mainly transported as Mn^{2+} and not as organic complexes. When it does move, Mn is preferentially translocated to the meristematic tissues (1). Translocation of Mn within the plant is greater when supplied to the leaf as a chelate (EDTA) than as inorganic Mn even though the chelate may not be taken up (68). Manganese toxicity can close stomates and decrease transpiration rates (76).

Interactions

Iron-manganese interactions occur at two levels: at uptake, where Fe hampers Mn uptake, and at the metabolic level, where Mn inactivates Fe metabolic activity by decreasing Fe concentration in plants (77). The uptake of Mn by plants can be reduced by high levels of available Fe, Cu, or Zn (1). Iron deficiency can be caused by high Mn supply decreasing Fe uptake and competing for binding sites. Manganese toxicity can also be responsible for deficiencies of Mg and Ca by the same mechanisms as for Fe. The crinkle leaf symptoms of Mn toxicity in beans is actually a Ca deficiency since Mn inhibits Ca translocation to the shoot apex (2,78).

COPPER

Uptake

The Cu content of most plants is 2–20 mg/kg dry weight, which is about one tenth the Mn content (1). The critical Cu level is at the bottom of that range, being 3 mg/kg for young subterranean clover (79). Copper uptake is metabolically mediated and in strongly inhibited by other divalent transition metals, especially Zn (1). Copper is strongly bound to organic matter in soils and to soluble organics in soil solution. It is likewise bound to ligands in plants in the xylem sap. Plant tolerance to Cu toxicity is through immobilization of Cu in cell walls, exclusion or restriction of Cu uptake, compartmentation of Cu in insoluble complexes and enzyme adaptation to high Cu levels (2).

Enzymes

Copper functions in plants by catalyzing terminal oxidation, which is a function not carried out by Fe (2). Both atoms of molecular oxygen are reduced by reactions catalyzed by Cu. These oxidases are cytochrome oxidase and other enzymes including ascorbic acid oxidase and polyphenyl oxidase. Both Cu and Zn are found in the enzyme superoxidase dismutase which is found in all aerobic organisms. It catalyzes the reaction $O_2^- + O_2^- + 2H^+ \rightarrow O_2 + H_2O_2$. The enzyme system contains two Cu and two Zn atoms. Superoxidase is a highly reactive free radical which is extremely detrimental to cells (1). Under Cu deficiency the activity of these oxidative proteins decreases rapidly. Copper and Fe are both important in the cytochrome oxidase system. Copper is also important in ascorbate oxidase, phenolase, laccase, and amine oxidases. Reduced phenolase activity that is due to Cu deficiency causes delay in flowering and maturation (2). The desaturation and the hydroxylation of fatty acids are catalyzed by Cu enzymes (1). The biosynthesis of diamine oxidase is controlled by Cu concentration in the leaves of clover (80). Copper deficiency causes dysfunction of phenolase and laccase in the oxidation of phenols which are precursors to lignin biosynthesis (2).

There are three types of copper in plants. The first is "blue Cu," which is coordinated by at least one cysteine residue and two N ligands in the protein. The enzymes involved are plastocyanin and six others. The second type is "nonblue Cu," which is in a coordination that differs from the first type. The enzymes here are amine oxidase and galactooxidase. The third type is "nonblue and nonparamagnetic." This is a two-electron acceptor in oxidation processes. The enzyme associated with the third type is tyrosinase (phenol o-monooxygenase), which oxidizes monophenols to orthodiphenyls and further to o-quinones and water. There are multicopper proteins containing all three types. They catalyze the monooxygenase reaction $2AH_2 + O_2 \rightarrow 2A + 2H_2O$, transferring four electrons from the substrate and four protons from the solvent to molecular oxygen giving water (74).

Carbohydrate and Protein Synthesis

More than half the Cu found in plants is localized in the chloroplasts bound to plastocyanin, a component of the electron transport chain linking the two photochemical systems of photosynthesis. In Cu-deficient plants photosynthesis decreases (1,2). The Cu in the chloroplasts constitute up to 70% of that found in the leaf. Copper may play a part in the synthesis or stability of chlorophyll. Copper participates in both protein and carbohydrate metabolism. In Cu-deficient plants protein synthesis is reduced causing soluble amino nitrogen compounds to build up (1). Copper-deficient plants have lower soluble carbohydrate during the vegetative stage which increase during the reproductive stage (2). Copper is a factor in DNA and RNA synthesis, since low levels of DNA are observed in Cu-deficient tissues. Nitrogen compounds build up both as a result of Cu enzymes being reduced in activity and by the low synthesis of DNA and RNA. There is a specific requirement for Cu in symbiotic nitrogen fixation (1). Copper toxicity increases mono- and disaccharide concentrations (20).

Translocation

Copper readily moves as anionic complexes in the xylem sap from the older to the younger leaves. Copper goes to the grain quite easily under sufficiency conditions but is immobile when deficient (1). Copper deficiency results in stomatal closure, lower water potentials, lower leaf conductances, and wilting. The wilting comes from structural weakness and reduced lignification of cell walls and xylem vessels. Stomatal closure is caused by inadequate levels of ATP which requires Cu for its production (81).

Interactions

Copper is considered to be somewhat free from competitive effects, except that uptake is strongly inhibited by Zn and, as indicated below, it interacts with P (1). Increased P induces a Cu deficiency which is due to a dilution effect and to depressing the Cu absorption (79). Varvel (82) indicated that at low Cu supply, P is high in the plant, whereas at high Cu, P is low in the plant. The Cu affected P uptake or its translocation.

Morphology

Copper deficiency affects grain, seed, and fruit formation more than vegetative growth. The critical deficiency level in plants ranges from 3 to 5 mg/kg dry weight. Typical deficiency symptoms are necrosis of the apical meristem and wilting of young leaves. Copper deficiency in legumes causes reduced nodulation and N_2 fixation. The critical toxicity level is above 20–30 mg/kg dry weight and may include Fe deficiency as a side effect. The initial response to Cu toxicity is reduced root elongation and damage to root cell membranes. Lateral root formation is enhanced (2). Further Cu deficiency symptoms are a decrease in growth (20) and a delay in maturity (81). Copper deficiency can result in irregular pollen development and reduced fertility. The development of the tapetum is often abnormal (83). Low water supply can decrease Cu concentrations in barley (20).

IRON

Uptake

The chemical activities of both Fe(II) and Fe(III) are very low in soil solution regardless of the total Fe content of the soil, and Fe(III) is usually reduced before uptake by plants. Transport in soils and inside plants is in chelated forms with various ligands (2) or as the Fe(II) form (1). Iron chlorosis is most often caused by insufficient mobilization and uptake of Fe. The mechanisms available to plants to assist in uptake include acidification of the rhizosphere, morphological changes like an increase in root hairs, and physiological changes such as proton release, chelator release, and increased reducing capacity at the plasma membrane of the roots (84).

Plants have two different strategies for mobilizing Fe to enable uptake. Dicots and nongramineous species are considered to be "strategy I" and produce no exudates, but release protons to lower the rhizosphere pH. The grasses are "strategy II" plants which release highly specific Fe chelator exudates but generally do not reduce Fe(III) (84,85). Some researchers have found that strategy II grasses not only produce exudates but are capable of acidifying their root environment (86). Strategy I plants develop specialized cells that release protons and sometimes reductants. Strategy II plants produce and excrete phytosiderophores, which chelate Fe, and the whole molecule is taken up by the plants that produce them. Iron(II) is essential to strategy I plants and Fe(III) is essential for strategy II plants (87).

Root exudates of barley were found to increase Fe uptake from calcareous soils by 10–20 times over the check. This increase was due to phytosiderophores and not microbial chelators, which were found to be somewhat more effective (88). Certain Fe-inefficient plants (strategy I sorghum) release caffeic acid as a major reductant (89). The uptake of Fe, the reduction of ferric Fe, and proton excretion (strategy I) occur in the young part of the root, less than 3 cm from the tip (90). The root effects on changing the pH, reducing capacity, and exudates affects not only Fe but also Zn and P (3). The reductants released by sorghum have been used in an attempt to measure Fe uptake efficiency (91). Higher-than-normal levels of HCO_3^- inhibit Fe absorption by limiting the ability of the roots to release reducing compounds (92). It has been shown that C4 plants require a higher Fe supply than C3 plants. The difference is in the ability of the roots to absorb Fe (93).

Enzymes

Iron functions that have been most studied are the enzyme systems having heme structures that function as the prosthetic groups. Iron functions much as Mg does in the porphyrin structure of chlorophyll. These Fe enzymes include catalase, peroxidase, cytochrome oxidases, and various cytochromes (1). The iron-containing proteins are considered in two groups: hemoproteins and iron-sulfur proteins. Of the hemoproteins, the most well known are cytochromes which make up the redox systems in chloroplasts and mitochondria. Others are catalase and peroxidases. Both systems decrease in activity when Fe is deficient. Of the iron-sulfur compounds the best known is ferredoxin, which acts as an electron transmitter in a number of basic metabolic processes. Other Fe-containing enzymes are riboflavin, which accumulates in the roots of Fe-deficient plants, and aconitase, which catalyzes isomerization of citrate to isocitrate causing organic acids to accumulate in the roots of Fe-deficient plants (2). A third group of Fe enzymes consists of superoxide dismutase and many less characterized Fe proteins. Superoxide dismutase is associated with Cu and Zn as one type, Fe as another, and Mn as another in three distinct groups. All three types of dismutases catalyze conversion of the superoxide radical (O_2^-) to hydrogen peroxide and oxygen (74).

The enzyme activity of Fe compounds can decrease with Fe deficiency. The catalase activity increases as Fe supply increases, but there is no relation with the leaf concentration of Fe (94). Iron levels in plants modify the total superoxide dismutase activity (71). However, ATPase activity is not affected by Fe deficiency (95).

Carbohydrate and Protein Synthesis

Iron deficiency chlorosis is caused by iron's role in the biosynthesis of chlorophyll. Thus there is a good correlation between the level of leaf Fe and chlorophyll content (1,2). Also, Fe deficiency causes reduced protein synthesis, which in turn inhibits chorophyll synthesis. This effect can be readily related to the fact that up to 80% of the Fe in leaves is in the chloroplasts (2).

Iron in the chloroplasts is in the form of phytoferrin (Fe storage), various cytochromes (electron transport), and ferredoxin (S compounds and electron transport). In Fe deficiency, ferredoxin becomes limiting affecting chloroplast biosynthesis (Fig. 5) (96). Chlorophyll is formed from precursors such as α-ketoglutaric acid or glutamate. This pathway may be activated by Fe related to aconitase activity and/or the formation of ferrodoxin (97). Ultraviolet radiation is capable of reducing Fe(III) to Fe(II) through a chlorotic leaf. This process may be important in establishing an "active" Fe fraction on the leaf (98).

Chlorophyll proteins and electron transport increases with increased Fe supply. Photosystem I is affected more by Fe than is photosystem II (99). Iron affects galactolipid and thylakoid protein synthesis. Iron deficiency most diminishes the chlorophyll protein associated with photosystem I. The number of polypeptides increases under Fe-deficient conditions (100).

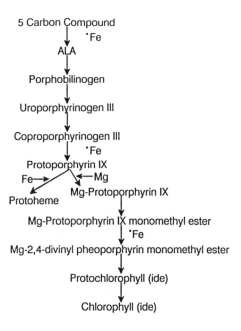

Figure 5   Iron involvement in chlorophyll biosynthesis. (From Ref. 96.)

Translocation

Iron is reduced to Fe(II) at the root surface and translocated to the leaf through the xylem in a combined form. In the leaf it is used as indicated above for chlorophyll formation as well as for the functioning of various enzyme systems (97). When Fe solutions are sprayed on plants, Fe is absorbed more as an inorganic salt than as a chelate (EDTA) (69).

Interactions

Since Fe uptake is active, there are competitive effects with other cations such as Mn, Zn, Cu, Ca, Mg, and K. Copper and Zn are known to displace Fe from chelate complexes which can be important both inside the plant and in the soil. The extent of Fe deficiency caused by heavy metals follows the stability sequence: Cu > Ni > Co > Zn > Cr > Mn (1). Phosphorus fertilization has been shown to increase P and Fe uptake in tomato on both alluvial and calcareous soils (101). An Fe–Mn interaction occurs on two levels: at uptake where Fe hampers Mn uptake and at the metabolic level where Mn inactivates Fe metabolic activity (77). Iron deficiency affects nitrate reductase and thus nitrogen nutrition. The plant needs a minimum of Fe to activate nitrate reductase (102). Iron deficiency in sunflower was found to increase the ferricyanide reduction activity and decrease K uptake, but in Fe-deficient wheat ferricyanide activity decreased and K uptake increased (95).

ZINC

Uptake

Zinc is taken up as a divalent ion and is not oxidized or reduced as is the case for some other transition metals. It is transported in the xylem as a free ion. The critical Zn deficiency level is 15–20 mg/kg dry weight for most plants. Since Zn plays a specific role in fertilization, grain and seed yield are affected more than total dry matter production by deficiency. Zinc toxicity causes inhibition of root elongation and chlorosis of young leaves. High Zn supply may cause Fe deficiency because of their similar ionic radii. The critical toxicity range for Zn is 400–500 mg/kg dry weight (2). There is disagreement in the literature as to whether Zn uptake is active or passive. Mengel and Kirkby (1) indicate that the evidence points to its being active. Zinc is taken up at a steady-state rate which is typical for active uptake and metabolic inhibitors reduce uptake.

Enzymes

Zinc is a metal component of enzymes or acts as a functional, structural, or regulatory cofactor of a large number of enzymes (2). In enzyme systems, Zn resembles Mn and Mg in that they both bind and assist the conformation between enzyme and substrate. The most widely known reaction that is catalyzed by Zn is that of carbonic anhydrase ($H_2O + CO_2 = H^+ + HCO_3^-$). This enzyme is localized in chloroplasts. The reaction acts to protect proteins from denaturation from local pH changes associated with hydrogen ion pumps. Other Zn enzymes are dehydrogenases (glutamic acid dehydrogenase, lactic acid dehydrogenase, alcohol dehydrogenase) and proteinases and peptidases. Zinc is also involved with nitrogen metabolism. The RNA and ribosome content of cells can be decreased due to zinc deficiency. Zinc deficiency influences IAA and auxins causing them to be lower in concentration thus reducing the growth rate. Zinc is a constituent of the enzyme superoxidase dismutase, which protects aerobic organisms from attack by O_2^- radicals. Zinc deficiency–induced changes in carbohydrate metabolism are not responsible for growth retardation or visible symptoms of Zn deficiency (1,2). Low Zn levels in the plant causing low carbonic anhydrase activity can be used to decide cases of latent Zn deficiency (103). Even though carbonic anhydrase is lower with Zn deficiency, it has been shown that it has no effect on photosynthesis (104).

Carbohydrate and Protein Synthesis

Zinc is involved either directly or indirectly in starch formation, since Zn deficiency has been shown to lower starch concentrations (1). The influence of Zn on protein synthesis leads to large amounts of tryptamine in barley under Zn-deficient conditions (105). Zinc deficiency increases root exudation of K, amino acids, sugars, and phenolics. Zinc has a role in membrane integrity and thus also root exudation. This effect is independent of the role of Ca (106). The permeability of the plasma membrane is increased in Zn-deficient plants giving a 3-, 5-, and 2.5-fold increase in the root cell leakage of K, NO_3, and organic compounds, respectively. Zinc directly affects the integrity of the plasma membrane by interfering with the O_2^- generation by a membrane-bound NADPH oxidase (107). In cotton, bean, and tomato, Zn deficiency enhances superoxide radical (O_2^-) generation and impairs detoxification by O_2^- and H_2O_2 leading to elevated levels of oxygen-derived oxidizing species and thus to increased peroxidation of membrane lipids (108).

Translocation

The form in which Zn is translocated from the root to the shoot is not known and the mobility is not great. Zinc is not bound to stable ligands as for Cu and Fe; neither is Zn transported as a citrate complex. Zinc accumulates in the root especially when the supply is high and is immobile in the older leaves. The rate of mobility in younger leaves is depressed under Zn-deficient conditions (1). Zinc is transported in the phloem to the roots, should they become low in Zn. The absence of Zn in the roots may affect rhizosphere pH and possibly membrane activity (109). The P status of the plant can influence Zn transport, as high P levels increase Zn translocation (110).

Interactions

Zinc interacts mostly with P and in an indirect way with K. Zinc deficiency leads to loss of membrane integrity allowing K to leak out of the roots (106). A high supply of P causes Zn deficiency by disrupting metabolic processes. It has been found that Zn is not precipitated as $Zn_3(PO_4)_2$ in the plant. Phosphorus-induced Zn-deficient corn accumulates high levels of Fe and to a lesser extent Mn (1). The P–Zn interaction is in the soil to some extent but is also in the plant. Zinc affects the P metabolism of the roots and increases the permeability of the plasma membranes of the root cells to P (2). The uptake rate of P in severely Zn-deficient plants is increased by a factor of 2–3, whereas the uptake rates of K, Ca, and NO_3 decrease. In Zn-deficient plants a feedback mechanism in the shoots is impaired which controls the P uptake by roots and the transport of P from roots to shoots. As a result, toxic concentrations of P can accumulate in the leaves (111). Although P treatment increases Zn uptake, Zn treatment has no effect on P uptake (110). Severe Zn deficiency increases the P concentration in the old leaves to 3–4% and gives symptoms of both Zn deficiency and P toxicity. Mild Zn deficiency has only a transient effect on increasing the P absorption rate (112). The P content of wheat was found to decrease as the Zn rate increased. Zinc uptake was shown to be higher at flowering than at tillering (113). The P/Zn ratio as an indicator of Zn supply is a good measure only for extremely Zn-deficient plants. It is not useful, however, for diagnosing latent Zn deficiency (102).

BORON

Uptake

Boron is taken up by plants in the form of undissociated boric acid and is present in the plant at the physiological pH in the same form. It has not yet been determined as to whether uptake is active or passive (1,2). There is no evidence that B is an enzyme component. Its functions are primarily extracellular and related to lignification and xylem differentiation (2).

The lower critical level for B in cotton is about 15 mg/kg in the young leaves (114). Boron is known for its extremely narrow sufficiency range, making it easy to overfertilize for B. The critical toxicity concentration in barley shoots is 50–70 mg/g dry weight (115). Boron deficiency in cauliflower (hollow stem) is found with B levels in the young leaves at or below 48 mg/kg dry weight (116). Boron plays an important role in the storage of assimilates in corn. The critical corn ear-leaf B concentration for yield is 8.4–25.3 mg/kg depending on the hybrid (117). Soybeans require 12 mg B/kg dry weight to achieve 90% of their maximal growth rate. This is a functional B requirement for leaf expansion and gives a more consistent and accurate critical value for diagnosis of B deficiency than by correlation with shoot dry matter (118).

Carbohydrate and Protein Synthesis

Boron is involved in uracil synthesis, which is an essential component of RNA. Protein synthesis is disrupted when RNA is in low supply and proteins that are essential for growth are not formed. Uracil is also a precursor of uridine disphosphate glucose, an essential coenzyme in the formation of sucrose. Since sucrose is the most important sugar transport form, B deficiency inhibits transport of assimilates, which has actually been observed experimentally. Boron deficiency reduces nucleic acid metabolism and protein synthesis by reducing the rate of incorporation of P into nucleotides. Boron deficiency decreases the synthesis of cytokinins, but other auxins often accumulate under B-deficient conditions. The accumulation of auxins and phenols may be associated with leaf necrosis in B deficiency (1).

The growth of roots is decreased by boron deficiency, and IAA oxidase activity in roots is increased. Under deficient conditions the protein content decreases and soluble nitrogen compounds, particularly nitrate, accumulate. Boron deficiency causes an accumulation of phenolics, which is responsible for the metabolic changes and cell damage in boron-deficient tissue (2).

Boron deficiency reduces the incorporation of glucose into polysaccharides as well as the total cellulose content of cell walls. Boric acid may directly or indirectly influence polysaccharide synthesis by forming complexes with mannose. Boron may affect the deposition of cell wall material by altering membrane properties (119). Both P and glucose uptake have reduced under B-deficient conditions (120). Boron deficiency and toxicity cause lower chlorophyll levels and net photosynthesis (121). [See Figure 6 for an overview of the influence of B in plants (122).]

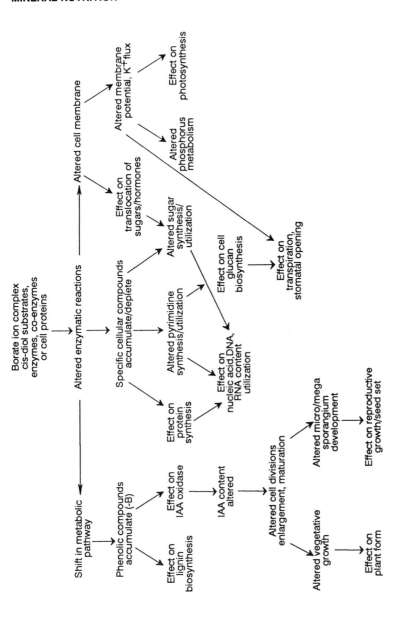

Figure 6 Hypothetical plant response to boron. (From Ref. 122.)

Translocation

Boron is relatively immobile in plants; thus B content often increases from the lower to the upper portion of the plant. The transport of B correlates with the rate of transpiration. Boron deficiency begins at the growing points as with Ca, and B is absent from phloem sap (1). Even though B is relatively immobile in the phloem, it can be remobilized under B-deficient conditions (116). Boron deficiency causes the reduction of stomatal conductance, which is not affected by B toxicity (121).

Interactions

There are few important interactions between B and other mineral elements (2). High B supply to cotton results in low uptake of Zn, Fe, and Mn, whereas the uptake of Cu increases (123). Boron deficiency increases Ca uptake but inhibits Ca translocation to the upper leaves and induces abnormal changes of the Ca metabolism of the cell wall (56). As a result, increased Ca supply accentuates the negative yield effects of B deficiency (124).

Morphology

As indicated above, B deficiency disturbs the development of meristematic tissues including the root tips and upper plant parts (1). Boron deficiency affects pollen germination and pollen tube growth (2). Boron treatment to deficient soybeans was found to increase the number of lateral pods per plant and increases the total seed weight per plant (124).

MOLYBDENUM

Uptake

Although Mo is a metal, it occurs as an oxyanion, MoO_4^{2-} and is absorbed by plants as the molybdate ion (1,2). Uptake is most likely active and the mode in which it is transported in the plant is unknown. However, it may be by MoO_4^{2-}, MoS–amino acid complex, or as a molybdate complex with sugars or other polyhydroxy compounds. Molybdenum is only moderately mobile. Older leaves become chlorotic first. Molybdenum deficiency restricts growth and flower formation. The concentration in plant tissue is usually 1 mg/kg (1). The plant requirement for Mo is lower than for any other mineral element. The critical deficiency levels range from 0.1 to 1.0 mg/kg dry weight. In legumes that rely on N fixation, Mo deficiency can cause N deficiency, since Mo is a component of nitrogenase. Molybdenum is often deficient in legumes grown in very acidic soils high in iron oxides, which strongly absorb Mo.

In acid soils the absorption of Mo is similar to the absorption of P on iron oxides. There is a wide variation between deficiency and toxicity (up to 10^4) as compared to a factor of 10 or less for B or Mn. Molybdenum uptake rates are very slow by soybean during the first 4 weeks, so that the plant must rely on seed Mo during that time (2). Thus, in Mo-deficient soils, the seeds are often treated with Mo to alleviate deficiency. Rhizosphere acidification by plant roots extruding H^+ may change the rhizosphere and serve to increase the availability of Mo. Molybdenum functions in the plant are related to valency changes from VI to V to IV as a metal component of enzymes (125).

Enzymes

Molybdenum is an essential component of two major enzymes, nitrogenase and nitrate reductase, both of which depend on an Mo valency change for their function. Nitrogenase is made up of two enzyme complexes, one of which contains both Fe and Mo in a ratio of about 9:1. Molybdenum deficiency can lead to N deficiency in legumes. Nitrate reductase catalyzes the reduction of NO_3^- to NO_2^-. When Mo-deficient plants are supplied with nitrate, it accumulates in the tissue accompanied by a decrease in the levels of soluble amino compounds. Plants grown with NH_4-N may not need Mo (1). There is a growth response in legumes to Mo on low-Mo soils if the N supply is low. Molybdenum can be substituted for N in low-Mo, low-N soils (2). Molybdenum has a stabilizing influence on the structure of nitrate reductase, protecting it from the effects of salinity in corn under salinity stress (126). Molybdenum is stored as Mo carriers in low molecular weight fractions that are cofactors of enzymes. One fraction is associated with xanthine oxidase and nitrate reductase activity (127).

Interactions

Since Mo uptake is as an oxyanion, ions of a similar type can interfere with uptake such as SO_4^{2-}, but it has been found that $H_2PO_4^-$ actually enhances uptake (1). Inorganic P increases the in vitro activity of NADH–nitrate reductase in barley leaves. The response to nitrate concentrations is specific for the presence of inorganic P. There is an interaction between an Mo site of the enzyme and inorganic P resulting in an alteration of the properties on the nitrate reductase molecule (128).

REFERENCES

1. K. Mengel and E. A. Kirby (1982). *Principles of Plant Nutrition*, 3rd ed., International Potash Institute, Bern.

2. H. Marschner (1986). *Mineral Nutrition of High Plants*, Academic Press, London.
3. H. Marschner, V. Romheld, and I. Cakmak (1987). *J. Plant Nutr.*, *10*:1175.
4. R. K. Misra, A. M. Alston, and A. R. Dexter (1988). *Plant Soil*, *107*:11.
5. D. Fohse, N. Claassen, and A. Jungk (1988). *Plant Soil*, *110*:101.
6. S. Adalsteinsson and P. Jensen (1988). *Physiol. Plant.*, *72*:271.
7. S. A. Barber (1988). *J. Plant Nutr.*, *10*:1309.
8. A. L. Fredeen, R. I. Madhusudana, and N. Terry (1989). *Plant Physiol.*, *89*:225.
9. M. A. Porter and G. M. Paulsen (1983). *Agron. J.*, *75*:303.
10. R. T. Besford (1979). *J. Sci. Food Agric.*, *30*:281.
11. J. R. Caradus and R. W. Snaydon (1987). *J. Plant Nutr.*, *10*:287.
12. I. M. Rao, J. Abadia, and N. Terry (1987). *Progress in Photosynthesis Research*, *Vol. 3* (J. Biggins, ed.), Martinus Nijhoff, Dordrecht, p. 751.
13. D. W. Israel and T. W. Rufty, Jr. (1989). *Crop Sci.*, *28*:954.
14. J. W. Radin and M. P. Eidenbock (1986). *Plant Physiol.*, *82*:869.
15. F. S. Chapin, III and I. F. Wardlaw (1988). *J. Exp. Bot.*, *39*:165.
16. N. S. Murali and A. H. Teramura (1985). *Physiol. Plant.*, *63*:413.
17. J. W. Radin and M. A. Mathews (1989). *Plant Physiol.*, *89*:264.
18. M. E. Sumner and M. P. W. Farina (1986). *Adv. Soil Sci.*, *5*:201.
19. P. Y. Yen, W. P. Inskeep, and R. L. Westerman (1988). *J. Plant Nutr.*, *11*:6.
20. M. N. El-Shourbagy, A. Wallace, and W. L. Berry (1989). *Soil Sci.*, *147*:432.
21. J. R. Caradus and R. W. Snaydon (1988). *J. Plant Nutr.*, *11*:277.
22. J. R. Caradus and R. W. Snaydon (1988). *J. Plant Nutr.*, *11*:289.
23. L. J. Graham, D. G. Blevins, and H. C. Minor (1986). *Plant Physiol.*, *82*:1008.
24. R. T. Leonard (1985). *Potassium in Agriculture* (R. D. Munson, ed.), Am. Soc. Agron. Inc., Madison, WI, p. 327.
25. T. Ohno and D. L. Grunes (1985). *Soil Sci. Soc. Am. J.*, *49*:685.
26. H. Kuhlmann and P. B. Barraclough (1987). *Zeit. Pflanz. Boden.*, *150*:24.
27. I. E. Widders and O. A. Lorenz (1982). *J. Am. Soc. Hort. Sci.*, *107*:960.
28. U. Luttage and D. T. Clarkson (1989). *Prog. Bot.*, *50*:51.
29. R. G. Win Jones and A. Pollard (1983). *Inorganic Plant Nutrition. Encyclopedia of Plant Physiology*, *Vol. 15* (A. Lauchili and R. L. Bieleski, eds.), Springer-Verlag, New York, p. 528.
30. K. Mengel (1985). *Potassium in Agriculture* (R. D. Munson, ed.), Am. Soc. Agron. Inc., Madison, WI, p. 397.
31. C. H. Suelter (1985). *Potassium in Agriculture* (R. D. Munson, ed.), Am. Soc. Agron. Inc., Madison, WI, p. 369.
32. S. C. Huber (1985). *Potassium in Agriculture* (R. D. Munson, ed.), Am. Soc. Agron. Inc., Madison, WI, p. 369.
33. D. G. Blevens (1985). *Potassium in Agriculture* (R. D. Munson, ed.), Am. Soc. Agron. Inc., Madison, WI, p. 413.

34. P. A. Pier and G. A. Berkowitz (1987). *Plant Physiol.*, *85*:655.
35. T. C. Hsiao and. A. Lauchli (1986). *Adv. Plant Nutr.*, *2*:281.
36. R. Pinton, Z. Varanini, A. Maggioni, and H. Frick (1987). *J. Plant Nutr.*, *10*:1975.
37. L. Erdei and M. R. Dhakal (1988). *Plant Soil*, *111*:171.
38. D. W. Dibb and W. R. Thompson, Jr. (1985). *Potassium in Agriculture* (R. D. Munson, ed.), Am. Soc. Agron. Inc., Madison, WI, p. 515.
39. A. A. Ali, M. Ikeda, and Y. Yamada (1987). *Soil Sci. Plant Nutr.*, *33*:585.
40. R. Stelzer, J. Kuo, and H. W. Koyro (1988). *J. Plant Physiol.*, *132*:671.
41. H. Berringer and F. Nothdurf (1985). *Potassium in Agriculture* (R. D. Munson, ed.), Am. Soc. Agron. Inc., Madison, WI, p. 351.
42. S. Klinguer, J. Martin-Tanguy, and C. Martin (1986). *Plant Physiol.*, *82*:561.
43. H. Smal, C. S. Kvien, M. E. Sumner, and A. S. Csinos (1989). *J. Plant Nutr.*, *12*:37.
44. P. A. Moore, Jr. and W. H. Patrick, Jr. (1989). *Soil Sci. Soc. Am. J.*, *53*:816.
45. R. K. Dela-Fuente, P. M. Tang, and C. C. Guzman (1986). "Plant growth substances 1985," Proceedings of the 12th International Conference on Plant Growth Substances, Berlin, p. 227.
46. G. Paliyath and B. W. Poovaiah (1988). *Plant Cell Physiol.*, *29*:67.
47. B. W. Poovaiah (1985). *Hort. Sci.*, *20*:347.
48. D. Marme (1983). *Inorganic Plant Nutrition. Encyclopedia of Plant Physiology*, Vol. 15 (A. Lauchli and R. L. Bieleski, eds.), Springer-Verlag, New York, p. 599.
49. H. Matsumoto and G. C. Chung (1988). *Plant Cell Physiol.*, *29*:1279.
50. S. Inanaga, T. Yoshida, T. Hoshino, and T. Nishihara (1988). *Plant Soil*, *106*:263.
51. H. M. Murtadha, J. W. Maranville, and R. B. Clark (1988). *Agron. J.*, *80*:125.
52. A. A. Mohamed, I. H. El-Sokkary, and T. C. Tucker (1987). *J. Plant Nutr.*, *10*:699.
53. F. M. Hons and K. D. Aljoe (1985). *Commun. Soil Sci. Plant Anal.*, *16*:349.
54. H. W. Scherer, J. E. Leggett, J. L. Sims, and P. Krasaesindhu (1987). *J. Plant Nutr.*, *10*:67.
55. J. S. Bailey (1989). *J. Plant Nutr.*, *12*:1019.
56. T. Yamauchi, T. Hara, and Y. Sonoda (1986). *Plant Soil*, *93*:223.
57. B. J. Scott and A. D. Robson (1990). *Aust. J. Agric. Res.*, *41*:499.
58. Z. Rengel and D. L. Robinson (1990). *Soil Sci. Soc. Am. J.*, *54*:785.
59. S. H. Duke and H. M. Reisenauer (1986). *Sulfur in Agriculture* (M. A. Tabatabai, ed.), Am. Soc. Agron. Inc., Madison, WI, p. 123.
60. J. F. Thompson, I. K. Smith, and J. T. Madison (1986). *Sulfur in Agriculture* (M. A. Tabatabai, ed.), Am. Soc. Agron. Inc., Madison, WI, p. 57.
61. P. S. Sukhija, V. Randhawa, K. S. Dhillon, and S. K. Munshi (1987). *Plant Soil*, *103*:261.
62. M. J. Archer (1987). *Aust. J. Plant Physiol.*, *14*:239.

63. A. Bottacin, K. Smith, B. J. Miflin, P. R. Shewry, and S. W. J. Bright (1985). *J. Exp. Bot.*, *36*:140.
64. S. L. Castle and P. J. Randall (1987). *Aust. J. Plant Physiol.*, *14*:503.
65. J. H. Skerritt, P. W. Lew, and S. L. Castle (1988). *J. Exp. Bot.*, *39*:723.
66. T. M. Hago and M. A. Salama (1987). *Exp. Agric.*, *23*:93.
67. K. Ohki (1987). *J. Plant Nutr.*, *10*:1583.
68. T. Horiguchi (1987). *Soil Sci. Plant Nutr.*, *33*:595.
69. M. Ferrandon and A. R. Chamel (1988). *J. Plant Nutr.*, *11*:247.
70. N. C. Uren and H. M. Reisenauer (1988). *Advances in Plant Nutrition* (B. Tinker and A. Lauchli, ed.), Praeger, New York, p. 79.
71. E. O. Leidi, M. Gomez, and L. A. del Rio (1987). *J. Plant Nutr.*, *10*:261.
72. D. O. Wilson, F. C. Boswell, K. Ohki, M. B. Parker, L. M. Shuman, and M. D. Jellum (1982). *Crop Sci.*, *22*:948.
73. K. Ohki, F. C. Boswell, M. B. Parker, L. M. Shuman, and D. O. Wilson (1979). *Agron. J.*, *71*:233.
74. G. Sandman and P. Boger (1983). *Inorganic Plant Nutrition. Encyclopedia of Plant Physiology*, Vol. 15 (A. Lauchli and R. L. Bieleski, eds.), Springer-Verlag, New York, p. 563.
75. R. E. Wilkinson and K. Ohki (1988). *Plant Physiol.*, *87*:841.
76. R. Suresh, C. D. Foy, and J. R. Weidner (1987). *J. Plant Nutr.*, *10*:749.
77. T. Zaharieva, D. Kasabov, and V. Romheld (1988). *J. Plant Nutr.*, *11*:1015.
78. C. Hecht-Buchholz, C. A. Jorns, and P. Keil (1987). *J. Plant Nutr.*, *10*:1103.
79. D. J. Reuter, A. D. Robson, J. F. Loneragan, and D. J. Tranthim-Fryer (1981). *Aust. J. Agric. Res.*, *32*:283.
80. E. Delhaize, M. J. Dilworth, and J. Webb (1986). *Plant Physiol.*, *82*:1126.
81. A. Casimiro (1987). *Plant Response to Stress. Functional Analysis in Mediterranean Ecosystems* (J. D. Tenhunen et al., eds.), Springer-Verlag, Berlin, p. 459.
82. G. E. Varvel (1984). *J. Plant Nutr.*, *11*:1623.
83. A. W. Jewell, B. G. Murray, and B. J. Alloway (1988). *Plant Cell Environ.*, *11*:273.
84. V. Romheld and H. Marschner (1986). *Adv. Plant Nutr.*, *2*:155.
85. J. C. Brown and V. D. Jolley (1988). *J. Plant Nutr.*, *11*:1077.
86. J. H. Bennett, N. J. Chatterton, and P. A. Harrison (1988). *J. Plant Nutr.*, *11*:1099.
87. A. Wallace (1990). *Hort. Sci.*, *25*:838.
88. F. Awad, V. Romheld, and H. Marschner (1988). *J. Plant Nutr.*, *11*:701.
89. D. B. McKenzie, L. R. Hossner, and R. J. Newton (1987). *J. Plant Nutr.*, *10*:15.
90. R. L. Chaney and P. F. Bell (1987). *J. Plant Nutr.*, *10*:963.
91. D. B. McKenzie, L. R. Hossner, and R. J. Newton (1985). *J. Plant Nutr.*, *8*:847.
92. S. M. Dofing, E. J. Penas, and J. W. Maranville (1989). *J. Plant Nutr.*, *12*:797.

93. G. S. Smith, I. S. Cornforth, and H. V. Henderson (1984). *New Pytologist*, *97*:543.
94. E. O. Leidi, M. Gomez, and M. de la Guardia (1986). *J. Plant Nutr.*, *9*:1239.
95. L. Erdei, A. Berczi, and A. Szabo-Nagy (1989). *Biochem. Physiol. Pflanz.*, *1984*:235.
96. G. W. Miller and J. C. Pushnik (1983). *Utah Sci.—Utah Agric. Exp. Stn.*, *44*:98.
97. G. W. Miller, J. C. Pushnik, and G. W. Welkie (1984). *J. Plant Nutr.*, *7*:1.
98. J. C. Pushnik, G. W. Miller, V. D. Jolly, J. C. Brown, T. D. Davis, and A. N. Barnes (1987). *J. Plant Nutr.*, *10*:1183.
99. J. N. Nishio, J. Abadia, and N. Terry (1985). *Plant Physiol.*, *78*:296.
100. J. N. Nishio, S. E. Taylor, and N. Terry (1985). *Plant Physiol.*, *77*:705.
101. A. S. Ismail, A. A. Orabi, and M. A. Mostafa (1985). *Plant Soil*, *83*:323.
102. D. P. Pandey (1989). *J. Plant Nutr.*, *12*:375.
103. A. Rahimi and A. Schropp (1984). *Zeit. Pflanz. Boden.*, *147*:572.
104. K. Ohki (1976). *Physiol. Plant.*, *38*:300.
105. H. Takaki and S. Arita (1986). *Soil Sci. Plant Nutr.*, *32*:433.
106. I. Cakmak and H. Marschner (1988). *J. Plant Physiol.*, *132*:356.
107. I. Cakmak and H. Marschner (1988). *Physiol. Plant.*, *73*:182.
108. I. Cakmak and H. Marschner (1988). *J. Exp. Bot.*, *39*:1449.
109. J. F. Loneragan, G. J. Kirk, and M. J. Webb (1987). *J. Plant Nutr.*, *10*:1247.
110. A. A. Orabi, T. El-Kobbia, and A. I. Fathi (1985). *Plant Soil*, *83*:317.
111. H. Marschner and I. Cakmak (1986). *Physiol. Plant.*, *68*:491.
112. M. J. Webb and J. F. Loneragan (1988). *Soil Sci. Soc. Am. J.*, *52*:1676.
113. M. A. Farah and M. F. Soliman (1986). *Agrochimie*, *30*:419.
114. W. E. Sabbe and L. J. Zelinski (1990). *Soil Testing and Plant Analysis*, 3rd ed. (R. L. Westerman, ed.), Soil Science Society of America, Madison, WI, p. 484.
115. M. M. Riley (1987). *J. Plant Nutr.*, *10*:2109.
116. B. J. Shelp and V. I. Shattuck (1987). *J. Plant Nutr.*, *10*:143.
117. A. Mozafar (1987). *J. Plant Nutr.*, *10*:319.
118. G. J. Kirk and J. F. Loneragan (1988). *Agron. J.*, *80*:758.
119. H. Goldbach and A. Amberger (1986). *J. Plant Physiol.*, *123*:263.
120. H. Goldbach (1985). *J. Plant Physiol.*, *118*:431.
121. P. D. Petracek and C. E. Sams (1987). *J. Plant Nutr.*, *10*:2095.
122. W. M. Dugger (1983). *Inorganic Plant Nutrition. Encyclopedia of Plant Physiology*, Vol. 15 (A. Lauchli and R. L. Bieleski, eds.), Springer-Verlag, New York, p. 628.
123. G. A. El-Gharably and W. Bussler (1985). *Zeit. Pflanzen. Boden.*, *148*:681.
124. M. K. Schon and D. G. Blevins (1987). *Plant Physiol.*, *84*:969.
125. J. A. Raven, A. A. Franco, E. L. de Jesus, and J. Jacob-Neto (1990). *New Phytologist*, *114*:369.

126. P. M. Safaraliev, N. P. L'vov, and V. L. Kretovich (1987). *Soviet Plant Physiol.*, *34*:250.

127. R. Vunkova-Radeva, J. Schiemann, R. R. Mendel, G. Salcheva, and D. Georgieva (1988). *Plant Physiol.*, 87:533.

128. Y. Oji, Y. Ryoma, N. Wakiuchi, and S. Okamoto (1987). *Plant Physiol.*, *83*:472.

6

Isoprenoid Biosynthesis

Robert E. Wilkinson

University of Georgia
Griffin, Georgia

INTRODUCTION

The isoprenoid biosynthetic system (Fig. 1) produces a wide variety of compounds requisite to metabolic and physiological functions in autotrophic plants. These include, but are not limited to, the phytol chain of chlorophyll, carotenes, sterols, quinones, xanthophylls, abscissic acid (ABA), and gibberellic acid (GA) (Fig. 1). Inhibition or diminution in production of these products induces specific phenological symptoms. Corn (*Zea mays* L.) seedlings that are white when grown in full sunlight are sometimes the result of a blockage in carotene synthesis. These mutants are pale green when grown in very diffuse light. Other mutants are pale yellow, which results from a chlorophyll biosynthesis inhibition without a carotene synthesis block. Similar mutants have been utilized in GA studies to establish the biosynthesis chain. Axiomatically, herbicide or any other stress that influences general or specific reactions results in specific symptoms. Then biochemical evaluation of specific steps can lead to a defined mechanism of action.

The basic enzymology, compartmentation, and activity of these systems are subject to intensive experimentation. Excellent reviews and texts on explicit biochemistry are available (1–5). Specificity of cation requirements were published (refs. 5,6, and references therein) and revealed the similarities of action and compartmentation as an explanation for the

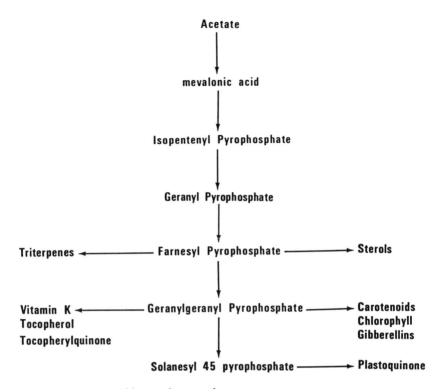

Figure 1   Isoprenoid biosynthetic pathway.

differences of biosynthetic products from what were previously presumed to be relatively analogous enzyme systems.

ISOPRENOID CARBON CHAINS

The basic isoprenoid carbon chain, as defined herein, runs from acetate to geranylgeranyl pyrophosphate (GGPP) (Fig. 1). Acetate is highly water-soluble whereas geranylgeraniol is highly lipophilic. Thus, the addition of pyrophosphate to the carbon chains alters their hydrophilic/lipophilic balance and makes them available in the milieus of the enzymes (1).

Several herbicides have been shown to influence Ac-CoA synthesis (7,8). This would induce overall changes of chlorophyll and carotenes, and possibly result in stunted plants. Manganese has been shown to be required for mevalonic kinase (9), phytoene synthetase (6,10), and *ent*-kaurene synthase (11). Additionally, excess Mn^{2+} concentrations are deleterious to all three of these enzymatic reactions, so that one form of response to acid soil stress is decreased chlorophyll, β-carotene, and GA syntheses resulting in stunted plants. The synthesis of GGPP (C_{20}) is accomplished by successive additions of isopentenyl pyrophosphate (IPP) (C_5). These additional reactions had been accepted as a repetitive addition of C_5 IPP units until a C_{20} unit was attained (Fig. 1). If such were the case, an excess Mn^{2+} concentration influence would be a general overall decrease in reactivity with accumulation at the C_{20} terminal product (GGPP). But in five sorghum cultivars ranging in acid soil stress sensitivity/tolerance (see acid soil stress chapter) at least four different patterns of ^{14}C incorporation and accumulation from $[^{14}C]IPP$ were found (Figs. 2–6). First, when based on mg protein/ml in the reaction mixture, the total quantity of $[^{14}C]IPP$ incorporated showed differences in response to Mn^{2+} concentration (Figs. 2–6). Maximum $[^{14}C]IPP$ incorporation was dependent on Mn^{2+} concentration at 1 μM Mn^{2+}

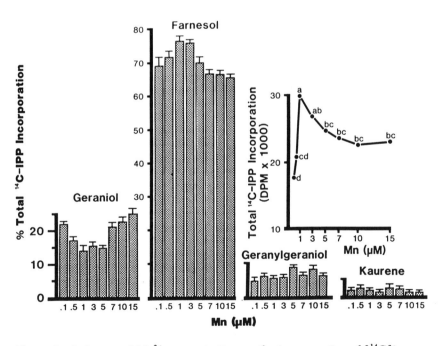

Figure 2   Influence of Mn^{2+} concentration on the incorporation of $[^{14}C]$isopentenyl pyrophosphate into isoprenoid intermediates in in vitro cell-free enzyme system from etiolated coleoptiles of Funk G522DR sorghum. Each treatment contained 10 μM Mg^{2+}. Values on a line followed by the same letter are not significantly different at the 5% level.

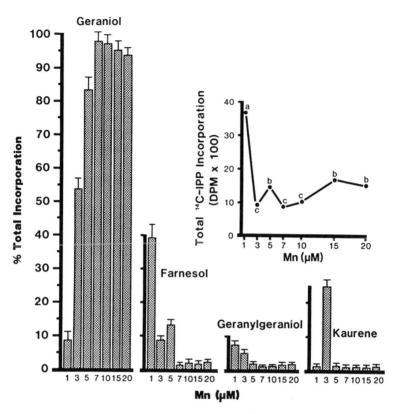

Figure 3 Mn^{2+} concentration influence on the incorporation of $[^{14}C]$isopentenyl pyrophosphate into isoprenoid intermediates in in vitro cell-free enzyme systems that etiolated coleoptiles of TAM428 sorghum. Each treatment contained $10 \mu M$ Mg^{2+}. Points on a line followed by the same letter are not significantly different at the 5% level.

(Funk G522DR– and TAM 428–sensitive) (Figs. 2 and 3, respectively); maximum incorporation of $[^{14}C]IPP$ was at 10 μM Mn^{2+} in SC599 (intermediate) and SC283 (tolerant) (Figs. 4 and 5, respectively); and $[^{14}C]IPP$ incorporation was maximized at 15 μM Mn^{2+} in SC574 (tolerant) (Fig. 6).

Second, in some sorghum cultivars Mn^{2+} concentration influenced the quantity of the intermediate isoprenoid products accumulated. Funk G522DR (Fig. 2) shows a "normal" Mn^{2+} concentration influence. Farnesyl pyrophosphate (FPP) (C_{15}) accumulated and the concentrations of the other intermediate isoprenoid products were not greatly changed by Mn^{2+} concentration. However, TAM 428 showed a massive accumula-

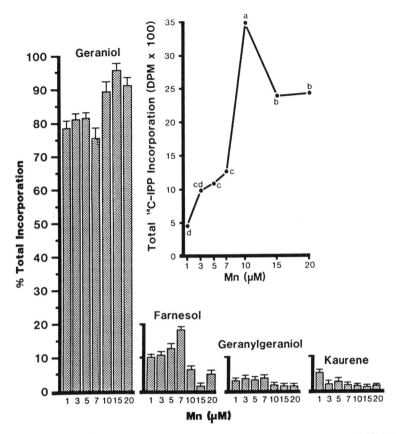

Figure 4 Mn^{2+} concentration influence on the incorporation of [^{14}C]isopentenyl pyrophosphate into isoprenoid intermediates in in vitro cell-free enzyme systems that etiolated coleoptiles of SC599 sorghum. Each treatment contained $10\,\mu M$ Mg^{2+}. Points on a line followed by the same letter are not significantly different at the 5% level.

tion of geranyl pyrophosphate (GPP) (C_{10}) (Fig. 3) that was influenced by Mn^{2+} concentration. In SC599, GPP accumulated without a massive influence of Mn^{2+} concentration (Fig. 4), whereas in SC283, GPP accumulated as the Mn^{2+} concentration increased and the FPP concentration decreased (Fig. 5). And, in SC574 (Fig. 6) GPP and FPP both accumulated while GGPP (C_{20}) decreased as the Mn^{2+} concentration increased. Thus, in sorghum there is more than a single enzyme in this addition sequence from C_{10} to C_{20}. This is substantiated by recent publication (12) of two different enzymes (i.e., IPP isomerase and GGPP synthase) in this biosynthetic system wherein Mg^{2+} and Mn^{2+} (2:1) were added for

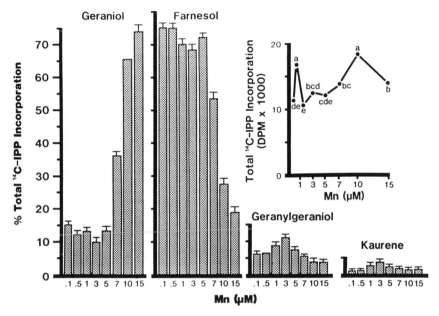

Figure 5   Influence of Mn^{2+} concentration (+ 10 μM Mg^{2+}) on incorporation of [^{14}C]isopentenyl pyrophosphate into isoprenoid intermediates by cell-free enzyme systems from etiolated coleoptiles of SC574 sorghum. Points on a line followed by the same letter are not significantly different at the 5% level.

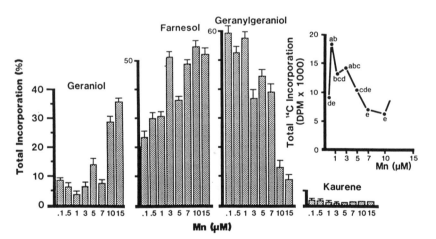

Figure 6   Influence of Mn^{2+} concentration (+ 10 μM Mg^{2+}) on incorporation of ^{14}C-isopentenyl pyrophosphate into isoprenoid intermediates by cell-free enzyme systems from etiolated coleoptiles of SC283 sorghum. Points on a line followed by the same letter are not significantly different at the 5% level.

activity. Consequently, several reactions in sorghum are highly susceptible to control by environment and/or herbicides.

STEROLS

Condensation of two C_{15}, i.e., FPP (Fig. 7), forms squalene, which is a
precursor to plant sterols (1). Sterols and steryl esters are requisite membrane components that are present in small proportions. Their exact
function has not been defined (1).

 Plant sterols stabilize phospholipid membranes. Stretched or perturbed membranes have been implicated in ion retention (2), and abscissic acid (ABA) perturbs the basic membrane so that permeability is
greatly increased (13–18). Plant sterols have been shown to inhibit ABA-
induced perturbations in phospholipid layers (19). Thus, at least an indication of the activity of both sterols and ABA have been reported in
membrane stability. Since membranes are the ubiquitous sequestering
systems that segregate enzymes into several separate functional units
(i.e., chloroplasts, mitochondria, microsomes, tonoplast, cytosol) as well
as separating and controlling the responses of the individual cells from
the contiguous ecosystem, maintenance of the function of those membranes is of prime importance to the existence and continued function of
the plant as a whole. Addition of this concept to those of differential
absorption serves as a clear pattern to explain the massive influence of
sterol synthesis inhibitors on the growth of plant pathogens which exist in
such restricted and protected environments. Several of these pathogens
have been shown to lack or have limited sterol synthesis capacity and
must obtain sterols from their host (20–23). In other fungi, inhibition of
sterol synthesis results in lack of growth (24,25).

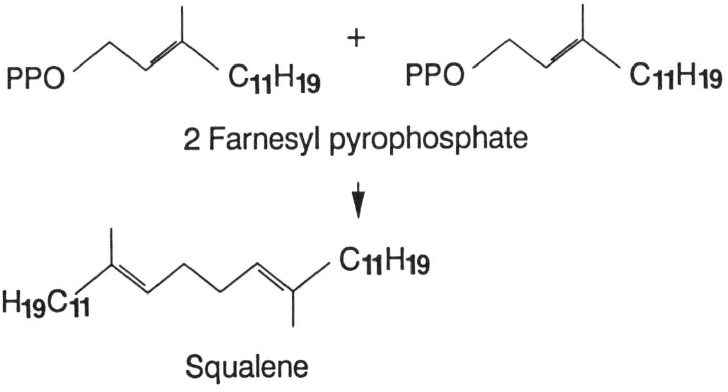

Figure 7 Condensation of two farnesyl pyrophosphate molecules to squalene.

β-CAROTENE SYNTHESIS

Production of β-carotene from GGPP (Fig. 8) involves condensation, dehydrogenation, and cyclization reactions. Phytoene synthesis is believed to be very similar to squalene synthesis except that the former is $2C_{20} \rightarrow C_{40}$ while the latter is $2C_{15} \rightarrow C_{30}$ (1). Recently, compartmentation and cofactor requirements have explained the differences in activity of these two biosynthetic systems (6). Squalene synthetase is located in the endoplasmic reticulum and utilizes Mg^{2+} as a cofactor (5 and references therein). Phytoene synthetase is located in chloroplasts and requires Mn^{2+} as a cofactor (6). Also, there appears to be differences in biosynthetic activity between leukoplasts, chromoplasts, and chloroplasts at different stages of development (5). Different herbicides have been shown to be active at each of these steps (26,27). When norflurazon [4-chloro-5(methylamino)-2-(3-(trifluoromethyl)phenyl)-3(2H)-pyridazinone]–treated plants are grown in full sunlight, the final result is albino plants caused by the photoxidation of chlorophyll when the carotenes are not present to function as free electron scavengers (28,29). Thus, norflurazon ($>0.2~\mu M$) inhibition of phytoene dehydrogenation was correlated with norflurazon concentration. At lower norflurazon ($<0.2~\mu M$, or $\leqslant 64$ ppbw) concentrations, GGPP accumulated (Figs. 8 and 9) (30). GGPP was also accumulated when EPTC [S-ethyldipropylthiocarbamate] in-

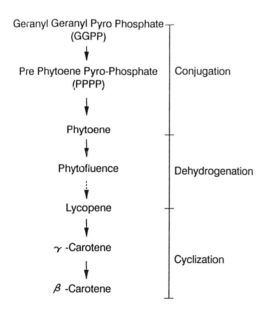

Figure 8   Conversion of geranylgeranyl pyrophosphate to β-carotene.

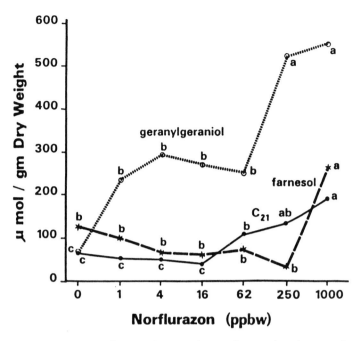

Figure 9 Accumulation of geranylgeranyl pyrophosphate at low norflurazon concentrations. (Ref. 33.)

hibited *ent*-kaurene synthetase (Fig. 10) (31). When applied concomitantly, the extra GGPP produced by EPTC treatment partially reversed the low concentration of norflurazon (< 0.2 μM) inhibition of phytoene synthetase and EPTC reversed the activity of low norflurazon concentrations (32). Since phytoene synthetase was reversed by excess substrate, but phytoene dehydrogenation has never been shown to be reversed by excess substrate and phytoene accumulates under norflurazon (> 0.2 μM) applications, the reactivities of these two enzyme systems are quite different.

Phyotene dehydrogenation can be partially reversed by concomitant applications of norflurazon-type carbon skeletons of no reactivity (Figs. 11 and 12) (33). This takes the form of competition of nonactive and active inhibitors for a site of activity and was shown to occur for two other inactive norflurazon-type compounds as well (33).

GIBBERELLIC ACID PRECURSOR

GGPP is cyclized to *ent*-kaurene (Fig. 13) (34,35). After GA_{12ald} the carbon skeleton is committed to GA biosynthesis (34,35). Prior to the syn-

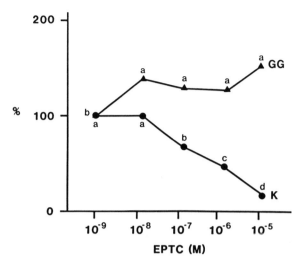

Figure 10   Influence of S-ethyldipropylthiocarbamate (EPTC) on *ent*-kaurene synthesis.

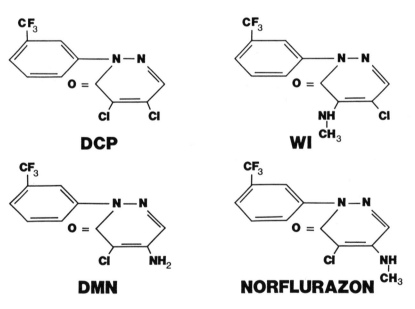

Figure 11   Structures of norflurazon, desmethyl norflurazon (DMN) wrong isomer (WI), and dichloropyrimidozinone (DCP). (From Ref. 33.)

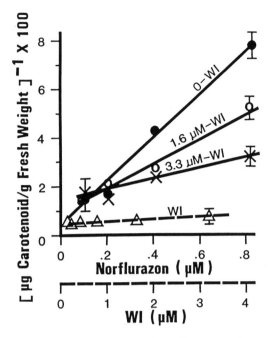

Figure 12 Dixon plot of the influence of the wrong isomer (WI) on the caro-tenogenic inhibition of norflurazon in wheat. Top abscissa: norflurazon (0.1, 0.2, 0.4, 0.8 μM) + WI (■ = 0; ● = 1.6; × = 3.3 μM). Bottom abscissa: influence of WI (△) without norflurazon.

thesis of GA_{12} aldehyde, the first GA produced in many biological systems, *ent*-kaurene is oxidized (Fig. 13) (34,35). These reaction steps have been shown to be inhibited by several chemicals with accompanying accumulation of the substrate and depletion of subsequent products (36). Since each of these, primarily grass herbicides, induce a cessation of coleoptile growth, the phenology and biochemistry of herbicide response are similar. However, as yet, very little information has been published on any possible reactivities of these herbicides on the metabolism of GA_{12}-aldehyde to GA_1 (Fig. 14).

Corn and sorghum have been shown to have the same GA biosynthetic pathway (37–40). Also, photoperiodically controlled steps have been documented (41). Thus environmental parameter influences can induce additional variability. However, the morphological responses of sorghum to metolachlor [2-chloro-*N*-(2-ethyl-6-methylphenyl)-*N*-(2-methoxy-1-methylethyl)acetomide], a selective herbicide applied to the soil prior to the emergence of the sorghum, and acid soil stress are identi-

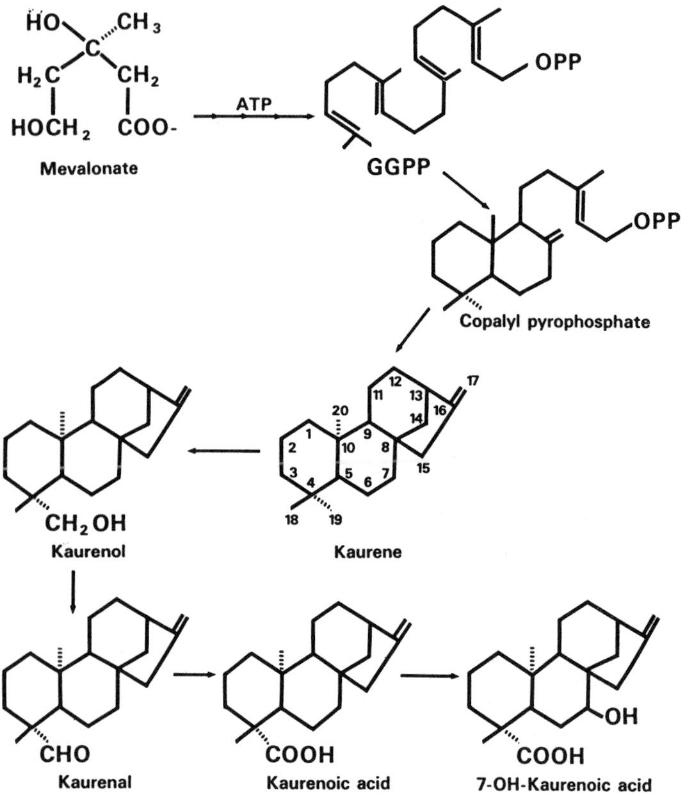

Figure 13 Conversion of geranylgeranyl pyrophosphate (GGPP) to 7-OH-kaurenoic acid.

cal. The coleoptiles are static, primary leaves often do not rupture the coleoptile or remain furled, and growth is inhibited (42). Metolachlor has been shown to inhibit GA precursor synthesis (36). Other chemical antidotes (safeners) are applied to the seed and prevent the action of metolachlor on the sorghum (36). These herbicide safeners also alleviate the response of sorghum seedlings to acid soil stress (43). This similarity of response indicated a common reaction mechanism. Data in Figures 2–6 show major Mn^{2+} influences in isoprenoid biosynthesis and these are corroborated by previous reports of excess Mn^{2+} activity in GA precursor biosynthesis (9,11,44).

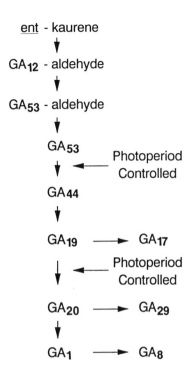

Figure 14   Conversion of *ent*-karuene to GA₁ in corn (*Zea mays* L.).

QUINONES

Several quinone systems are present in plants and in their respective sites of activity all are requisite (45–47). Chloroplasts utilize quinone systems for electron transfer (46). Mitochondria use a separate quinone system for such transfer (47). All of these components are produced by the isoprenoid biosynthetic system. When EPTC was applied to wheat (*Triticum aestivum* L. em Thell) increased quantities of various quinones were developed (48). Two factors were possible. Either quinone, chlorophyll, and carotene syntheses were continued at an unaltered rate while fresh weight was decreased, or there was an increased synthesis of these components (48). Demonstration of the GGPP accumulation with EPTC treatment showed a pathway wherein accumulated quinone, chlorophyll, and carotene precursors could result in increased biosynthesis of these products.

Abscissic acid (ABA) has been suggested to be produced by violax-anathin metabolism (1); and ABA is a plant growth regulator (PGR) whose action has generated multiple reviews and hypotheses on modes of action (3,4). This wide variability in responses only shows that this essential group of compounds has multiple activities that depend on (or are triggered by) physiological conditions. Obviously, differences in growth stage, environment, and species will have a major influence on specific reactivities.

SUMMARY

The isoprenoid biosynthetic system produces multiple components that are requisite for the growth of autotrophic plants. Interruption of the production of any component will have major implications to the growth of the plant. Economic applications have taken the form of pesticides which inhibit the synthesis (or activity) of specific isoprenoid components. However, imbalances in mineral nutrition are also a location of response. Only a few examples have been listed. This is an extremely important biosynthetic system whose details of compartmentation, enzyme activities, and cofactors are a quest of continued research by a large number of experts and highly proficient researchers. This chapter has attempted to document some of the many interactions of environmental parameters in the biosynthesis of this group of essential compounds. Complete examination of all known interactions would require multivolume compendia and could never be completed.

REFERENCES

1. J. W. Porter and S. L. Spurgeon (1981). *Biosynthesis of Isoprenoid Compounds*, Vols. 1 and 2, John Wiley and Sons, New York.
2. E. Zeiger (1983). *Annu. Rev. Plant Physiol.*, 34:441.
3. D. C. Walton (1980). *Annu. Rev. Plant Physiol.*, 31:453.
4. J. A. D. Zeevaart and R. C. Creelman (1988). *Annu. Rev. Plant Physiol.*, 39:439.
5. H. Kleinig (1989). *Annu. Rev. Plant Physiol.*, 40:39.
6. O. Dogbo, A. Laferrier, A. D'Harlingue, and B. Camara (1988). *Proc. Natl. Acad. Sci. USA*, 85:7054.
7. R. E. Wilkinson and T. H. Oswald (1987). *Pest. Biochem. Physiol.*, 28:38.
8. R. E. Wilkinson (1988). *Pest. Biochem. Physiol.*, 32:25.
9. T. T. Tchen (1958). *J. Biol. Chem.*, 233:1100.
10. B. Maudinas, M. L. Buckholtz, C. Papastephanou, S. S. Katiyar, A. V. Briedis, and J. W. Porter (1977). *Arch. Biochem. Biophys.*, 180:354.
11. J. Knotz, R. C. Coolbaugh, and C. A. West (1977). *Plant Physiol.*, 60:81.
12. O. Dogbo and B. Camara (1987). *Biochim. Biophys. Acta*, 920:140.

13. W. Stillwell, B. Brengle, P. Hester, and S. R. Wassall (1988). *Biochemistry*, 28:2798.
14. S. R. Wassall, P. Hester, and W. Stillwell (1985). *Biochim. Biophys. Acta*, 815:519.
15. W. Stillwell, B. Brengle, D. Belcher, and S. R. Wassall (1987). *Phytochemistry*, 26:3145.
16. W. Stillwell and P. Hester (1984). *Z. Pflanzenphysiol.*, 114:65.
17. W. Stillwell and P. Hester (1984). *Phytochemistry*, 23:2187.
18. C. Harkers, W. Hartung, and W. Gunter (1986). *J. Plant Physiol.*, 122:385.
19. W. Stillwell, Y. F. Cheng, and S. R. Wassall (1990). *Biochim. Biophys. Acta*, 1024:345.
20. T. A. Chen and R. E. Davis (1979). *The Mycoplasmas*, Vol. 3 (R. F. Whitcomb and J. G. Tully, eds.), Academic Press, New York, p. 65.
21. C. J. Chang and T. A. Chen (1982). *Science*, 215:1121.
22. C. J. Chang and T. A. Chen (1983). *J. Bacteriol.*, 153:452.
23. C. J. Chang and R. E. Wilkinson (1984). *Structure, Function, and Metabolism of Plant Lipids* (P. A. Siegenthaler and W. Eichersberger, eds.), Elsevier, Amsterdam, p. 357.
24. M. Taton, P. Benveniste, and A. Rahier (1987). *Pest. Sci.*, 21:269.
25. R. S. Burden, G. A. Carter, T. Clark, D. T. Cooke, S. J. Croker, A. H. B. Deas, P. Hedden, C. S. James, and J. R. Lenton (1987). *Pest. Sci.*, 21:253.
26. S. M. Ridley (1981). *Carotenoid Chemistry and Biochemistry* (G. Britton and T. W. Goodwin, eds.), Pergamon Press, New York, p. 539.
27. G. Sandmann, I. E. Clarke, P. M. Bramley, and P. Böger (1984). *Z. Naturforsch. C: Biosci.*, 39:443.
28. J. S. Schiff, F. X. Cunningham, Jr., and M. S. Green (1981). *Carotenoid Chemistry and Biochemistry* (G. Britton and T. W. Goodwin, eds.), Pergamon Press, New York, p. 329.
29. N. I. Krinsky (1966). *Biochemistry of Chloroplasts* (T. W. Goodwin, ed.), Academic Press, New York, p. 423.
30. R. E. Wilkinson (1985). *Pest. Biochem. Physiol.*, 23:370.
31. R. E. Wilkinson (1983). *Pest. Biochem. Physiol.*, 19:321.
32. R. E. Wilkinson (1989). *Pest. Biochem. Physiol.*, 33:257.
33. R. E. Wilkinson (1987). *Pest. Biochem. Physiol.*, 28:381.
34. J. E. Graebe, D. T. Dennis, C. D. Upper, and C. A. West (1965). *J. Biol. Chem.*, 240:1847.
35. G. W. M. Barendse (1975). *Gibberellins and Plant Growth* (H. N. Krishnamoorthy, ed.), John Wiley and Sons, New York, p. 65.
36. R. E. Wilkinson (1989). *Crop Safeners for Herbicides: Development, Uses, and Mechanism of Action* (K. K. Hatzios and R. E. Hoagland, eds.), Academic Press, New York, p. 221.
37. B. O. Phinney and C. Spray (1982). *Pesticide Chemistry: Human Welfare and Environment*, Vol. 2 (J. Minamoto and P. C. Kearney, eds.), Pergamon Press, New York, p. 81.
38. C. Spray, B. O. Phinney, P. Gaskin, S. J. Gilmour, and J. MacMillan (1984). *Planta*, 160:464.

39. S. B. Rood, D. M. Bruns, and S. J. Smienk (1988). *Can. J. Bot.*, *66*:1101.
40. S. B. Rood, K. M. Larson, L. N. Mander, H. Abe, and R. P. Pharis (1986). *Plant Physiol.*, *82*:330.
41. S. J. Gilmour, J. A. D. Zeevaart, L. Schwenen, and J. E. Graebe (1986). *Plant Physiol.*, *82*:190.
42. R. E. Wilkinson (1981). *Pestic. Biochem. Physiol.*, *16*:63.
43. R. E. Wilkinson and R. R. Duncan (1990). *Abstracts: Int. Conf. on Herbicide Safeners: Chemistry, Mode of Action, and Applications*, Budapest, August 12–15, p. 32.
44. R. E. Wilkinson and K. Ohki (1988). *Plant Physiol.*, 87:841.
45. G. E. W. Wolstenholme and C. M. O'Connor (1961). *Quinones in Electron Transport*, J. & A. Churchill Ltd., London.
46. A. Trebst and M. Avron (1977). *Photosynthesis I. Encyclopedia of Plant Physiology*, New Series, Vol. 5, Springer-Verlag, New York.
47. R. Donce and D. A. Day (1985). *Higher Plant Cell Respiration. Encyclopedia of Plant Physiology*, New Series, Vol. 18, Springer-Verlag, New York.
48. R. E. Wilkinson (1978). *Pest. Biochem. Physiol.*, *8*:208.

7

Whole-Plant Response to Salinity

Michael C. Shannon, Catherine M. Grieve, and Leland E. Francois

U.S. Salinity Laboratory, ARS, USDA
Riverside, California

INTRODUCTION

Our world is dissolving around us. As a consequence of rain, the earth's
mantle is slowly being solubilized and washed into the oceans. Salts of
sodium, calcium, magnesium, chloride, sulfate, carbonate, and
numerous other elements are formed as water flows from soils into rivers,
lakes, and, finally, the sea. The water and dissolved salts are essential to
plant growth, but water reuse and high evaporation rates in arid or
semiarid regions concentrate the salt as the general phenomenon of salination occurs. Throughout the range of saline ecosystems, from mountain
to sea, diverse plant species abound, each the result of an accumulation
of inherent adaptive characters and mechanisms.

Salinity is one of the most severe and insidious limiters of crop
growth because it is intricately meshed with water and nutrient uptake
into the plant and because its effects at low and moderate concentrations
are so ubiquitous. Plant growth is adversely affected when specific ion
concentrations exceed their thresholds and become toxic. Salts may also
reduce plant growth by significantly reducing the water potential or by
interfering with nutrient uptake. The concentrations at which these
effects take place differ with the genetic capacity of the species, growth
stage, environmental interactions, and ion species. Native plant species,
from mangroves to salt bush, have adapted to an incredibly wide range

199

of saline environments. Plants inhabit oceans, saline marshes, semiarid deserts, and saline seeps. They have adapted to high sodium environments in oceans and alkaline flats, to habitats high in magnesium or potassium, and to areas with high chloride, sulfate, or carbonate. But crop plants with a few exceptions, grow best with fairly low concentrations of salts around their roots. In general, salt concentrations higher than 2600 mg/L ($\approx 4\,dS/m$) can decrease yields of many crops. Salts interfere with nutrient uptake, decrease water potential outside the plant, and may cause ion toxicities upon entrance into plant.

Historically, mismanagement in irrigated agriculture and the salt sensitivity of many crop species have contributed to the decline of both Old and New World civilizations (219 and included references). Despite this legacy, an increasing proportion of the world's food supply is produced in semiarid regions with irrigation. In many countries, drainage water from irrigation systems in addition to effluent from industrial sites contribute to increased river and groundwater pollution at record rates. Our population numbers continue to rise as supplies of high-quality water are diminished.

This chapter deals with the effects of salinity on higher plants, i.e., those with vascular systems. The focus will be on the plant system and its immediate interfaces between the soil-water and the atmosphere. External factors that can be manipulated to limit the quantity or quality of the plant's saline environment will be discussed where appropriate. The salt-coping mechanisms of the plant will be examined from root to shoot in context with the myriad of mechanisms for salt tolerance that have evolved throughout the higher plant kingdom.

THE SALINE SOIL ENVIRONMENT

Plants are most often exposed to salinity through their root systems. The principal components of soil water salinity that confront the root are sodicity and specific ion concentrations, but the degree to which these affect plant growth are determined by genotype, climatic conditions, soil properties, irrigation, and cultural practices (16,205).

Soil Salinity

Salt concentration at the soil–root interface increases dramatically as soil water is depleted through evaporation and transpiration (18). Thus, increased irrigation frequency is beneficial under saline cropping conditions (8). Shorter irrigation intervals minimize the concentrating effect of evapotranspiration on soil salinity (25,123).

Most irrigation waters contain more salts than are removed by the crop, so that continued irrigation without leaching will progressively salinize the root zone. Water in excess of consumptive use (evapotranspiration) must therefore be applied to leach the residual salts. In addition, soils must be sufficiently permeable to allow the extra water needed for leaching to infiltrate in a reasonable time. To achieve the necessary drainage and salinity control, it is usually necessary to grow crops for which evapotranspiration is sufficiently less than attainable infiltration.

Evidence indicates that plants respond primarily to the soil salinity in that part of the root zone with the highest total water potential (25,85). With more frequent irrigations, this zone corresponds primarily to the upper part of the root zone, where soil salinity is influenced by the salinity of the irrigation water. With infrequent irrigations, the zone of maximum water uptake becomes larger as the plant extracts water from increasingly saline solutions at greater depths.

Sodicity

Sodic soils, formerly called alkali soils, contain excess exchangeable sodium, with 15% or more of the cation exchange sites in the soil occupied by Na^+ (241). These soils may be either saline or nonsaline, depending on the concentration of Ca^{2+} and/or Mg^{2+} present in the soil solution. In sodic, nonsaline soils, the high exchangeable Na^+ is balanced by low exchangeable Ca^{2+} and Mg^{2+}. The resultant Ca^{2+} and/or Mg^{2+} deficiencies, rather than Na^+ toxicity, are frequently the cause of poor growth among nonwoody species. In contrast, saline-sodic soils contain higher concentrations of Ca^{2+} and Mg^{2+} and may therefore remain nutritionally adequate. With saline-sodic soils, salinity effects predominate, and the detrimental effects of sodicity on plant nutrition are usually absent. Rana (199) notes the crops adapted to alkali soils are usually tolerant of nonalkaline saline soils, but not the converse.

In addition to the nutritional imbalances encountered in sodic soils, the hydraulic conductivity and permeability of the soil to both water and air are significantly reduced. The deterioration of the soil physical condition caused by the high exchangeable Na^+ content can lead to flooding and waterlogging. The use of gypsum and the importance of Ca^{2+} in relation to sodic soils and their reclamation has been extensively reviewed by Oster (191) and Rengasamy (204). To alleviate the poor permeability of these soils, the electrolyte concentration in the soil water must be increased. This is accomplished by the addition of gypsum ($CaSO_4$), sulfuric acid, or acid-forming compounds to the soil or irrigation water (8). The acid and acid-forming compounds react with the soil lime ($CaCO_3$) to release Ca^{2+} into the soil solution.

Waterlogging

As indicated, waterlogging can result from too much irrigation or from sodicity-generated infiltration problems. Shallow water tables and excessive groundwater discharge also contribute to the worldwide problem of flooded or poorly drained soils. Waterlogged soils can have poor aeration, which may exacerbate the plant's sensitivity to salt stress and greatly increase salt uptake compared with nonwaterlogged conditions (251,253). Depletion of soil oxygen contributes to reduced shoot growth of salt-stressed tomato (*Lycopersicon esculentum*) (5) and maize (*Zea mays*) (65). Energy-dependent ion uptake, i.e., exclusion and extrusion processes in root membranes, is disrupted due to deprivation of the required energy sources from oxidative metabolism.

Inadequate drainage that results in a shallow water table can also increase the salt tolerance of some deep-rooted crops. Plants that can extract water directly from this source respond much differently than expected from the level of salinity in the soil, depending on the quality of the water.

SOLUTE AND WATER MOVEMENT IN ROOTS

Plants have developed anatomical and physiological adaptations that permit their survival and, in the case of halophytes, promote their best growth under saline conditions. The root is the plant organ in direct contact with the saline environment; therefore, its structure and function regulate ion uptake and transport. The root is the primary barrier to solute movement into the plant and as a result the ion concentrations delivered to the shoot are vastly different from those in the external medium.

Morphology and Anatomy

There are two ways for a solute to move from the outside solution into the root. The extracellular (apoplastic) pathway is a free diffusional continuity of the external solution with its properties modified by the fixed negative charges of the cell walls. The intracellular, or symplastic, route may be viewed as the cytoplasm of all the living cells in the plant linked by the plasmadesmata. In the recently matured region of the root, the apoplastic and symplastic systems appear to merge, at least for a short distance, at the endodermis with its heavily suberized and/or lignified Casparian strip. Although solutes must, in principle, pass through the endodermal cells as they move from the root cortex to the stele, there are a number of ways that the integrity of the endodermal barrier may be breached. Lateral roots develop from the pericycle in the stele and the endodermis divides to accommodate their elongation. The ring of cells at the lateral root–endodermis interface lacks Casparian bands. At this junction the apoplast in the cortex and stele may be continuous. Another "leaky" zone may be the root apical meristem where differentiation of the various cell types may not be complete. These areas are believed to be of major importance in the radial transport of Ca^{2+} and Mg^{2+} (171). The quantitative aspects of this apoplastic continuity have not been established (45,146); however, under saline conditions, the contribution to ion uptake may be significant. Undoubtedly additional processes that lower ion concentrations in the shoot are involved, otherwise flaws in the barrier would result in large increases of Na^+ and Cl^- in the leaves (177). Salinity-induced changes in morphology, anatomy, and ultrastructure as well as some of physiological implications of the altered growth patterns have been reviewed by Poljakoff-Mayber (196,197), Maas and Nieman (164), and Kramer (141).

The plant root system is composed of several types of roots. Therefore, it may be inappropriate to address a general response of the root system to salinity stress. In *Raphanus sativus*, the initiation of new laterals was more sensitive to 200 mM NaCl than extension of the tap roots (245). Such a response may be an adaptation whereby a predominately vertical system would develop under highly saline or sodic conditions whereas the roots would spread horizontally under nonsaline conditions (197).

In the roots of some NaCl-stressed plants, meristematic activity, particularly that associated with the vascular tissues, may be suppressed. As a result, the cross-sectional diameter of both the roots and the vascular cylinder may be reduced (*Zea mays* [198]; *Pisum sativum* [215]). In contrast, the stelar diameter of the halophyte, *Suaeda maritima*, increased in response to NaCl-stress (82). Salinity-induced growth inhibition may result from imbalances in phytohormone levels (171). Treatment of *P. sativum* with growth hormones largely offsets such detrimental effects of NaCl-salinity such as decreases in vascular activity, secondary xylem differentiation, and cross-sectional area (215).

Salinity stimulates the development of fully vacuolated epidermal and cortical root cells of both halophytes (e.g., *Suaeda maritima* [109] and *Salicornia europea* [170]), and glycophytes such as *Pisum sativum* (226), *Hordeum vulgare*, and *Phaseolus vulgaris* (179). In root apices of salt-stressed *H. vulgare*, the ground meristem was the most sensitive of the meristematic tissues to early vacuolation (131,132). Vacuolation occurs significantly closer to the root apex under saline conditions. As an adaptive mechanism, this response may regulate the internal concentration of solutes and protect the cytoplasm from toxic levels of ions by storing them in the vacuole. At the same time, early vacuolation may contribute to the plant's ability to survive the challenge of salinity by the inhibition of growth. Cells with a central vacuole have much smaller surface area/volume and cytoplasm/volume ratios than slightly vacuolated cells. These differences in cell geometry would result in slower rates of solute uptake, or of solute synthesis, per unit cell volume and would result in reduced plant growth (177).

Salinity often promotes suberization of the hypodermis and endodermis with the formation of a well-developed Casparian strip in a region closer to the root apex than in the nonsaline roots. This effect has been observed in roots of *Suaeda maritima* (109), *Puccinellia peisonia* (232), and *Citrus* species (248). While early differentiation was not evident in pea (*Pisum sativum*) roots, NaCl stress stimulated lignification of tracheary elements in the conducting tissue (226). Early differentiation of the Casparian strip would provide a barrier that minimizes free ion influx through the apoplast to the root xylem and finally into the transpirational stream.

Cell walls of roots grown in the presence of NaCl often are unevenly thickened and convoluted, e.g., parenchyma cells surrounding the xylem vessels of *Plantago coronopus*. The increased plasmalemmal area that results from the corrugations may enhance ion transport between the vessels and parenchyma and the remainder of the stele and may prevent NaCl-overloading in the root cells (112). However, thickened convolutions together with bladder-shaped hairs also developed on the outer peripheral walls of epidermal and xylem parenchyma cells of *Atriplex hastata*. Rather than a specific response to salinity, this anatomical change proved to be an Na-induced response to Fe deficiency (140,141). In NaCl-stressed pea roots, the thicker, larger cells contained five times as much cell wall substances and twice as much protein per cell than did the controls (224).

Lawton et al. (149) compared the root anatomy of two mangrove species: *Avicennia marina* possesses glands on the leaves that excrete salt, whereas *Bruguiera gymnorrhiza* has no salt glands and appears to be a salt excluder. Differences were apparent in the length and thickness of the root cap, the development and positioning of the Casparian strip, and the differentiation of the vascular tissues. Suberization of the endodermal and hypodermal cells of the salt excluder starts close behind the root tip, and this adaptation may function to partially block the apoplastic pathway.

Water Relations

Water comprises over 80% of the weight of most plant tissues and is required as a physiological solvent, a transport medium for nutrients, an evaporative coolant, and a pressure source to support form and function and drive the growth process. The amount of water inside a plant at any given time is relatively small compared to the quantities that pass from root to shoot throughout its life cycle. Actively growing tomato plants transpire 30–50 times as much water as may be accumulated in fresh weight during their first 6–7 weeks growth (220).

As water passes through the plant in response to its potential gradients, salts are filtered through the biological membranes where they are exposed to a variety of active and passive transport mechanisms. Some ions are accumulated as useful nutrients, whereas others may be excluded or sequestered as wastes. The mechanisms for ion transport vary widely (99), but the final result is that the ionic concentration at which normal cellular physiological processes occur is very limited. Despite this, the concentrations at which specific ions become harmful to plant growth vary over several orders of magnitude. For example, boron is toxic to some plants at soil water concentrations as low as 0.05 mol/m^3, whereas chloride concentrations of 200 mol/m^3 may not harm some plants.

Water extraction by plant roots is generally viewed as a passive process which responds only to differences in plant and soil water potentials (243). In contrast, ion absorption by plant roots cells is considered an active process requiring metabolic energy and generally following Michaelis–Menten enzyme kinetics (36,74). When the volume of water transpired is orders of magnitude greater than the volume of the plant, the relative permeability of roots to water and ions is a significant factor relating to salt tolerance. The theory of irreversible thermodynamics has been used to couple water and solute flow independent of specific mechanisms of ion transport and the exact location of barriers to water or ion transport (52,78,79). The root solution is assumed to be separated from the xylem solution by an effective semipermeable membrane with active ion transport capabilities. The transpiration process is dependent on a hydraulic gradient and its associated flow resistance, and an osmotic component (54,56). The latter component depends on the quality of the semipermeable membrane (reflection coefficient), the diffusive resistance of the membrane (osmotic permeability coefficient), and a strongly temperature-dependent rate of metabolically driven ion uptake (active transport). This model describes a relationship between water flux and hydraulic gradient that is more complicated than a passive water uptake model, but provides an analytic expression that relates salt loading in the shoot to transpiration, root zone salinity, and the hydraulic properties of the root system (55).

Ion Exclusion, Uptake, Transport

The regulation of ion transport involves mineral nutrition, osmotic adjustment, and pH regulation (99). Specific ions associated with high salinities may have either direct or indirect effects on metabolism and may also disrupt mineral uptake and transport. Ion exclusion and accumulation are prominent features of salt-tolerant plants. Some tolerance to salinity has been reported as the result of the enhanced ability of particular genotypes to actively or passively exclude ions. Schubert and Läuchli (212) attributed high salt tolerance in a maize genotype to the ability of the plasmalemma in the root epidermis and cortex to passively exclude Na^+ more effectively than the sensitive genotypes. Ion sequestering capacity in the shoot will be discussed in a later section.

Ionic effects associated with high salinity are both general and specific. General ionic effects are the result of the increased ionic strength of the soil water, but ionic effects also may interfere with the normal mechanisms by which plants take up nutrients by changing physical properties of the cell wall and surface chemistry near cell walls and membranes. Fertility interactions related to salinity composition are a primary concern and have been reviewed extensively (101,192). Interactions between chloride and nitrate, and chloride and phosphate are important in both their nutritional and toxic aspects. The specific major ions associated with high salinity may also damage metabolic processes directly. For instance, high sodium concentrations relative to other salts disrupt root permeability to ions by displacing calcium in the plasma membrane.

The importance of plant Ca status, particularly under sodic conditions, has been the topic of recent reviews (101,148). Several morphological and anatomical changes have been observed in the roots of glycophytes when grown in cultures containing high Na^+ and low Ca^{2+} concentrations. For example, the apices of *Pisum sativum* roots showed successive regions of curvature, constriction, and thickening. Swelling was caused by lateral root initials whose growth had been arrested (224,226). When additional Ca^{2+} was included in the growth media, this anomolous growth did not occur (225). Growth parameters including root length, cell elongation, radial cell expansion, rate of cell production (144 and literature cited therein), and cellular stability of cotton (*Gossypium hirsutum*) roots (97) as well as the growth of maize root epidermal cells (275) depend on the Na^+/Ca^{2+} ratio in the media.

Osmotic Adjustment and Solute Synthesis

The observations that water balance is maintained within the plant and that metabolic activities are sensitive to high salt concentrations have led to the general conclusion that osmotic potentials within the cytoplasm are maintained by "compatible solutes" while inorganic salts accumulate inside the vacuole (30,177,258). For example, oxidative phosphorylation in mitochondria of both *Pisum sativum* and *Suaeda maritma* is sensitive to high sodium (80). Organic solutes which accumulate to high concentrations with little effect on plant metabolism occur in a wide variety of alga and higher plants and include varied forms of polyols, glycerol, mannitol, sucrose, proline, and glycinebetaine (260,261).

Some halophytes are adapted to absorb ions from their root environment to use as an osmoticum. Thus, physiological desiccation is avoided, plant osmotic potential is increased, and water balance is maintained. Generally, absorbed salts are sequestered in vacuoles which comprise the largest volume of most plant cells. Osmotic adjustment within the cytoplasm is maintained by synthesis of compatible solutes, some of which have deleterious effects on metabolism and growth at high concentrations (259). Nonhalophytes, or glycophytes, have adapted a different strategy. They generally exclude salts or sequester salts within roots and stems. Thus, osmotic adjustment is more dependent on compatible solute synthesis. Under salinity stress, the dry weight–to–fresh weight ratio of many glycophytes will increase as a result of osmotic adjustment.

Osmotic adjustment strategies of both halophytes and glycophytes have significant physiological maintenance costs determined by the extent and efficiency of organic synthesis, ion transport, and repair and alterations of cell structures and components (213,229,265). It has been estimated that the energy costs associated with the use of salts as an osmoticum may be a magnitude lower than those solely dependent on the synthesis of organic solutes (201).

Cell Walls and Membranes

Cell walls and membranes provide the key metabolically active interfaces between the plant and its environment. In this capacity membranes play a significant role in separating toxic ion concentrations within the plants from the sensitive metabolic machinery responsible for growth and reproduction. Membranes participate in the ion discrimination process by selective partitioning at the interface of the root cell wall and plasma membrane and at vacuoles, by actively exporting ions back into the soil water, by inhibiting apoplastic or symplastic ion transport, and by providing discrimination in transport processes at the level of tissues and organelles. The cell wall is a source of static change and its influence on the interpretation of the kinetics of potassium uptake was shown to be significant (53). One way in which salinity may disrupt root growth is by decreasing ATP-dependent proton expulsion at the root plasmalemma (228). Thus, cell wall acidification necessary for wall loosening leading to expansion growth might be interrupted or delayed. The observations that membrane vesicles isolated from roots of tomato and *Plantago* grown in saline media have lower ATPase activities is indirect evidence for this (75,104), but is in conflict with other evidence that indicates that root plasma membrane ATPase activity and proton pumping is not diminished by salt stress (114,234). Recently, it has been demonstrated that salinity levels that inhibit maize root growth do not prevent cell wall acidification in the growing zone of the root (275).

Measurements of osmotically-induced backflow in maize root apices indicate that salinity increases hydraulic conductivity (76). It was also found that added Ca mitigated both the effects of NaCl salinity and the observed increase in hydraulic conductivity. Salinity stress also lowers root hydraulic conductivity in *Citrus* but not as much in salt-tolerance genotypes as in sensitive ones (274). Using the coupled water and solute transport model of Dalton et al. (56), Shalhevet et al. (216) found that salinity had no appreciable influence on root permeability for tomato and sunflower (*Helianthus annuus*).

Stein (230) summarized the evidence that indicates that most cell membranes are highly structured bimolecular layers, containing primarily of cholesterol, phospholipid, and protein molecules. The phospholipid molecules, which form the two surfaces of a membrane, are electrically charged. This charge, together with pores physically smaller than the ions, is ultimately responsible for salt-sieving by biological membranes. The sign of the change can vary due to the amophoteric nature of the phospholipid molecules and is thus pH-dependent (185). Membranes will vary in their ability to impede the motion of ions and uncharged molecules through them. The existing theories that describe transport (without convection) can be classified as (a) mechanical-electrical, (b) thermodynamic, and (c) molecular. They are comprehensively discussed by Helfferich (120) and Lakshminarayanaiah (145). From a molecular point of view, Parsegian (194) investigated the energy of an ion crossing a low dialectic membrane and showed that the electrical influence of "channels" and "carriers" were the only significant factors that could lower the energy state of an ion. This energy is given in terms of the dielectric properties of the membrane. Other transport properties that depend on the value of this electrical parameter are outlined in the references above. Some measurements along this line have been made by Chloupek (42,43), Dainty et al. (51), Dvorak et al. (67), and Chernohorska et al. (41).

Ion Pumps

Membrane-bound enzymes, such as ATPases and pyrophosphatases, function as pumps to develop electrochemical proton (H^+) gradients between cellular compartments (75,236). Such gradients result in a protonmotive force that can be used to drive endergonic processes through uniports by means of the generated electrical potential difference or through coupling of the H^+ reflux to the movement of solutes. Such coupling involves H^+ symports or specific antiports of cations, anions, sugars, or amino acids (170). These pumps are important because they are energy-consumptive and can significantly modify the rate of transport of many solutes, which in turn affects composition of cellular compartments and thus the metabolic processes within.

Various forms of ATPases are located in plasma membrane, tonoplast, Golgi, mitochondria, and several other membranes. Isoforms of the enzyme also exist in specific membranes isolated from different tissues. In tomato, seven isoforms of plasma membrane ATPase have been raised with immunological antibodies; three of these isoforms exist in the root (77). Root plasma membrane ATPases have been cytologically located at the epidermis and around the elements of the xylem, to include parenchyma and companion cells (99).

Studies concerning the effect of salinity stress on the plasma membrane H^+-ATPase of either glycophytic or halophytic species are limited. Among citrus genotypes differing in salt tolerance, Douglas and Walker (62) found that, in addition to lipid differences, the best Cl excluder had the highest plasma membrane ATPase activity. The Arrhenius activation energy and thermotropic phase transition temperature of this enzyme were found to be significantly lower in salt-tolerant genotypes than in the enzyme isolated from sensitive genotypes. Braun et al. (34) reported that growing *Atriplex nummularia*, a halophyte, in hydroponic culture containing 400 mol/m^3 NaCl enhanced the H^+-translocating, Mg^{2+}-ATPase activity of sealed microsomal vesicles isolated from roots. Erdei et al. (75) examined Mg^{2+}-ATPase activity of crude microsomal fractions isolated from roots of *Plantago maritima* (a salt-tolerant species) and *P. coronopus* (a moderately salt-sensitive species) grown in the presence of salt. Mg^{2+}-ATPase activity isolated from roots of *P. maritima* decreased when NaCl levels in the hydroponic culture exceeded 150 mol/m^3. In the case of *P. coronopus*, Mg^{2+}-ATPase activity was reduced approximately 30% and 45% when plants were hydroponically cultured in a medium containing 75 and 150 mol/m^3 NaCl, respectively. The H^+-ATPase activity of sealed microsomal vesicles isolated from roots of cotton, a moderately salt-tolerant glycophyte, was not altered in a medium containing 75 mol/m^3 NaCl (114). Crude microsomal fractions were used in most of the above studies; hence whether the observed effects were specific for plasma membrane ATPase is uncertain.

Lipids

Lipids form the bimolecular layer in which membrane proteins are partially imbedded and thus have a direct role in the selective permeability of the membrane either as a direct barrier to water-soluble substances or through their effects on adjacent protein conformation. Various components of lipids have been associated with salt tolerance in a number of species and there is a significant body of evidence to indicate the involvement of sterol. For example, Kuiper (142) found low levels of free sterol in a salt-sensitive grape (*Vitis* sp.) variety and subsequent studies indicated a correlation between sterol level and tolerance in bean (*Phaseolus vulgaris*), barley (*Hordeum vulgare*), and sugarbeet (*Beta vulgaris*) (233). In *Citrus*, the capacity to exclude chloride was directly correlated with an increase in the ratio of more planar to less planar free sterol (61). Hybridization studies between genotypes indicated that an inverse correlation existed between the ratio of phospholipid to free sterol in control roots and the inability of the genotype to exclude chloride uptake into the shoot (63). In cell cultures it has been reported that a salt-tolerant tobacco (*Nicotiana tabacum*) cell line lost its tolerance when it was treated with an inhibitor that affected sterol structure (115).

Not all of the data on the effects of salt and drought on lipid composition are in agreement. Simonds and Orcutt (221) noted that among five genotypes of *Zea mays* there were only minor increases in the stigmasterol-to-sitosterol ratios of roots exposed to stress with polyethylene glycol. But under field conditions, drought decreased total lipid and phospholipid content of maize seedlings without changing either the degree of unsaturation of the fatty acids or the ratios of more planar to less planar sterol (180). The decrease in phospholipid raised the sterol-to-phospholipid ratio threefold. Moreover, Brown and DuPont (37) reported only a slight increase in stigmasterol in barley roots exposed to 100 mol/m^3 NaCl and no changes in the proportions of the individual phospholipid, glycolipid, and sterol fractions. Clearly, there is a need to ascertain the difference between adaptive and degenerative lipid changes in plants.

Other Organelles

Werker et al. (250) observed condensation of chromatin in the nuclei of epidermal and cortical cells of *H. vulgare* grown in the presence of 192 mol/m^3 NaCl. Metaphase chromosomes isolated from meristematic cells of *Vicia faba* root tips showed progressive disorganization as NaCl in the incubation media increased from 300 to 900 mol/m^3 (143). Mitochondrial damage, including the reduction of the numbers and length of the cristae, the presence of both vacuoles and myeline figures, was observed in root epidermal and pericyclic cells of a salt-sensitive ecotype of *Agrostis stolonifera* (223) and maize (269). This response appears to depend on the plant species, age of the tissue, salt type, and degree of salt stress. Salinity (either NaCl or Na_2SO_4) resulted in an increase in the numbers of high cristate mitochondria in maize (135) and barley (250).

Hypertrophy of the Golgi apparatus and the lumen of the endoplasmic reticulum in response to salinity has been noted in a salt-tolerant ecotype of *A. stolonifera* (223) and maize (269).

There appears to be a wide range of tolerance to salinity among the various *Rhizobium* species. While some strains of *Rhizobium meliloti* can survive soil water salinities greater than seawater (46 dS/m), most strains of *R. japonicum* grow poorly at salinities of 12 dS/m (11). Studies comparing various *Rhizobium* species report the salt tolerance of *R. meliloti* > *R. trifolii* > *R. leguminosarum* > *R. japonicum* (11,231). Salt effects on rhizobia appeared to be ion-specific, with chloride ions of Na^+, K^+, and Mg^{2+} being more toxic than corresponding sulfate ions (70,262). In addition, Mg ions inhibited growth at a much lower concentration than did Na^+ or K^+ (31,71).

Most *Rhizobium* species are relatively unaffected by soil salinity levels less than the yield tolerance threshold reported for most leguminous crops. At soil salinities in excess of the threshold, the ability of rhizobia to survive and fix nitrogen may be severely reduced (9,28,242). This is particularly important, since legumes that are already weakened by salinity stress will be deprived of essential nitrogen fertilization as well.

In *Casuarina* species grown in the presence of 0–200 mol/m^3 NaCl, the rate of symbiotic N_2 fixation remained fairly constant. Over a wide range of NaCl concentrations, the root/shoot and the nodule/shoot ratios remained the same as the inoculated but unsalinized controls (181,203). Since some rhizobia can withstand large increases in salinity, they must be able to regulate and adjust their internal solute concentration. Osmoregulation in *Rhizobium* grown at high external salt concentrations involves the accumulation of organic and/or inorganic solutes. While some strains respond to salt stress by increasing their intracellular K^+ levels (264), others accumulate organic compounds such as amino acids, betaine, and carbohydrates in their cytoplasm (130,133).

ROLE OF THE SHOOT

Salinity reduces plant growth as a function of total electrolyte concentration, soil water content, and soil matric potential; thus both osmotic and ionic influences are involved. The osmotic effects of salinity are a result of increased ion concentrations at the root–soil water interface which create lower water potential. The earliest response of a nonhalophyte to salinity is that leaves grow more slowly as a result of reduced cell enlargement and metabolism. It is possible that root water deficits elicit a signal that leads to decreased leaf expansion. When plants are grown under high root pressures, water stress may occur for a short period until osmotic adjustment takes place, but shoot water deficit does not limit growth, even at relatively high salinity (178).

Whole-plant response to excess salts in the root zone or on leaf surfaces (from ocean sprays or irrigation) is quantitatively dependent on salt concentration, time of exposure, and salt composition. Salt sensitivity also varies greatly with growth stage, species, variety, and ecotype, but some general trends can be noted. Plants are often salt-tolerant during germination, become more sensitive during the emergence and young seedling stages, and become more tolerant through the reproductive stage with the exception of anthesis. Salinity generally decreases plant growth at low concentrations and is lethal at higher concentrations. Salt-affected plants appear darker green and are stunted, with shorter and fewer internodes. Some species may develop succulence or a rosette growth habit.

Salt Loading

Many tolerant species restrict translocation of Na^+ and/or Cl^-. The uptake and translocation of ions into the shoot relative to water balance and growth are critical to ion loading in the shoot and salt tolerance (83,178). In addition, specific ions may affect metabolic pathways (102,164). This could be of special significance to the plant if the regulation of pathways involved in the supply of essential metabolites such as ATP or carbohydrates is altered.

Salt loading into the shoot is an important part of the toxic effect of salinity. Salt accumulation in the shoot is generally a one-way process; very little salt, with the exception of essential nutrient transport, returns via the phloem to the root. Plant species have developed a variety of mechanisms to adapt and adjust to the problem of maintaining positive hydraulic gradients for growth and development while at the same time limiting the total import of inorganic solute. One mechanism for the control of excessively high salt content in photosynthetically active leaf tissue is the differential distribution of ions to various plant organs. In certain cereal species, Na is preferentially accumulated in the sheaths or culms, while Na^+ concentration in the lamina remains relatively low. Alternatively, Na^+ may be localized in the lower leaves that are shed when the salt concentration reaches a critical level (e.g., rice [oryzasativa] 167).

Recently, Boursier and Läuchli (32) have found that *Sorghum bicolor* has the capacity to sequester Cl^- preferentially in sheaths, particularly in adaxial epidermal cells; whereas Cl^- concentrations in photosynthetically active mesophyll and bundle sheath cells remained lower. Other species have increased tolerance which is achieved by selectively directing ions into older leaves or specialized salt glands and away from the growing and photosynthetically active tissues (238).

Long-term salinity effects may be due to the accumulation of specific ions in the leaves (178). Accumulated ions may be sequestered in vacuoles, in older leaves with lower metabolic activity, or even in salt glands. With time, there is a simple overloading of niches that do not affect growth or metabolism; a slow poisoning of the tissues occurs. As external salt concentration increases, ion flux rate to the shoot may decrease due to reduced transpiration rate; however, growth rates continue to decline (178). That carbohydrate pools accumulate during this period as a result of utilization suggests that CO_2 fixation is not limiting.

Morphology and Anatomy

Morphological and anatomical changes induced by salt stress may be beneficial for plant survival and growth (196). For example, light leaf coloration decreases leaf temperature in hot, arid environments. Some species adapted to such environments have abaxial stomates and/or high mesophyll resistance to reduce transpiration water loss. Rice (*Oryza sativa*), although salt-sensitive, is considered a reclamation crop on saline lands because it has a shallow roots and can be grown in flooded fields if good-quality water is available. Susceptibility to leaf injury in rice depends more on leaf characteristics and rate of ion absorption than on tolerance to soil salinity (84,268,270). In other crops the shoot/root ratio may decrease at low to moderate salinity to provide the plant a larger surface area at which water can be collected (148,164). In a split-root study, root growth of tomatoes grown in soil-filled pots was less sensitive to salinity than the tops (193). It is interesting, however, that reductions in root growth in compartments with high salinity were compensated by increased growth in those portions of the root system in the less saline environments.

Structural changes that occur in salt-stressed glycophytes may be a measure of their salt tolerance or, alternatively, may be an indication of the structural disorganization that is associated with reduced osmotic potential, water stress, unfavorable ion balances, nutritional disturbances, the presence of toxic ions in the external media, or a combination of these factors. Udovenko and Tsibkovskaya (240) observed the salt-induced damage to cellular ultrastructure of leaf tissue was followed by a period of repair and, finally, adjustment and adaptation to the saline conditions. In this series of experiments, organelles in the leaf mesophyll cells of barley plants showed significant changes and disorganization within 2 hr after NaCl (150 mol/m^3) salination; the number of cristae per mitochondria and the number of grana per chloroplast decreased. These organelles as well as the Golgi apparatus and the endoplasmic reticulum underwent swelling–contraction cycles. Evidence of lysis and the presence of autophagic vacuoles within 5–10 days after salination suggested that cellular repair processes were in progress. After 15–25 days, cellular adjustment to the saline conditions occurred and the ultrastructure of the salt-stressed leaves was similar to that of the nonsaline controls.

In response to 50 mol/m^3 NaCl stress, companion cells and xylem parenchyma in leaf veins of *Trifolium alexandrinum* were differentiated into transfer cells with characteristic wall ingrowths (256). Phloem transfer cells were also identified that appeared to be involved in export of Na from the leaf. Disorganization of these cells occurred when their internal Na$^+$/K$^+$ ratio exceeded 4 even though the rest of the leaf was still healthy (256). Leaf mesophyll cells from wheat seedlings grown in 100 mol/m^3 NaCl exhibited increases in the volume fractions of cytoplasm, chloroplasts, and cell walls; the volume fraction of the vacuole decreased (113). The changes in these compartments were in the opposite direction of those reported for *S. maritima* (107).

The number of mitochondria per mesophyll cell section of *S. maritima* increased with salinity. In addition the microbodies increased in diameter, absolute surface area, and absolute volume (107).

The organelle most affected by salinity is the chloroplast (196). Photosynthetic activity of chloroplasts is dependent on optimal ion concentration and compartmentation and osmotic adjustments to regulate volume. Under drought conditions chloroplast volume decreases, Mg^{2+} concentrations increase, and photophosphorylation is inhibited (33,272). Under salt stress, water deficits are not usually involved and chloroplasts maintain ion homeostasis with only slight changes in Na$^+$ and K$^+$ (210). Photosynthetic activity, even in halophytes, is maintained through the selective accumulation of K$^+$ and organic solutes in the chloroplasts (44,58,206).

In salt-tolerant cell cultures of alfalfa (*Medicago sativa*), the expression of genes associated with photosynthesis is altered (255). Several gene transcripts in both nuclear and chloroplast DNA increase as a result of salt adaptation. There is evidence that photosystem II is involved in this salt-induced adaptation (254).

Salt-induced modifications include swelling within the grana loculi and frets, accumulation of lipid droplets, and changes in the size and number of grana. Damage to the chloroplasts in the salt-sensitive rice cultivar "Amber" included loss of the envelope and disorganization of the grana (81). In response to 340 mol/m^3 NaCl, chloroplasts isolated from the mesophyll cells of *Suaeda maritima* leaves showed increases in width, length, absolute volume, and absolute surface area and decreases in both the grana number and the number of lipid globules per chloroplast as well as in the fractional volume of the cell occupied by the chloroplasts. While the number of chloroplasts per cell did not change significantly with salinity, the number of chloroplasts that contained starch increased and the presence of the starch grains changed the shape of the chloro-plasts in the salt-stressed cells. The number of chloroplasts positioned around the intercellular spaces also increases in response to salinity (107).

Lipids in leaves may also reflect salt tolerance differences. Among five halophytes, the lipid composition of leaves did not vary significantly, but high linoleic content of the fatty acids was closely correlated with salt tolerance rankings (273). Recently, Genard et al. (96) found that reduc-ing the NaCl content of the culture medium induced reductions in the total lipid contents of *Suaeda maritima* leaves and that salt shock induced lipid synthesis. Similarly, drought stress of three cucumber (*Cucumis sativus*) genotypes resulted in an increase in total lipids and phospholipids on a fresh weight basis in leaf tissues and also led to a greater degree of fatty acid unsaturation (38).

Reproductive Growth and Development

Salinity often affects the timing of the growth stages of certain cereal crops so that ear emergence, anthesis, and grain maturity typically occur earliest in those plants that have been exposed to the most saline condi-tions. This response has been reported for wheat (*Triticum aestivum*) (159,202), sorghum (*Sorghum bicolor*) (92), oats (*Avena sativa*) (Fran-cois, unpublished observations), triticale (*X Triticosecale*) (93). Barley (*Hordeum vulgare*) (202) and rye (*Secale cereale*) (91) appear to be exceptions to this generalization; the time to anthesis and grain matura-tion was virtually unaffected by salinity.

After imbibition and germination, two distinct stages of wheat apex development are evident: leaf primordia initiation is followed by initia-tion of spikelet primordia. Each stage proceeds at a constant but different rate. The final number of primordia initiated during each stage is the product of the initiation rate and the length of that phase. The effects of salinity on the timing of wheat development are largely based on data collected at the U.S. Salinity Laboratory during 1988–1990. The dura-tion of spikelet primordia initiation decreases with increasing salinity; the rate is unchanged. As a result the number of spikelets, an important yield determinant, is reduced.

Salinity significantly alters the patterns of grain distribution along the spike and within the spikelets. The mainstem spikes of salt-stressed plants are shorter; fewer spikelets are produced and the number of ker-nels per spike is reduced (159). Depending on cultivar and the level of stress, weight of individual kernels may increase, particularly those in the two basal florets of the central spikelets. Thus, mainstem yield may not change, or may even increase in response to salinity as the decrease in kernel number may be completely compensated by an increase in kernel weight. (See also section "Cereal Crops.")

Salinity causes a significant reduction in the flowering intensity, fruit retention, final fruit number, and size of orchard crops [e.g., *Prunus salicina* (122) and *Citrus sinensis* (90,126,129)]. From a study with *Spergularia* species, Okusanya and Ungar (188) concluded that salinity can either delay or promote flowering depending on the species and the nutrient level. Reproductive effort (dry weight of reproductive parts/total above-ground biomass) of *Atriplex triangularis* was unaffected by salinity; plants allocated an average of 28% of the above-ground biomass to seed and bract production regardless of the salt treatment (64).

Shoot/Root Ratio

Partitioning of assimilates between shoot and root in response to salinity is also influenced by the genotype, resource (nutrients, water) availability, plant age, and other edaphic and environmental variables. Root growth of glycophytes is generally affected less by salinity than either vegetative shoot growth or fruit and seed production (164). Depending on the species, the level of salinity stress, and the composition of the external solution, root growth may be stimulated [e.g., *Cynodon* sp. (66), *Chloris gayana* (244)], inhibited [e.g., *Pisum sativum* (215), *Sorghum bicolor*, *S. halepense* (263), and *Gossypium hirsutum* (138)], or unaffected [e.g., *Hordeum vulgare* (57)]. Likewise, the type of salt has a profound influence on root growth. At low salinity, Weimberg et al. (249) found root growth of *S. bicolor* was stimulated by KCl and K_2SO_4, was unaffected by Na_2SO_4, and was inhibited by NaCl. Supplemental Ca protects cotton roots from the adverse effects of Na only when the Ca is added prior to NaCl salination (50,144).

When salinity inhibits shoot growth more than root growth, the resultant decrease in shoot/root ratio (S/R) leads to better exploitation of soil moisture and nutrients for the support of a plant with a relatively smaller leaf area (164). The S/R of C_4 turfgrasses, whether expressed in terms of dry weight, surface area, or length, decreased with increasing NaCl stress (66,169,175). Shoot growth in these grasses decreased linearly as salt stress increased, whereas root growth initially increased and then declined. Decreases in S/R in response to increasing salinity have also been reported for cotton (128), *Suaeda maritima* (266), *Leucaena leucocephala* (100), mesquite (*Prosopis juliflora*) (134), maize (211), wheat and barley (202). The S/R of the halophytic C_4 grass *Distichlis spicata* was not affected by NaCl stress (250 and 500 mol/m³) when the plants were grown under high illumination (137). In contrast, shoot growth of a number of species was less inhibited by salinity than was root growth, and as a consequence the S/R increased in response to increasing salt stress [e.g., *Sorghum bicolor* and *S. halepense* (263), as well as a number of flowering annual ornamental plants (59)]. In both of these studies, the adverse effects of NaCl-stress on whole-plant response may have been alleviated by the low Na^+/Ca^{2+} ratio in the root media.

Growth stimulation of the halophytes *Atriplex* (190) and *Spergularia* (189) by salinity is associated with an increase in S/R ratio inasmuch as photosynthate is allocated to develop additional photosynthetic tissue. Osmond et al. (190) noted that salinity had little effect on the rate of CO_2 fixation by *Atriplex* over the range of salt concentrations and further postulated that growth response was controlled principally by the partitioning of assimilates.

The influence of plant nutrient status on S/R ratio of the halophytes *Spergularia marina*, *S. rupicola*, *S. rubra* (189), and *Lavatera arborea* (187) has been studied. In response to increasing sea salt concentrations, root growth was severely reduced and the S/R ratio of these species increased. The addition of supplemental nutrients (Ca^{2+}, $H_2PO_4^-$, NO_3^-) to the 50% sea salt treatment increased the uptake of K^+ and Mg^{2+}, improved plant growth, particularly that of the root system, and decreased the S/R ratio.

The S/R ratios of barley, salinized with irrigation water containing 1:1 mixtures of NaCl and $CaCl_2$, decreased during the early stages of growth. As the crop matured, however, this trend was reversed, and finally, 116 days after sowing, the S/R ratio of the control plants was indistinguishable from that of the salt-stressed plants (4).

Reduced shoot growth in response to salinity is the consequence of reduction in the number of leaves formed on the main axis and an inhibition of lateral bud initiation and growth. In this connection, the most serious limitation of grain yield in salt-stressed cereals is the reduction in the tillering capacity that leads to a decrease in the number of spikes per unit area.

Succulence

Succulence, defined as a high water content per unit leaf surface area, is a typical morphological response to salinity in dicotyledonous species but is rarely observed in the monocots. This adaptation generally minimizes the effects of excessive salt concentrations in leaf tissue. Increased succulence may also have beneficial effects on CO_2 exchange by increasing internal surface area per unit leaf surface over which CO_2 diffusion can occur (151). The function, anatomy, and ecophysiological implications of succulence have been reviewed by Gibson (98) and Lüttge and Smith (155).

Increases in cell size, associated with succulence, may be the result of increased cell wall extensibility in concert with increased turgor pressure

(177). The thickness of both the epidermal cell walls and the cuticle also increases. Salinity-induced changes in the amount, composition, and structure of the epicuticular waxes are important in the reduction of cuticular transpiration rates [*Arachis hypogaea* (200) and *Suaeda maritima* (108)] and in the prevention of salt spray damage to coastal plants by reduction of leaf surface wettability [*Dactylis glomerata* and *Agrostis stolonifera* (3,172)].

Salt Glands

Perhaps the most elaborate adaptation to salinity is the formation of salt-excreting structures (salt glands, bladders, microhairs) on leaf or stem epidermal surfaces. Terrestrial halophytes can survive and complete their life cycles at optimum salt concentrations of 1.2–30 g/L salt in their rooting medium. Halophytes that lack a powerful salt exclusion mechanism in their root systems frequently have a process for desalination of leaf parenchyma by active excretion (149). The excretion of even small quantities of salt may contribute to the survival of plants challenged by salt stress. Salt-excreting mechanisms are normally highly selective for Na^+ and Cl^-, but against the nutrient ions. Thus, the ratios of Na^+/K^+, Cl^-/NO_3^-, and $Cl^-/H_2PO_4^-$ may be reduced to normal physiological levels (246). The function, physiology, and ultrastructure of salt excretion has been extensively reviewed (121,150,237,238).

THE AERIAL ENVIRONMENT

Natural Environments

The influence of factors in the plant's aerial environment significantly affects salt tolerance. Most crops can tolerate greater salt stress when the weather is cool and humid than when it is hot and dry. Magistad et al. (167), working with equivalent soil salinities, showed that crops grown in a coastal climate (cool and humid) consistently produced higher yields than those grown in a desert climate (hot and dry). Hoffman and Rawlins (127) reported that the salt tolerance of kidney beans grown with cool temperatures and high relative humidity was more than double that obtained with high-temperature/low-humidity conditions.

These factors also affect the expression of specific salt injury symptoms. Fruit crops and woody plants, susceptible to leaf injury by excess chloride or sodium accumulation, often develop leaf necrosis with the onset of hot, dry weather in late spring or early summer (27). Ehlig (68) reported similar results with grapes, which showed no leaf injury symptoms during cool, cloudy spring weather even though the leaves contained levels of chloride considered to be toxic.

Although high humidity has been shown to consistently improve growth and water balance under salt stress (182,208), temperature is believed to be the dominate factor in plant response to saline conditions (2). Other studies have confirmed that temperature influences plant salt tolerance to a greater degree than does relative humidity (127,153).

Light intensity has also been implicated in growth reduction caused by salinity. While some studies have shown that growth depression from salinity is generally greater under higher than under low light intensities (35,95,183), the salt tolerance of muskmelon (*Cucumis melo*) (173) and faba bean (*Vicia faba*) (119) increases at high irradiation. Toxicity symptoms are frequently observed on citrus leaves growing on the south side of trees, in response to higher light intensities, whereas leaves on the north side may remain symptom-free (17).

It is likely that at least part of the reduction in plant growth on saline media is due to increased transpiration, since high temperature, low relative humidity, and high irradiance are conditions that favor a high rate of transpiration. This may explain why some crops grown under field conditions are more salt-sensitive than when grown in the greenhouse.

Agromanagement and Pollutants

Some crops, when irrigated by sprinkling, experience foliar injury and subsequent yield reductions that may not occur when they are surface- or drip-irrigated with water of similar quality (24,111). The injury is the result of foliar absorption of salts that accumulate on the leaf as the sprinkled water evaporates from the leaf surface. Numerous factors affect the amount of salt accumulated by leaves, including the leaf age, shape, angle, and position on the plant, the type and concentration of salt, the ambient temperature and humidity, and the length of time the leaf remains wet. In addition, the leaf surface properties, such as a waxy cuticular layer or pubescence, may restrict ion absorption.

Francois and Clark (89) reported a linear increase in Na^+ and Cl^- concentration in grape leaves when sprinkled with saline water. When Cl^- is readily absorbed directly by the leaves, the advantage of chloride-resistant grape rootstocks that reduce Cl^- uptake by the roots is of little benefit with sprinkler irrigation.

Susceptibility to foliar injury varies considerably among crop species. A comparative study by Maas et al. (158) with 11 herbaceous species revealed wide differences in the rates of Na^+ and Cl^- absorption when the plants were sprinkled with saline water. Leaves of deciduous fruit trees appear to absorb Na^+ and Cl^- even more readily than the herbaceous crops (69). Citrus leaves absorbed these ions more slowly, but over time ion accumulation may be sufficient to cause severe leaf burn (111).

If sprinkler irrigation must be used, then good water management is essential. Since foliar injury is related more to frequency of sprinkling than duration (89,157), infrequent heavy irrigations should be applied rather than frequent light irrigations. Slowly rotating sprinklers that allow drying between cycles should be avoided, since this increases the wetting–drying frequency. Sprinkling should be done at night or early morning when evaporation is less. Sprinkling on hot, dry, windy days is not recommended. In general, poorer quality water can be used for surface-applied irrigation than can be used for sprinkler irrigation.

Ozone, a major air pollutant, decreases the yield of some oxidant-sensitive crops more under nonsaline than saline conditions (39,124,125,163). This aberration has the tendency to make many crops grown in air-polluted regions appear to be more salt-tolerant than they really are. This salinity–ozone interaction may be agronomically important in air-polluted areas. However, the increased ozone tolerance induced by salinity may be more than offset by the detrimental effects of salinity on the harvestable product (124,125,186).

In contrast to ozone, higher CO_2 concentrations in the atmosphere have been shown to increase the salt tolerance of bean, maize, and tomato (73,214). This increased tolerance is believed to be the result of an increased rate of photosynthesis (174).

CROP SALT TOLERANCE

An extensive range of genetic variability has been described with which higher plants adjust and adapt to saline environments. In agriculture, salinity problems are confronted by changing management techniques within the feasibility of economic and material resources and through crop selection. Crops such as sugarbeet, asparagus (*Asparagus officinalis*), cotton, and barley are relatively salt-tolerant compared to different types of beans, berries (*Rubus* sp.), and fruit trees (156). Most commercially grown cultivars are developed under nonsaline conditions and not bred to endure salt stress. Therefore, their relative tolerances to salinity are often similar and difficult to measure. In addition, many cultivars developed in the past were derived from a narrow genetic base and so possessed similar traits. Cultivars more recently developed are from a much more diverse genetic base and therefore may possess a wider range of salt tolerance.

Among the crop species that show some intraspecific diversity in salt tolerance are Bermuda grass (*Cynodon dactylon*), bromegrass (*Bromus inermis*), creeping bentgrass (*Agrostis stolonifera*), rice, wheat, barley, soybean (*Glycine max*), berseem clover (*Trifolium alexandrinum*), squash (*Cucurbita pepo melopepo*), muskmelon, and strawberry (*Fragaria* sp.). While cotton and sugarcane also show significant cultivar differences, these differences occur only at high salinity where yields are below commercially acceptable levels (14,26).

Yield response curves have been developed for many crops (161). The two-piece linear response model for these curves consists of the threshold and slope. The threshold is the maximum salinity that can be applied without causing a reduction in yield below that obtained under similar but nonsaline conditions. It is a function of environmental conditions and the adaptive ability of the crop to physiologically adjust to salinity without yield reduction. This does not mean that growth is not reduced or that visual symptoms of salt stress do not exist. Only economic yield is unaffected. Plant adaptive responses may include an increase in respiration rate, improved water use efficiency, and alterations in the partitioning of available photosynthate. The slope is the percent yield decrease per unit increase in salinity beyond the threshold and probably reflects the direct effects of salt loading into the shoot, but the physiological determinants of either parameter have not been well defined and represent a needed area of research. The threshold and slope data for different crop species are presented in terms of the electrical conductivity of the saturated-soil extract, EC_e, at 25°C with units of decisiemens per meter. Absolute tolerances vary, depending on climate, cultural practices, soil conditions, and specific salts.

Vegetable Crops

Vegetable crops tend to fall into the salt-sensitive category. The only notable exceptions are asparagus, red beet (*Beta vulgaris*), and zucchini squash. Since most vegetables are salt-sensitive, the choice of land and/or irrigation water where they can be successfully grown is severely restricted.

Salt affects the growth of vegetables, directly or indirectly, by a reduction in the osmotic potential of the growth media (164). In addition, certain ions may interfere with normal nutrition. Some lettuce (*Lactuca sativa*) cultivars often develop calcium deficiency symptoms when sulfate levels in the soil are too high. Excessive calcium may restrict the uptake of potassium, which may be a factor in reduced yields of bean and carrot (*Daucus carota*) (13). However, with most vegetable crops, the osmotic effect greatly predominates and nutritional effects are either absent or of decidedly secondary importance.

Under marginal conditions of salinity many vegetables are stunted and exhibit a reduction in growth rate without showing other visible injury symptoms (116). At high salinity levels, some vegetables do exhibit pronounced injury symptoms at the later stages of growth. Bean leaves develop a marginal chlorosis-necrosis with an upward cupping of the leaves (20). Onions (*Allium cepa*) have also been shown to develop a leaf necrosis (21).

High levels of exchangeable sodium frequently restrict vegetative growth because of the unfavorable physical conditions associated with sodic soils. Most vegetable crops appear to be at least moderately tolerant to exchangeable sodium. Bean plants, however, are sensitive to nutritional factors in sodic soils and may be severely affected, even before the soil physical condition is impaired.

Most vegetable crops produced on saline soils are not of prime market quality. This is seen in such diverse ways as smaller fruit size of tomatoes and peppers (*Capsicum annuum*) (13), reduced petiole length of celery (*Apium graveolens*) (94), and misshapen potatoes (*Solanum tuberosum*) (29). However, not all quality effects are detrimental. The flavor of carrots (13) and asparagus (87) is enhanced by a measurable increase in sugar content when grown under saline conditions. Likewise, a number of studies (1,136,207,217) have shown that total soluble solids in tomatoes are significantly increased as salt stress is increased. Unfortunately, this gain in quality is more than offset by lower yields.

Cereal Crops

Most of the major cereal crops exhibit high tolerance to soil salinity. In this group are sorghum, wheat, triticale, rye, oats, and barley. The only exceptions are maize and rice (156).

Regardless of the overall salt tolerance, all cereals tend to follow the same sensitivity or tolerance pattern in relation to their stage of growth. The seedling or early vegetative stage appears to be the most sensitive, with subsequent stages showing increased tolerance. This phenomenon has been reported for sorghum (166), wheat (165), barley (7), maize (162), and rice (195). The other cereal crops, although not tested but with similar growth patterns, would also be expected to show sensitivity at the early vegetative stage of growth.

Since the life cycle of cereals is an orderly sequence of developmental events, salinity stress can have a significant effect on the developmental process that is occurring at a particular time. The sequence of events has been separated into three distinct but continuous developmental phases (160). In the first phase, which encompasses the early vegetative growth stage, leaf and spikelet primordia are initiated, leaf growth occurs, and tiller buds are produced in the axils of the leaves. High soil salinity at this time will reduce the number of leaves per culm, the number of spikelets per spike, and the number of tillers per plant (159,160). Differentiation of the terminal spikelet signals completion of this phase.

During phase II, the tillers grow, mainstems and tiller culms elongate, and the final number of florets is determined (139). Salinity stress during this phase may affect tiller survival and reduce the number of functional florets per spikelet. This phase ends with anthesis. Carpel fertilization and grain filling occur during the final phase (139). At this time, salinity affects seed number and seed size.

The references previously cited in this section show that the effect of salinity on spikelet and tiller number established during phase I have a greater influence on final seed yield than the effects exerted on yield components in the latter two phases.

Forage Crops

Forage crops fall into two broad salt tolerance categories. Most of the grasses belong to the tolerant group, with the majority of legumes in the sensitive group. Exceptions to this generalization are meadow foxtail (*Alopecurus pratensis*), lovegrass (*Eragrostis* sp.), and orchardgrass (*Dactylis glomerata*), which are moderately sensitive to salt stress, and birdsfoot trefoil (*Lotus corniculatus tenuifolium*), and the sweet clovers (*Melilotus* sp.), which are moderately tolerant (161).

Many of the forage grasses possess the same growth habit as the cereal grasses and, like the cereals, are more sensitive to salinity during the early seedling stage of growth (168). However, unlike the cereals, many of the grasses are maintained in a perpetual vegetative stage of growth from continued grazing or mowing. Therefore it appears that these grasses, once beyond the early seedling stage and well established, are less sensitive to soil salinity.

Because of their fibrous roots, grasses alone or in combination with forage legumes are frequently used in the reclamation of saline and sodic soils to restore good soil structure (12). Under nonirrigated conditions, those grasses, which accumulate significantly high concentrations of Na^+ and Cl^- in their shoots, may be used to restore soil structure and also to remove these ions from the soil profile (209). Grasses used for this purpose may be unfit for animal feed because of the high salt content (12).

Clovers are the predominate legume of pastures and are frequently grown in combination with various grass species. However, salt-sensitive clovers tend to die out on saline soils as the more tolerant grass becomes the predominate vegetation. Loss of the clover from the pasture mixture significantly reduces the nutritional value of the pasture (176).

The salt tolerance of clovers (252) and alfalfa (222) is highly dependent on the stage of growth at which salinity is first imposed.

The salt tolerance of alfalfa has been reported to be closely associated with Cl^- accumulation in the leaves (184,222). Salt-affected plants are characterized initially by a dark green leaf coloration and reduced leaf size (184), followed by a general reduction in plant size (15).

Although the salt tolerance of alfalfa appears to be dependent on a salt exclusion mechanism (184), no consistent correlation seems to exist between salt tolerance and salt exclusion for legumes in general (147). There appears to be sufficient evidence that the genetic variability which exists among the grass and legume species and cultivars offers the possibility of developing higher salt tolerant strains (88,147,168,252,271).

Tree and Vine Crops

With the exception of date palm (*Phoenix dactylifera*), and a few other species believed to be moderately tolerant, most fruit trees are relatively sensitive to salinity. Stone fruits, citrus, and avocado (*Persea americana*) have all shown specific sensitivity to foliar accumulations of chloride and sodium. The accumulation of these ions to harmful levels, as well as the general osmotic growth inhibition, contributes to the reduction in tree growth and fruit yield.

Different cultivars and rootstocks absorb chloride and sodium at different rates so tolerance can vary considerably within a species. However, in the absence of specific ion effects, the tolerance of these crops can be expressed as a function of the concentration of total soluble salts or the osmotic potential of the soil solution.

Some of the more sensitive fruit crops may accumulate toxic levels of sodium, chloride, or both over a period of years from soils that would be classified as nonsaline and nonsodic (6,19).

Injury by sodium can occur at concentrations as low as 5 mol/m^3 in the soil solution (156). However, injury symptoms, which are characterized as tip, marginal, and/or interveinal necrosis, may not appear for considerable time after exposure to salinity. Initially, the sodium is thought to be retained in the sapwood of the tree. With the conversion of sapwood to heartwood, the sodium is released and then translocated to the leaves causing leafburn (22). This may partly explain why stone fruits and grapes appear to be more sensitive to salinity as the plants grow older. With succeeding years, the chloride and sodium accumulate more rapidly in the leaves, causing leaf burn to develop earlier and with increasing severity (122).

Chloride toxicity in woody species is generally more severe and observed on a wider range of species than is sodium toxicity. What differences do exist among species, varieties, or rootstocks in susceptibility to chloride usually reflect the capability of the plant to prevent or retard chloride accumulation in the plant tops.

Recent studies have shown that sodium accumulation in plum (*Prunus domestica*) leaves did not significantly increase until the leaves were already severely damaged by chloride accumulation (40,122). These studies indicate that when chloride and sodium are present in the soil solution, chloride is the primary damaging ion on stone fruits. Sodium accumulation only occurs after the leaf membranes have already been damaged.

The initial symptom of excess chloride accumulation in fruit crops is leaf tip necrosis developing into marginal necrosis. With citrus, a chlorosis and bronzing of the leaves occur without a well-defined necrosis. As chloride continues to accumulate, the effects become more severe with premature leaf drop, complete defoliation, twig dieback, and, in extreme cases, death of the tree or vine (19,118).

Growth and yield reduction may occur with woody fruit species in the absence of specific ion toxicity. Francois and Clark (90), working with "Valencia" orange (*Citrus sinensis*), reported a 50% reduction in fruit yield from salinity with no visible leaf injury symptoms. However, it is generally believed that growth and yield of most woody fruit crops suffer from both osmotic effects and toxicities caused by chloride or sodium accumulation (19).

The tolerance of many fruit tree and vine crops can be significantly improved by selecting rootstocks that restrict Cl^- and/or Na^+ accumulation. Although citrus is not considered to be very salt-tolerant, there are differences in salt tolerance among the various rootstocks (17,48,218). These differences are attributed to salt exclusion and particularly to chloride exclusion (47,247). Citrus apparently excludes chloride from shoots, not by sequestering it in the root but by restricting its entry into and/or movement within the roots. The chloride concentration differences found in leaves and to a lesser extent in stems emphasize pronounced rootstock differences in root to shoot transport of chloride (247). The scion appears to have no major influence on chloride transport from the roots to the shoot (10).

Differences among rootstocks are much greater for chloride accumulation than for sodium and there appears to be no correlation between chloride tolerance and sodium tolerance (47). These differences are due to the existence of apparent separate mechanisms that operate to limit or regulate the transport of sodium or chloride to the leaves (103).

The chloride tolerance range for avocado rootstocks is much narrower than for citrus. In addition, because of the wide variation among varieties of the same rootstock, the rootstock tolerances tend to overlap (72). However, it is generally agreed that the average chloride tolerance of West Indian is greater than that of Guatemalan which is greater than that of Mexican (46,72,105). The general pattern for sodium accumulation with avocado rootstocks tend to follow that for chloride accumulation and, like chloride, shows differences among varieties on the same rootstock (105,106).

Cold hardiness has been implicated with the salt tolerance of citrus and avocado rootstocks. Wutscher (257) reported that citrus rootstocks, which have good chloride-excluding characteristics, tend to be relatively cold-hardy. For some citrus species, a short-term, moderate salt stress has been shown to enhance cold hardiness in seedlings by modifying growth, water relations, and mineral nutrition (235).

In contrast to citrus, the more salt-tolerant avocado rootstocks, such as West Indian and West Indian–Guatemalan hybrids, are the least cold-tolerant. Likewise, the salt-sensitive Mexican is the most cold-tolerant rootstock (49).

Chloride toxicity has been the principal limiting factor for grapevines grown on their own roots. However, a significant reduction in chloride accumulation has been shown to occur in chloride-sensitive scions when grown on "Dog Ridge" or "1613-3" rootstocks (23). The salt tolerance of these two rootstocks would probably be limited by soil osmotic effects long before chloride reached toxic levels.

Ornamentals

In contrast to crop species that produce a marketable product, the salt tolerance of ornamental trees and flowers is determined by the aesthetic value of the plant species. Injury or loss of leaves or flowers due to salt stress is unacceptable even though growth may be unaffected. A significant growth reduction might be acceptable and possibly desirable for some species as long as they appear healthy and attractive.

The type of injury seen on woody ornamentals and trees is similar to damage recorded for fruit trees and vines. A number of reports have shown that, although some woody ornamentals and trees accumulate sodium, the salt tolerance of these species is closely associated with the ability to limit chloride uptake and accumulation (60,86,239).

In northern climates, where NaCl and/or $CaCl_2$ are used as deicing salts, typical salt injury symptoms occur on roadside trees. These trees are subjected to both soil salinity from runoff and saline spray from passing automobiles. Although salt spray is thought to be the more detrimental of the two modes of deposition (110,152), the soil salinity effects may be cumulative and that, over a period of years, may result in a slow but progressive decline of the trees.

A limited number of floricultural plants have been tested for salt tolerance. While chrysanthemum (*Chrysanthemum* sp.), carnation

(*Dianthus caryophyllus*), and stock (*Matthiola* sp.) are considered to be moderately tolerant to salt stress, aster (*Aster* sp.), poinsettia (*Euphorbia pulcherrima*), gladiolus (*Gladiolus xhortulanus*), azalea (*Rhododendron* sp.), gardenia (*Gardenia* sp.), gerbera (*Gerbera jamesonii*), amaryllis (*Amaryllis* sp.), and African violet (*Saintpaulia* sp.) are considered to be somewhat sensitive (59,117,227). Like other ornamental species, the aesthetic value of floral plants is the determining factor for salt tolerance.

CONCLUSIONS

The plant is a complex system of organelles, cells, tissues, and organs all of which are subject to salinity stress. Unlike a chain, the weakest link in terms of salt sensitivity does not determine the salt tolerance of the whole. A myriad of mechanisms and systems operate to filter and protect sensitive components of the plant from the effects of salt. Such mechanisms may start beyond the physical barriers of the plant cell itself. As an example, electrical charge properties may influence the relative abilities of ions to approach the plasmalemma (146,234). Other mechanisms may operate to pump specific ions back out of the cytosol or into vacuoles. Mangroves probably have the most highly adapted membranes for efficiently filtering salt.

Ions may be sequestered further along the pathway of water flow into the xylem by specialized parenchyma cells. In the shoot, individual leaves may serve as reservoirs for toxic ions. The meristematic regions of the shoot and fruiting bodies seem to be the most protected from the accumulation of toxic ions. Location of meristems within relatively humid leaf sheath environments may be a protective mechanism against ion accumulation. Certainly, the delay of xylem development is also a contributing factor. Specialized structures such as salt glands or salt hairs exist in some halophytes adapted to highly saline environments. Many other components of the shoot, such as stomates, leaf mesophyll structure, epicuticular waxes, leaf coloration, and leaf shape, may play significant roles in both water relations and, as a result, salinity stress.

Science has come a long way in identifying the particular components of salinity stress and in describing the vast amount of variation among different species and genera. At the same time, rapid advances in the areas of molecular biology have provided new and exciting tools for fingerprinting and manipulating genomes. But, as yet, these tools have only been useful in further descriptions of the plant system. In order to advance beyond our present state of the art, two main research efforts are needed. A clearer understanding of the ways in which the morphological and physiological salt response mechanisms are integrated within a plant must be developed, and the control mechanisms that trigger plant adjustment to salinity must be identified. The development of knowledge in these areas will pave the way for the eventual design and development of salt-tolerant crop species suited to specific environments within particular agromanagement systems.

REFERENCES

1. P. Adams and L. C. Ho (1989). Effects of constant and fluctuating salinity on the yield, quality and calcium status of tomatoes. *J. Hort. Sci.*, *64*:725–732.
2. S. M. Ahi and W. L. Powers (1938). Salt tolerance of plants at various temperatures. *Plant Physiol.*, *13*:767–789.
3. I. Ahmad and S. J. Wainwright (1976). Ecotype differences in leaf surface properties of *Agrostis stolonifera* from salt marsh, spray zone and inland habitats. *New Phytol.*, *76*:361–366.
4. S. Al-Khafaf, A. Adnan, and N. M. Al-Asadi (1990). Dynamics of root and shoot growth of barley under various levels of salinity and water stress. *Agric. Water Manage.*, *18*:63–75.
5. G. M. Aubertin, R. W. Rickman, and J. Letey (1968). Differential salt-oxygen levels influence plant growth. *Agron. J.*, *60*:345–349.
6. A. D. Ayers, D. G. Aldrich, and J. J. Coony (1951). Sodium and chloride injury of Fuerte avocado leaves. *Calif. Avocado Soc. Yearbook*, pp. 174–178.
7. A. D. Ayers, J. W. Brown, and C. H. Wadleigh (1952). Salt tolerance of barley and wheat in soil plots receiving several salinization regimes. *Agron. J.*, *44*:307–310.
8. R. S. Ayers and D. W. Westcot (1985). Water quality for agriculture. FAO Irrigation and Drainage Paper 29, Rev. 1 FAO, United Nations, Rome.
9. V. Balasubraminiam and S. K. Sinha (1976). Effects of salt stress on growth, nodulation and nitrogen fixation in cowpea and mungbean. *Physiol. Plant.*, *36*:197–200.
10. M. H. Behboudian, E. Torokfalvy, and R. R. Walker (1986). Effect of salinity on ionic content, water relations and gas exchange parameters in some citrus scion-rootstock combinations. *Sci. Hort.*, *28*:105–116.
11. T. Bernard, J. A. Pocard, B. Perroud, and D. LeRudulier (1986). Variations in the response of salt-stressed *Rhizobium* strains to betaines. *Arch. Microbiol.*, *143*:359–364.
12. L. Bernstein (1958). Salt tolerance of grasses and forage legumes. *USDA Info. Bull.*, 194.
13. L. Bernstein (1959). Salt tolerance of vegetable crops in the west. *USDA Agric. Info. Bull.*, 205.
14. L. Bernstein (1964). Salt tolerance of plants. *USDA Agric. Info. Bull.*, 283.

15. L. Bernstein (1964). Effect of salinity on mineral composition and growth of plants. *Plant Anal. Fert. Prob.*, 4:25–44.

16. L. Bernstein (1967). Quantitative assessment of irrigation water quality. *Am. Soc. Test. Water. Spec. Tech. Pub.*, 416:51–65.

17. L. Bernstein (1969). Salinity factors and their limits for citrus culture. *Proc. First Int. Citrus Symp.*, 3:1779–1782.

18. L. Bernstein (1974). Crop growth and salinity. *Drainage for Agriculture* (J. van Schilfgaarde, ed.). *Agronomy*, 17:39–54. (Am. Soc. Agron.)

19. L. Bernstein (1980). Salt tolerance of fruit crops. *USDA Agric. Info. Bull.*, 292.

20. L. Bernstein and A. D. Ayers (1951). Salt tolerance of five varieties of greenbeans. *Proc. Am. Soc. Hort. Sci.*, 57:243–248.

21. L. Bernstein and A. D. Ayers (1953). Salt tolerance of five varieties of onions. *Proc. Am. Soc. Hort. Sci.*, 62:367–370.

22. L. Bernstein, J. W. Brown, and H. E. Hayward (1956). The influence of rootstock on growth and salt accumulation in stone-fruit trees and almond. *Proc. Am. Soc. Hort. Sci.*, 68:86–95.

23. L. Bernstein, C. F. Ehlig, and R. A. Clark (1969). Effect of grape rootstocks on chloride accumulation in leaves. *J. Am. Soc. Hort. Sci.*, 94:584–490.

24. L. Bernstein and L. E. Francois (1973). Comparison of drip, furrow, and sprinkler irrigation. *Soil Sci.*, 115:73–86.

25. L. Bernstein and L. E. Francois (1973). Leaching requirement studies: Sensitivity of alfalfa to salinity of irrigation and drainage waters. *Proc. Soil Sci. Soc. Am.*, 37:931–943.

26. L. Bernstein, L. E. Francois, and R. A. Clark (1966). Salt tolerance of N. Co. varieties of sugarcane I. Sprouting, growth, and yield. *Agron. J.*, 58:489–493.

27. L. Bernstein, L. E. Francois, and R. A. Clark (1972). Salt tolerance of ornamental shrubs and ground covers. *J. Am. Soc. Hort. Sci.*, 97:550–556.

28. L. Bernstein and G. Ogata (1966). Effect of salinity on nodulation, nitrogen fixation, and growth of soybeans and alfalfa. *Agron. J.*, 58:201–203.

29. E. C. Blodgett and R. S. Snyder (1946). Effect of alkali salts on shape and appearance of Russett Burbank potatoes. *Am. Potato J.*, 23:425–430.

30. L. J. Borowitzka and A. D. Brown (1974). The salt relations of marine and halophilic species of the unicellular green alga *Dunaliella*. *Arch. Microbiol.*, 96:37–52.

31. J. L. Botsford (1984). Osmoregulation in *Rhizobium meliloti*: Inhibition of growth by salts. *Arch. Microbiol.*, 137:124–127.

32. P. Boursier and A. Läuchli (1990). Growth responses and mineral nutrient relations of salt-stressed sorghum. *Crop Sci.*, 30:1226–1233.

33. J. S. Boyer (1976). Water deficits and photosynthesis. *Water Deficits and Plant Growth*, Vol. 4, Academic Press, New York, pp. 153–190.

34. Y. Braun, M. Hassidim, H. R. Lerner, and L. Reinhold (1986). Studies on H^+-translocating ATPase in plants of varying resistance to salinity. 1.

Salinity during growth modulates the proton pump in the halophyte *Atriplex nummularia*. *Plant Physiol.*, *81*:1050–1056.

35. R. Brouwer (1963). The influence of the suction tension of the nutrient solution on growth transpiration and diffusion pressure deficit of bean leaves (*Phaseolus vulgaris*). *Acta Bot. Neerl.*, *12*:248–260.

36. R. Brower (1953). Water absorption by the roots of *Vicia faba* at various transpiration strengths. 2. Crucial relation between suction tension, resistance and uptake. *Proc. K. Ned. Akad. Wet.*, *56*:129–136.

37. D. J. Brown and F. M. DuPont (1989). Lipid composition of plasma membranes and endomembranes prepared from roots of barley (*Hordeum vulgare* L.). Effects of salt. *Plant Physiol.*, *90*:955–961.

38. H. A. M. Bulder, W. R. van der Leij, E. J. Speek, P. R. van Hasselt, and P. J. C. Kuiper (1989). Interactions of drought and low temperature stress on lipid and fatty acid composition of cucumber genotypes differing in growth response at suboptimal temperature. *Physiol. Plant.*, *75*:362–368.

39. A. Bytnerowicz and O. C. Taylor (1983). Influence of ozone, sulfur dioxide, and salinity on leaf injury, stomatal resistance, growth, and chemical composition of bean plants. *J. Environ. Qual.*, *12*:397–405.

40. P. B. Catlin, G. J. Hoffman, R. M. Mead, R. S. Johnson, and L. E. Francois (1992). Responses of mature plum trees to salinity-fourth to sixth years (in preparation).

41. J. Cernohorska, M. Dvorak, and E. M. Wiedenroth (1989). Electrical conductivity and capacitance of root tissues in different conditions of energetic metabolism. *Structural and Functional Aspects of Transport in Roots* (B. C. Loughman, O. Gasparikova, and J. Kolek, eds.), Kluwer Academic, Boston, pp. 93–95.

42. O. Chloupek (1972). The relation between electric capacitance and some other parameters of plant roots. *Biol. Plant*, *14*(3):227–230.

43. O. Chloupek (1977). Evaluation of the size of a plant's root system using its electrical capacitance. *Plant Soil*, *48*:525–532.

44. W. S. Chow, M. C. Ball, and J. M. Anderson (1990). Growth and photosynthetic responses of spinach to salinity: Implications of K^+ nutrition for salt tolerance. *Aust. J. Plant Physiol.*, *17*:563–578.

45. D. T. Clarkson (1988). Movements of ions across roots. *Solute Transport in Plant Cells and Tissues* (D. A. Baker and J. L. Hall, eds.), Longman, London, pp. 251–305.

46. W. C. Cooper (1951). Salt tolerance of avocados on various rootstocks. *Texas Avocado Soc. Yearbk.*, pp. 24–28.

47. W. C. Cooper (1961). Toxicity and accumulation of salts in citrus trees on various rootstocks in Texas. *Proc. Fla. State Hort. Soc.*, *74*:95–104.

48. W. C. Cooper, B. S. Gorton, and C. Edwards (1951). Salt tolerance of various citrus rootstocks. *Proc. Rio Grande Valley Hort. Soc.*, *5*:46–52.

49. W. C. Cooper, A. Peynado, N. Maxwell, and G. Otey (1957). Salt tolerance and cold-hardiness tests on avocado trees. *J. Rio Grande Valley Hort. Soc.*, *11*:67–74.

50. G. R. Cramer, A. Läuchli, and E. Epstein (1986). Effects of NaCl and

CaCl$_2$ on ion activities in complex nutrient solutions and root growth of cotton. *Plant Physiol.*, *81*:792–797.

51. J. Dainty, P. C. Croghan, and D. S. Fensom (1963). Electro-osmosis, with some applications to plant physiology. *Can. J. Bot.*, *41*:953–966.

52. F. N. Dalton (1972). A physical-mathematical model describing the simultaneous transport of water and solutes across root membranes. PhD thesis, University of Wisconsin, Madison.

53. F. N. Dalton (1984). Dual pattern of potassium transport in plant cells: a physical artifact of a single uptake mechanism. *J. Exp. Bot.*, *35*:1723–1732.

54. F. N. Dalton and W. R. Gardner (1978). Temperature dependence of water uptake by plants. *Agron. J.*, *20*:404–406.

55. F. N. Dalton and J. A. Poss (1990). Water transport and salt loading: A unified concept of plant response to salinity. Symposium on Scheduling of Irrigation for Vegetable Crops Under Field Condition. *Acta Hort.*, *1*:187–193.

56. F. N. Dalton, P. A. C. Raats, and W. R. Gardner (1975). Simultaneous uptake of water and solutes by plant roots. *Agron. J.*, *67*:334–339.

57. R. Delane, H. Greenway, R. Munns, and J. Gibbs (1982). Ion concentration and carbohydrate status of the elongating leaf tissue of *Hordeum vulgare* growing at high external NaCl. 1. Relationship between solute concentration and growth. *J. Exp. Bot.*, *33*:557–573.

58. G. Demming and K. Winter (1986). Sodium, potassium, chloride and proline concentrations of chloroplasts isolated from a halophyte, *Mesembryanthium crystallinum* L. *Planta*, *168*:421–426.

59. D. A. Devitt and R. L. Morris (1987). Morphological response of flowering annuals to salinity. *J. Am. Soc. Hort. Sci.*, *112*:951–955.

60. M. A. Dirr (1976). Selection of trees for tolerance to salt injury. *J. Arboric.*, *2*:209–216.

61. T. J. Douglas and R. R. Walker (1983). 4-Desmethylsterol composition of citrus rootstocks of different salt exclusion capacity. *Physiol. Plant.*, *58*:69–74.

62. T. J. Douglas and R. R. Walker (1984). Phospholipid, free sterol and adenosine triphosphatase of plasma membrane-enriched preparations from roots of citrus genotypes differing in chloride exclusion ability. *Physiol. Plant.*, *62*:51–58.

63. T. J. Douglas and R. R. Walker (1985). Phospholipid, galactolipid and free sterol composition of fibrous roots from citrus genotypes differing in chloride exclusion capacity. *Plant Cell Environ.*, *8*:693–699.

64. D. R. Drake and I. A. Ungar (1989). Effects of salinity, nitrogen, and population density on the survival, growth, and reproduction of *Atriplex triangularis* (Chenopodiaceae). *Am. J. Bot.*, *76*:1125–1135.

65. M. C. Drew and A. Läuchli (1985). Oxygen-dependent exculsion of sodium ions from shoots by roots of *Zea mays* (cv. Pioneer 3906) in relation to salinity damage. *Plant Physiol.*, *79*:171–176.

66. A. E. Dudeck, S. Singh, C. E. Giordano, T. A. Nell, and D. B. McCon-

nell (1983). Effects of sodium chloride on *Cynodon* turfgrasses. *Agron. J.*, 75:927–930.

67. M. Dvorak, J. Cernohorska, and K. Janacek (1981). Characteristic of current passage through plant tissue. *Biol. Plant,* 23:306–310.

68. C. F. Ehlig (1960). Effects of salinity on four varieties of table grapes grown in sand culture. *Proc. Am. Soc. Hort. Sci.*, 76:323–331.

69. C. F. Ehlig and L. Bernstein (1959). Foliar absorption of sodium and chloride as a factor in sprinkler irrigation. *Proc. Am. Soc. Hort. Sci.*, 74:661–670.

70. E. A. E. Elsheikh and M. Wood (1989). Response of chickpea and soybean Rhizobia to salt: osmotic and specific ion effects to salt. *Soil Biol. Biochem.*, 21:889–895.

71. E. A. A. Elsheikh and M. Wood (1990). Salt effects on survival and multiplication of chickpea and soybean rhizobia. *Soil Biol. Biochem.*, 22:343–347.

72. T. W. Embleton, M. Matsumura, W. B. Storey, and M. J. Garber (1962). Chlorine and other elements in avocado leaves as influenced by rootstock. *Proc. Am. Soc. Hort. Sci.*, 80:230–236.

73. H. Z. Enoch, N. Zieslin, Y. Biran, A. H. Halevy, M. Schwartz, B. Kessler, and D. Shimshi (1973). Principles of carbon dioxide nutrition research. *Acta Hort.*, 32:97–110.

74. E. Epstein (1976). Kinetics of ion transport and the carrier concept. *Encyclopedia of Plant Physiology. Transport in Plants. 2. Part B. Tissues and Organs* (U. Lüttge and M. G. Pitman, eds.), Springer-Verlag, New York, pp. 70–94.

75. L. Erdei, B. Stuiver, and P. J. C. Kuiper (1980). The effect of salinity on lipid composition and on activity of Ca^{2+}- and Mg^{2+}-stimulated ATPase in salt-sensitive and salt-tolerant *Plantago* species. *Physiol. Plant.*, 49:315–319.

76. D. Evlagon, I. Ravina, and P. Neumann (1990). Interactive effects of salinity and calcium on hydraulic conductivity, osmotic adjustment, and growth in primary roots of maize seedlings. *Israel J. Bot.*, 39:239–247.

77. N. N. Ewing, L. W. Wimmers, D. J. Meyer, R. T. Chetelat, and A. E. Bennett (1990). Molecular cloning of tomato plasma membrane H^{+}-ATPase. *Plant Physiol.*, 94:1874–1881.

78. E. L. Fiscus (1975). The interaction between osmotic and pressure induced water flow in plant roots. *Plant Physiol.*, 55:917–922.

79. E. L. Fiscus and P. J. Kramer (1975). Liquid phase resistance to water flow in plants. *What's New in Plant Physiology?* 7(3):1–4.

80. T. J. Flowers (1974). Salt tolerance in *Suaeda maritima* (L.) Dum.: A comparison of mitochondria isolated from green tissues of *Suaeda* and *Pisum. J. Exp. Bot.*, 101:101–110.

81. T. J. Flowers, E. Duque, M. A. Hajibagheri, T. P. McGonigle, and A. R. Yeo (1985). The effect of salinity on leaf ultrastructure and net photosynthesis of two varieties of rice: further evidence for a cellular component of salt-resistance. *New Phytol.*, 100:37–43.

82. T. J. Flowers, M. A. Hajibagheri, and N. J. W. Clipson (1986). The

mechanism of salt tolerance in halophytes. *Annu. Rev. Plant Physiol.*, 28:89–121.

83. T. J. Flowers and A. R. Yeo (1986). Ion relations of plants under drought and salinity. *Aust. J. Plant Physiol.*, 13:75–91.

84. T. J. Flowers and A. R. Yeo (1988). Ion relations of salt tolerance. *Solute Transport in Plant Cells and Tissues* (D. A. Baker and J. L. Hall, eds.), John Wiley and Sons, New York, pp. 392–416.

85. L. E. Francois (1981). Alfalfa management under saline conditions with zero leaching. *Agron. J.*, 73:1042–1046.

86. L. E. Francois (1982). Salt tolerance of eight ornamental tree species. *J. Am. Soc. Hort. Sci.*, 107:66–68.

87. L. E. Francois (1987). Salinity effects on asparagus yield and vegetative growth. *J. Am. Soc. Hort. Sci.*, 112:432–436.

88. L. E. Francois (1988). Salinity effects on three turf bermudagrasses. *Hortscience*, 23:706–708.

89. L. E. Francois and R. A. Clark (1979). Accumulation of sodium and chloride in leaves of sprinkler-irrigated grapes. *J. Am. Soc. Hort. Sci.*, 104:11–13.

90. L. E. Francois and R. A. Clark (1980). Salinity effects on yield and fruit quality of "Valencia" orange. *J. Am. Soc. Hort. Sci.*, 105:199–202.

91. L. E. Francois, T. J. Donovan, K. Lorenz, and E. V. Maas (1989). Salinity effects on rye grain yield, quality, vegetative growth, and emergence. *Agron. J.*, 81:707–712.

92. L. E. Francois, T. Donovan, and E. V. Maas (1984). Salinity effects on seed yield, growth, and germination of grain sorghum. *Agron. J.*, 76:741–744.

93. L. E. Francois, T. J. Donovan, E. V. Maas, and G. L. Rubenthaler (1988). Effect of salinity on grain yield and quality, vegetative growth, and germination of triticale. *Agron. J.*, 80:642–647.

94. L. E. Francois and D. W. West (1982). Reduction in yield and market quality of celery caused by soil salinity. *J. Am. Soc. Hort. Sci.*, 107:952–954.

95. J. Gale, H. C. Kohl, and R. M. Hagan (1967). Changes in the water balance and photosynthesis of onion, bean and cotton plants under saline conditions. *Physiol. Plant.*, 20:408–420.

96. H. Genard, J. Le Saos, and J. Boucaud (1988). [Effect of NaCl on foliar lipid content and composition in *Suaeda maritima* (L.) Dum. var. *Macrocarpa* Moq. with special reference to galactolipids and phosphatidylglycerol.] *C. R. Acad. Sci. Paris*, 306:75–80.

97. C. J. Gerard and E. Hinojosa (1973). Cell wall properties of cotton roots as influenced by calcium and salinity. *Agron. J.*, 65:556–560.

98. A. C. Gibson (1982). The anatomy of succulence. *Crassulacean Acid Metabolism* (I. P. Ting and M. Gibbs, eds.), American Society of Plant Physiologists, Rockville, MD, pp. 1–17.

99. A. D. M. Glass (1983). Regulation of ion transport. *Annu. Rev. Plant Physiol.*, 34:311–326.

100. J. Gorham, O. S. Tomar, and R. G. Wyn Jones (1988). Salinity-induced

changes in the chemical composition of *Leucaena leucocephala* and *Sesbania bispinosa*. *J. Plant Physiol.*, *132*:678–682.

101. S. R. Grattan and C. M. Grieve (1992). Mineral element acquisition and growth response of plants grown in saline environments. *Agric. Ecosyst. Environ.* *38*:275–300.

102. H. Greenway and R. Munns (1980). Mechanisms of salt tolerance in non-halophytes. *Annu. Rev. Plant Physiol.*, *31*:149–190.

103. A. M. Grieve and R. R. Walker (1983). Uptake and distribution of chloride, sodium and potassium ions in salt-stressed citrus plants. *Aust. J. Agric. Res.*, *34*:133–143.

104. J. W. Gronwald, C. G. Suhayda, M. Tal, and M. C. Shannon (1990). Reduction in plasma membrane ATPase activity of tomato roots by salt stress. *Plant Sci.*, *66*:145–153.

105. A. R. C. Haas (1950). Effect of sodium chloride on Mexican, Guatelmalan, and West Indian avocado seedlings. *Calif. Avocado Soc. Yearbk.*, *35*:153–160.

106. A. R. C. Haas (1952). Sodium effects on avocado rootstocks. *Calif. Avocado Soc. Yearbk.*, *37*:159–166.

107. M. A. Hajibagheri, T. J. Flowers, and J. L. Hall (1985). Cytometric aspects of the leaves of the halophyte *Suaeda maritima*. *Physiol. Plant.*, *64*:365–370.

108. M. A. Hajibagheri, J. L. Hall, and T. J. Flowers (1983). The structure of the cuticle in relation to cuticular transpiration in leaves of the halophyte *Suaeda maritima* (L.). *Dum. New Phytol.*, *94*:125–131.

109. M. A. Hajibagheri, A. R. Yeo, and T. J. Flowers (1985). Salt tolerance in *Suaeda maritima* (L.) Dum. Fine structure and ion concentration in the apical region of roots. *New Phytol.*, *99*:331–343.

110. R. Hall, G. Hofstra, and G. P. Lumis (1972). Effects of deicing salts on eastern white pine: foliar injury, growth suppression and seasonal changes in foliar concentrations of sodium and chloride. *J. For. Res.*, *2*:244–249.

111. R. B. Harding, M. P. Miller, and M. Fireman (1958). Absorption of salts by citrus leaves during sprinkling with water suitable for surface irrigation. *Proc. Am. Soc. Hort. Sci.*, *71*:248–256.

112. D. M. R. Harvey, R. Stelzer, R. Brandtner, and D. Kramer (1985). Effects of salinity on ultrastructure and ion distribution in roots. *Physiol. Plant.*, *66*:328–338.

113. D. M. R. Harvey and J. R. Thorpe (1986). Some observations on the effects of salinity on ion distributions and cell ultrastructure in wheat leaf mesophyll cells. *J. Exp. Bot.*, *37*:1–7.

114. M. Hassidim, Y. Braun, H. R. Lerner, and L. Reinhold (1986). Studies on H^+-translocating ATPase in plants of varying resistance to salinity. 2. K^+ strongly promotes development of membrane potential in vesicles from cotton roots. *Plant Physiol.*, *81*:1057–1061.

115. S. Hata, K. Shirata, and H. Takagishi (1987). Effect of alteration of sterol composition on the salt sensitivity of a tobacco cell suspension culture. *Plant Cell Physiol.*, *28*:959–961.

116. H. E. Hayward (1955). Factors affecting the salt tolerance of horticultural

crops. *Rept. XIV Int. Hort. Congr.*, Wageningen, Netherlands, pp. 385–399.

117. H. E. Hayward and L. Bernstein (1958). Plant-growth relationships on salt-affected soils. *Bot. Rev.*, *24*:584–635.
118. H. E. Hayward, E. M. Long, and R. Uhvits (1946). Effect of chloride and sulfate salts on the growth and development of the Elberta peach on Shalil and Lovell rootstocks. *USDA Tech. Bull.*, 922.
119. H. M. Helal and K. Mengel (1981). Interaction between light intensity and NaCl salinity and their effects of growth, CO_2 assimilation, and photosynthate conversion in young broad beans. *Plant Physiol.*, *67*:999–1002.
120. F. Helfferich (1962). *Ion Exchange*, McGraw-Hill, New York.
121. A. E. Hill and B. S. Hill (1976). Elimination processes by glands: mineral ions. *Encyclopedia of Plant Physiology. Transport in Plants 2. Part B. Tissues and Organs* (U. Lüttge and M. G. Pitman, eds.), Springer-Verlag, New York, pp. 225–243.
122. G. J. Hoffman, P. R. Catlin, R. M. Mead, R. S. Johnson, L. E. Francois, and D. Goldhamer (1989). Yield and foliar injury responses of mature plum trees to salinity. *Irrig. Sci.*, *10*:215–229.
123. G. J. Hoffman, J. A. Jobes, and W. J. Alves (1983). Response to tall fescue to irrigation water salinity, leaching fraction, and irrigation frequency. *Agric. Water Manage.*, *7*:439–456.
124. G. J. Hoffman, E. V. Maas, and S. L. Rawlins (1973). Salinity-ozone interactive effects on yield and water relations of pinto bean. *J. Environ. Qual.*, *2*:148–152.
125. G. J. Hoffman, E. V. Maas, and S. L. Rawlins (1975). Salinity-ozone interactive effects on alfalfa yield and water relations. *J. Environ. Qual.*, *4*:326–331.
126. G. J. Hoffman, J. D. Oster, E. V. Maas, J. D. Rhoades, and J. van Schilfgaarde (1984). Minimizing salt in drain water by irrigation management—Arizona field station studies with citrus. *Agric. Water Manage.*, *9*:61–78.
127. G. J. Hoffman and S. L. Rawlins (1970). Design and performance of sunlit climate chambers. *Trans. Am. Soc. Agric. Eng.*, *13*:656–660.
128. G. J. Hoffman, S. L. Rawlins, M. J. Garber, and E. M. Cullen (1971). Water relations and growth of cotton as influenced by salinity and relative humidity. *Agron. J.*, *63*:822–826.
129. H. Howie and J. Lloyd (1989). Response of orchard "Washington navel" orange, *Citrus sinensis* (L.) Osbeck to saline irrigation water. 2. Flowering, fruit set and fruit growth. *Aust. J. Agric. Res.*, *40*:371–380.
130. S. S. T. Hua, V. Y. Tsai, G. M. Lichens, and A. T. Noma (1982). Accumulation of amino acids in Rhizobium sp. strain WR1001 in response to NaCl salinity. *Appl. Environ. Microbiol.*, *44*:135–140.
131. C. X. Huang and R. F. M. van Stevenick (1988). Effect of moderate salinity on patterns of potassium, sodium and chloride accumulation in cells near the root tip of barley: Role of differentiating metaxylem vessels. *Physiol. Plant.*, *73*:525–533.
132. C. X. Huang and R. F. M. van Steveninck (1990). Salinity induced struc-

tural changes in meristematic cells of barley roots. *N. Phytol.*, *115*:17–22.

133. J. F. Imhoff (1986). Osmoregulation and compatible solutes in eubacteria. *FEMS Microbiol. Rev.*, *39*:57–66.

134. W. M. Jarrell and R. A. Virginia (1990). Response of mesquite to nitrate and salinity in a simulated phreatic environment: water use, dry matter and mineral nutrient accumulation. *Plant Soil*, *125*:185–196.

135. C. D. John and A. Läuchli (1980). Metabolic adaption in mature roots of salt-stressed *Zea mays*. *Ann. Bot.*, *46*:395–400.

136. J. A. Jobes, G. J. Hoffman, and J. D. Wood (1981). Leaching requirement for salinity control. 2. Oat, tomato, and cauliflower. *Agric. Water Manage.*, *4*:393–407.

137. P. R. Kemp and G. L. Cunningham (1981). Light, temperature and salinity effects on growth, leaf anatomy and photosynthesis of *Distichlis spicata* (L.) Greene. *Am. J. Bot.*, *68*:507–516.

138. L. M. Kent and A. Läuchli (1985). Germination and seedling growth of cotton: salinity-calcium interactions. *Plant Cell Environ.*, *8*:155–159.

139. E. J. M. Kirby (1988). Analysis of leaf, stem and ear growth in wheat from terminal spikelet stage to anthesis. *Field Crop Res.*, *18*:127–140.

140. D. Kramer (1983). The possible role of transfer cells in the adaptation of plants to salinity. *Physiol. Plant.*, *58*:549–555.

141. D. Kramer (1984). Cytological aspects of salt tolerance in higher plants. *Salinity Tolerance in Plants: Strategies for Crop Improvement* (R. C. Staples and G. H. Toenniessen, eds.), John Wiley and Sons, New York, pp. 3–15.

142. P. J. C. Kuiper (1968). Lipids in grape roots in relation to chloride transport. *Plant Physiol.*, *43*:1367–1371.

143. N. Kume and K. Maruyama (1986). Effects of NaCl on *Vicia* chromosomes studied by scanning electron microscopy. *J. Electron Microsc.*, *35*:288–291.

144. E. Kurth, G. R. Cramer, A. Läuchli, and E. Epstein (1986). Effects of NaCl and $CaCl_2$ on cell enlargement and cell production in cotton roots. *Plant Physiol.*, *82*:1102–1106.

145. N. Lakshminarayanaiah (1984). *Equations of Membrane Biophysics*, Academic Press, New York.

146. A. Läuchli (1976). Apoplasmic transport in tissues. *Transport in Plants. 2. Part B. Tissues and Organs* (U. Lüttge and M. G. Pitman, eds.), Springer-Verlag, New York, pp. 3–34.

147. A. Läuchli (1984). Salt exclusion: An adaptation of legumes for crops and pastures under saline conditions. *Salinity Tolerance in Plants: Strategies for Crop Improvement* (R. C. Staples and G. H. Toenniessen, eds.), John Wiley and Sons, New York, pp. 171–187.

148. A. Läuchli and E. Epstein (1990). Plant responses to saline and sodic conditions. *Agricultural Salinity Assessment and Management* (K. K. Tanji, ed.), Am. Soc. Civil Eng. Manuals and Reports on Engineering Practice No. 71. ASCE, New York, pp. 113–137.

149. J. R. Lawton, A. Todd, and D. K. Naidoo (1981). Preliminary investigations into the structure of roots of the mangroves, *Avicennia marina* and

Bruguiera gymnorrhiza, in relation to ion uptake. *New Phytol.*, 88:713–722.

150. N. Liphschitz and Y. Waisel (1982). Adaptation of plants to saline environments: salt excretion and glandular structure. *Contributions of the Ecology of Halophytes: Tasks for Vegetative Science*, Vol. 2 (D. N. Sen and K. S. Rajpurohit, eds.), W. Junk The Hague, pp. 187–214.

151. D. J. Longstreth and P. S. Nobel (1979). Salinity effects on leaf anatomy: consequences for photosynthesis. *Plant Physiol.*, 63:700–703.

152. G. P. Lumis, G. Hofstra, and R. Hall (1973). Sensitivity of roadside trees and shrubs to aerial drift of deicing salts. *Hortscience*, 8:475–477.

153. O. R. Lunt, J. J. Oertli, and H. C. Kohl, Jr. (1960). Influence of environmental conditions on the salinity tolerance of several plant species. *7th Int. Congr. Soil Sci.*, Madison, WI, pp. 560–570.

154. U. Lüttge (1975). Salt glands. *Ion Transport in Plant Cells and Tissues* (D. A. Baker and J. L. Hall, eds.), North-Holland, Amsterdam, pp. 335–376.

155. U. Lüttge and J. A. C. Smith (1984). Structural, biophysical, and biochemical aspects of the role of leaves in plant adaptation to salinity and water stress. *Salinity Tolerance in Plants: Strategies for Crop Improvement* (R. C. Staples and G. H. Toenniessen, eds.), John Wiley and Sons, New York, pp. 125–150.

156. E. V. Maas (1990). Crop salt tolerance. *Agricultural Salinity Assessment and Management* (K. K. Tanji, ed.), Am. Soc. Civil Eng. Manuals and Reports on Engineering Practice No. 71, ASCE, New York, pp. 262–304.

157. E. V. Maas, R. A. Clark, and L. E. Francois (1982). Sprinkler-induced foliar injury to pepper plants: effects of irrigation frequency, duration and water composition. *Irrig. Sci.*, 3:101–109.

158. E. V. Maas, S. R. Grattan, and G. Ogata (1982). Foliar salt accumulation and injury in crops sprinkled with saline water. *Irrig. Sci.*, 3:157–168.

159. E. V. Maas and C. M. Grieve (1990). Spike and leaf development in salt-stressed wheat. *Crop. Sci.*, 30:1309–1313.

160. E. V. Maas and C. M. Grieve (1992). Salt tolerance of plants at different stages of growth. *Proc. Int. Conf. Tando Jam*, Pakistan, Jan. 7–11, 1990 (in press).

161. E. V. Maas and G. J. Hoffman (1977). Crop salt tolerance—current assessment. *J. Irrig. Drainage Div.*, ASCE, 103(IR2):115–134.

162. E. V. Maas, G. J. Hoffman, G. D. Chaba, J. A. Poss, and M. C. Shannon (1983). Salt sensitivity of corn at various growth stages. *Irrig. Sci.*, 4:45–57.

163. E. V. Maas, G. J. Hoffman, S. L. Rawlins, and G. Ogata (1973). Salinity-ozone interactions on pinto bean: Integrated response to ozone concentration and duration. *J. Environ. Qual.*, 2:400–404.

164. E. V. Maas and R. H. Nieman (1978). Physiology of plant tolerance to salinity. In *Crop Tolerance to Suboptimal Land Conditions* (G. A. Jung, ed.), ASA Spec. Publ. 32:277–299.

165. E. V. Maas and J. A. Poss (1989). Salt sensitivity of wheat at various growth stages. *Irrig. Sci.*, 10:29–40.

166. E. V. Maas, J. A. Poss, and G. J. Hoffman (1986). Salinity sensitivity of

sorghum at three growth stages. *Irrig. Sci.*, *7*:1–11.

167. O. C. Magistad, A. D. Ayers, C. H. Wadleigh, and H. G. Gauch (1943). Effect of salt concentration, kind of salt, and climate on plant growth in sand cultures. *Plant Physiol.*, *18*:151–166.

168. N. E. Marcar (1987). Salt tolerance in the genus *Lolium* (ryegrass) during germination and growth. *Aust. J. Agric. Res.*, *38*:297–307.

169. K. B. Marcum and C. L. Murdock (1990). Growth responses, ion relations, and osmotic adaptations of eleven C4 turfgrasses to salinity. *Agron. J.*, *82*:892–896.

170. E. Marre and A. Ballarin-Denti (1985). The proton pumps of the plasmalemma and the tonoplast of higher plants. *J. Bioenerg. Biomemb.*, *17*:1–21.

171. H. Marschner (1986). *Mineral Nutrition of Higher Plants*, Academic Press, London.

172. T. McNeilly, M. Ashraf, and C. Veltkamp (1987). Leaf micromorphology of sea cliff and inland plants of *Agrostis stolonifera* L., *Dactylis glomerata* L. and *Holcus Ianatus* L. *New Phytol.*, *106*:261–269.

173. A. Meiri, G. J. Hoffman, M. C. Shannon, and J. A. Poss (1982). Salt tolerance of two muskmelon cultivars under two radiation levels. *J. Am. Soc. Hort. Sci.*, *107*:1168–1172.

174. A. Meiri and Z. Plaut (1985). Crop production and management under saline conditions. *Plant Soil*, *89*:253–271.

175. M. J. Meyer, M. A. L. Smith, and S. L. Knight (1989). Salinity effects on St. Augustine grass: a novel system to quantify stress response. *J. Plant Nutr.*, *12*:893–908.

176. F. B. Morrison (1956). *Feeds and Feeding*, Morrison, Ithaca, NY.

177. R. Munns, H. Greenway, and G. O. Kirst (1983). Halotolerant eukaryotes. *Physiological Plant Ecology. 3. Responses to the Chemical and Biological Environment* (O. L. Lange, P. S. Nobel, C. B. Osmond, and H. Ziegler, eds.), Springer-Verlag, Berlin, pp. 59–135.

178. R. Munns and A. Termaat (1986). Whole-plant responses to salinity. *Aust. J. Plant Physiol.*, *13*:143–160.

179. H. Nassery and R. L. Jones (1976). Salt-induced pinocytosis in barley and bean. *J. Exp. Bot.*, *27*:358–367.

180. F. Navari-Izzo, M. F. Quartacci, and R. Izzo (1989). Lipid changes in maize seedling in response to field water deficits. *J. Exp. Bot.*, *215*:675–680.

181. B. H. Ng (1987). The effects of salinity on growth, nodulation and nitrogen fixation of *Casuarina equisetifolia*. *Plant Soil*, *103*:123–125.

182. R. H. Nieman and L. L. Poulsen (1967). Interactive effects of salinity and atmospheric humidity on the growth of bean and cotton plants. *Bot. Gaz.*, *128*:69–73.

183. R. H. Nieman and L. L. Poulsen (1971). Plant growth suppression on saline media: interactions with light. *Bot. Gaz.*, *132*:14–19.

184. C. L. Noble, G. M. Halloran, and D. W. West (1984). Identification and selection for salt tolerance in Lucerne (*Medicago sativa* L.). *Aust. J. Agric. Res.*, *35*:239–252.

185. H. E. Oberlander (1966). The relative importance of active and passive processes in ion translocation across root tissue. *Limiting Steps in Ion Uptake by Plants from Soil*, Int. Atomic Energy Agency, Vienna, pp. 101–120.

186. G. Ogata and E. V. Maas (1974). Interactive effects of salinity and ozone on growth and yield of garden beet. *J. Environ. Qual.*, 2:518–520.

187. O. T. Okusanya and T. Fawole (1985). The possible role of phosphate in the salinity tolerance of *Lavatera arborea*. *J. Ecol.*, 73:317–322.

188. O. T. Okusanya and I. A. Ungar (1983). The effect of time and seed production on the germination response of *Spergularia marina*. *Physiol. Plant.*, 59:335–341.

189. O. T. Okusanya and I. A. Ungar (1984). The growth and mineral composition of three species of *Spergularia* as affected by salinity and nutrients at high salinity. *Am. J. Bot.*, 71:439–447.

190. C. B. Osmond, O. Bjorkman, and D. J. Anderson (1980). Physiological processes in plant ecology: Toward a synthesis with *Atriplex*. *Ecology Studies*, Vol. 36, Springer-Verlag, New York.

191. J. D. Oster (1982). Gypsum usage in irrigated agriculture: a review. *Fert. Res.*, 3:73–89.

192. A. L. Page and A. C. Chang (1990). Deficiencies and toxicities of trace elements. *Agricultural Salinity Assessment and Management* (K. K. Tanji, ed.), Am. Soc. Civil Eng. Manuals and Reports on Engineering Practice No. 71. ASAE, New York, pp. 138–160.

193. I. Papadopoulos and V. V. Rendig (1983). Tomato plant response to soil salinity. *Agron. J.*, 75:696–700.

194. A. Parsegian (1969). Energy of an ion crossing a low dielectric membrane: solutions of four relevant electrostatic problems. *Nature*, 221:884–886.

195. G. A. Pearson and L. Bernstein (1959). Salinity effects at several growth stages of rice. *Agron. J.*, 51:654–657.

196. A. Poljakoff-Mayber (1975). Morphological and anatomical changes inplants as a response to salinity stress. *Plants in Saline Environments* (A. Poljakoff-Mayber and J. Gale, eds.), Springer-Verlag, New York, pp. 97–117.

197. A. Poljakoff-Mayber (1988). Ecological-physiological studies on the responses of higher plants to salinity and drought. *Sci. Rev. Arid Zone Res.*, 6:163–183.

198. M. U. Rahman and S. A. Kayani (1988). Effects of chloride type of salinity on root growth and anatomy of corn (*Zea mays* L.). *Biologie*, 34:123–131.

199. R. S. Rana (1985). Breeding for salt resistance: concept and strategy. *Int. J. Trop. Agric.*, 111(4):236–254.

200. G. Rao, S. K. M. Basha, and G. R. Rao (1981). Effect of sodium chloride salinity on amount and composition of epicuticular wax and cuticular transpiration rate in peanut *Arachis hypogaea*. *Indian J. Exp. Bot.*, 19:880–881.

201. J. A. Raven (1985). Regulation of pH and generation of osmolarity in vascular plants: a cost-benefit analysis in relation to efficiency of use of energy, nitrogen and water. *New Phytol.*, 101:25–77.

202. H. M. Rawson (1986). Gas exchange and growth in wheat and barley grown in salt. *Aust. J. Plant Physiol.*, *13*:475–489.

203. P. Reddell, R. C. Foster, and G. D. Bowen (1986). The effects of sodium chloride on growth and nitrogen fixation in *Casuarine obesa* Miq. *New Phytol.*, *102*:397–408.

204. P. Rengasamy (1987). Importance of calcium in irrigation with saline-sodic water—a viewpoint. *Agric. Water Manage.*, *12*:207–219.

205. J. D. Rhoades (1972). Quality of water for irrigation. *Soil Sci.*, *113*:177–184.

206. S. P. Robinson and W. J. S. Downton (1985). Potassium, sodium and chloride ion concentrations in leaves and isolated chloroplasts of the halophyte *Suaeda australis* R. Br. *Aust. J. Plant Physiol.*, *12*:471–449.

207. D. W. Rush and E. Epstein (1981). Breading and selection for salt tolerance by the incorporation of wild germplasm into a domestic tomato. *J. Am. Soc. Hort. Sci.*, *106*:699–704.

208. M. Salim (1989). Effects of salinity and relative humidity on growth and ionic relations of plants. *New Phytol.*, *113*:13–20.

209. G. R. Sandhu and K. A. Malik (1975). Plant succession—a key to the utilization of saline soils. *The Nucleus*, *12*:35–38.

210. G. Schroppel-Meier and W. M. Kaiser (1988). Ion homeostasis in chloroplasts under salinity and mineral deficiency. 1. Solute concentrations in leaves and chloroplasts from spinach plants under NaCl or $NaNO_3$ salinity. *Plant Physiol.*, *87*:822–827.

211. S. Schubert and A. Läuchli (1986). Na^+ exclusion, H^+ release, and growth of two different maize cultivars under NaCl salinity. *J. Plant Physiol.*, *126*:145–154.

212. S. Schubert and A. Läuchli (1990). Sodium exclusion mechanisms at the root surface of two maize cultivars. *Plant Soil*, *123*:205–209.

213. M. Schwarz and J. Gale (1981). Maintenance respiration and carbon balance of plants at low levels of sodium chloride salinity. *J. Exp. Bot.*, *32*:933–941.

214. M. Schwarz and J. Gale (1984). Growth response to salinity at high levels of carbon dioxide. *J. Exp. Bot.*, *35*:193–196.

215. R. C. Setia and S. Narang (1985). Interactive effects of NaCl salinity and growth regulators on vascular tissue differentiation in pea (*Pisum sativum* L.) roots. *Phytomorphology*, *35*:207–211.

216. J. Shalhevet, E. V. Maas, G. J. Hoffman, and G. Ogata (1976). Salinity and the hydraulic conductance of roots. *Physiol. Plant.*, *38*:224–232.

217. J. Shalhevet and B. Yaron (1973). Effect of soil and water salinity on tomato growth. *Plant Soil*, *39*:285–292.

218. J. Shalhevet, B. Yaron, and U. Horowitz (1974). Salinity and citrus yield—an analysis of results from a salinity survey. *J. Hort. Sci.*, *49*:15–27.

219. M. C. Shannon (1987). *Salinity—An Environmental Constraint on Crop Productivity*, 4th Aust. Agron. Soc. Conf., La Trobe Univ., Melbourne, Australia, Aug. 24–17, 1987.

220. M. C. Shannon and C. L. Noble (1990). Genetic approaches for developing economic self-tolerant crops. *Agricultural Salinity Assessment and Management* (K. K. Tanji, ed.), Am. Soc. Civil Eng. Manuals and Reports on Engineering Practice No. 71, ASCE, New York, pp. 161–185.

221. J. M. Simonds and D. M. Orcutt (1988). Free and conjugated desmethyl-sterol composition of *Zea mays* hybrids exposed to mild osmotic stress. *Physiol. Plant.*, 72:395–402.

222. D. Smith, A. K. Dobrenz, and M. H. Schonhorst (1981). Response of alfalfa seedling plants to high levels of chloride-salts. *J. Plant Nutr.*, 4:143–174.

223. M. M. Smith, M. J. Hodson, H. Opik, and S. J. Wainwright (1982). Salt-induced ultrastructural damage to mitochondria in root tips of a salt-sensitive ecotype of *Agrostis stolonifera*. *J. Exp. Bot.*, 33:886–895.

224. M. Solomon, R. Ariel, M. J. Hodson, A. M. Mayer, and A. Poljakoff-Mayber (1987). Ion absorption and allocation of carbon resources in excised pea roots grown in liquid medium in absence or presence of NaCl. *Ann. Bot.*, 59:387–398.

225. M. Solomon, R. Ariel, A. M. Mayer, and A. Poljakoff-Mayber (1989). Reversal by calcium of salinity-induced growth inhibition in excised pea roots. *Israel Bot.*, 38:65–69.

226. M. Solomon, E. Gedalovich, A. M. Mayer, and A. Poljakoff-Mayber (1986). Changes induced by salinity to the anatomy and morphology of excised pea roots in culture. *Ann. Bot.*, 57:811–818.

227. C. Sonneveld, and W. Voogt (1983). Studies on the salt tolerance of some flower crops grown under glass. *Plant Soil*, 74:41–52.

228. R. M. Spanswick (1981). Electrogenic ion pumps. *Annu. Rev. Plant Physiol.*, 32:267–289.

229. S. J. Stavarek and D. W. Rains (1985). Effect of salinity on growth and maintenance costs of plant cells. *Cellular and Molecular Biology of Plant Stress* (J. L. Key and T. Kosuge, eds.), Alan R. Liss, New York, pp. 161–185.

230. W. P. Stein (1967). *The Movement of Molecules Across Cell Membranes*, Academic Press, New York.

231. J. Steinborn and R. J. Roughley (1975). Toxicity of sodium and chloride ions to Rhizobium spp. in broth and peat culture. *J. Appl. Bact.*, 39:133–138.

232. R. Stelzer and A. Läuchli (1977). Salt and flooding tolerance of *Puccinellia peisonis*. 2. Structural differentiation of the root in relation to function. *A. Pflanzenphysiol.*, 84:95–108.

233. C. E. E. Stuiver, P. J. C. Kuiper, and H. Marschner (1978). Lipids in bean, barley and sugar beet in relation to salt resistance. *Physiol. Plant.*, 42:124–128.

234. C. G. Suhayda, J. L. Giannini, D. P. Briskin, and M. C. Shannon (1990). Electrostatic changes in *Lycopersicon esculentum* root plasma membrane resulting from salt stress. *Plant Physiol.*, 93:471–478.

235. J. P. Syvertsen and G. Yelenosky (1988). Salinity can enhance freeze toler-

ance of citrus rootstock seedlings by modifying growth, water relations, and mineral nutrition. *J. Am. Soc. Hort. Sci.*, *113*:889–893.

236. H. Sze (1985). H$^+$-translocating ATPase: Advances using membrane vesicles. *Annu. Rev. Plant Physiol.*, *36*:175–208.

237. W. W. Thomson (1975). The structure and function of salt glands. *Plants in Saline Environments* (A. Poljakoff-Mayber and J. Gale, eds.), Springer-Verlag, New York, pp. 118–146.

238. W. W. Thomson, C. D. Faraday, and J. W. Oross (1988). Salt glands. *Solute Transport in Plant Cells and Tissues* (D. A. Baker and J. L. Hall, eds.), Longman, Essex, pp. 498–537.

239. A. M. Townsend (1980). Response of selected tree species to sodium chloride. *J. Am. Soc. Hort. Sci.*, *113*:889–893.

240. G. V. Udovenko and N. S. Tsibkovskaya (1983). Dynamics of changes in ultrastructure of cellular organelles in barley leaves under conditions of salinization. *Sov. Plant Physiol.*, *30*:397–405.

241. United States Salinity Laboratory Staff (1954). *Diagnosis and Improvement of saline and Alkali Soils.* USDA Handb. 60. 160 p. U.S. Government Printing Office, Washington, D.C.

242. R. G. Upchurch and G. H. Elkan (1977). Comparison of colony morphology, salt tolerance and effectiveness in *Rhizobium japonicum. Can. J. Microbiol.*, *23*:1118–1122.

243. T. H. van den Honert (1948). Water transport in plants as a catenary process. *Disc. Faraday Soc.*, *3*:146.

244. Y. Waisel (1985). The stimulating effects of NaCl on root growth of Rhodes grass (*Chloris gayana*). *Physiol. Plant.*, *64*:519–522.

245. Y. Waisel and S. W. Breckle (1987). Differences in responses of various radish roots to salinity. *Plant Soil*, *104*:191–194.

246. Y. Waisel, A. Eshel, and M. Agami (1986). Salt balance of leaves of the mangrove *Avicennia marina. Physiol. Plant.*, *67*:67–72.

247. R. R. Walker and T. J. Douglas (1983). Effect of salinity on uptake and distribution of chloride, sodium and potassium ions in citrus plants. *Aust. J. Agric. Res.*, *34*:145–153.

248. R. R. Walker, M. Sedgley, M. A. Blesing, and T. J. Douglas (1984). Anatomy, ultrastructure and assimilate concentrations of roots of citrus genotypes differing in ability for salt exclusion. *J. Exp. Bot.*, *35*:1481–1494.

249. R. Weimberg, H. R. Lerner, and A. Poljakoff-Mayber (1984). Changes in growth and water-soluble solute concentration in *Sorghum bicolor* stressed with sodium and potassium salts. *Physiol. Plant.*, *62*:472–480.

250. E. Werker, H. R. Lerner, R. Weimberg, and A. Poljakoff-Mayber (1983). Structural changes occurring in nuclei of barley root cells in response to a combined effect of salinity and ageing. *Am. J. Bot.*, *70*:222–225.

251. D. W. West (1978). Water use and sodium chloride uptake by apple trees. 2. The response to soil oxygen deficiency. *Plant Soil*, *50*:51–65.

252. D. W. West and J. A. Taylor (1981). Germination and growth of cultivars of *Trifolium subterraneum* L. in the presence of sodium chloride salinity. *Plant Soil*, *62*:221–230.

253. D. W. West and J. A. Taylor (1984). Response of six grape cultivars to the combined effects of high salinity and rootzone waterlogging. *J. Am. Soc. Hort. Sci.*, *109*:844–851.

254. I. Winicov (1990). Gene expression in salt tolerant alfalfa cell cultures and salt tolerant plants regenerated from these cultures. *Progress in Plant Cellular and Molecular Biology* (H. J. J. Nijkamp, L. H. W. van der Plas, and J. van Aartijk, eds.), Kluwer, Dordrecht, The Netherlands, pp. 142–147.

255. I. Winicov and J. D. Button (1991). Accumulation of photosynthesis gene transcripts in response to sodium chloride by salt-tolerant alfalfa cells. *Planta*, *183*:478–483.

256. E. Winter (1982). Salt tolerance of *Trifolium alexandrinum* L. 3. Effects of salt on ultrastructure of phloem and xylem transfer cells in petioles and leaves. *Aust. J. Plant Physiol.*, *9*:239–250.

257. H. K. Wutscher (1979). Citrus rootstocks. *Hort. Rev.*, *1*:237–269.

258. R. G. Wyn Jones (1981). Salt tolerance. *Physiological Processes Limiting Plant Productivity* (C. B. Johnson, ed.), Butterworths, London, pp. 271–292.

259. R. G. Wyn Jones (1984). Phytochemical aspects of osmotic adaptation. *Phytochemical Adaptation to Stress* (B. N. Timmermann, C. Steelink, and F. A. Loewus, eds.), Plenum Press, New York, pp. 55–78.

260. R. G. Wyn Jones and J. Gorham (1983). Osmoregulation. *Encyclopedia of Plant Physiology*, New Series, Vol. 12C, Physiological Plant Ecology, III (O. L. Lange, P. S. Nobel, C. B. Osmond, and H. Ziegler, eds.), Springer-Verlag, Berlin, pp. 35–58.

261. R. G. Wyn Jones, R. Storey, R. A. Leigh, N. Ahmad, and A. Pollard (1977). A hypothesis on cytoplasmic osmoregulation. *Regulation of Cell Membrane Activity in Plants* (E. Marre and O. Ciferri, eds.), Elsevier, Amsterdam, pp. 121–136.

262. N. K. Yadav and S. R. Vyas (1973). Salts and pH tolerance of rhizobia. *Folia Microbiol.*, *18*;242–247.

263. Y. W. Yang, R. J. Newton, and F. R. Miller (1990). Salinity tolerance in *Sorghum*. Whole plant response to sodium chloride in *S. bicolor* and *S. halepense*. *Crop Sci.*, *30*:775–781.

264. M. M. Yelton, S. S. Yang, S. A. Edie, and S. T. Lim (1983). Characterization of an effective salt-tolerant, fast-growing strain of *Rhizobium japonicum*. *J. Gen. Microbiol.*, *129*:1537–1547.

265. A. R. Yeo (1983). Salinity resistance: physiologies and prices. *Physiol. Plant.*, *58*:214–222.

266. A. R. Yeo and T. J. Flowers (1980). Salt tolerance in the halophyte *Suaeda maritima* L. Dum: evaluation of the effect of salinity upon growth. *J. Exp. Bot.*, *31*:1171–1183.

267. A. R. Yeo and T. J. Flowers (1982). Accumulation and localisation of sodium ions within the shoots of rice (*Oryza sativa*) varieties differing in salinity stress. *Physiol. Plant.*, *56*:343–348.

268. A. R. Yeo and T. J. Flowers (1983). Varietal differences in the toxicity of sodium ions in rice leaves. *Physiol. Plant.*, *59*:189–195.

269. A. R. Yeo, D. Kramer, A. Läuchli, and J. Gullasch (1977). Ion distribution in salt-stressed mature *Zea mays* roots in relation to ultrastructure and retention of sodium. *J. Exp. Bot.*, *28*:17–29.

270. A. R. Yeo, M. E. Yeo, S. A. Flowers, and T. J. Flowers (1990). Screening of rice (*Oryza sativa* L.) genotypes for physiological characters contributing to salinity resistance, and their relationship to overall performance. *Theor. Appl. Genet.*, *79*:377–384.

271. V. B. Youngner, O. R. Lunt, and F. Nudge (1967). Salinity tolerance of seven varieties of creeping bentgrass, *Agrostis palustris* Huds. *Agron. J.*, *59*:335–336.

272. H. M. Younis, G. Weber, and J. S. Boyer (1983). Activity and conformational changes in chloroplast coupling factor induced by ion binding: formation of magnesium-enzyme-phosphate complex. *Biochemistry*, *22*:2505–2512.

273. M. Zarrouk and A. Cherif (1983). Teneur en lipides de halophytes et resistance au sel. *Z. Pflanzenphysiol.*, *112*:373–380.

274. M. Zekri and L. R. Parsons (1989). Growth and root hydraulic conductivity of several citrus rootstocks under salt and polyethylene glycol stresses. *Physiol. Plant.*, *77*:99–106.

275. I. Zidan, H. Azaizeh, and P. M. Neuman (1990). Does salinity reduce growth in maize root epidermal cells by inhibiting their capacity for cell wall acidification? *Plant Physiol.*, *93*:7–11.

8

Symbiotic Nitrogen Fixation

John G. Streeter

The Ohio State University
Wooster, Ohio

INTRODUCTION

Nitrogen is the element that most often limits the growth and yield of crop plants. Although the earth contains an enormous quantity of elemental nitrogen in its rocks and air, only a tiny fraction of this nitrogen is in a chemical form that can be utilized directly by plants (63). Plants have essentially only two options for acquisition of their nitrogen. The first is from mineral forms of nitrogen (generally nitrate) from the soil solution and, in general, this nitrogen comes from chemical fertilizers.

The second option is biological nitrogen fixation. Although numerous microorganisms are capable of the synthesis of ammonium from dinitrogen gas, bacteria known as rhizobia are of greatest agricultural importance because of the associations formed with legume plants. Although this process is restricted to certain crop plants, it is of enormous importance because it is environmentally "friendly"; little of the nitrogen fixed escapes to the soil and the process permits the "delivery" of this scarce nutrient to farms in developing countries where fertilizer nitrogen may not be available. Improvements in biological nitrogen fixation would also have significant economic impact on U.S. agriculture, estimated at as much as $4.5 billion per year (65).

This chapter focuses on symbiotic nitrogen fixation for two reasons. The first is the importance of the process to worldwide food production

245

just discussed. The second reason is that legume nodules are complex plant organs subject to many biological and environmental constraints. Effects of environmental stresses are much more acute for nitrogen fixation than for the uptake and assimilation of mineral nitrogen. As a result, much more work has been done in an attempt to understand stress effects on nitrogen fixation than on the other nitrogen input pathway.

There has been a relatively small amount of work on the effects of salinity on symbiotic nitrogen fixation; a few sample studies are those of Paul Singleton (57,58). In contrast to the paucity of work on effects of salinity on nodule performance, there has been considerable effort on the effects of salinity on survival of rhizobia in the soil. Some of the literature on osmoregulatory responses of rhizobia to salinity has been discussed in Chapter 7, and the subject will not be considered further here.

Most of the work on stress effects on nodule function has focused on temperature stress and drought stress. These two topics constitute the bulk of this review. In addition, although this book emphasizes plant stress, some studies on the effects of desiccation on rhizobia in soil will be briefly considered because this subject is a component of general influence of water availability on symbiotic nitrogen fixation. Finally, because this book is for the nonspecialist, I will devote some space to a short description of the structural features of nodules which are pertinent to an understanding of stress effects.

EFFECTS OF TEMPERATURE AND DESICCATION ON THE SURVIVAL OF RHIZOBIA

Although the surface layers on soil may reach or surpass 40°C (32), there have been few studies on temperature stress per se. In an early major study, Marshall (40) established that fast-growing species such as *Rhizobium meliloti* are more susceptible to temperature stress that the slow-growing species such as *R. lupini* and *Bradyrhizobium japonicum*. This is consistent with the fact that most fast-growing species infect temperate legumes while the slow growers generally infect tropical legumes. Marshall also established that clays such as montmorillonite have a protective effect for the bacteria. In addition to differences among species, differences between strains of *B. japonicum* have been found in culture (44) and in the soil (32). The basis for these interesting differences in temperature tolerance among species and strains of rhizobia have apparently not been pursued.

There has been much more effort regarding the effects of water stress on rhizobia and only a portion of this literature is reviewed here. Many of the responses are similar to those for temperature stress. For example, there are differences between fast- and slow-growing types, the former being more susceptible to desiccation (8). In addition, clays again have a protective effect, but for desiccation, clay did not have a protective effect for the slow-growing species (8).

B. *japonicum* has been divided into serogroups, i.e., serologically related strains, and there are discernible differences between serogroups in water stress tolerance (38). Also, there are major differences in stress tolerance among strains of R. *leguminosarum* biovar *trifolii* (21) and strains of B. *japonicum* (3). To make matters even more complex, there are interactions between B. *japonicum* strains and soil texture classes (39), so that predicting the response of any given strain in any given environment is difficult.

There have been a few attempts to understand the basis for these highly variable effects of desiccation. Because rhizobia are prolific producers of extracellular polysaccharides and because these slimy materials are known to play a role in desiccation tolerance of other gram-negative bacteria, the role of extracellular polysaccharide in the protection of rhizobia has been tested. However, in neither R. *meliloti* nor R. *leguminosarum* bv *trifolii* was extracellular polysaccharide production found to be important in protecting cells against desiccation (9,41). Under some circumstances organic carbon content of the soil was found to be related to survival of rhizobia in dry soils, so perhaps some organic material other than polysaccharides are important in protection against desiccation (10).

As typical gram-negative bacteria, rhizobia are susceptible to desiccation, and the above studies uniformly illustrate the drastic decline in rhizobial numbers which may occur under conditions of soil drying. But options for improving the survival of rhizobia remain to be elucidated.

DEVELOPMENT AND STRUCTURE OF LEGUME NODULES

Rhizobia which come into contact with root hairs of the appropriate legume host enter the root hair and the subtending cells by way of a structure called an infection thread. Simultaneous to this infection event, the cortical cells of the root begin to divide rapidly and a swelling of the root occurs in a few days. Vascular bundles connected to the root vascular system extend into the developing nodule. As the infection thread grows through the central cells of the developing nodule, bacteria in the thread are released into membranous sacks in the cytoplasm of these cells.

The final structure is a spherical or cylindrical growth on the root which may even exceed the diameter of the root. An example for soybean

[*Glycine max* (L.) Merr.] is shown in Figure 1. The earliest nodules form on the primary or tap root, but nodule formation continues throughout plant growth period resulting in most nodules being formed on lateral roots.

Nodule structure at the level of tissues is illustrated in Figure 2. An important feature of nodules which is pertinent to this chapter is that vascular bundles are not continuous around the periphery of the nodule. That is, each bundle comes to a "dead end," with the tips of the xylem

Figure 1 Nitrogen-fixing nodules formed on the roots of soybean by the bacterium *Bradyrhizobium japonicum*. The rough appearance of the nodules is due to the presence of lenticels on the exterior surface of the nodules. Magnification is about 5×.

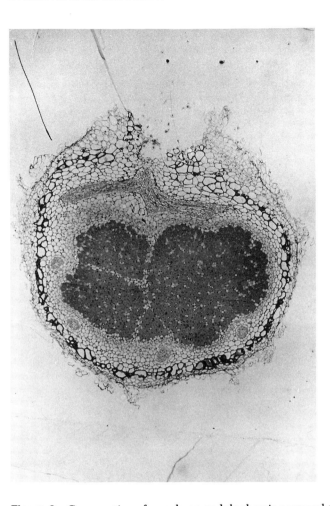

Figure 2 Cross-section of a soybean nodule showing general tissue regions. The interior portion (darker cells) is the infected zone; most of these cells are packed with bacteria ("bacteroids"), but there are some uninfected cells (lighter) in this tissue zone. Exterior to the infected cells is the nodule cortex. The dark layer consists of sclerenchyma cells which give the nodule structural strength. Outside of the sclerenchyma layer, the cells are loosely organized and have large intercellular spaces. Inside of the sclerenchyma layer, vascular bundles are imbedded in the cortex. These generally appear as circles of cells, but in this section a portion of a vascular trace has been sectioned longitudinally at the top of the picture which also represents the point at which the nodule was originally attached to the root. Magnification is about 50 ×.

and phloem traces being surrounded by endodermis (68). This vascular arrangement is important because it means that export of the nitrogenous products of fixation in the xylem is dependent on the import of water in the phloem (67). To give this even more emphasis, we can assert that the water status of the nodule is dependent on water import in the phloem. There are some theoretical arguments that phloem import cannot account for the entire water requirement of nodules (52), and there is some recent evidence for the symplastic transport of water through the cortical parenchyma of nodules (64). Thus, some as yet undefined portion of the nodule water requirement may be satisfied by nonvascular water transport, but there is general agreement that most of the water transport into nodules is via the phloem.

Before continuing with the main body of this chapter, some definitions are provided for the reader who is not familiar with the literature on symbiotic nitrogen fixation. Additional details and references for these terms can be found in a recent review (62).

Bacteroids and Cytosol

Bacteroid is the term given to the bacterial symbiont; it refers to the highly differentiated form of the organism which is also active in the fixation of gaseous nitrogen. Cytosol is an operational term given to the supernatant following high-speed centrifugation of a nodule extract. The cytosol mainly represents the host cytoplasm, but it also contains various soluble components from mitochondria, vacuoles, nuclei, plastids, etc.

Acetylene Reduction Assay

Nitrogenase, the enzyme complex which converts N_2 ($N \equiv N$) to ammonium in bacteroids, is also active with several other substrates. One of these, acetylene ($CH \equiv CH$), provides a convenient assay for nitrogenase activity. The product ethylene ($CH_2 = CH_2$) cannot be further metabolized by bacteroids or the host and is rapidly excreted from nodules. Ethylene can easily be separated from acetylene by gas chromatography and the rate of ethylene formation provides an estimate of nitrogenase activity.

Relative Efficiency

Another nitrogenase substrate, the H^+ ion, is converted to H_2 in a wasteful side reaction. All nitrogenases studied so far use a portion of the electrons supplied to support this reaction. In contrast, the reduction of acetylene consumes all electrons supplied to the enzyme and thus provides a convenient estimate of total electron flux through the enzyme complex.

The difference between electron flux coupled to H_2 generation and electron flux for acetylene reduction provides an estimate of relative efficiency. Relative efficiency is defined as $RE = 1 - (H_2 \text{ evolution}/C_2H_2$ reduction). This value can be relatively easily measured using isolated bacteroids or intact nodules, and, as we will see, the relative efficiency of nodules is influenced by a variety of factors.

Diffusion Barrier and Gas Permeability

In the inner cortex of nodules there is a layer of cells which has no intercellular spaces and constitutes a major barrier to the diffusion of gases. The main impact of this barrier is probably to limit the diffusion of O_2 to the interior regions of the nodule. There is evidence indicating that the resistance to diffusion provided by the barrier is variable. The resistance of the barrier can be measured and it is influenced by a wide variety of treatments and environmental factors.

EFFECTS OF TEMPERATURE STRESS ON NODULE FUNCTION

Effects of Low Temperature

Although there has been relatively little interest in effects of low temperature, the need to optimize legume performance in cold climates has prompted some studies. In an early study, no effect of growing *Vicia faba* at 10°C on nodule structure was noted, but cold temperatures significantly delayed nodule development and increased nodule size (volume/nodule) (22). As would be expected, legume plants grown at temperatures below optimum have lower specific acetylene reduction activity (18,22,53), and the increase in nodule volume may be a response which compensates for the depression in specific fixation activity. In studies reminiscent of those on rhizobia in soils, negative effects of cold temperatures on nodules are dependent on *Rhizobium* strain and on plant species (46,53).

A few attempts have been made to understand mechanisms underlying the effects of cold on nodule function, but these results are not in complete agreement. Rennie and Kemp (53) reported lower relative efficiency of *Phaseolus vulgaris* nodules at lower temperatures. But other more detailed studies with *Lotus pedunculatus* nodules (46) and with *Glycine max* nodules (36) indicated that relative efficiency increases with a lowering of temperature. The explanation for the difference in response may be related to plant species, but the issue remains unresolved. It should be emphasized that the effect of cold on the CO_2/C_2H_4 ratio (i.e., nodule respiration/nitrogenase activity) also differed between *Lotus pedunculatus* and *Glycine max* (36,46). It is important to note that in all of these studies nodule respiration, acetylene reduction activity, and hydrogen evolution were all decreasing substantially in response to temperature lowering; thus, the ratios represent differences in the relative *rate* of decline, and small differences among species in the decline rate may lead to large differences among species in these ratios.

Diurnal Cycles and Temperature Coupling

Following the development of the acetylene reduction assay, continuous assays allowed for the analysis of diurnal cycles of nitrogenase activity in nodules, and these were generally interpreted in terms of the availability of photosynthate to support symbiotic fixation (30,42). However, more detailed studies with soybean showed that most or all of the diurnal variation in acetylene reduction could be accounted for by variation in temperature (55). In fact, recent studies with field-grown soybeans showed high correlations between soil temperature and acetylene reduction activity ($R^2 = 0.85$) even when changes in soil temperature were manipulated to be out of phase with changes in ambient temperature (15,70). Thus, as effects of extreme temperatures are considered, it is important to remember that nodule activity and temperature are closely and positively coupled within some "normal" range of temperatures.

Of course, the optimum temperature for nodule activity does vary among legumes, with temperate legumes having lower optima (23–25°C) than the optima for tropical legumes (30–32°C), as would be expected (23,37,54,56,66). Studies on the temperature optima for white clover nodules (*Trifolium repens* L.) are especially interesting because they included a comparison of nitrate-fed plants and plants totally dependent on nitrogen fixation. These studies show that fixation is more sensitive to temperature than nitrate assimilation and that nitrate assimilation dominates at the extremes of high and low root temperatures (23,37).

Effects of High Temperature

As we would anticipate from results already described in other sections, the negative effects of high temperature vary depending on plant species (11,69), on plant cultivar (45), and on strain of *Rhizobium* (44,45). To complicate matters even further, LaFavre and Eaglesham (34) found that differences among rhizobial strains in response to high temperature were not the same in culture as in the symbiotic state. Still another complicating factor is that the application of starter nitrogen partially alleviated subsequent negative effects of high temperature on soybean nodule function (35).

A somewhat surprising result is that the nodule formation process is much more temperature-tolerant than nodule function (26,34). For example, for *Phaseolus vulgaris*, nodule formation was unaffected at temperatures as high as 38°C (26). Another important consideration is whether high temperature is constant or is lowered for the dark period. For example, Date and Roughly (11) found two *Trifolium* species to be tolerant of a day/night 30°C/25°C temperature regime, but growth and nitrogen fixation were severely impaired with a day/night 30°C/30°C temperature treatment. Obviously, in reading this literature on temperature stress effects, it is important to consider carefully the protocol involved in each experiment.

Attempts to understand mechanisms underlying the effects of high temperature on nodule function have included analysis of carbohydrate fractions in nodules, including total nonstructural carbohydrate (66) or individual sugars (27). In neither of these studies did the evidence indicate that carbohydrate deprivation might explain the temperature effects. Hernandez-Armenta et al. (27) did note that there were major changes in the structure of nodules transferred to 38°/22°C for 6 days, principally the apparent destruction of the membranes surrounding the bacteroids.

Effects of high temperature on relative efficiency have also been reported for *Pisum sativum* and *Vigna unguiculata*, namely, relative efficiency declined with increasing temperature for both legumes (7,51). In spite of the negative effect of high temperature on the relative efficiency of nitrogenase, the overall energy cost of nitrogen fixation in *Trifolium repens* nodules was found to be relatively insensitive to temperature over a range from 5°C to 32°C (54). This conclusion was based on the close coupling of respiration and nitrogenase activity, so that the CO_2/C_2H_4 ratio was largely invariant over the temperature range studied (54).

GENERAL EFFECTS OF WATER STRESS ON NODULE FUNCTION

A logical place to begin this section is with the evidence that responses of legume nodules to temperature stress and to water stress are interactive. Pankhurst and Sprent found for both *Phaseolus vulgaris* and *Glycine max* that water-stressed nodules have a lower optimum temperature for nitrogen-fixing activity (48). The first detailed studies of the effects of water stress were reported for soybean nodules by Sprent in 1971, and this work emphasized the changes in nodule structure, especially the possibility for rupture of plasmodesmata because of the shrinkage of cytoplasm in the uninfected cells (59). Based on her early studies, Sprent concluded that "water supply was the major environmental factor affecting nitrogen fixation" (60).

The nitrogen fixation activity of nodules is largely independent of transpiration rate (16) and is not influenced by the supplying of water to the nodule surface (31). These findings are consistent with the "dead end" nature of the vascular bundles in the nodule cortex, discussed in an earlier section, and they emphasize that the water status of the nodule is dependent on water import in the phloem. Thus, water potentials in nodules on stressed plants may be as low as those in leaves, and nitrogenase activity in soybean nodules was linearly related to water potential over a wide range of values (6). In spite of the great sensitivity of nitrogenase to water stress, recovery from stress may occur in a matter of hours, as long as nitrogenase is not completely destroyed by the stress treatment (4,49).

Legumes have two options for acquisition of nitrogen and the effects of stress on nitrate-grown plants vs. plants entirely dependent on nitrogen fixation have been compared. Using three cycles of moderate water stress on *Trifolium subterraneum*, DeJong and Phillips (14) found that growth of nitrate-dependent plants was depressed slightly less than those dependent on N_2 fixation. Soybean plants grown with nitrate and treated with continuous, moderate water stress had greater root development, higher transpiration rates, and lower threshold values of leaf water potential for stomatal closure than nonstressed plants (43). Plants dependent on nitrogen fixation and subjected to water stress responded somewhat differently, having retarded root development, lower transpiration rates, and stomatal closure at higher leaf water potentials than nonstressed plants. These authors concluded that nitrate-dependent plants adopted strategies to tolerate stress whereas nodulated plants developed strategies to avoid stress (43).

POSSIBLE MECHANISMS UNDERLYING WATER STRESS EFFECTS

Carbohydrate Deprivation

Because water stress has long been known to inhibit photosynthesis, it is reasonable to predict that the sensitivity of nitrogen fixation in legume nodules to water stress may be mediated via effects on carbohydrate supply. The early reports of Huang et al. (29) supported this view, showing roughly coincident declines in photosynthesis and acetylene reducation activity of soybean plants. Furthermore, the inhibition of acetylene reduction activity at low water potentials could be partially reversed by supplying higher concentrations of CO_2 to shoots (30). The overall results of Huang et al. led them to conclude that "inhibition of shoot photosynthesis accounted for the inhibition of nodule acetylene reduction at low water potentials" (30).

Subsequent studies have led to different conclusions. More detailed analyses of the time course of photosynthesis, transpiration, and nodule activity show that these processes are not well correlated during the onset or recovery from water stress (2,17). In a recent study, continuation of a low level of nodule activity during a stress treatment which eliminated photosynthesis of soybean plants was found (17). Application of a localized water stress lowered nitrogenase activity in cowpea nodules by 67%, but the sugar content of nodules was lowered by only 26% (33). Also, analysis of carbohydrate composition of soybean nodules showed higher concentrations of sucrose in stressed nodules than in nonstressed nodules (20). These authors concluded that carbohydrate utilization in nodules (transport or metabolism) might limit nodule activity during stress but that supply of carbohydrate to the nodule should not be the cause of lowered activity (20).

Studies on the energy requirement for symbiotic fixation also support the notion that carbohydrate deprivation is not a cause of decreased fixation activity under stress. Studies on *Cicer arietinum*, *Trifolium subterraneum*, and *Medicago sativa* show that although both nodule respiration and acetylene reduction activity are depressed under conditions of water stress, the ratio of CO_2/C_2H_4 was greater under stress conditions (1,13,28). Thus, the energy cost for symbiotic N_2 fixation increased under stress conditions, in the case of *Medicago sativa*, as much as sevenfold (1). When considered relative to the knowledge that carbohydrate concentrations in stressed nodules are higher than those in nonstressed nodules, this makes it unlikely that carbon utilization was seriously impaired under stress. The most logical overall conclusion seems to be that the depression of nitrogenase under conditions of water stress is not caused by a lack of reduced carbon or by impairment of carbohydrate utilization.

Restriction in Oxygen Diffusion

The effect of supplemental oxygen was first reported by Pankhurst and Sprent who found that increasing the O_2 concentration around detached soybean nodules completely restored rates of respiration and nitrogenase activity following moderate water stress (47). In contrast to the response of intact nodules, acetylene reduction activity of nodule slices, nodule breis, and isolated bacteroids was similar in fully turgid and water-stressed nodules. This led to the conclusion that a major effect of stress was to restrict the diffusion of gas (specifically O_2) through the cortex of the nodules (47). In general, these effects have been confirmed by other groups for other legumes (12,19,25). One difference is that bacteroids isolated from *Vicia faba* nodules following mild stress were found to be more sensitive to changes in O_2 than bacteroids isolated from control nodules (25).

Recently two detailed studies have been published, one using *Trifolium subterraneum* and the other using *Glycine max;* both of these studies were designed to compare the impact of decreased supply of photosynthate to nodules with the impact of changes in resistance to gas diffusion. Both groups concluded that an increase in diffusion resistance best explains the effects of nodule dehydration during whole-plant water stress (12,19). Unfortunately, the mechanisms which control the variable barrier to gas diffusion in nodules are not yet known.

Other Possible Mechanisms

Sprent and Gallacher (61) showed that ethanol accumulates in water-stressed nodules and suggested that the net effect of the anaerobiosis could be a lowering of energy charge in stressed nodules. This was actually confirmed by Patterson et al. (50) who showed that energy charge in soybean nodules declined concurrently with acetylene reduction activity during the onset of water stress. However, during a recovery phase, energy charge values quickly returned to control levels whereas acetylene reduction activity recovered only slowly. The authors concluded that the availability of ATP was probably not the key to explaining the effects of stress (50).

Increases in intracellular pH and decreases in leghemoglobin content of nodules have led other workers to suggest that there may be an increase in proteolytic activity in water-stressed *Vicia faba* nodules (24). This suggestion was supported by actual measurements of increased alkaline protease activity in nodule cytosol and by changes in nodule ultrastructure. These changes occurred over periods of several days; thus, although they may represent important long-term effects, it seems unlikely that they can explain effects which can be measured in periods of stress lasting only a few hours. Finally, Becana et al. (5) showed that the cytosolic enzymes involved in ammonium assimilation in *Medicago sativa* nodules were not significantly perturbed in water-stressed nodules.

SUMMARY AND CONCLUSIONS

Legume nodules have a complex structure which is apparently required to accommodate the specialized chemical and physical conditions necessary to promote and sustain nitrogenase activity in the bacteroids. Delicate balances in import and export of solutes, in influx and efflux of gases, and in host and bacteroid metabolism seem to have been established and seem to be required to permit efficient symbiotic nitrogen fixation. Thus, it is perhaps not surprising that this system is easily perturbed by environmental stresses.

Although the literature is not clear on many points, it is clear that temperature and water stress may have rapid and large negative effects on fixation. In the "real world" environment, temperature stress may be less important. That is, except in desert and some tropical settings, soil temperatures probably rarely reach inhibitory levels after the crop canopy is established, assuming that precipitation is sufficient to provide some evaporative cooling. There are clear differences between temperature optima for tropical and temperature legume plants and for the rhizobia which infect them. Thus, the symbiotic system appears to have evolved to accommodate temperature differences among environments. This suggestion is also supported by the positive correlation between temperature and nitrogenase activity in nodules over a "reasonable" range of expected soil temperatures.

Water status of soils is probably much more variable and, therefore, water stress is probably a much more common determinant of fixation activity in the field. The main short-term effect of dehydration appears to be to increase the resistance to gas diffusion in nodules. More specifically, nodule function may generally be oxygen-limited and the decrease in gas diffusion in stressed nodules appears to accentuate this oxygen starvation. It is not clear how this problem might be circumvented, but future studies should focus on mechanisms which may underlie the control of the cortical diffusion barrier.

ACKNOWLEDGMENT

Support was provided by state and federal funds appropriated to the Ohio Agricultural Research and Development Center and The Ohio State University. I thank Seppo Salminen and Peter Hartel for suggestions regarding organization and content of the manuscript.

REFERENCES

1. J. Aguirreolea and M. Sánchez-Díaz (1989). CO_2 evolution by nodulated roots in *Medicago sativa* L. under water stress. *J. Plant Physiol.* 134:598–602.

2. S. L. Albrecht, J. M. Bennett, and K. J. Boote (1984). Relationship of nitrogenase activity to plant water stress in field-grown soybeans. *Field Crops Res.*, 8:61–71.

3. R. K. Al-Rashidi, T. E. Loynachan, and L. R. Frederick (1982). Desicca-

tion tolerance of four strains of *Rhizobium japonicum*. *Soil Biol. Biochem.*, *14*:489–493.

4. P. M. Aparicio-Tejo, M. F. Sánchez-Díaz, and J. I. Peña (1980). Nitrogen fixation, stomatal response and transpiration in Medicago sativa, Trifolium repens and T. subterraneum under water stress and recovery. *Physiol. Plant.*, *48*:1–4.

5. M. Becana, P. M. Aparicio-Tejo, and M. F. Sánchez-Díaz (1984). Effects of water stress on enzymes of ammonia assimilation in root nodules of alfalfa (*Medicago sativa*). *Physiol. Plant.*, *61*:653–657.

6. J. M. Bennett and S. L. Albrecht (1984). Drought and flooding effects on N_2 fixation, water relations, and diffusive resistance of soybean. *Agron. J.*, *76*:735–740.

7. H. Bertelsen (1985). Effect of temperature on H_2 evolution and acetylene reduction in pea nodules and in isolated bacteroids. *Plant Physiol.*, *77*:335–338.

8. H. V. A. Bushby and K. C. Marshall (1977). Water status of rhizobia in relation to their susceptibility to desiccation and to their protection by montmorillonite. *J. Gen. Microbiol.*, *99*:19–27.

9. H. V. A. Bushby and K. C. Marshall (1977). Some factors affecting the survival of root-nodule bacteria on desiccation. *Soil Biol. Biochem.*, *9*:143–147.

10. W.-L. Chao and M. Alexander (1982). Influence of soil characteristics on the survival of *Rhizobium* in soils undergoing drying. *Soil Sci. Soc. Am. J.*, *46*:949–952.

11. R. A. Date and R. J. Roughly (1986). Effects of root temperature on the growth and nitrogen fixation of *Trifolium semipilosum* and *Trifolium repens*. *Exp. Agric.*, *22*:133–147.

12. A. G. Davey and R. J. Simpson (1990). Nitrogen fixation by subterranean clover at varying stages of nodule dehydration. I. Carbohydrate status and short-term recovery of nodulated root respiration. *J. Exp. Bot.*, *41*:1175–1187.

13. A. G. Davey and R. J. Simpson (1990). Nitrogen fixation by subterranean clover at varying stages of nodule dehydration. II. Efficiency of nitrogenase functioning. *J. Exp. Bot.*, *41*:1189–1197.

14. T. M. DeJong and D. A. Phillips (1982). Water stress effects on nitrogen assimilation and growth of *Trifolium subterraneum* L. using dinitrogen or ammonium nitrate. *Plant Physiol.*, *69*:416–420.

15. R. F. Dennison and T. R. Sinclair (1985). Diurnal and seasonal variation in dinitrogen fixation (acetylene reduction) rates by field-grown soybeans. *Agron. J.*, *77*:679–684.

16. R. deVisser and H. Poorter (1984). Growth and root nodule nitrogenase activity of *Pisum sativum* as influenced by transpiration. *Physiol. Plant.*, *61*:637–642.

17. A. Djekoun and C. Planchon (1991). Water status effect on dinitrogen fixation and photosynthesis in soybean. *Agron. J.*, *83*:316–322.

18. S. H. Duke, L. E. Schrader, C. A. Henson, J. C. Servaites, R. D. Vogel-

zang, and J. W. Pendleton (1979). Low root temperature effects on soybean nitrogen metabolism and photosynthesis. *Plant Physiol.*, *63*:956–962.

19. J.-L. Durand, J. E. Sheehy, and F. R. Minchin (1987). Nitrogenase activity, photosynthesis and nodule water potential in soybean plants experiencing water deprivation. *J. Exp. Bot.*, *38*:311–321.

20. R. J. Fellows, R. P. Patterson, C. D. Raper, and D. Harris (1987). Nodule activity and allocation of photosynthate of soybean during recovery from water stress. *Plant Physiol.*, *84*:456–460.

21. J. Fuhrmann, C. B. Davey, and A. G. Wollum II (1986). Desiccation tolerance of clover rhizobia in sterile soils. *Soil Sci. Soc. Am. J.*, *50*:639–644.

22. A. Fyson and J. I. Sprent (1982). The development of primary root nodules on *Vicia faba* L. grown at two temperatures. *Ann. Bot.*, *50*:681–692.

23. A. J. Gordon, J. H. MacDuff, G. J. A. Ryle, and C. E. Powell (1989). White clover N_2-fixation in response to root temperature and nitrate. *J. Exp. Bot.*, *40*:527–534.

24. V. Guérin, D. Pladys, J.-C. Trinchant, and J. Rigaud (1991). Proteolysis and nitrogen fixation in faba-bean (*Vicia faba*) nodules under water stress. *Physiol. Plant.*, *82*:360–366.

25. V. Guérin, J.-C. Trinchant, and J. Rigaud (1990). Nitrogen fixation (C_2H_2 reduction) by broad bean (*Vicia faba* L.) nodules and bacteroids under water-restricted condition. *Plant Physiol.*, *92*:595–601.

26. R. Hernandez-Armenta, H. C. Wien, and A. R. J. Eaglesham (1989). Maximum temperature for nitrogen fixation in common bean. *Crop Sci.*, *29*:1260–1265.

27. R. Hernandez-Armenta, H. C. Wien, and A. R. J. Eaglesham (1989). Carbohydrate partitioning and nodule function in common bean after heat stress. *Crop Sci.*, *29*:1292–1297.

28. R. S. Hooda, I. S. Sheoran, and R. Singh (1990). Partitioning and utilization of carbon and nitrogen in nodulated roots and nodules of chickpea (*Cicer arietinum*) grown at two moisture levels. *Ann. Bot.*, *65*:111–120.

29. C.-Y. Huang, J. S. Boyer, and L. N. Vanderhoef (1975). Acetylene reduction (nitrogen fixation) and metabolic activities of soybean having various leaf and nodule water potentials. *Plant Physiol.*, *56*:222–227.

30. C.-Y. Huang, J. S. Boyer, and L. N. Vanderhoef (1975). Limitation of acetylene reduction (nitrogen fixation) by photosynthesis in soybean having low water potentials. *Plant Physiol.*, *56*:228–232.

31. D. J. Hume, J. G. Criswell, and K. R. Stevenson (1976). Effects of soil moisture around nodules on nitrogen fixation by well watered soybeans. *Can. J. Plant Sci.*, *56*:811–815.

32. A. S. Kennedy and A. G. Wollum II (1988). Enumeration of *Bradhyrhizobium japonicum* in soil subjected to high temperature: Comparison of plate count, most probable number and fluorescent antibody techniques. *Soil Biol. Biochem.*, *20*:933–937.

33. R. Khanna-Chopra, K. R. Koundal, and S. K. Sinha (1984). A simple technique of studying water deficient effects on nitrogen fixation in nodules without influencing the whole plant. *Plant Physiol.*, *76*:254–256.

34. A. K. LaFavre and A. R. J. Eaglesham (1986). The effects of high tempera-
    tures on soybean nodulation and growth with different strains of bra-
    dyrhizobia. *Can. J. Microbiol.*, *32*:22–27.

35. A. K. LaFavre and A. R. J. Eaglesham (1987). Effects of high temperatures
    and starter nitrogen on the growth and nodulation of soybean. *Crop Sci.*,
    27:742–745.

36. D. B. Layzell, P. Rochman, and D. T. Canvin (1984). Low root tempera-
    tures and nitrogenase activity in soybean. *Can J. Bot.*, *62*:965–971.

37. J. H. MacDuff and M. S. Shanoa (1990). N_2 fixation and nitrate uptake by
    white clover swards in response to root temperature in flowing solution cul-
    ture. *Ann. Bot.*, *65*:325–335.

38. R. L. Mahler and A. G. Wollum II (1980). Influence of water potential on
    the survival of Rhizobia in a Goldsboro loamy sand. *Soil Sci. Soc. Am. J.*,
    44:988–992.

39. R. L. Mahler and A. G. Wollum II (1981). The influence of soil water
    potential and soil texture on the survival of *Rhizobium japonicum* and
    Rhizobium leguminosarum isolates in the soil. *Soil Sci. Soc. Am. J.*,
    45:761–766.

40. K. C. Marshall (1964). Survival of root-nodule bacteria in dry soils exposed
    to high temperatures. *Aust. J. Agric. Res.*, *15*:273–281.

41. P. Mary, D. Ochin, and R. Tailliez (1986). Growth status of rhizobia in
    relation to their tolerance to low water activities and desiccation stresses.
    Soil Biol. Biochem., *18*:179–184.

42. H. J. Mederski and J. G. Streeter (1977). Continuous, automated acetylene
    reduction assays using intact plants. *Plant Physiol.*, *59*:1076–1081.

43. M. I. Minguez and F. Sau (1989). Responses of nitrate-fed and nitrogen-
    fixing soybeans to progressive water stress. *J. Exp. Bot.*, *40*:497–502.

44. F. Munevar and A. G. Wollum II (1981). Growth of *Rhizobium japonicum*
    strains at temperatures above 27°C. *Appl. Environ. Microbiol.*, *42*:272–
    276.

45. F. Munevar and A. G. Wollum II (1982). Response of soybean plants to
    high root temperature as affected by plant cultivar and *Rhizobium* strain.
    Agron. J., *74*:138–142.

46. C. E. Pankhurst and D. B. Layzell (1984). The effect of bacterial strain and
    temperature changes on the nitrogenase activity of *Lotus pedunculatus* root
    nodules. *Physiol. Plant.*, *62*:404–409.

47. C. E. Pankhurst and J. I. Sprent (1975). Effects of water stress on the
    respiratory and nitrogen-fixing activity of soybean root nodules. *J. Exp.
    Bot.*, *26*:287–304.

48. C. E. Pankhurst and J. I. Sprent (1976). Effects of temperature and oxygen
    tension on the nitrogenase and respiratory activities of turgid and water-
    stressed soybean and French bean root nodules. *J. Exp. Bot.*, *27*:1–9.

49. S. Pararajasingham and D. P. Knievel (1990). Nitrogenase activity of
    cowpea (*Vigna unguiculata* (L.) Walp.) during and after drought stress.
    Can. J. Plant Sci., *70*:163–171.

50. R. P. Patterson, C. D. Raper, Jr., and H. D. Gross (1979). Growth and

specific nodule activity of soybean during application and recovery of a leaf moisture stress. *Plant Physiol.*, *64*:551–556.

51. R. M. Rainbird, C. A. Atkins, and J. S. Pate (1983). Effect of temperature on nitrogenase functioning in cowpea nodules. *Plant Physiol.*, *73*:392–394.

52. J. A. Raven, J. I. Sprent, S. G. McInroy, and G. T. Hay (1989). Water balance of N₂-fixing root nodules: can phloem and xylem transport explain it? *Plant Cell Environ.*, *12*:683–688.

53. R. J. Rennie and G. A. Kemp (1981). Dinitrogen fixation in *Phaseolus vulgaris* at low temperatures: interaction of temperature, growth stage, and time of inoculation. *Can. J. Bot.*, *60*:1423–1427.

54. G. J. A. Ryle, C. E. Powell, M. K. Timbrell, and A. J. Gordon (1989). Effect of temperature on nitrogenase activity in white clover. *J. Exp. Bot.*, *40*:733–739.

55. L. E. Schweitzer and J. E. Harper (1980). Effect of light, dark, and temperature on root nodule activity (acetylene reduction) of soybeans. *Plant Physiol.*, *65*:51–56.

56. T. R. Sinclair and P. R. Weisz (1985). Response to soil temperature on dinitrogen fixation (acetylene reduction) rates by field-grown soybeans. *Agron. J.*, *77*:685–688.

57. P. W. Singleton (1983). A split-root growth system for evaluating the effect of salinity on components of the soybean *Rhizobium japonicum* symbiosis. *Crop Sci.*, *23*:259–262.

58. P. W. Singleton and B. B. Bohlool (1983). Effect of salinity on the functional components of the soybean—*Rhizobium japonicum* symbiosis. *Crop Sci.*, *23*:815–818.

59. J. I. Sprent (1971). The effects of water stress on nitrogen-fixing root nodules. I. Effects on the physiology of detached soygean nodules. *N. Phytol.*, *70*:9–17.

60. J. I. Sprent (1972). The effects of water stress on nitrogen-fixing root nodules. IV. Effects on whole plants of *Vicia faba* and *Glycine max*. *N. Phytol.*, *72*:603–611.

61. J. I. Sprent and A. Gallacher (1976). Anaerobiosis in soybean root nodules under water stress. *Soil Biol. Biochem.*, *8*:317–320.

62. J. G. Streeter (1991). Transport and metabolism of carbon and nitrogen in legume nodules. *Adv. Bot. Res.*, *18*:129–187.

63. J. G. Streeter and A. L. Barta (1984). Nitrogen and minerals. *Physiological Basis of Crop Growth and Development* (M. B. Tesar, ed.), American Society of Agronomy, Madison, WI.

64. J. G. Streeter and S. O. Salminen (1992). Evidence supporting a nonphloem source of water for export of solutes in the xylem of soybean root nodules. *Plant Cell Environ.* *15*:735–741.

65. L. W. Tauer (1989). Economic impact of future biological nitrogen fixation technologies on United States agriculture. *Plant Soil*, *119*:261–270.

66. K. M. Trang and J. Giddens (1980). Shading and temperature as environmental factors affecting growth, nodulation, and symbiotic N₂ fixation by soybeans. *Agron. J.* *72*:305–308.

67. K. B. Walsh (1990). Vascular transport and soybean nodule function. III. Implications of a continual phloem supply of carbon and water. *Plant Cell Environ.*, *13*:893–901.

68. K. B. Walsh, M. E. McCully, and M. J. Canny (1989). Vascular transport and soybean nodule function: nodule xylem is a blind alley, not a throughway. *Plant Cell Environ.*, *12*:395–405.

69. G. J. Waughman (1977). The effect of temperature on nitrogenase activity. *J. Exp. Bot.*, *28*:949–960.

70. P. R. Weisz and T. R. Sinclair (1988). Soybean nodule gas permeability, nitrogen fixation and diurnal cycles in soil temperature. *Plant Soil*, *109*:227–234.

9

Plant Response Mechanisms
to Soil Compaction

M. J. Vepraskas

North Carolina State University
Raleigh, North Carolina

INTRODUCTION

Compaction occurs when soil is compressed by the wheels of agricultural implements or rainfall. During the compression, soil particles are rearranged and pressed closer together so that large pores or macropores are filled in or crushed (1,2). The overall result of compaction is that the soil's total porosity is reduced. Compaction affects plant growth by reducing the amount of soil air available to roots, and by increasing the pressures that roots and seedlings must exert to penetrate the soil (3,4). A soil is said to be "compacted" when it contains dense layers that reduce crop yields as a result of poor seedling emergence, localized waterlogging that causes restricted aeration, or restricted root growth that results in the plant not obtaining adequate moisture or nutrients from the soil. Knowing the extent to which these conditions actually reduce yields in a given field is difficult to measure. Alternatively, a compacted soil can be considered one in which soil loosening may increase yields.

Compaction is a widespread problem, but its economic impact is difficult to quantify. It can affect soils throughout the United States, and compacted soils probably occur on all continents where agriculture is practiced. Gill (5) estimated that the loss in crop value in 1964 due to soil compaction in the United States was $1.2 billion annually. Such losses are probably greater today because the weights of agricultural implements have increased since 1964 and caused compaction to be more extensive.

The objectives of this chapter are (a) to review how compaction can be measured and characterized, (b) to describe how plant roots respond to compacted soil layers, and (c) to describe how compacted soil layers affect crop yield.

MEASUREMENTS OF SOIL COMPACTION

Compaction has been assessed primarily by measuring three soil properties: bulk density, porosity, and mechanical impedance. These properties are interrelated but each has advantages for evaluating compaction.

Bulk Density

Bulk density is the mass of oven-dry soil per unit bulk volume of soil (6). The term "bulk volume" includes the volume of both solids and pores in a given mass of soil. The bulk volume is usually determined before the soil sample is oven-dried at 105°C. Volumes of "field-moist" samples are measured if the soil material does not shrink appreciably on drying. For shrinking soils, the soil material can be equilibrated to a specific soil water potential.

Bulk density has been considered the most direct quantitative measure of compaction (7). For soils where pronounced volume change does not occur by wetting and drying, the bulk density value remains constant until the soil is compressed further, or is loosened by tillage or by the growth of roots, worms, or other organisms. Measurement of bulk density is simple but labor-intensive (8).

Porosity

Porosity is the total volume of pores per unit bulk volume of soil. The compaction process reduces the soil porosity by expelling some of the air and water that occupies soil pores (9). Chancellor (7) considered porosity to be the most meaningful property to use when discussing compaction because it is related to the total soil volume available to plant roots, water, and air.

Porosity can be estimated from bulk density using the equation:

$$P = 1 - \frac{BD}{PD} \qquad [1]$$

where P is porosity, BD bulk density, and PD the average density of the soil particles. Most mineral soils have an average PD of approximately

2.65 g/cm^3, which is also the average density of quartz, the most abundant soil mineral. Porosity can also be measured directly on soil samples that are saturated with water so that water fills all pores. The volume of water held in the soil is then determined by the soil's loss in weight following oven-drying.

Equation [1] indicates that as bulk density increases porosity decreases. This is not the whole story because as bulk density increases, it is primarily the larger pores or "macropores" that are destroyed first. These pores are important because they allow roots to grow through the soil with little resistance to their elongation. Large cracks near the soil surface allow seedlings to emerge with little resistance to their growth.

There is not universal agreement on the definition of macropores or their dimensions (10). From a functional standpoint, Greenland (11) defined "transmission pores" (a type of macropore) as having equivalent cylindrical diameters (ecd) from 50 to 500 μm and "fissures" as having ecd's >500 μm. Transmission pores are large enough to allow some roots to grow through them, and to allow movement of water or air into the soil. The effect of increasing bulk density on the percentage of pores of different diameters is shown in Table 1. Pores having ecd's >100 μm were reduced as bulk density increased from 1.24 to 1.52 g/cm^3.

Mechanical Impedance

Mechanical impedance is the physical resistance a soil offers to the elongation of roots or soil probes being pushed into the soil (12). It is usually determined by measuring the force required to push a rigid rod (a penetrometer) through the soil. The penetrometer usually has a cone-shaped tip, but flat-tipped penetrometers are also used (13). Mechanical impedance is often expressed as a pressure or as a "cone index" by divid-

Table 1 Volume of Pores for Different Equivalent Diameters Found in a Sandy Loam Soil Compacted to Bulk Densities

Pore diameter (μm)	Bulk density (g/cm^3)	
	1.24	1.52
	cm^3/cm^3	cm^3/cm^3
> 1200	0.02	0.01
100–200	0.15	0.01
6–100	0.24	0.24
< 6	0.10	0.14
Total	0.51	0.40

Source: Modified from Ref. 4.

ing the force that is required to move the rod through the soil by the basal area of the cone (13).

Mechanical impedance measurements are assumed to approximate the amount of pressure that elongating roots must exert to move through the soil. Such measurements are believed by some researchers to be the most appropriate ones to make to assess whether soils are compacted to the point that root elongation is slowed or stopped (4,12). While this is true in theory, the pressure that resists the elongation of a growing root through soil is difficult to measure exactly (14). Root tips have small diameters, are lubricated by polysaccharide gels, and can twist and turn to avoid obstructions. Rigid rods pressed into the soil only crudely simulate root growth. Nevertheless, penetrometer measurements are useful in identifying zones of different mechanical impedance under field conditions (15,16).

The cone index (CI) values measured for a given soil layer will vary with the kind of penetrometer used. Taylor and Ratliff (15) reported that CI values determined with a blunt-tipped or pocket penetrometer were approximately two thirds the magnitude of CI values determined with a cone penetrometer. Cassel (16) and Bengough and Mullins (17) cited data to show that, for a given soil, the CI values measured with cone penetrometers will generally decrease as the basal area of the cone increases.

For a given soil, the mechanical impedance is related to the soil's bulk density and water content. In general, mechanical impedance will increase as the soil becomes drier or as its bulk density increases. A general relationship among mechanical impedance, bulk density, and water content, which was obtained using equations reported by Ehlers et al. (18), is shown in Figure 1. The relationships shown pertain to a particular soil, but the trends apply to most soils. It is important to note that a high mechanical impedance value (e.g., 3 MPa) may be due to either a high bulk density or a low water content. Because mechanical impedance varies with water content, a soil layer's mechanical impedance will vary through the season as the soil wets and dries, even if bulk density remains constant.

Compact Layers Found Under Field Conditions

A form of compaction that occurs in some agricultural soils is a dense horizontal layer with a bulk density value that is greater than overlying or underlying soil layers. Such a "pan-like" layer is usually densest below the normal depth of moldboard plowing (16). It forms when a tractor wheel rolls along a furrow such that the wheel rides on top of the subsurface layer being compacted. When the furrow is filled with soil during

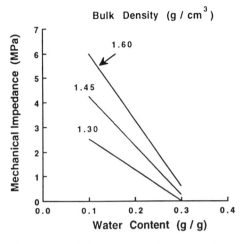

Figure 1   Mechanical impedance as a function of water content and bulk density. Data pertain to the silt loam soil studied by Ehlers et al. (18), but the general relationships can be applied to many soils.

the next pass of the tractor during the plowing operation, the compacted base of the furrow is buried below the Ap horizon. Over a number of years, the compacted furrow bottoms basically form a continuous layer across a field because farmers do not plow the field in exactly the same pattern each year.

These compacted layers are called tillage pans, plow pans, traffic soles, or plow soles among other terms (6). They are common in agricultural fields that are plowed at the same depth every year, and are most prevalent in sandy soils (19). Yields can be reduced in soils containing pans because the pans slow downward extension of roots and thus decrease the plant's access to subsoil moisture.

PRINCIPLES OF ROOT GROWTH

Compacted soils frequently affect plant growth by slowing the growth of roots through the soil and thereby reducing the size of the root system (12). Roots elongate as a result of the cell division in the apical meristem and subsequent expansion of these cells a few micrometers behind the meristematic region which lies within 1 mm of the root tip. The root systems of both monocotlydons and dicotlydons develop from apical meristems (4,20). The mechanisms that plant roots use to respond to compacted soils relate primarily to the elongation of root tips through soil.

Root Anatomy

The anatomy of root tips will be reviewed briefly by focusing on those parts of the root that have a direct effect on how roots can grow through compacted soil. This discussion was based on the reviews of Russell (4) and Kramer (20) which discuss root anatomy in detail. The basic components of a root are shown in Figure 2.

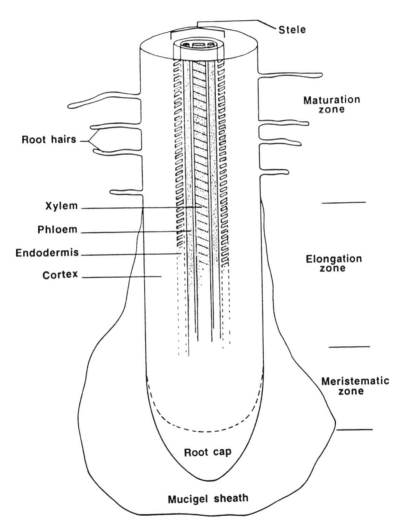

Figure 2   Major anatomical components of a root tip. Features are not drawn to scale. (Modified from Ref. 4.)

Root Cap and Mucilage

The root cap is a unique group of cells at the root tip. The cap has at least three functions (21): (a) to protect the apical meristem or zone of cell production; (b) to create a channel for the root to grow into; and (c) to produce a slime or mucigel that protects the elongating root tip. The cap may also provide a stimulus that enables roots to respond to gravity.

The root cap probably experiences the most abrasion of any part of the root as it is forced through the soil. The cap continually produces cells within it, and these cells are pushed to the edge of the cap where they slough off as a result of the abrasion. For a given root the size of the root cap remains fairly constant due to the continual production and loss of cap cells. Cells in the cap apparently can be completely replaced within 6–9 days in some monocotyledons (22). The dimensions of root caps vary with plant species and range in the axial direction from 270 to 500 μm for pea (*Pisum sativum* L.) and corn (*Zea mays* L.), respectively. Root caps as long as 2700 μm have been found in some desert plants that are adapted to growth in dry soils (23).

The mucilage secreted by the root cap lubricates the cap and other portions of the root. It also protects the root from desiccation and increases the root–soil contact. The mucilage, which contains organic acids and polysaccharides (23), may reduce the mechanical impedance the root experiences by reducing the surface tension of the soil water. This would have an effect on mechanical impedance that is similar to wetting the soil.

Vascular Stele

The major conducting organs of the root are contained inside the vascular stele. Within the stele, the xylem elements conduct water and solutes through the root to the stem, while the phloem elements conduct solutes and metabolites from the leaves to the roots. These elements are contained within a cylinder of cells that form the pericycle. The stele—consisting of the pericycle, xylem, and phloem—is flexible and can be bent through virtually any angle as long as the xylem and phloem are not constricted and do not lose their conducting capacity. If the stele loses its conductive ability, then that portion of the root between the tip and the constriction will die.

Cortex

The stele is separated from the soil by the endodermis, cortex, and epidermis. Cortical cells, which extend from the single layer of endodermal cells adjacent to the pericycle to the single layer of epidermal cells at the root's surface, are thin-walled and contain intercellular, or free, spaces that permit movement of water, solutes, and gases. Under ideal conditions, cortical cells expand into the shape of a cylinder. In some cases the cortex will conform to the shape of the void through which the root is elongating, and can assume oval or even triangular cross-sections without impairment of root function (4). The innermost cells of the cortex form the endodermis that abuts the stele.

A complete cortex apparently is not essential for roots to function. With time, the cortical cells lying external to the endodermis can collapse by desiccation or attack by microorganisms. The root's conductive capacity may not be affected as long as the endodermis remains intact. In poorly aerated environments, air- conducting voids (arenchyma) occur within the cortex of plants adapted to live under such conditions. Roots of mesophytic plants such as corn and sunflower (*Helianthus annus* L.) also can form arenchyma by breakdown of cortical cells (24,25). Thus, the cortex is a malleable component of roots.

Root Hairs

Root hairs are single cells that in most species extend from the surface of the epidermis. Their location along a root can be influenced by the environment, but they typically occur within 5–10 mm of the root tip. Root hairs are thought to be capable of anchoring roots as the growing tips extend into the soil.

Root Laterals

Lateral, or branch, roots can form on most roots, and these generally compose the greatest lengths of a root system. The location of lateral roots around a major root axis is controlled by the arrangement of the conducting elements in the stele. The number of laterals formed per centimeter of root axis varies with plant species, but it is strongly influenced by the environment. The lateral roots are formed on the edge of the stele within the endodermis. When the soil moistens, laterals are initiated and extend into the wet areas. On the other hand, when an elongating root meets an obstruction around which it must grow, the root will bend and grow in one direction while lateral roots are initiated and extend into favorable soil on the convex side of the bent root axis. Deflection of the root tip results in the increased initiation of lateral roots (26).

Diameters and Relative Lengths of Roots

Soil compaction does not affect root growth as long as continuous macropores are present that have diameters greater than the diameters of the roots. The distribution of root diameters were reported in detail for perennial ryegrass (*Lolium perenne* L.) by Scholefield and Hall (27). The diameters of all roots examined ranged from approximately 90 to 990

μm. The modal diameter for root axes was approximately 665 μm, while the modal diameter for first-order laterals was 165 μm. While the distribution of roots for a given size class may vary among plant species, this range of root diameters approximates those found for plants other than ryegrass (28,29). Lyford (29) reported that most roots in the forest floors of hardwood stands have diameters between 200 and 500 μm, with root tips being 100–200 μm in diameter. While roots having larger diameters ($\gg$ 1 mm) are found in forests or crops, the cumulative lengths of roots having diameters < 1 mm compose most of the root systems.

The root diameters described above refer to a "whole root" or one containing fully expanded cortical cells. Root diameters are not constant but change in response to environmental factors. Huck et al. (30) showed that root diameters can decrease during periods of moisture stress, with the change probably being caused by desiccation of cortical cells. Root diameters can be reduced by desiccation to 60% or more of the fully turgid diameter.

Growth of a Root System

In monocotyledons the main root axes originate from the seed (termed seminal roots) or from the base of the stalk (termed nodal roots) (4). First-order lateral roots develop from nodal and seminal roots while second-order laterals develop on laterals of the first order. The lengths of these different root types for 4-week-old barley (*Hordeum vulgare* L.) are shown in Table 2. In dicotyledons, a single taproot is clearly the dominant root early in the growth of the plant; however, primary lateral roots and their branches can quickly account for most of the root system's length. Yorke and Sagar (31) showed that the taproot dominated the length of the root system of peas for the first 8 days of growth, but by 20

Table 2 Diameter and Lengths of Roots Determined for Barley Grown for 4 Weeks in Solution Culture

Root type	Diameter (μm)	Length (cm) Total	Length (cm) Single root
Axes: seminal	400	420	65
nodal	700	380	20
Laterals: first order	200	4900	2.5
second order	< 100		

Source: Ref. 4.

days it constituted only 5% of the total length of the root system. The lateral roots also can extend as deeply as the taproots in some cases. Thus, the volume of soil explored by dicots and monocots may be similar when development of the root system has reached its maximum extent for a given soil. For both types of plants, root dry matter is greatest near the surface and typically decreases exponentially with depth. More than 90% of the root dry matter can be found within 30 cm of the soil surface for both monocot and dicot plants, but this can be influenced by environmental conditions (32,33).

Under ideal conditions, the roots of many plants tend to grow outward in relatively straight lines until deflected by objects they cannot penetrate (20). The first-formed roots tend to grow downward, apparently in response to the pull of gravity; however, later roots do not always show this tendency and can grow laterally.

When root growth is unrestricted, the depth of rooting and extension of laterals increase progressively until flowering (34). The total length of the root system often declines during reproductive growth as the rate of death of old roots exceeds the rate of initiation and growth of new roots. Examples of maximum rooting depths are shown in Table 3 (35). The roots of many monocots and dicots used for crop production extend to depths of >1.5 m when their growth is not impeded. Lateral spread is variable but is probably related to row spacing.

Table 3 Depths and Lateral Spread of Roots for Common Agricultural Crops

	Rooting depth (cm)		Lateral spread of roots (cm)
Plant	Maximum	Working[a]	
Monocotyledons			
Corn (*Zea mays* L.)	188	180	100
Sorghum (*Sorghum bicolor* L.)	180	180	60
Wheat (*Triticum aestivum* L.)	200	150	15
Dicotyledons			
Cotton (*Gossypium hirsutum* L.)	200	180	—
Soybean (*Glycine max* L.)	225	200	50
Alfalfa (*Medicago sativum* L.)	610	300	15
Peanut (*Arachis hypogaea* L.)	250	150	—

[a] Depth of water extraction.
Source: Ref. 35.

ROOT GROWTH THROUGH COMPACTED SOILS

Root Growth Through Rapid Pores

The ways roots grow through soil is considered first by describing how roots grow through rigid pores. In this case, the root cannot expand the pore, so that whether a root can or cannot penetrate the pore is determined by the size of the opening (36). The basic process was illustrated in detail by the experiments of Scholefield and Hall (27) who grew perennial ryegrass over wire mesh screens filled with perlite for periods of 8 weeks. The screens had pore openings ranging from 76 to 712 μm. Ryegrass roots that penetrated pores that were $\geqslant 315$ μm had a maximum thickness below the mesh of 880 μm; however, their root caps and steles had diameters that were < 315 μm. Photographs showed that the cortex had expanded radially both above and below the mesh, but that it had failed to expand within the mesh itself. The ratio of the diameter of the whole root to the root cap was 2.77, and the ratio of the whole root to the stele was 3.00. Thus, the smallest pore diameter that a root could pass through was one third the diameter of the whole root which had expanded cortical cells. The root stele was apparently able to conduct water, solutes, and metabolites despite the restricted cortex. When a root cap could not penetrate a pore, the root tip was deflected above the mesh, and production of lateral roots was associated with the deflection. These experiments were repeated using the roots of corn, oats (*Avena sativa* L.), wheat (*Triticum aestirum* L.), and barley with virtually the same results.

Scholefield and Hall (27) also grew ryegrass roots down glass capillary tubes that had an inside diameter of 450 μm. Roots were constricted within the capillaries but still elongated at a reduced rate as compared to unconstricted roots. The growth was accompanied by copious production of mucigel from the root cap. This suggested that when roots are constricted and elongation has slowed, the production of mucigel will increase.

Reaction of Roots to Mechanical Stresses

The rate at which an elongating root can grow through soil is controlled by the soil conditions. When large pores (e.g., > 315 μm in diameter) are absent, roots must enlarge existing soil pores or create their own channels by displacing soil particles. When soil particles are displaced, the root growth rate is strongly influenced by the soil's mechanical impedance. Mechanical impedance values for soil vary widely, and their magnitude ranges from the degree that root growth is slowed slightly to the extreme case where root growth essentially ceases.

The mechanism that roots use to grow through soil was studied by Abdalla et al. (37) who suggested the process shown in Figure 3. When root growth is not impeded by the soil's mechanical impedance, the root cap is pushed forward through the soil by axial expansion of cells in the zone of elongation. In this case the root has a fairly constant diameter along its length. When the cap encounters a mechanical impedance that stops or retards its growth, the cells in the zone of elongation expand radially to increase the root's diameter. As the root diameter increases, pressure is exerted against the soil to cause a reduction in the soil's mechanical impedance at the root tip. In some cases, the radial expansion may even form cracks in the soil that the tip can grow into. If the root's radial expansion reduces the soil's mechanical impedance at the root cap sufficiently, the root will advance into the soil. The increase diameter resulting from radial expansion is relatively permanent. If the advancing root tip encounters soil with a lower mechanical impedance, the root diameter in the zone of elongation will develop to its previous narrower dimensions. Thus, a single root can have different diameters along its length if it has grown through a soil whose mechanical impedance is spatially variable.

The theory of Abdalla et al. (37) concurs with observations that roots will thicken radially near the tip when their growth is impeded and would explain why pressures experienced by the root cap may be less than those measured at the tips of penetrometers. It has been widely

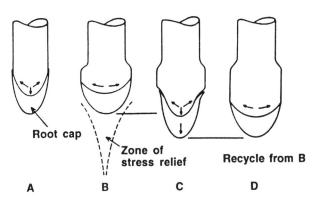

Figure 3 Schematic showing Abdalla et al.'s (37) theory of radial expansion that increases the root's diameter when it elongates into a soil that mechanically impedes its growth. (A) Root elongates normally when mechanical impedance is low. (B) When root cap is impeded, cortex will expand radially to relieve stress at root cap. (C) Root tip elongates following radial expansion. (D) Process continues if soil mechanical impedance remains high.

observed that root diameter will increase when the root growth is slowed by a soil's high mechanical impedance (38–41). The radial expansion occurs primarily by radial rather than axial enlargement of cortical cells. The final volume of the cortical cells that expand radially can be greater than those that expand axially (4,26). The basic process appears to have the effect of the root prying the soil open to enable passage of the root cap, much like the action of a wedge being used to split a log. Hettiaratchi and Fergusson (42) termed this process of root extension inverse peristalsis. As shown by their analysis of soil mechanics, this inverse peristalic mode of growth would permit roots to extend into soil that has mechanical impedances that are approximately five times greater than the root tip could overcome by being pushed only straightforward. Bengough and Mullins (17) reviewed the results of at least five experiments that showed that the pressures resisting the movement of penetrometers through soil were about two to nine times greater than the pressures resisting the penetration of root tips. In fact, the maximum pressures that root tips can exert varies by species, but the largest value recorded is 1.3 MPa (4). Mechanical impedance values that are capable of stopping root elongation are as high as 7 MPa (43).

Relationship Between Soil Mechanical Impedance and Root Growth

The relationship between root growth and mechanical impedance was determined initially by growing roots in soil cores for short periods of 3–18 days and then measuring both the soil's mechanical impedance with a penetrometer and root length (44,45). Taylor and coworkers (15,44,45) were among the first to evaluate the effect of mechanical impedance on root growth, and the results of one study are shown in Figure 4. While there are some differences among species, in general root growth rates decrease sharply for mechanical impedance values between 1 and 3 MPa, and while influenced by the kind of penetrometer used, root growth rates generally are very low when a soil's mechanical impedance is ≥ 3 MPa (46,47). The results shown in Figure 4 have been verified for field conditions as well (48–51). An example for tobacco (*Nicotiana tabacum* L.) is shown in Figure 5. The relationship between root abundance and mean cone index was approximately linear, with few roots being found when the mean cone index was > 3 MPa.

Bulk density values have been used to estimate when a soil will have mechanical impedance values that greatly retard root growth. Root-limiting bulk density values vary with soil textural class and increase as sand percentage increases (52–54). Daddow and Warrington (53) estimated that root-limiting bulk densities for the clay, clay loam, sandy loam, and loamy sand textural classes were approximately 1.40, 1.50,

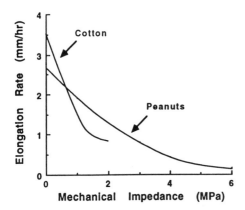

Figure 4 Root elongation rate as a function of mechanical impedance. Roots were allowed to elongate for 40–80 hr through a loamy sand contained in plastic chambers. (From Ref. 15.)

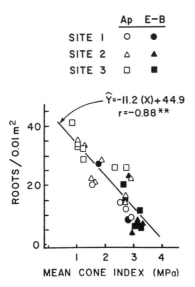

Figure 5 Relationship between tobacco root concentration and mechanical impedance for three Paleudults. Roots were observed in the field of subsoiled and nonsubsoiled plots at approximately 70 days after transplanting. Mechanical impedance (cone index) was determined weekly for two horizons (Ap and E-B) and the values were averaged for approximately a 5-week period prior to the time roots were observed. (From Ref. 51.)

1.65, and 1.75 g/cm^3, respectively. The root-limiting bulk densities for the loamy sand and sandy loam textural classes correspond closely to the bulk densities of root-limiting tillage pans in Typic and Arenic Paleudults that had mechanical impedance values of > 3 MPa (54).

Interactions Among Mechanical Impedance, Aeration, Aggregation, and Plant-Available Water

The rates at which roots grow through soil are affected by factors other than mechanical impedance, including aeration, aggregation, and plant-available water (55). These factors interact with one another and affect the relationship between mechanical impedance and root growth rate (56–58).

Aeration Effects

Oxygen is needed by the root for respiration and to maintain cell division in the apical meristems. The amount of oxygen a root absorbs depends on the diffusion rate of oxygen through the soil and the rate of oxygen consumption during root respiration. When the amount of oxygen in the soil falls to zero, root growth stops because cell division and expansion cease (24). However, root growth slows before all oxygen has been exhausted. In addition, the permeability of the roots to water is reduced at low oxygen levels, and eventually the root loses the ability to control water movement across it (24).

Tackett and Pearson (59) showed that oxygen percentage clearly affected root penetration of a sandy loam soil that was compacted into cores to bulk densities ranging from 1.3 to 1.9 g/cm^3. Mechanical impedance was not measured directly, but it can be assumed to have increased with increasing bulk density. Cotton (*Gossypium hirsutum* L.) was grown in the cores for approximately 4 days, and then root length was measured. Oxygen percentages of the soil air were controlled during the experiments by pumping gas mixtures containing the desired oxygen percentage into the cores. At a bulk density of 1.3 g/cm^3, the soil did not impede root growth until the oxygen percentage decreased from 21% to 5%. At a bulk density of 1.5 g/cm^3, root growth began to decrease when oxygen levels were reduced to 10%, and root penetration decreased sharply at lower oxygen levels. At bulk densities of $\geqslant$ 1.6 g/cm^3, root penetration was greatly reduced due to high mechanical impedance values, and the effect of oxygen percentage decreased until it had a negligible effect at a bulk density of 1.9 g/cm^3. These results showed that at moderate bulk densities (i.e., 1.5–1.6 g/cm^3) root penetration of soil could be slowed when oxygen percentages decreased to less than 10%. At the higher bulk densities, mechanical impedance appeared to be the dominant factor affecting root growth.

Aggregation and Plant-Available Water

As noted earlier, mechanical impedance affects root growth only when the root must displace soil particles to either elongate or increase its diameter. Where macropores or cracks are present, soil mechanical impedance does not control root elongation if roots can grow quickly along the large voids. Cracks or voids large enough for roots to grow through frequently occur between peds or structural aggregates found in many soils. An example of the effect of structural cracks on root growth is illustrated in Figure 6. The soil shown is not compacted because the pores between the aggregates are inteconnected. The lower portion of the soil may be compacted where the soil structure is massive and no interconnected macropores are present. The soil mechanical impedance may be high enough to prevent roots from growing into the massively structured horizon. To compact horizons with granular structures, the soil aggregates would have to be crushed and compressed by intensive tillage

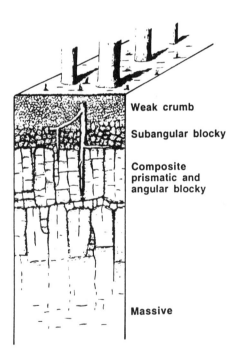

Weak crumb

Subangular blocky

Composite prismatic and angular blocky

Massive

Figure 6 Schematic showing the effects of aggregation on root growth. Cracks or macropores occur between aggregates (peds) of various types. The cracks offer little resistance to root growth. (From Ref. 60.)

operations. Heavy vehicles moving over wet soil when mechanical impedance is low would have the same effect.

Aggregation affects root penetration of compact soil layers under field conditions. Vepraskas and Wagger (61) examined corn root distribution and yield response to deep tillage in eight soils (Typic and Arenic Paleudults) containing dense pan-like layers. The size of aggregates in the pans was measured by weighing the amounts of aggregates (or peds) retained on sieves that had a range of mesh sizes. The average diameter of aggregates was reported as the ped mean weight diameter (62). The relationship of root penetration of dense pan layers to the ped mean weight diameter is shown in Figure 7A. Less than 20% of the corn roots were found below the tillage pans where the pans had ped mean weight diameters less than approximately 2 mm. In fields where tillage pans had ped mean weight diameters of >2 mm, approximately 30% of the corn roots occurred below the tillage pans. The pans in all eight fields had mean mechanical impedance values between 3.1 and 4.1 MPa that were determined for a 4- to 6-week period prior to the time roots were observed. Subsoiling significantly increased corn yields where ped mean weight diameters were <2 mm in the tillage pans.

Corn roots were apparently able to penetrate tillage pans that had ped mean weight diameters of >2 mm by following the cracks or planes of weakness between the aggregates as shown in Figure 6. This interpretation is oversimplified because the aggregated pan layers had lower amounts of plant available water (Figure 7B). Root growth through the aggregated tillage pans was probably also enhanced by the roots having access to more plant-available water in addition to the roots following planes of weakness. It has been shown that root growth rates are related to soil water potential (20,55). Under field conditions, the effects of aggregation and available soil water on root growth cannot be clearly separated. The effects of aggregation on root growth have been studied intensively and more information can be found elsewhere (63–67).

COMPACTION EFFECTS ON YIELD

Compaction affects yield primarily by reducing the amount of water and nutrients a plant obtains (68,69). The degree to which compaction affects yield has been difficult to define due to two basic factors. First, the amount and distribution of rain falling during the growing season (particularly near flowering) determines whether or not compaction affects yields (69,70). In some years an "ideal" rainfall distribution can result in compaction having no effect on yield. Second, the relationship between rainfall and yield in a compacted soil differs among soils. In coarse-

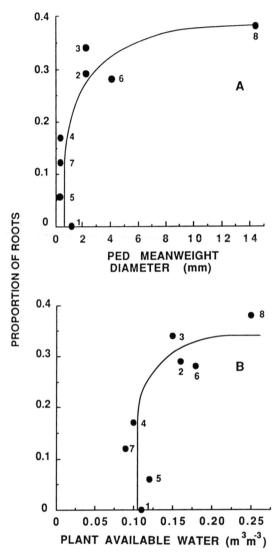

Figure 7 Proportion of corn roots found below dense tillage pans at 77–100 days after planting for eight paleudults in relation to (A) ped mean weight diameter (aggregate size) of the pans and (B) plant- available water held in the pan. (From Ref. 50.)

textured soils compaction can reduce yields during dry years, while in the fine-textured soils compaction seems to reduce yields during wet years (69,70). Because soil properties determine how compaction affects yield on a given soil, this subject will be reviewed by considering coarse-textured and fine-textured soils separately.

Coarse-Textured Soils

These soils have Ap horizons with textural classes of sand, loamy sand, or sandy loam. Such soils are found extensively in the Gulf and Atlantic Coastal Plains of the United States (19). Tillage pans are common features in such soils, and the pans appear to form under the standard agricultural practices used on the sandy soils where agricultural vehicles apply loads that are < 4.5 Mg/axle. Compaction experiments have usually been done on soils where pans have been formed, and the experimental methods compare the effects of different kinds of tillage in their ability to loosen compact layers and increase yields. The compact pan layers affect yields by reducing root penetration into the subsoil (Fig. 8). Most roots occur in the Ap horizons before flowering. In coarse-textured soils, the amount of plant-available water held in the Ap horizon is low, and it can supply available water to the plant for about 7 days without rainfall

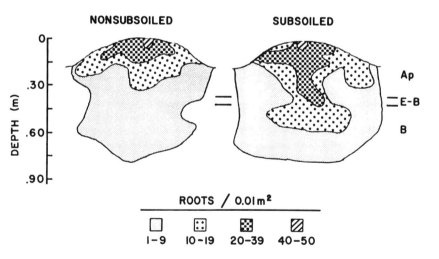

Figure 8 Mean distribution of tobacco roots across four rows found for two tillage treatments at 69 days after transplanting. The paleudult had a tillage pan in the E-B horizon which was restricting roots in the nonsubsoiled treatment. Deeper root distribution in the subsoiled treatment was produced by ripping the pan. (From Ref. 51.)

(68). During dry years, moisture stress occurs frequently when pans are present, and plant growth is slowed. Subsoiling is a common practice used to increase yields when pans are present. The subsoiler creates slits in the pans and increases the rate at which roots grow into the B horizon where they extract available water.

Corn grain yields have been increased by more than 120% in some dry years when subsoiling has been used (61). Yields of tobacco, soybean, and cotton also can be increased by subsoiling, but the maximum yield increases are generally lower than those for corn, being generally within 30–40% of the yields found for the nonsubsoiled treatments (69,71).

Kamprath et al. (69) examined the effect of subsoiling on the grain yield of soybeans in coarse-textured soils with tillage pans. When the pan was subsoiled, the soybean roots extracted more moisture from below the pan (depths >30 cm). During dry years this increased moisture utilization found with subsoiling increased the leaf area index to nearly three times that found for plots where the pan was not ripped. Despite the large increase in leaf area, grain yield was increased by only 29% over that found for the conventional tillage treatment. By considering the rainfall distribution over a 50-year period, Vepraskas (74) estimated that subsoiling would increase tobacco yields on the average in 7 out of every 10 years on Coastal Plains soils with tillage pans. In soils where the depth to the B horizon was ≥46 cm, the probability of a yield increase from subsoiling rose to 9 out of every 10 years. Where the B horizon was <38 cm deep, yield increases from subsoiling occur about every 4 out of 10 years.

Subsoiling is most beneficial when it is used every year as the standard tillage practice. The subsoiler slits tend to collapse within a year, and the beneficial effects of subsoiling may be lost after 3 years if the operation is not repeated (75,76).

Fine-Textured Soils

Fine-textured soils include those with clay and clay loam textural classes in their Ap horizons. Compaction experiments on these soils are done differently than with coarse-textured soils because the fine-textured soils are normally not compacted by standard agricultural practices used at the experimental site. Therefore, a compaction treatment is imposed to compress the soil surface using repeated passes of agricultural vehicles, some of which have been exceptionally heavy (77,78). Results of compaction experiments on fine-textured soils were summarized by Voorhees (70) and Hakansson et al. (78).

Experiments by Lindemann et al. (79) on a clay loam soil (Typic Haplaquoll) in Minnesota illustrate the effects of "surface" compaction on soybean yields for wet and dry years. Compaction treatments were imposed 2 weeks before planting by driving a tractor over each plot from zero to three times to compact the soil to a depth ≤30 cm. Treatments were repeated over a 2-year period. Prior to planting, plots were tilled to a depth of 5 cm with a cultivator harrow to prepare the seedbed.

The effect of the tractor passes on bulk density are shown in Table 4. In 1976 most of the compaction occurred during the initial pass of the tractor, but in 1977 additional compaction was produced by the second or third passes. Bulk density values for the 15- to 25- cm depth were similar to those shown. Compaction did not affect bulk density below a depth of 25 cm.

More than twice the amount of precipitation fell in 1977 compared to 1976 and the difference was reflected in the yields of soybean seed (Table 4). In the drier year, 1976, compaction tended to increase yields, but the differences were not statistically significant. Yields on plots receiving three tractor passes were approximately 15% higher than on plots receiving no tractor traffic. In the wetter year, 1977, the trend was for compaction to decrease yields. Compaction increased taproot nodulation and taproot acetylene reduction activity during the drier year, but reduced plant growth and nodulation in the wetter year. Yield results similar to these also were reported by Johnson et al. (80) who studied similar soils over a 5-year period.

The reasons why compaction reduces yields on fine-textured soils during wet years have not been determined exactly. The soils on which the yield reductions are observed tend to be imperfectly drained, and

Table 4 Compaction of a Clay Loam Soil (5- to 15-cm depth) by Different Numbers of Tractor Passes, and the Resulting Soybean Seed Yields for a 2-Year Period

Number of passes	Bulk density (g/cm^3)		Seed yield (Mg/ha)	
	1976	1977	1976 (327 mm)[b]	1977 (704 mm)
0	1.16a[a]	1.25a	1.98	4.12
1	1.26b	1.46b	2.16	4.11
2	1.25b	1.55c	2.11	3.96
3	1.28b	1.54c	2.28	3.85

[a] Mean values followed by the same letter within a column were not significantly different ($p = 0.05$). Yield data were not significantly different.
[b] Number in parentheses is cumulative rainfall for the months April through September
Source: Ref. 79.

during wet years the high soil water contents may reduce aeration. Increased loss of nitrogen by denitrification and colder temperatures have also been suggested as factors causing yield reductions (81,82).

The reasons for compaction increasing yields during dry years on fine-textured soils also are not completely understood, but appear to be related to improved water uptake by the roots (70,80). Kooistra et al. (83) showed that the degree of root–soil particle contact increased as soil bulk density increased. Thus, in dry years, compaction may increase the volume of soil from which roots extract their water. The rate at which soils transmit water to plant roots likely increases as particles are pressed closer together (20).

SUMMARY

Soil compaction occurs whenever agricultural implements or rainfall impact compress soil. During the compaction process, soil bulk density increases as the particles are pushed closer together and macropores are crushed. While soil mechanical impedance increases as a soil dries, the impedance increases to greater levels as bulk density increases. Dense tillage pans are a common form of compact layers that occur below the normal depth of plowing. They are prevalent in nonaggregated sandy soils.

Compaction can affect plant growth by reducing the rate at which roots grow through soil. Reduced growth rates occur as mechanical impedance increases. When their growth is impeded, roots expand radially near their tip. This sideways expansion tends to reduce the mechanical impedance at the root tip. Also the root cap may increase its production of mucigel when root growth is impeded. Soil loosening by tillage is probably beneficial when mechanical impedance values are $\geqslant 3$ MPa. Lower impedances also may impede root growth when aeration is low.

Compaction can reduce crop yields, but the magnitude of the yield reduction varies with rainfall. In sandy soils, compaction reduces yields during dry years. The crop experiences moisture stress because these soils hold little plant-available water. In fine- textured soils, the effect of compaction on reducing yields appears to be most prevalent during wet years. The reasons for these yield reductions are not completely understood. At present, compaction problems probably occur in the coarse-textured soils more often than in other soils, such that deep tillage to loosen compact layers is routinely used by growers.

REFERENCES

1. W.L. Harris (1971): *Compaction of Agricultural Soils* (D.K. Barnes, ed.), Am. Soc. Agric. Eng., St. Joseph, Michigan, p. 9.

2. B.D. Soane, P.S. Blackwell, J.W. Dickson, and D.J. Painter (1980/81): *Soil Till. Res.*, *1*:207.

3. G.R. Bathke, D.K. Cassel, W.R. Hargrove, and P.M. Porter (1992): *Soil Sci.*, *154*:316.

4. R.S. Russell (1977): *Plant Root Systems*, McGraw-Hill (UK) Ltd., London.

5. W.R. Gill (1971): *Compaction of Agricultural Soils* (D.K. Barnes, ed.), Am. Soc. Agric. Eng., St. Joseph, Michigan, p. 431.

6. *Glossary of Soil Science Terms* (1975): Soil Sci. Soc. Am., Madison, Wisconsin.

7. W.J. Chancellor (1977): *Univ. California, Div. Agric. Sci. Bulletin 1881.*

8. G.R. Blake and K.H. Hartge (1986): *Methods of Soil Analysis*, Part 1 (A. Klute, ed.), Am. Soc. Agron., Madison, Wisconsin, p. 363.

9. D. Hillel (1980): *Fundamentals of Soil Physics*, Academic Press, New York.

10. R.J. Luxmoore, P.M. Jardine, G.V. Wilson, J.R. Jones, and L.W. Zelazny (1990): *Geoderma*, *46*:139.

11. D.J. Greenland (1977): *Phil. Trans. Royal Soc. London. B.*, *281*:193.

12. H.D. Bowen (1981): *Modifying the Root Environment to Reduce Crop Stress* (G.F. Arkin and H.M. Taylor, eds.), Am. Soc. Agric. Eng., St. Joseph, Michigan, p. 21.

13. J.M. Bradford (1986): *Methods of Soil Analysis*, Part 1 (A. Klute, ed.), Am. Soc. Agron., Madison, Wisconsin, p. 463.

14. G.M. Whiteley, W.H. Utomo, and A.R. Dexter (1981): *Plant Soil*, *61*:351.

15. H.M. Taylor and L.F. Ratliff (1969): *Soil Sci.*, *108*:113.

16. D.K. Cassel (1982): *Predicting Tillage Effects on Soil Physical Properties and Processes* (P.W. Unger and D.M. Van Doren, eds.), ASA Special Publ. No. 44, Am. Soc. Agron. and Soil Sci. Soc. Am., Madison, Wisconsin, p. 45.

17. A.G. Bengough and C.E. Mullins (1990): *J. Soil Sci.*, *41*:341.

18. W. Ehlers, U. Kopke, F. Hesse, and W. Bohm (1983): *Soil Till. Res.*, *3*:261.

19. D.K. Cassel (1981): *Crop Reactions to Water and Temperature Stresses in Humid, Temperate Climates* (C.D. Raper, Jr., and P.J. Kramer, eds.), Westview Press, Boulder, Colorado, p. 167.

20. P.J. Kramer (1983): *Water Relations of Plants*, Academic Press, New York.

21. P.W. Barlow (1975): *The Development and Function of Roots* (J.G. Torrey and D.T. Clarkson, eds.), Academic Press, New York, p. 21.

22. J. Glinski and J. Lipiec (1990): *Soil Physical Conditions and Plant Roots*, CRC Press, Boca Raton, Florida.

23. M. Luxova and M. Ciamporova (1992): *Physiology of the Plant Root System* (J. Kolek and V. Kozinka, eds.), Kluwer Academic, Dordrecht, Netherlands, p. 46.

24. M.C. Drew (1992): *Soil Sci.*, *154*:259.

25. M. Kawase (1979): *Am. J. Bot.*, *66*:183.

26. M.J. Goss and R.S. Russell (1980): *J. Exp. Bot.*, *31*:577.

27. D. Scholefield and D.M. Hall (1985): *Plant Soil*, *85*:153.

28. R. Tippkotter (1983): *Geoderma*, *29*:355.

29. W.H. Lyford (1975): *The Development and Function of Roots* (J.G. Torrey and D.T. Clarkson, eds.), Academic Press, New York, p. 179.
30. M.G. Huck, B. Klepper, and H.M. Taylor (1970): *Plant Physiol.*, 45:529.
31. J.S. Yorke and G.R. Sagar (1970): *Can. J. Bot.*, 48:699.
32. W.R. Jordan (1983): *Crop-Water Relations* (I.E. Teare and M.M. Peet, eds.), John Wiley and Sons, New York, p. 213.
33. W.K. Robertson, L.C. Hammond, J.T. Johnson, and K.J. Boote (1980): *Agron. J.*, 72:548.
34. D.B. Mengel and S.A. Barber (1974): *Agron. J.*, 66:341.
35. H.M. Taylor and E.E. Terrell (1982): *Handbook of Agricultural Productivity*, Vol. 1 (M. Rechcigl, ed.), CRC Press, Boca Raton, Florida, p. 185.
36. L.K. Wiersum (1957): *Plant Soil*, 9:75.
37. A.M. Abdalla, D.R.P. Hettiaratchi, and A.R. Reece (1969): *J. Agric. Eng. Res.*, 14:236.
38. B.G. Richards and E.L. Greacen (1986): *Aust. J. Soil Res.*, 24:393.
39. S.A. Materechera, A.R. Dexter, A.M. Alston, and J.M. Kirby (1992): *Plant Soil*, 143:85.
40. S.A. Materahera, A.R. Dexter, and A.M. Alston (1991): *Plant Soil*, 135:31.
41. A.G. Bengough and C.E. Mullins (1991): *Plant Soil*, 131:59–66.
42. D.R.P. Hettiaratchi and C.A. Ferguson (1973): *J. Agric. Eng. Res.*, 18:309.
43. C.J. Gerard, P. Sexton, and G. Shaw (1982): *Agron. J.*, 74:875.
44. H.M. Taylor and H.R. Gardner (1963): *Soil Sci.*, 96:153.
45. H.M. Taylor, G.M. Roberson, and J.J. Parker (1966): *Soil Sci.*, 102:18.
46. B.W. Veen and F.R. Boone (1990): *Soil Till. Res.*, 16:219.
47. A.R. Dexter (1987): *Plant Soil*, 98:303.
48. D.W. Grimes, R.J. Miller, and P.L. Wiley (1975): *Agron. J.*, 67:519.
49. G. Ide, G. Hofman, C. Ossemerct, and M. van Ruymbeke (1984): *Soil Till. Res.*, 4:419.
50. M.J. Vepraskas and M.G. Wagger (1989): *Soil Sci. Soc. Am. J.*, 53:1499.
51. M.J. Vepraskas and G.S. Miner (1986): *Soil Sci. Soc. Am. J.*, 50:423.
52. C.A. Jones (1983): *Soil Sci. Soc. Am. J.*, 47:1208.
53. R.L. Daddow and G.E. Warrington (1983): *U.S.D.A. Forest Serv. Rep. WSDA-TN-00005*, Fort Collins, Colorado, p. 9.
54. M.J. Vepraskas (1988): *Soil Sci. Soc. Am. J.*, 52:1117.
55. H.M. Taylor (1983): *Limitations to Efficient Water Use in Crop Production* (H.M. Taylor, W.R. Jordon, and T.R. Sinclair, eds.), Am. Soc. Agron., Madison, Wisconsin, p. 87.
56. B.W. Eavis (1972): *Plant Soil*, 36:613.
57. B.C. Warnaars and B.W. Eavis (1972): *Plant Soil*, 36:623.
58. M.J. Vepraskas, G.S. Miner, and G.F. Peedin (1986): *Soil Sci. Soc. Am. J.*, 50:1541.
59. J.L. Tackett and R.W. Pearson (1964): *Soil Sci. Soc. Am. Proc.*, 28:600.
60. E.A. FitzPatrick (1980): *Soils: Their Formation, Classification, and Distribution*, Longman, London.
61. M.J. Vepraskas and M.G. Wagger (1990): *Soil Sci. Soc. Am. J.*, 54:849.
62. W.D. Kemper and R.C. Rosenau (1986): *Methods of Soil Analysis*, Part 1 (A. Klute, ed.), Am. Soc. Agron., Madison, Wisconsin, p. 425.

63. V.E. Nash and V.C. Baligar (1974): *Plant Soil, 41*:81.
64. A.R. Dexter (1978): *J. Soil Sci., 29*:102.
65. G.M. Whiteley and A.R. Dexter (1984): *Plant Soil, 77*:141.
66. R.K. Misra, A.R. Dexter, and A.M. Alston (1986): *Plant Soil, 94*:59.
67. M.V. Braunack and Dexter (1989): *Soil Till. Res., 14*:259.
68. R.B. Campbell, D.C. Reicosky, and C.W. Doty (1974): *J. Soil Water Cons., 29*:220.
69. E.J. Kamprath, D.K. Cassel, H.D. Gross, and D.W. Dibb (1979): *Agron. J., 71*:1001.
70. W.B. Voorhees (1987): *Soil Till. Res., 10*:29.
71. M.J. Vepraskas, G.S. Miner, and G.F. Peedin (1987): *Agron. J., 79*:141.
72. S. Gameda, G.S.V. Raghavan, E. McKeyes, and R. Theriault (1987): *Soil Till. Res., 10*:123.
73. L. Gaultney, G.W. Krutz, G.C. Steinhardt, and J.B. Liljedahl (1982): *Trans. Am. Soc. Agric. Eng., 25*:563.
74. M.J. Vepraskas (1988): *Soil Sci. Soc. Am. J., 52*:229.
75. C. van Ouwerkerk and P.A.C. Raats (1986): *Soil Till. Res., 7*:273.
76. G. Ide, G. Hofman, C. Ossemerct, and M. van Ruymbeke (1987): *Soil Till. Res., 10*:213.
77. W.B. Voorhees, W.W. Nelson, and G.W. Randall (1986): *Soil Sci. Soc. Am. J., 50*:428.
78. I. Hakansson, W.B. Voorhees, P. Elonen, G.S.V. Raghavan, B. Lowery, A.L.M. van Wijk, K. Rasmussen, and H. Riley (1987): *Soil Till. Res., 10*:259.
79. W.C. Lindemann, G.E. Ham, and G.W. Randall (1982): *Agron. J., 74*:307.
80. J.F. Johnson, W.B. Voorhees, W.W. Nelson, and G.W. Randall (1990): *Agron. J., 82*:973.
81. J.P. Graham, P.S. Blackwell, J.V. Armstrong, D.G. Christian, K.R. Howse, C.J. Dawson, and A.R. Butler (1986): *Soil Till. Res., 7*:189.
82. S.D. Logsdon, R.R. Allmaras, W.W. Nelson, and W.B. Voorhees (1992): *Soil Till. Res., 23*:95.
83. M.J. Kooistra, D. Schoonderbeek, F.R. Boone, B.W. Veen, and M. van Noordwijk (1992): *Plant Soil, 139*:119.

10

Plant Response to Flooding

S. R. Pezeshki

Louisiana State University
Baton Rouge, Louisiana

INTRODUCTION

Soil flooding affects plants in numerous ways. Under normal soil moisture conditions with ample aeration, roots experience the same oxygen level as that of atmospheric level, i.e., 20.9 kPa (1). When a soil is waterlogged, pore spaces are filled with water preventing gas exchange between soil and atmosphere due to low O_2 diffusion in water. The dissolved O_2 in floodwater will be depleted shortly, within hours to days, depending on the temperature and respiration activity of plant and microorganisms, leading to anaerobic (oxygen-free) conditions (2). Depending on the level of flood tolerance, morphological and anatomical changes may be followed by leaf chlorosis, leaf senescence, inhibition of growth, partial injury, and death.

For flood-sensitive species, the rapid depletion of soil oxygen following flooding is the beginning of a process which, if continued, will result in severe injury because plant cells require oxygen for normal metabolism. In flood-tolerant plants, O_2 deficiency will lead to a switching of their metabolism from aerobic to anaerobic pathways resulting in a much less efficient system of energy production. The mechanisms involved in flood injury and their potential interactions are not completely understood. However, several physiological processes are affected at the cell, process, and whole-plant levels. Lack of oxygen affects cell membranes, water relations, mineral nutrition, growth regulator production and allocation, photosynthesis, respiration, and carbohydrate allocation. Most of these processes may also be affected by soil phytotoxins such as H_2S, which are byproducts of flooding.

This chapter reviews the literature concerning flood response mechanisms in plants including morphological, anatomical, and biochemical changes which occur due to soil oxygen deficiency. Although plant responses are the main focus of this chapter, brief references will be made to some aspects of soil (especially rhizosphere) changes due to oxygen depletion.

EFFECTS OF FLOODING ON SOILS

Soil waterlogging begins a chain of reactions in physical, chemical, and biological processes, in soils with significant implications for plant functioning (see Refs. 2 and 3 for reviews). Physical processes include such important aspects as restriction of gas exchange and depletion of molecular O_2. Oxygen is needed for root respiration and soil microorganisms. Thus ample supply of O_2 is essential for normal root functioning. Flooding dramatically reduces soil O_2 as floodwater fills soil pore spaces replacing air, thus restricting soil–air gas exchange. Oxygen diffusion in gasfilled pores is approximately 10,000 times greater than in water-filled pores. The limited supply of O_2 in floodwater is rapidly depleted by roots and soil microorganisms. The restriction of gas exchange and subsequent depletion of O_2 results in a series of chemical changes in soil including accumulation of such gases as CO_2, methane, N_2, and H_2 (2). Depending on the soil type, the bulk of soil is depleted of O_2 within a short period and aerobic microorganisms are replaced by anaerobic ones dominated by bacteria (4). This process is followed by denitrification; reduction of iron, manganese, and sulfate; and a changing soil pH and redox potential (Eh).

Major changes in soil physicochemical properties are best reflected in reductions in soil redox potential (5). Soil redox potential decreases (becomes more negative) in response to flooding (2,3). Upon depletion of O_2, anaerobic microorganisms begin to utilize other terminal electron acceptors for respiration. In a typical series of reduction, NO_3^- is reduced to NO_2^-, followed by reductions of Mn^{4+} to Mn^{2+}, Fe^{3+} to Fe^{2+}, SO_4^{2-} to S^{2-}, and accumulations of acetic and butyric acids produced by microbial metabolism (2,3).

Several phytotoxic compounds are found in flooded soils. Anaerobic bacteria such as *Desulfovibrio* and *Desulfotomaculum* reduce sulfate forming hydrogen sulfide, which is assumed to be the most damaging compound. Other phytotoxic compounds include ethanol, reduced forms of Fe and Mn, fatty acids, and cyanogenic compounds. These compounds may accumulate in flooded soils at levels that can cause injury to plants (3,6). For instance, inhibitory effects of H_2S on cytochrome oxidase is disruptive to aerobic respiration and excess cytosolic Fe and Mn are harmful to enzymatic structures (7). Soil flooding could result in death of the existing root system due to the enhanced activity of *Phytophthora* fungi (8,9) as was documented for overirrigated citrus trees (10). The resultant effects of soil physicochemical changes on plants are a series of complex responses which, depending on the species level of flood tolerance and the plant's developmental stage at the onset of flooding, cause various symptoms ranging from mild responses to severe injury and death of plants.

EFFECTS OF FLOODING ON PLANTS

Water Relations

Flooding affects plant water relations. Flooded roots have slower water flux compared to the roots under normal (aerated) condition (11–13). Stomatal closure is among the early responses of plants to soil flooding (14,15). Flood-induced stomatal closure has been reported for bean (*Phaseolus vulgaris* L.) (16), tobacco (*Nicotiana tabacum* L.), tomato (*Lycopersicon esculentum* Mill.) (17), sunflower (*Helianthus annus* L.) (18,21), bell pepper (*Capsicum annuum* L.) (19), wheat (*Triticum aestivum* L.) (20), cotton (*Gossypium hirsutum* L.) (21), and numerous other species (15,22). Mechanisms involved in flood-induced stomatal closure are complex and in most instances are yet to be determined (23).

Leaf dehydration may be caused by a decrease in root permeability under flooded conditions (24,25). While development of water stress may be species-specific (26), in most cases initial stomatal closure occurs without significant changes in plant water status (27–31). The satisfactory maintenance of water status in most instances may not reflect a sustained root conductivity but is probably due to rapid stomatal closure. For instance, in flooded oak (*Quercus* spp.) seedlings, there was a significant reduction in water loss for which even a slow water absorption rate by roots would compensate sufficiently (31). Similar changes in water relations were found for tomato plants by Bradford and Hsiao (12). In addition, further evidence documented for tomato plants clearly shows that flood-stressed roots influence stomatal conductance and gas exchange by mechanisms not involving leaf water potential (32,33). In many crop species, stomatal closure is not due to leaf dehydration. In wheat, stomatal closure occurred due to a low soil oxygen diffusion rate (20). Similar findings have been reported for other crop species including sunflower and cotton (34). The rapid responses in the leaves caused by root inundation suggest that transfer of substances produced under anaerobic conditions and/or fluctuations in the level of growth regulators transported from roots through the transpiration stream are the causes of shoot responses. Fluctuations in the level of root-synthesized growth regulators, including cytokinins (CKs) (33) and gibberellins (GAs) (35), can cause growth regulator imbalances in the shoots. Growth regulator effects may also include changes in ethylene production (36,37) and concentration of abscisic acid (ABA) (25,38).

Wilting in response to flooding has been observed in many species including tomato (39), tobacco (18), maize (*Zea mays* L.) (40), broad bean (*Vicia faba* L.) (41), and alfalfa (*Medicago sativa* L.) (42). The response has been attributed to an increased resistance to water flow in flooded roots which is presumed to last a short time before root death (12). The wilting in most cases is alleviated by stomatal closure (12,17). Depending on the air temperature, humidity, the extent of stomatal closure and root resistance to water movement, wilting may occur at various intensities (43). The possibility of guard cell turgor loss due to potassium ion efflux has been proposed (23). This is based on existing data that show reduction in potassium content of flood-stressed leaf tissues in several species, including tobacco (44,45), and decrease in potassium uptake in some species, including maize, in response to O_2 stress (44).

Flood-induced stomatal closure may also be related to ABA buildup in leaves (46). Rapid increase in leaf tissue ABA content has been found in several species in response to flooding (25,38,47). Such ABA increases have been implicated in stomatal closure and inhibition of leaf growth under hypoxic conditions. Whether ABA buildup is due to translocation from roots is not known. It is known, however, that exogenous ABA application causes stomatal closure (48,49). ABA application to tomato plants simulated stomatal responses found under flooded conditions (50). The ABA buildup may interfere with various metabolic processes and/or ion transport between guard cells and surrounding cells as well as guard cell starch formation (23,46).

Ethylene may also be involved in stomatal response to flooding when it is found at elevated levels (12,36), although this effect may be species-specific. The effects have been attributed to the changes in guard cell membranes which interfere with water and/or ion efflux or elevated CO_2 concentrations resulting from decreased photosynthesis (36). In peanut (*Arachis hypogaea* L.), stomatal closure has been attributed to elevated ethylene; however, it has no apparent effects on stomatal response in several species including maize and tomato (36,50,51). Virtually all plant growth regulators have been implicated in the observed stomatal responses to some degree. For instance, reduced production of GA, auxins (IAA), and CK and enhanced production of ABA and 1-aminocyclopropane-1-carboxylic acid have been reported in several species in response to flooding (47,52,53). Under flooded conditions, reduced root production and supply of CK and GA may affect stomatal behavior in a complex interaction scheme which may also involve ABA. However, as was pointed out by Reid and Bradford (52), these hypotheses are based on limited data and require additional investigation.

In long-term flooding, stomatal reopening may occur (54,55). The degree of resumption of normal stomatal functioning after floodwaters recede is dependent on the duration of flooding and is faster after a short rather than a long period of flooding (55). Patterns of stomatal reopening were, however, correlated with production of adventitious roots (28,56). Kozlowski (22) hypothesized that a combination of increased root surface area, increased aerobic respiration, and oxidation of the rhizosphere may account for the physiological functioning of adventitious roots. If stomatal reopening occurs under flooded conditions and in the absence of adventitious roots, then the reduction in root water uptake capacity may become an important factor affecting plant functioning through water stress.

Based on the general patterns of stomatal behavior under short-term and prolonged flooding, there is little doubt that oxygen stress in the roots produces signals, via transpiration stream, to the leaves, affecting stomatal behavior as has been documented for tomato (12) and some woody species (28,56). In prolonged flooding, a close correlation was found between adventitious root development and stomatal reopening (56), further supporting the communicating signal theory between root and shoot.

Photosynthesis

Flood-induced stomatal closure is accompanied by a decrease in photosynthetic activity as shown in numerous species including tomato, wheat, and pepper (19,57,58). In bean, net photosynthesis was reduced to near zero in response to 7 days of flooding. Flooding for 1 day reduced photosynthesis 17% as compared to control conditions and caused reduction in plant dry weight (59). Flood-sensitive plants show drastic and rapid reduction of net photosynthesis in response to flooding. For instance, net photosynthesis was reduced >90% in citrus seedlings within 10 days (60). In cherrybark oak (*Quercus falcata* var. *pagodaefolia* Ell.) seedlings, soil flooding reduced net photosynthesis to zero within 3 days (31). Photosynthesis in pepper dropped substantially within 3 days of flooding (Fig. 1). Similarly, many woody species show reduc-

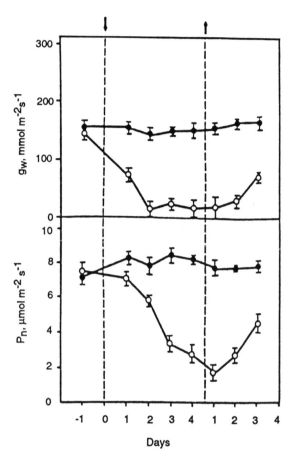

Figure 1 Effect of flooding on mean daily stomatal conductance (g_w) and net photosynthesis (p_n) for flooded (open circles) and control (closed circles) *Capsicum annuum* plants. (From Ref. 19.)

tion of net photosynthesis in response to soil hypoxia (26,30,31,54). This is a common response among species found in various flood tolerance categories ranging from least tolerant to most tolerant. Figure 2 represents the photosynthetic responses of seedlings of woody species under a broad range of experimental conditions (soil type, duration of flooding, etc.) under laboratory or greenhouse conditions. As depicted in Figure 2, in a majority of species studied, soil O_2 deficiency adversely affects plant gas exchange to various degrees. A major difference found, however, is that net photosynthesis in most tolerant species begins to recover rapidly following the initial reduction (Fig. 3). However, under

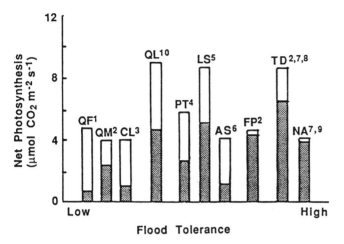

Figure 2   Net photosynthetic response of seedling of various tree species to flooding showing the drained (control) plants (open bars) vs. the flooded plants (closed bars). Data represent the responses of seedlings under a broad range of experimental conditions (duration, intensity of reductions, soil, types etc.). QF, *Quercus falcata* var. pagodaefolia; QM, *Q. michauxii*; QL, *Q. lyrata*; TD, *Taxodium distichum*; LS, *Liquidambar styraciflua*; FP, *Fraxinus pennsylvanica*; AS, *Acer saccarinum*; CI, *Carya illinoensis*; PT, *Pinus taeda*. Numbers refer to each respective reference as follows: 1, Pezeshki and Chambers (30): 2, Pezeshki and Chambers (62); 3, Smith and Ager (281); 4, Pezeshki (282); 5, Pezeshki and Chambers (31); 6, Peterson and Bazzaz (283): 7, Pezeshki (284); 8, Pezeshki et al. (285); 9, Pezeshki et al. (286); 10, Pezeshki (unpublished). (From Ref. 287.)

continuous flooded conditions, little or no recovery is found in sensitive species. Flood-sensitive species apparently are not capable of resuming normal photosynthetic functions under hypoxic conditions. Therefore, continuous flooding causes progressive stomatal closure and a decrease of photosynthetic activity. In birch (*Betula papyrifera* Marsh.), stomata remained closed during 2 weeks of flooding (61) while in overcup oak (*Q. macrocarpa* Michx.) and sycamore (*Platanus occidentalis* L.), stomatal closure persisted for 30 days—the entire length of flooding (29,37). However, in highly flood-tolerant species such as baldcypress [*Taxodium distichum* (L.) Rich.] and greenash (*Fraxinus pennsylvanica* Marsh.), stomatal reopening and increased photosynthetic activity has been observed (62). There is a wide range of inter- and intraspecies differences in photosynthetic response of plants to the flooding. For the most part, however, the mechanisms involved are unknown. Reduction in photosynthesis has been attributed to both stomatal closure and metabolic (nonsto-

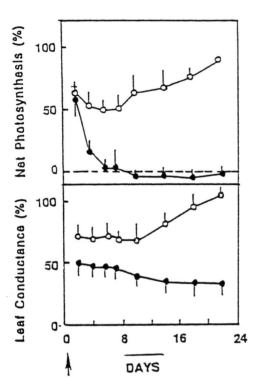

Figure 3 Time–course responses of net photosynthesis and stomatal conductance in *Taxodium distichum* (open circles) and *Quercus falcata* var. *pagodaefolia* (closed circles) in response to flooding. Data are presented as percentage of control for each species. Each point is the mean for 12 measurements. Bars represent SE. Arrow signifies treatment initiation. (From Pezeshki, unpublished.)

matal) processes. Stomatal closure causes diffusional limitation of photosynthesis since there is a close correlation between stomatal conductivity and photosynthetic activity (63,64). The explanation involving stomatal (diffusional) limitations can account for the differences partially. However, metabolic factors have been implicated and their potential role will be discussed.

Nonstomatal inhibition of photosynthetic capacity has been reported in bean (16), tomato (32,33), and bell pepper (19). While the adverse effects of flooding on photosynthetic capacity is known, the mechanisms involved have not been identified; thus much speculation remains in the literature. For instance, ethylene has been implicated in the observed photosynthetic declines. While the effects may be due to loss of photosynthetic capacity of mesophyll (65), the specific site of action for ethylene's effects on photosynthesis has yet to be determined. Many factors may contribute to the reduction in photosynthetic capacity. Chlorophyll degeneration occurs in some species following flooding (66). Leaf chlorophyll content in barley (*Hordeum vulgare* L.) was reduced by flooding (67). There is a close correlation between leaf tissue chlorophyll content and photosynthesis (68). The metabolic processes affected may include reduction in activity of photosynthetic enzymes. The activity of these enzymes is highly sensitive to changes in environmental conditions (69–71). The effect of flood stress on photosynthetic apparatus and protein syntheses has not been documented adequately. Regulation of protein turnover is an important factor in leaf development and function (72). Chlorophyllase (chlorophyllchlorophyllidohydrolase, EC 3.1.1.14) catalyzes the first step in carboxylation of chlorophyll (71). The enzyme has been implicated in leaf senescence during stress in plants including soybean and barley (69,71,73). Plant stresses such as temperature, salinity, and so forth generally adversely affect photosynthetic efficiency through their effects on PEPC (EC 4.1.1.31) and/or rubisco (EC 4.1.1.39) (70,74). Rubisco composes up to 70% of total leaf soluble protein as shown in barley (75). The leaf tissue rubisco concentration increases during the leaf expansion phase (76), followed by a plateau and finally by degradation by proteolytic enzymes to nitrogen and carbon as shown in wheat (77). Stress factors enhance rubisco degradation in potato (*Solanum* spp.) (78) as well as reducing its activity in soybean (*Glycine max* L.) (79). Rubisco activity appears to be highly sensitive to changes in environmental conditions such as temperature (80), drought (71), and air pollutants (81). In response to drought, rubisco activity showed a significant decrease within 5 days of drought initiation in soybean (71). Rubisco is also sensitive to air pollutants such as ozone as was shown in alfalfa (81).

The relationship between gas exchange and flood-induced changes in photosynthetic metabolites is poorly understood. For instance, as mentioned before, rubisco is an important enzyme in photosynthetic process; it catalyzes carboxylation yielding two molecules of 3-phosphoglycerate and oxygenation producing one molecule of 3-phosphoglycerate and one molecule of 2-phosphoglycolate (82). Rubisco has been studied in much photosynthetic research (82,83). Several studies have shown a direct relationship between maximum rate of carbon assimilation and activity and kinetics of rubisco in rice, wheat, and other species (83–89). Since activity of this enzyme is critical for photosynthetic carbon fixation, flood-induced reductions of net photosynthesis may be partially caused by changes in the status of rubisco activity. In flood-stressed plants, reductions in rubisco activity may provide the signals leading to the initial decline of photosynthetic activity. If hypoxia results in reduction in the pool of leaf active rubisco, it will result in reduced net photosynthesis (50). It is possible that in highly flood-tolerant plants the recovery of rubisco activity may lead to the recovery of net photosynthesis.

Other factors restricting photosynthetic activity include premature senescence of leaves and CK deficiency (50). In tomato plants, photosynthetic reduction in response to flooding was attributed to ABA accumulation and CK depletion, which resulted in reduced ribulose bisphosphate regeneration capacity of leaves (50). A hypothesis that emerged earlier proposes that leaf response to root stress is due to interruptions in transport of substances such as CK which were presumed to be produced solely in roots (50,90,91). Subsequent research, however, showed that many of these substances are also produced in the leaves (92,93). Neuman et al. (94) demonstrated that root O_2 deficiency reduced CK fluxes in xylem sap of bean; however, bulk leaf tissue content of CK was not affected. Their finding and that of Smit et al. (95) questioned the proposed hypothesis, at least as it related to the effects of flooding on CK concentrations in leaves. Nevertheless, they pointed out that the compartmentalization and/or active pools of CK in the leaves may be the critical factor rather than bulk CK. The question raised requires additional research including data on compartmentalization of CK as well as mechanisms and sites of CK action (94,95).

Soil flooding and the resultant excess soil sulfide is known to inhibit growth of various marsh species (96–98). The soluble sulfide species including H_2S are toxic to the roots of flood-tolerant plants such as rice (99,100). The O_2 diffusion from aerial parts to the roots and the subsequent sulfide oxidation in the rhizosphere has been considered as a major mechanism allowing a high level of sulfide tolerance in some species (101). However, oxygen release from young rice seedlings was inhibited when exposed to an H_2S concentration of 0.2 mg/l, therefore limiting ability of roots to oxidize the rhizosphere (102).

Sulfate uptake from the root medium as well as its translocation and accumulation in foliage has been documented for several species. For instance, Carlson and Forrest (103) and Pearson and Havill (104) using different techniques demonstrated that sulfide was taken up by roots and subsequently oxidized to sulfate. Total S^{2-} in rice foliage increased 12.1% in response to sulfide treatment (104). The inhibitory effects of hypoxia and elevated sulfide concentrations on leaf photosynthetic capacity has been demonstrated in marsh species, *Panicum hemitomon* Shult. and *Spartina patens* (Ait.) Muhl. (Fig. 4). The adverse effects of sulfide on photosynthesis of Lemnaceae species has been reported by Takemoto

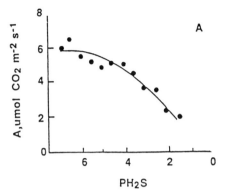

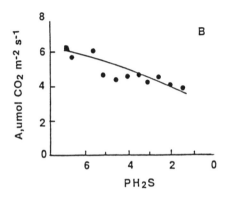

Figure 4 The relationship between net carbon assimilation (A), μmol $CO_2/m^2/$ sec and pH_2S levels in *Panicum hemitomon* (A) and *Spartina Patens* (B). P − log concentration of H_2S in moles per liter (e.g., p3 = 0.034 g/L or 3.4 × 10^{-3} $\mu g/$ L). (From Ref. 288.)

and Nobel (105). The photosynthetic capacity decreases because of alterations in activity of photosynthesis enzymes (106–108), disruption of light reactions (109), and/or photophosphorylation (110).

Soil flooding also influences translocation of various photosynthetic products. Reduced rates of assimilate translocation to roots has been reported in flood-stressed soybean (111). In squash (*Cucurbita melopepo* L.), oxygen deficiency resulted in 35–45% reduction in rate of assimilate movement within 30–40 min (112). Similar disruption of assimilate translocation has been reported in other species (113). Carbohydrate allocation patterns and translocation rate appear to be critical for hypoxia tolerance. For example, survival of rice roots was improved by approximately 40 hr in response to glucose application (114). The mechanisms involved in changes in assimilate translocation are poorly understood. Some products of anaerobic respiration, believed to be toxic, have been identified as potential causes (115). Nevertheless, additional data are needed to fill the gap in our understanding of mechanisms involved.

Nutrition

It is well established that soil flooding results in reduced foliage concentrations of N, P, and K (116–118). Under flooded conditions, mineral nutrition is affected by many factors including the soil's initial nutrient status, changes in soil physicochemical characteristics induced by flooding, plant developmental/physiological status, and flood tolerance capabilities (15). Oxygen deficiency causes inhibition of nutrient uptake and transport in flood-sensitive species (119) due to the death of the root system. Oxygen stress may change the permeability of cell membranes in the roots causing nutrient leaching (120). The effects may also be due to the inefficiency of anaerobic metabolism in providing adequate energy for active ion uptake (121,122). However, ion uptake may continue through passive means (123).

In flood-sensitive species, nitrogen concentration and total N content of tissue decreases in response to flooding (23). The reduction in N has been attributed, at least partially, to the rapid volatilization and loss through denitrification occurring in soil under flooded conditions (124). Reduced N concentration in several corn cultivars was attributed to the decline in soil nitrate availability (125). In flooded wheat, however, inhibition of N uptake was the cause of nitrogen deficiency rather than nitrogen availability (118,121,123). Inhibition of N uptake, enhanced senescence, and N remobilization within the plant have been reported in barley under oxygen stress (126,127).

Plant P content decreases in response to flooding presumably due to the suppressed uptake capability of roots as found in barley (128). Depending on the preflood conditions of the soil, P availability may increase upon flooding (124). The increase may be reflected in rise in plant tissue P content early in flooding as reported in maize (125). Nevertheless, P concentration and total content is reduced in extended flooding (23).

Potassium uptake is also reduced in flood-sensitive plants in response to flooding as reported for maize (129). Tissue content of Ca and Mg is less affected by flooding than content of N, P, and K (23). While the uptake of Ca and Mg may be metabolically controlled, available data do not establish a close correlation between active uptake mechanisms and plant Ca and Mg accumulation at least for species such as wheat and bean (123,130). Sodium accumulates in flood-sensitive plants under flooded conditions as found in cotton and sunflower (131,132). Flooded cereal plants had greater Na concentration in shoots than control plants (116,133). Soil iron (Fe) and manganese (Mn) availability increases under flooded conditions due to the reduction processes (124). Tissue concentration of these elements increases in some species. For instance, increased Fe concentrations was found in maize (134) and in avocado (*Persea americana* Mill.) (135). Increased tissue Mn concentrations have been found in flooded maize plants (134). Many fewer data are available on the effects of flooding on plant uptake of Cu, Zn, Mo, and Bo (23).

In plants adapted to wetland conditions, ion uptake continues possibly due to the existence of an internal O_2 supply system (136). The development of adventitious roots and aerenchyma formation in flood-tolerant species facilitates oxygen transport from aerial tissues to the roots and root–soil interface (137,138). While these adaptations allow root functioning and nutrient uptake, depending on soil conditions, various nutritional deficiencies and toxicities may occur such as zinc deficiency in rice growing on alkaline soils (139). During prolonged flooding conditions, as soil reduction continues, pH decreases and zinc availability increases leading to increased tissue zinc concentration (140). Flooded rice plants had greater concentrations of N, P, Ca, and Fe whereas Mn and Mg did not change significantly (141–143). Under flooded conditions, ferric and manganic forms are reduced to ferrous and manganous forms which are soluble (124). Thus tissue Mn and Fe concentrations are greater than found in plants under nonflooded conditions. Leaf discoloration (bronzing) in rice due to high soluble ferrous iron has been reported when a soil characterized by low pH undergoes prolonged flooding (139). Such a high iron concentration may also interfere with P uptake in rice (137). Sulfide toxicity may occur due to soil chemical changes. The decline in redox potential begins a chain of reaction which includes reduction of sulfate to sulfide by anaerobic microorganisms (144). Sulfide is considered to be phytotoxic, however, and in most cases roots of flood-tolerant plants such as rice have the capability to oxidize sulfide thus avoiding or minimizing injury (139).

Growth Regulators

The growth regulator balance within a plant changes rapidly in response to flooding due to changes in synthesis and translocation patterns. Flooding increases ethylene, auxins (IAA), and ABA while reducing CK and GA (52). Shortly after flooding is initiated, xylem contents of growth regulators such as CK and GA are reduced substantially (89,145) perhaps due to the death of roots and/or reduced root metabolic activity (90). It is difficult to evaluate the potential effects of such reductions on shoots because the shoot can also produce these substances (122).

Oxygen deficiency enhances ethylene biosynthesis in the roots (53,146). This enhancement is apparently coupled with reduced gas leakage from roots to the rhizosphere further increasing ethylene concentrations in root tissues. When roots of tomato plants grown in aerated nutrient solution were subjected to N_2 gas treatment, the foliage content of ethylene increased (147,148). These experiments demonstrated that root hypoxia enhanced ethylene synthesis in the shoots. Experiments using split-root systems provided further evidence that a precursor of ethylene is transported from the stressed roots to the shoots via the transpiration stream (17,147,149). This precursor has been identified as 1-aminocyclopropane-1-carboxylic acid (ACC) in apple tissue (150).

The adverse effects of ethylene on plants have been debated. Evidence shows that under laboratory conditions, elevated ethylene did not damage plants (151). In maize roots, ethylene synthesis increased at 3 kPa oxygen pressure while the process is apparently stopped when conditions became anaerobic (146) further supporting previous findings that molecular O_2 is essential for ACC conversion to ethylene. A large ACC pool can be accumulated in hypoxic tissues which reacts with O_2 to produce ethylene (152). In flooded tomato plants, stimulation of ACC production in the roots and its conversion to ethylene in the leaves has been demonstrated (153). The mechanisms involved in flood-induced ACC synthesis is unknown (154) but it may involve IAA-induced formation of ACC synthase (52,155,156). While CK may enhance ACC and ethylene synthesis, ABA appears to have the opposite affect (157,158). As has been pointed out previously (52), all growth regulators are affected by root anaerobiosis to some degree; thus the results for ACC synthesis may depend on complex interactions among these compounds in the stressed tissue. The increase in ABA in flooded plants may also inhibit IAA movement (52). Hiron and Wright (25) reported that flood stress caused ABA buildup in leaves of bean plants. Similar findings were reported for other species (38,159). Further evidence for the potential importance of ABA in flood tolerance has been reported in maize (160). Exogenous application of ABA increased seedling tolerance of anoxia 10-fold. The mechanism of flood-induced ABA response is not known (52).

Flooding increased IAA in sunflower shoots substantially (161–163). IAA accumulation may result in enhanced ethylene production (155,164) while buildup of ethylene could disrupt IAA transport as shown in cotton (165). In tomato plants, GA concentration in various plant components was reduced rapidly in response to flooding (35,145). In sunflower, flooding reduced CK concentration in xylem sap (90). Signals from stressed roots may interfere causing inhibition of shoot GA and CK biosynthesis (52). In general, much work is needed to identify the mechanisms of responses for each growth regulator as well as the potential interactions among them (52,122).

Growth

The effects of soil O_2 deficiency on plant growth is determined by species as well as the stage of plant development at the time of stress initiation. Soil waterlogging influences numerous aspects of root and shoot physiology (15,166). Root exposure to O_2 stress produces signals which cause foliage responses. Signals are presumed to be in the form of specific compounds and/or changes in flow of substances between root and shoot, and accumulation of substances in the shoot which normally are transported to the roots (12,122,164). For instance, inhibitions of sugar transport from shoots or endosperm to the roots was reported for 4- to 6-day-old rice plants under anoxic conditions (167).

Moving floodwater causes less damage to the roots than stagnant water because of the higher dissolved O_2 content of moving water as compared to stagnant water. For example, wetland plants develop roots in moving water and in sediments which are not highly reduced. Root growth in these plants, however, is affected by highly reduced conditions and root penetration into reduced zones is prevented (168,169). Similar conditions are found in rice growing in paddy fields (170).

Root growth is an energy-dependent process. In soybean, 7 days of flooding resulted in rapid cessation of root growth and root death (171). The depletion of oxygen in roots is more rapid than in the soil water unless a substantial aerenchymatic tissue is present. Two major causes of O_2 depletion are the root tissue respiration and the limitation on O_2 diffusion caused by floodwater surrounding the roots. Such conditions may lead to complete tissue O_2 deprivation (anoxia). In numerous species, cessation of root elongation have been observed in response to hypoxia. Cotton and soybean grown in soil subjected to N_2 gas treatment showed taproot death after 3 and 5 hr, respectively (172). Seminal root in maize survived 70 hr of treatment (173). A short period of anaerobiosis may be sufficient to reduce root growth (131;132,174,175). For instance, within 1–2 hr of initiation of O_2 stress, root growth in soybean was reduced significantly (175). Pea and pumpkin roots survived anaerobiosis up to 9 and 20 hr, respectively (176). Adventitious roots in rice survived 7 days and root growth resumed following termination of O_2 deficiency (177). The threshold O_2 concentration at which root elongation is affected is believed to be around half of the concentration in the air (133,177,178). However, O_2 transport from shoots changes this threshold. If the supply of O_2 from the shoot is eliminated, O_2 pressure of 0.01–0.03 atm at 20°C causes cessation of root growth (122). Under flooded conditions, root functioning is affected rapidly because molecular O_2 is required as an electron acceptor for oxidative phosphorylation (7,179).

The oxygen concentration at which root respiration is limited has been defined as the critical oxygen pressure, COP_{ext} (180–183). In maize the COP_{ext} is 30 kPa for root tips at 25°C and 10 kPa for old roots (184). The importance of COP has also been implicated in other plant functioning such as root elongation and nutrient uptake. At an O_2 concentration of 0.015–0.050 atm, normal root fresh weight increases have been reported for tomato and soybean (133). When oxygen concentration falls below COP_{ext}, cellular oxygen deficiency (hypoxia) occurs in cells located away from the epidermis (7). Such hypoxic conditions are likely to be heterogeneous within the plant tissue. The COP_{ext}, however, may not necessarily reflect the COP_{int} since the O_2 transport system from leaves to roots (aerenchyma) may compensate for the low external oxygen (185). For instance, in wheat grown hydroponically with the root medium purged with N_2 gas treatment (1–2 kPa O_2), the O_2 concentration (13–16 kPa O_2) of the internal root gases was substantially greater than the external O_2 concentration (186).

Root growth is a function of cell division and elongation. In onion (*Allium cepa* L.) grown in O_2 concentrations ranging from 0.2 atm (air) to 0.02 atm, root growth was reduced from 21.3 to 8.8 mm/day primarily through the effects on the rate of production of new cells (187). Cessation of root growth is found under O_2 deficiency in most cases. However, cell expansion and division may continue albeit at slow rates in some species such as rice (177,188,189). Figure 5 depicts root elongation responses of maize to various levels of soil Eh (190). Root elongation is affected by O_2 deficiency in both flood-tolerant and flood-sensitive species (Fig. 6). In wheat roots, low ATP synthesis leads to cell injury and thus a loss of root elongation capability upon a return to aerobic conditions. In this species, >85% of roots lost their elongation potential after 9 hr of anoxia (191). Hypoxic pretreatment is important in allowing roots to acclimate to anoxic conditions as has been shown in wheat (191) and maize (192). In wheat roots pretreated under hypoxic conditions for 15–30 hrs, only 30% of roots lost their elongation potential (191).

Accumulation of high ethylene concentrations has been implicated in the reduction/cessation of root growth in several species such as white mustard (see ref. 122 for the pertinent review). However, in flood-

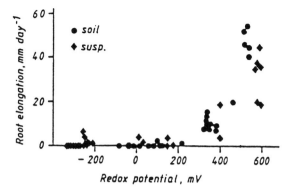

Figure 5 Relationship between redox potential (Eh, mV) in the root medium and the elongation rate of primary maize roots grown under different redox potential treatments in soil (closed circles) and suspension (closed diamonds). (From Ref. 190.)

tolerant plants such as rice, ethylene production in response to flooding is low (193). In addition, the concentration of reduced soil compounds such as NO_2^-, Mn^{2+}, Fe^{2+}, S^-, and other compounds may increase to the levels toxic to plants causing root injury.

Other root processes affected by hypoxia include ion relations, water relations, and growth regulator production. These changes are rapidly communicated to the shoots affecting shoot physiology (194). Flooding reduces shoot growth except in some flood-tolerant species such as lowland rice (195) and some aquatic species (196). The reduction in shoot growth has been attributed to the adverse effects of flooding on root aerobic respiration and disruption of translocation of root-produced metabolites to shoots (52). The decrease in GA and increase in ABA and ethylene may be responsible for reduced shoot growth as was found in tomato plants (35). However, the existing data are not sufficient to identify the role of individual growth regulators and the potential interactions among them in the reduction of shoot growth (52). In addition to growth regulators, toxins, nutrient deficiency, and water stress all have been implicated in the observed shoot responses (122) and their role requires further research and documentation.

Leaf growth and development is adversely affected by root hypoxia (197,198). For instance, in wheat, 83% reduction in leaf area has been reported (20). The effect on leaf growth is rapid, within minutes as reported in maize (199) which was primarily attributed to nutrient deficiency (see also Ref. 200). Ethylene buildup has been implicated in

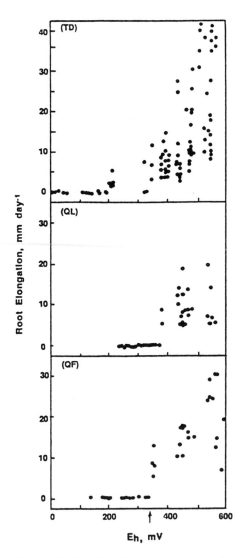

Figure 6 Relationship between root elongation of *Taxodium distichum* (TD), *Quercus falcata* var. *pagodaefolia* (QF), *Quercus lyrata* (QL), and soil redox potential (Eh, mV). Arrow signifies Eh at which oxygen disappears. (From Ref. 289.)

this response (122,201) but the exact mechanisms are yet to be determined.

Flooding results in overall reduction of biomass (20,123) in flood-sensitive and flood-tolerant species. In wheat, an initial increase in dry weight has been reported followed by a reduction in dry weight (118). Such a response has been attributed to reduced dark respiration (58) and shift in photosynthate allocation (202). In bean, flooding for 7 days reduced root, stem, and leaf dry weights (59). In flood-tolerant *Scirpus americanus* Pers. seedlings, flooding for 6 months resulted in significant reduction of height, above-ground and below-ground biomass. Height on average was reduced 23% and dry weight was reduced 33% as compared to the drained plants (203). Jones and Etherington (204) found adverse effects of waterlogging on growth and productivity of four grass species from Great Britain. Growth, fresh and dry weights were significantly reduced in poorly aerated solutions for both *Echinochloa oryzoides* (Ard.) Fritsch. and rice (205). In marsh grass *Sporobolus virginicus* (L.) Kunth. grown for 6 months under poorly aerated conditions, plant biomass decreased substantially; there was a 46% reduction in plant biomass as compared to aerated control (206). Significant changes are reported in below-ground/above-ground biomass ratio as the effects of anaerobiosis is more drastic on root systems (206). While a range of responses to flooding may occur in a given species, these changes nevertheless occur within the context of an overall reduction in biomass accumulation (122).

Other Responses

Plant foliage often responds rapidly to root oxygen deficiency. Such responses include epinasty, wilting, stem hypertrophy, and leaf senescence. These responses are injurious to plants, adversely affecting crop productivity while providing some means for improved survival (7).

Epinasty

Epinastic curvature of leaves in response to flooding is a well-known phenomenon which has been reported in numerous species including sunflower (207) and tomato (147,148). Nonuniform elongation of cells, i.e., the cells on the upper side of a petiole elongating at much faster rates than lower side cells, is the cause of epinasty (51,147,148,208–210). Ethylene has been implicated in epinasty as ACC is transported from roots to shoots where conversion to ethylene occurs in the presence of oxygen (53). Other growth regulators including IAA and GA have also been implicated (207,210). The importance of epinasty to avoidance/tolerance of flood stress is presumably due to the potential advantage to plants in allowing water conservation, thus avoiding development of water stress. However, the exact role of epinasty in flood stress remains unknown (122).

Hypertrophy

Hypertrophy or stem base swelling, caused by collapse or enlargement of cells in the cortex, has been reported in many species including tomato (18), maize (211), and sunflower (207,212). The role of floodwater in entrapping ethylene, thus promoting hypertrophy, has been proposed (212,213) as well as IAA accumulation (163). In sunflower hypocotyls, both IAA and ethylene may be involved in formation of hypertrophy (163). Ethylene and IAA stimulate enzymatic activity, which enhances cell wall weakening and cell collapse (214).

Leaf Chlorosis and Senescence

Flooding causes leaf chlorosis and premature senescence in many species (18,90). Several factors involved in this response including reduced CK levels (215), reduced GA levels (216), increased ABA and ethylene (217,218). Reduced GA and CK levels may promote deterioration of membrane affecting a range of leaf metabolism including synthesis of chlorophyll and protein (51,219). Leaf chlorosis in some plants has been attributed to an increase in tissue iron concentration (220) as shown in rice (221). Leaf chlorosis in barley was related to depressed nitrogen uptake (126). A similar condition has been reported in wheat (121). Flood-induced senescence has been observed in crop species such as sunflower (212), tomato and tobacco (18), barley (126), maize (199), and wheat (121). The causes of such response are not known but may involve nutrient deficiency and/or soil phytotoxins. Loss of capacity to take up nitrogen is believed to be a dominant factor in induction of leaf senescence in several species including barley and wheat (120,126).

Despite a great deal of progress made in identifying various edaphic and endogenous factors which control flood responses in plants, there are numerous areas requiring additional research. Research is needed in such areas as mechanisms controlling hypertrophy of lenticels, adventitious rooting, photosynthesis, and carbohydrate allocation in response to oxygen deficiency.

MECHANISMS OF FLOOD TOLERANCE

The tolerance of oxygen deficiency differs among various species ranging from hours to months. In species tolerant of flooding, flood tolerance is achieved through various means including morphological, anatomical, and metabolic adaptations.

Morphological/Anatomical Adaptations

Morphological responses to flooding include epinasty and hypertrophy phenomena, which have been discussed in previous sections. In addition to the morphological changes, flood-tolerant plants are capable of developing specific anatomical structural characteristics in order to avoid or delay root anoxia (222–225), allowing them to survive and function under stressful conditions. Such features include aerenchyma tissues and adventitious root development.

Oxygen may enter through leaves (137,226,227) and lenticels (228). In wetland plants, aerenchyma tissue is found in above-ground and below-ground portions characterized by large intercellular spaces. Rapid aerenchyma tissue development in response to flooding has been reported in some species such as sunflower (214,229). Increased ethylene-induced cellulase activity has been implicated in this process (229). Aerenchyma tissue development is a process caused by cell separation (schizogenous) or cortex breakdown (lysigenous) producing gas-filled lacunae (230,231). The process of lysis in maize root occurs primarily in cortex (149,232) and not in surrounding cells, suggesting site-specific actions for ethylene (7). Lysigenous aerenchyma formation in response to O_2 stress has been reported in cortex tissues 3–4 cm behind root tips in maize (149), wheat (117), and rice (233). Ethylene may be involved in cell lysis since ethylene concentration increases in response to O_2 deficiency (43). In some wetland species, low soil O_2 is needed for aerenchyma formation (222,234) while development of these systems in some species such as rice does not require oxygen stress, i.e., it is controlled genetically (235). For example, in rice, aerenchyma develops under well-aerated conditions while hypoxic conditions further enhance the process (233).

Aerenchyma tissue enhances root oxygenation in plants (236). It may also be found in the meristematic zone (230) where there is a great need for O_2 (237). The existence of aerenchyma improves radial O_2 leakage (233,238) and root growth in anaerobic environment as found in rice (230) and wheat (118). An oxidized area is found close to the root tips due to oxygen leakage to the rhizosphere. This layer is assumed to help detoxify the immediate zone surrounding the root tips (239). Mineral deposits around roots have been reported due to oxidation of Fe^{2+} to Fe^{3+} (225,240,241).

Development of adventitious roots on the submerged areas of stem or existing roots has been documented in several species (242,243). Adventitious root formation on flooded portions of stem or hypocytl has been reported in many crop species including tomato (18,35), sunflower (18), maize (244), and sugarcane (245). Removal of adventitious roots causes reduction of growth in herbaceous and crop plants (246). This flood-induced rooting is a rapid response (247–249) controlled by many factors including growth regulators (52,247–249). While the physiological significance of adventitious roots is not completely known and is under debate (250–252), it appears that these roots can replace the losses due to damaged functioning of original roots which occurs under flooded conditions (18) and thus improve plant capability for functioning (252). Adventitious roots also produce root-synthesized growth regulators such as GA and CK (52). The mechanisms involved in adventitious root formation are still subject to debate. The potential involvement of plant growth regulators (ethylene, auxin, cytokinins, etc.) and high concentrations of carbohydrates has been implicated by various workers (123,163,207,210,212,253) but requires additional experimentation.

Metabolic Adaptations

In addition to the morphological and anatomical adaptations, metabolic adaptations are critical for plant survival in anaerobic environments. The mechanisms involved are complex and include avoidance of accumulation of toxic compounds and maintenance of a continuous supply of carbohydrate. Under normal (aerated) conditions, oxidation of one molecule of sugar to CO_2 and water produces 38 molecules of ATP. However, under anaerobic conditions, the glycolytic pathway produces pyruvic acid from glucose, which is converted to CO_2 and ethyl alcohol. The process is catalyzed by cytoplasmic enzymes (254) with net conversion of two molecules of ADP to ATP from each glucose molecule thus slowing down ATP synthesis and affecting energy-dependent processes critical for plant functioning most of which require ATP such as protein synthesis (7). In barley roots subjected to O_2 stress, ATP concentration decreased by 88% in 15 min (255), which resulted in dramatic changes in the pH gradient across the cell membrane. ATP turnover in root tips is a rapid process under aerated conditions. For instance, ATP turnover was reported to be around 8 sec in aerated maize root tips (256).

There are two prominent theories of metabolic adaptations of plants to oxygen deficiency. Crawford (257,258) and McManmon and Crawford (259) proposed that flood tolerance in a given plant is dependent on the ability to avoid ethanol production in the glycolytic pathway thus maintaining a low level of ethanol biosynthesis. The mechanisms proposed were low alcohol dehydrogenase (ADH) activity and use of alternate end products such as lactate, malate, etc., thus avoiding the potential ethanol toxicity. This theory proposes that in flood-sensitive species glycolysis will be promoted in response to oxygen stress resulting in ethanol accumulation at faster rates than in flood-tolerant species. However, this hypothesis has been questioned (151,260–262). Anaerobic conditions result in ethanol production in both flood-tolerant and flood-sensitive species, e.g., in rice (136,263), and in flood-sensitive species, e.g., in pea (264). While anoxic or hypoxic conditions increase the activity of fermentative enzymes, ADH and pyruvate decarboxylase (PDC) (179,265), no relationship was found between ethanol content and the induction of ADH and PDC synthesis in maize and rice (266). Data so far appear to show that flood-tolerant plants produce ethanol, which is subsequently released to the outside solution or through the transpiration stream (151,263) eliminating the potential accumulation to toxic levels (151,263). Ethanol accumulation in flood-sensitive species is believed to cause death of root cells (258). Plants use several mechanisms in order to avoid ethanol buildup to toxic levels including transport from anaerobic tissue to aerated tissue (262,267), ethanol leakage to the surrounding areas, and rerouting of glycolytic intermediates to produce different end products such as malate, lactate, succinate, and alanine (267). As was pointed out by Kennedy et al. (268), the question of ethanol toxicity has not yet been completely resolved and is still subject to debate.

Davies (260) proposed an alternative hypothesis proposing that cytoplasmic pH determines the relative rates of ethanol and lactate synthesis. Cytosolic pH decreases in response to anaerobiosis leading to inhibition of lactate dehydrogenase (LDH) activity, and increased PDC activity which lead to enhanced ethanol synthesis. Roberts (269) found that in O_2-stressed maize root tips, cytoplasmic pH decreased from 7.3 to 6.8 during the first 20 min of treatment, stabilizing thereafter. The pH stabilization was attributed to increased PDC and ADH activity, reduced LDH activity and lactate transport to the vacuole. In hypoxic root tips of maize, no relationship was found between cytoplasmic acidification and lactic acid synthesis (270). There are studies which support the involvement of pH in anaerobic metabolism (271–273). The pH theory, however, has not been applicable in all plants including rice (268,274). For instance, Menegus et al. (271,274) found that anoxic conditions resulted in pH reduction in flood-sensitive plants such as wheat and barley while it increased cellular pH in flood-tolerant rice plants. They concluded that there is a reverse correlation between flood tolerance and cytoplasmic acidification. In addition, such response may serve as an early signal of O_2 stress in sensitive species whereas in flood-tolerant species mechanisms other than pH acidification may be involved.

The role of ADH in flood tolerance is still being debated. In maize, a certain level of ADH activity is required for survival under O_2 stress conditions (275) while in *Echinochloa* species, there is not an apparent correlation (276). Cobb and Kennedy (277) demonstrated a potential correlation between location of ADH activity within plants and flood tolerance.

For instance, in rice the majority of ADH activity was found in shoots while in flood-sensitive maize and pea much of the ADH activity was found in the roots. In 5- to 21-day-old soybean plants, the induction of ADH was organ-specific and was related to the age of the plant (278). The critical role of ADH in cytoplasmic pH regulation has been confirmed by the work of Roberts et al. (272,273). Changes in cytoplasmic acidity is injurious to cells causing cellular death in maize and pea (272). However, the mechanisms of cytoplasm injury due to low pH is not known (7).

Protein metabolism is also affected by O_2 stress. In flood-sensitive plants such as maize, oxygen deficiency results in synthesis of a few anaerobic stress proteins (ASPs) including ADH_1, ADH_2, glucose-6-phosphate isomerase, aldolase, sucrose synthase, cytosolic glyceraldehyde-3-phosphate dehydrogenase, PDC, and LDH (279). Synthesis of normal protein is halted leading to plant death (173). In contrast, in flood-tolerant *Echinochloa*, ASP synthesis in response to O_2 stress is accompanied by resumption of normal protein synthesis within 24 hr (268). In rice shoot, protein synthesis continued in anaerobic mitochondria in the absence of a functional respiratory chain under anoxic conditions (280). Additional data is needed to identify the mechanisms involved in synthesis of ASPs as well as normal protein synthesis in response to oxygen deficiency.

CONCLUSIONS

Our knowledge of oxygen stress effects on plants and the mechanisms involved have expanded significantly in recent years. The correlation between morphological, anatomical, and metabolic responses to the level of flood tolerance in plants has been established. Recent advances in techniques particularly at the molecular level allow substantial gain in understanding of gene regulation and plant functioning under oxygen-deficient environments.

There are, however, many speculations remaining in the literature which require clarification and numerous areas in need of further attention (see Refs. 7, 23, 52, and 122 for detailed reviews). These areas, to name a few, include mechanisms involved in the regulation of plant–water relations, gas exchange, carbohydrate allocation, adventitious rooting, growth regulators (biosynthesis, metabolism, and interactions), regulation of gene expression by oxygen deficiency, anaerobic protein synthesis and regulation, mechanisms involved in root acclimation to oxygen deficiency, and the potential role of reduced soil compounds in induction of flood injury to plants.

ACKNOWLEDGMENTS

The author is grateful to Dr. Mary Musgrave, Department of Plant Pathology and Crop Physiology, Louisiana State University, for helpful comments and critical review of this manuscript.

REFERENCES

1. E. W. Russell (1973): *Soil Conditions and Plant Growth*, Longmans, London.
2. F. N. Ponnamperuma (1984): *Flooding and Plant Growth* (T. T. Kozlowski, ed.), Academic Press, Orlando, p. 10.
3. R. P. Gambrell and W. H. Patrick, Jr. (1978): *Plant Life in Anaerobic Environments* (D. D. Hook and R. M. M. Crawford, eds.), Ann Arbor Science, Ann Arbor, Michigan, p. 375.
4. T. Yoshida (1978): *Soils and Rice*, International Rice Research Institute, Los Banos, Philippins, p. 445.
5. W. H. Patrick and R. D. DeLaune (1977): *Geoscience and Man*, *18*:131.
6. M. C. Drew and J. M. Lynch (1980): *Annu. Rev. Phytopathol.*, *18*:37.
7. M. C. Drew (1990): *Plant Cell Environ.*, *13*:681.
8. L. H. Stolzy, J. Letey, L. J. Klotz, and T. A. De Wolfe (1965): *Soil Sci.*, *99*:403.
9. L. H. Stolzy and R. E. Sojka (1984): *Flooding and Plant Growth* (T. T. Kozlowski, ed.), Academic Press, New York, p. 221.
10. S. J. Field (1982): Studies on the role of irrigation and soil water matrix potential on *phytophthora parasitica* root rot of citrus, Ph.D. dissertation, Univ. of California, Riverside.
11. P. J. Kramer and W. J. Jackson (1954): *Plant Physiol.*, *29*:241.
12. K. J. Bradford and T. C. Hsiao (1982): *Plant Physiol.*, *70*:1508.
13. J. D. Everard and M. C. Drew (1989): *J. Exp. Bot.*, *40*:95.
14. T. T. Kozlowski (1984): *Flooding and Plant Growth* (T. T. Kozlowski, ed.), Academic Press, New York, p. 1.
15. T. T. Kozlowski (1984): *Bioscience*, *34*:162.
16. H. Moldau (1973): *Photosynthetica*, *7*:1.
17. M. B. Jackson, K. Gales, and D. J. Campbell (1978): *J. Exp. Bot.*, *29*:183.
18. P. J. Kramer (1951): *Plant Physiol.*, *26*:722.
19. S. R. Pezeshki and F. J. Sundstrom (1988): *Scientia Hort.*, *35*:27.
20. R. E. Sojka, L. H. Stolzy, and M. R. Daufman (1975): *Agron. J.*, *67*:591.
21. R. E. Sojka and L. H. Stolzy (1980): *Soil Sci.*, *130*:350.
22. T. T. Kozlowski (1982): *For. Abstract.*, *43*:145.
23. T. T. Kozlowski and S. G. Pallardy (1984): *Flooding and Plant Growth* (T. T. Kozlowski, ed.), Academic Press, Orlando, p. 165.
24. P. J. Kramer (1940): *Am. J. Bot.*, *27*:216.
25. R. W. P. Hiron and S. T. C. Wright (1973): *J. Exp. Bot.*, *24*:769.
26. J. B. Zaerr (1983): *Forest Sci.*, *29*(1): 71.
27. J. S. Pereira and T. T. Kozlowski (1977): *Physiol. Plant.*, *41*:184.

28. A. R. Sena Gomes and T. T. Kozlowski (1980): *Physiol. Plant, 49*:373.
29. Z. C. Tang and T. T. Kozlowski (1982): *Can. J. Forest. Res., 12*:196.
30. S. R. Pezeshki and J. L. Chambers (1985): *Can. J. For. Res., 15*:371.
31. S. R. Pezeshki and J. L. Chambers (1985): *For. Sci., 31*:760.
32. K. J. Bradford (1983): *Plant Physiol., 73*:475.
33. K. J. Bradford (1983): *Plant Physiol., 73*:480.
34. R. E. Sojka and L. H. Stolzy (1981): *Calif. Agric., 35*:18.
35. D. M. Reid and A. Crozier (1971): *J. Exp. Bot., 22*:39.
36. J. E. Pallas and S. J. Kays (1982): *Plant Physiol., 70*:598.
37. Z. C. Tang and T. T. Kozlowski (1982): *Plant Soil, 66*:243.
38. B. Shaybany and G. C. Martin (1977): *J. Am. Soc. Hort. Sci., 102*:300.
39. W. T. Jackson (1956): *Am. J. Bot., 43*:496.
40. S. Szlovak (1975): *Acta Bot. Acad. Sci. Hung., 21*:167.
41. A. S. El-Beltagy and M. A. Hall (1974): *New Phytol., 73*:47.
42. B. D. Van't Woudt and R. M. Hagan (1957): *Drainage of Agricultural lands* (J. N. Luthin, ed), Am. Soc. Agric. Eng., Madison, Wisconsin, p. 514.
43. M. B. Jackson (1982): *Plant Growth Substances 1982* (P. R. Wareing, ed.), Academic Press, New York, p. 291.
44. L. C. Hammond, W. H. Allaway, and W. E. Loomis (1955): *Plant Physiol., 30*:155.
45. D. G. Harris and C. H. M. van Bavel (1957): *Agron. J., 49*:176.
46. T. A. Mansfield and R. J. Jones (1971): *Planta, 101*:147.
47. J. Zhang and W. J. Davies (1987): *J. Exp. Bot., 38*:649.
48. R. J. Jones and T. A. Mansfield (1970): *J. Exp. Bot., 21*:714.
49. P. E. Kriedemann, B. R. Loveys, G. I. Fuller, and A. C. Leopold (1972): *Plant Physiol., 49*:842.
50. K. J. Bradford (1982): *Plant Growth Substances 1982* (P. F. Wareing, ed.), Academic Press, New York, p. 599.
51. C. K. Pallaghy and K. Raschke (1972): *Plant Physiol., 42*:275.
52. D. M. Reid and K. J. Bradford (1984): *Flooding and Plant Growth* (T. T. Kozlowski, ed.), Academic Press, New York, p. 195.
53. K. J. Bradford and S. F. Yang (1980): *Plant Physiol., 65*:322.
54. D. L. Regehr, F. A. Bazzaz, and W. R. Boggess (1975): *Photosynthetica, 9*:52.
55. T. T. Kozlowski and S. G. Pallardy (1979): *Physiol. Plant, 46*:155.
56. A. R. Sena Gomes and T. T. Kozlowski (1980): *Plant Physiol., 66*:267.
57. L. H. Stolzy, O. L. Taylor, W. M. Dugger, Jr., and J. D. Mersereau (1964): *Soil Sci. Soc. Am. Proc., 28*:305.
58. E. M. Wiedenroth and J. Poskuta (1981): *Z. Pflanzenphysiol., 103*:459.
59. B. P. Singh, K. A. Tucker, J. D. Sutton, and H. L. Bharadwaj (1991): *Hort. Sci., 26*:372.
60. H. T. Phung and E. B. Knipling (1976): *Hort. Sci., 11*:131.
61. Z. C. Tang and T. T. Kozlowski (1982): *Physiol. Plant, 55*:415.
62. S. R. Pezeshki and J. L. Chambers (1986): *For. Sci., 32*:914.
63. P. Holmgren, P. G. Jarvis, and M. S. Jarvis (1965): *Physiol. Plant, 18*:557.

64. P. E. Kriedemann (1971): *Physiol. Plant, 24*:218.
65. G. E. Taylor and C. A. Gunderson (1988): *Plant Physiol., 86*:85.
66. R. R. Rodriguez and H. W. Gausman (1973): *J. Rio Grande Va. Hort. Soc., 27*:81.
67. A. V. Mikhailova (1977): *Biol. Nauki, 20*:104.
68. T. Keller and W. Koch (1964): *Plant Soil, 20*:116.
69. C. R. Barmore (1975): *Hort. Sci., 10*:595.
70. J. Berry and O. Bjorkman (1980): *Annu. Rev. Plant Physiol., 31*:491.
71. S. Majumdar, S. Ghosh, B. R. Glick, and E. B. Dumbroff (1991): *Physiol. Plant, 81*:473.
72. S. S. Thayer, H. T. Choe, A. Tang, and R. C. Huffaker (1989): *Plant Senescence: Its Biochemistry and Physiology* (W. W. Thompson, E. A. Northnagel, and R. C. Huffaker, eds.), Am. Soc. Plant Physiol., Rockville, Maryland, p. 71.
73. M. T. Rodriguez, M. P. Gonzales, and J. M. Linares (1987): *J. Plant Physiol., 129*:369.
74. S. Ghosh, S. Gepstein, B. R. Glick, J. J. Heikkila, and E. B. Dumbroff (1989): *Plant Physiol., 90*:1298.
75. B. L. Miller and R. C. Huffaker (1982): *Plant Physiol., 69*:58.
76. M. J. Dalling (1987): *Plant Senescence: Its Biochemstry and Physiology* (W. W. Thomson, E. A. Northnagel, and R. C. Huffaker, eds.), Am. Soc. Plant Physiologists, Rockville, Maryland, p. 54.
77. T. Mae, N. Kai, A. Making, and K. Ohira (1984): *Plant Cell Physiol., 25*:333.
78. M. S. Dann and E. J. Pell (1989): *Plant Physiol., 91*:427.
79. M. J. Lauer, S. G. Pallardy, D. C. Belvins, and D. D. Randall (1989): *Plant Physiol., 91*:848.
80. V. M. Oja, B. H. Rasulov, and A. H. Laisk (1988): *Aust. J. Plant Physiol., 15*:737.
81. E. J. Pell and N. S. Pearson (1983): *Plant Physiol., 73*:185.
82. G. Bowes (1991): *Plant Cell Environ., 14*:795.
83. G. H. Lorimer, M. R. Ballgen, and T. J. Andrews (1977): *Anal. Biochem., 78*:66.
84. P. F. Wareing, M. M. Khalifa, and K. J. Treharne (1968): *Nature, 220*:453.
85. R. G. Jensen and J. T. Bahr (1977): *Ann. Rev. Plant Physiol., 28*:379.
86. D. I. Dickman and J. C. Gordon (1975): *Plant Physiol., 56*:23.
87. A. Makino, T. Mae, and K. Ohira (1985): *Planta, 166*:414.
88. R. R. Evan (1986): *Planta, 167*:351.
89. I. E. Woodrow and J. A. Berry (1988): *Ann. Rev. Plant Physiol. Plant Molec. Biol., 39*:533.
90. D. J. Carr and D. M. Reid (1968): *The Biochemistry and Physiology of Plant Growth Substances* (F. Wightman and G. Setterfield, eds.), Runge, Ottawa, p. 1169.
91. W. J. Burrows and D. J. Carr (1969): *Physiol. Plant, 22*:1105.
92. C. M. Chen and B. Petschow (1978): *Plant Physiol., 62*:861.

93. Y. Koda and Y. Okazawa (1980): *Physiol. Plant,* 49:193.

94. D. S. Neuman, S. B. Rood, and B. A. Smit (1990): *J. Exp. Bot.,* 41:1325.

95. B. A. Smit and M. L. Stachowiak (1990): *Plant Physiol.,* 92:1021.

96. G. M. King, M. J. Klug, R. G. Wiegert, and A. G. Chalmers (1982): *Science,* 218:61.

97. A. Ingold and D. C. Havill (1984): *J. Ecol.,* 72:1043.

98. D. C. Havill, A. Ingold, and J. Pearson (1985): *Vegetatio,* 62:279.

99. A. Tanaka, R. P. Mulleriyawa, and T. Yasu (1968): *Soil Sci. Plant Nutr.,* 14:1.

100. A. I. Allam and J. P. Hollis (1972): *Phytopathology,* 62:634.

101. J. M. Teal and J. Kanwisher (1966): *J. Exp. Bot.,* 17:355.

102. M. M. Joshi, I. K. A. Ibrahim, and J. P. Hollis (1975): *Phytopathology,* 65:1165.

103. P. R. Carlson and J. Forrest (1982): *Science,* 216:633.

104. J. Pearson and D. C. Havill (1988): *J. Exp. Bot.,* 39:363.

105. B. K. Takemoto and R. D. Noble (1986): *New Phytol.,* 103:525.

106. S. G. Garsed (1981): *Envir. Pollut. Series A,* 24:883.

107. A. A. Khan and S. S. Malhotra (1982): *Phytochemistry,* 21:2607.

108. M. J. Dropff (1987): *Plant Cell Environ.,* 10:753.

109. K. Shimazaki and K. Sugahara (1980): *Plant Cell Physiol.,* 21:125.

110. A. R. Wellburn, C. Higginson, D. Robinson, and C. Walmsley (1981): *New Phytol.,* 88:223.

111. E. G. Zvareva and B. I. Bartkov (1976): *Fiziol. Biokhim. Rast.,* 8:204.

112. J. W. Sij and C. A. Swanson (1973): *Plant Physiol.,* 51:368.

113. F. A. Qureshi and D. C. Spanner (1973): *Planta,* 110:131.

114. T. Webb and W. Armstrong (1983): *J. Exp. Bot.,* 34:579.

115. D. R. Geiger and S. A. Savonick (1975): *Encyl. Plant Physiol. New Ser.,* 1:256.

116. J. Letey, L. H. Stolzy, and N. Valoras (1965): *Agron. J.,* 57:91.

117. M. C. Drew and E. J. Sisworo (1979): *New Physiologist,* 82:301.

118. M. C. T. Trought and M. C. Drew (1980): *Plant Soil,* 54:77.

119. M. G. Pitman (1976): *Encycl. Plant Physiol. New Ser.,* 2B:95.

120. C. J. Rosen and R. M. Carlson (1984): *Plant Soil,* 80:345.

121. M. C. T. Trought and M. C. Drew (1980): *Plant Soil,* 56:187.

122. M. B. Jackson and M. C. Drew (1984): *Flooding and Plant Growth* (T. T. Kozlowski, ed.), Academic Press, Orlando, FL, p. 47.

123. M. C. T. Trought and M. C. Drew (1980): *J. Exp. Bot.,* 31:1573.

124. F. N. Ponnamperuma (1972): *Adv. Agron.,* 24:29.

125. R. Singh and B. P. Ghildyal (1980): *Agron. J.,* 72:737.

126. M. C. Drew and E. J. Sisworo (1977): *New Phytol.,* 79:567.

127. M. C. Drew, E. J. Sisworo, and L. R. Saker (1979): *New Phytol.,* 82:315.

128. A. J. Leyson and R. W. Sheard (1974): *Can. J. Soil Sci.,* 54:463.

129. K. Lawton (1945): *Soil Sci. Soc. Am. Proc.,* 10:263.

130. M. C. Drew and O. Biddulph (1971): *Plant Physiol.,* 48:426.

131. J. Letey, L. H. Stolzy, G. B. Blank, and O. R. Lunt (1961): *Soil Sci.,* 92:314.

132. J. Letey, O. Lunt, L. M. Stolzy, and T. E. Szuszkiewicz (1961): *Soil Sci. Soc. Am. Proc.*, 25:183.
133. H. T. Hopkins, A. W. Specht, and S. B. Hendricks (1950): *Plant Physiol.*, 25:193.
134. R. Lal, and G. S. Taylor (1970): *Soil Sci. Soc. Am. Proc.* 34:246.
135. K. Slowik, C. K. Labanauskas, L. H. Stolzy, and G. A. Zentmyer (1979): *J. Am. Soc. Hortic. Sci.*, 104:172.
136. C. D. John, V. Limpinuntana, and G. Greenway (1974): *Aust. J. Pl. Physiol.*, 1:513.
137. W. Armstrong (1968): *Physiol. Plant*, 21:539.
138. M. P. Coutts and W. Armstrong (1976): *Tree Physiology and Yield Improvement* (M. G. R. Cannell and F. T. Last, eds.), Academic Press, New York, p. 361.
139. A. Tanaka and S. Yoshida (1970): *Int. Rice Res. Inst. Tech.*, Bulletin No. 10.
140. V. Pavanasasivam and J. H. Axley (1980): *Commun. Soil Sci. Plant Anal.*, 11:163.
141. C. S. Weeraratna (1975): *Indian J. Agric. Sci.*, 45:461.
142. A. Islam and W. Islam (1973): *Plant Soil*, 39:555.
143. R. S. Chahal, P. S. Tulla, and S. S. Khanna (1980): *J. Indian Soc. Soil Sci.*, 28:254.
144. W. Armstrong (1975): *Environment and Plant Ecology* (J. R. Etherington, ed.), Academic Press, New York, p. 181.
145. D. M. Reid, A. Crozier, and B. M. R. Harvey (1969): *Planta*, 89:376.
146. B. J. Atwell, M. C. Drew, and M. B. Jackson (1988): *Physiol. Plant*, 72:15.
147. M. B. Jackson and D. J. Campbell (1976): *New Phytol.*, 76:21.
148. K. J. Bradford and D. R. Dilley (1978): *Plant Physiol.*, 61:506.
149. M. C. Drew, M. B. Jackson, and S. C. Gifford (1979): *Planta*, 147:83.
150. D. O. Adams and S. F. Yang (1979): *Proc. Natl. Acad. Sci. USA*, 76:170.
151. M. B. Jackson, B. Herman, and A. Goodenough (1982): *Plant Cell Environ.*, 8:163.
152. N. Marrisen, W. A. Kannewoff, and L. H. W. pan der Plas (1991): *Physiol. Plant*, 82:465.
153. T. Wang and R. N. Arteca (1992): *Plant Physiol.*, 98:97.
154. Y. B. Yu and S. F. Yang (1980): *Plant Physiol.*, 66:281.
155. H. A. Imaseki, A. Watanabe, and S. Odawara (1977): *Plant Cell Physiol.*, 18:577.
156. Y. Yoshii, A. Watanabe, and H. Imaseki (1980): *Plant Cell Physiol.*, 21:279.
157. H. Yoshii and H. Imaseki (1981): *Plant Cell Physiol.*, 22:369.
158. T. A. McKeon, N. E. Hoffman, and S. F. Yang (1982): *Planta*, 155:437.
159. M. A. Hall, J. A. Kapuya, S. Sivakumaran, and A. John (1977): *Pest. Sci.*, 8:217.
160. S. Y. Hwang and T. T. van-Toai (1991): *Crop Sci.*, 97:593.
161. I. D. J. Phillips (1964): *Ann. Bot. N.S.*, 28:17.

162. I. D. J. Phillips (1964): *Ann. Bot. N.S.*, *28*:38.
163. R. L. Wample and D. M. Reid (1979): *Physiol. Plant*, *45*:219.
164. Y. B. Yu and S. F. Yang (1979): *Plant Physiol.*, *64*:1074.
165. E. M. Beyer and P. W. Morgan (1969): *Plant Physiol.*, *44*:1690.
166. R. Q. Cannel and M. B. Jackson (1981): *Modifying the Root Environment to Reduce Crop Stress* (G. F. Arkin and H. M. Taylor, eds.), Am. Soc. Agric. Eng., St. Joseph, Michigan, p. 141.
167. I. Waters, P. J. C. Kuiper, E. Watkin, and H. Greenway (1991): *J. Exp. Bot.*, *42*:1427.
168. W. Armstrong and D. J. Boatman (1967): *J. Ecol.*, *55*:101.
169. S. R. Pezeshki and R. D. DeLaune (1990): *Acta Oecologia*, *11*:377.
170. S. Aomine (1962): *Soil Sci.*, *94*:6.
171. C. D. Stanley, T. C. Kaspar, and H. M. Taylor (1980): *Agron. J.*, *72*:341.
172. M. G. Huck (1970): *Agron. J.*, *62*:815.
173. M. M. Sachs, M. Freeling, and R. Okimoto (1980): *Cell*, *20*:761.
174. L. J. Stolzy, D. D. Focht, and H. Fluhler (1981): *Flora (Jena)*, *171*:236.
175. F. T. Turner, J. W. Sij, G. N. McCauley, and C. C. Chen (1983): *Crop. Sci.*, *23*:40.
176. T. Webb (1982): Ph.D. dissertation, Univ. of Hull, UK.
177. H. A. Kordan (1976): *New Phytol.*, *76*:81.
178. F. T. Turner, C. C. Chen, and G. N. McCauley (1981): *Agron. J.*, *73*:566.
179. A. Bertani and I. Brambilla (1982): *Z. Pflanzenphysiol.*, *108*:282.
180. W. A. Cannon (1925): *Carnegie Inst.*, Publ. No. 368, Washington, DC.
181. L. J. Berry and W. E. Norris (1949): *Biochem. Biophys. Acta*, *3*:593.
182. D. M. Griffin (1968): *New Phytologist*, *67*:561.
183. W. Armstrong and T. Gaynard (1976): *Physiol. Plant*, *37*:200.
184. P. H. Saglio, F. Rancillac, F. Bruzas, and A. Pradet (1984): *Plant Physiol.*, *76*:151.
185. W. Armstrong and P. M. Beckett (1987): *New Physiol.*, *105*:221.
186. B. Erdmann and E. M. Wiedenroth (1988): *Ann. Bot.*, *62*:277.
187. J. F. Lopez-Saez, F. Gonzalez-Bernaldez, A. Gonzalez-Fernandez, and G. Garcia-Ferrero (1969): *Protoplasma*, *67*:312.
188. B. B. Vartapetian, I. M. Andreeva, and N. Nuritdinov (1978): *Plant Life in Anaerobic Environments* (D. D. Hook and R. M. M. Crawford, eds.), Ann Arbor Science, Ann Arbor, Michigan, p. 13.
189. B. J. Atwell, I. Waters, and H. Greenway (1982): *J. Exp. Bot.*, *33*:1030.
190. W. Stepniewski, S. R. Pezeshki, R. D. DeLaune, and W. H. Patrick, Jr. (1991): *Vegetatio*, *94*:47.
191. I. Waters, S. Morrell, M. Greenway, and T. D. Colmer (1991): *J. Exp. Bot.*, *42*:1437.
192. D. J. Hole, B. G. Cobb, P. S. Hole, and M. C. Drew (1992): *Plant Physiol.*, *99*:213.
193. H. Konings and M. B. Jackson (1979): *Z. Pflanzenphysiol.*, *92*:385.
194. M. B. Jackson and A. K. B. Kowalewska (1983): *J. Exp. Bot.*, *34*;493.
195. H. S. Ku, H. Suge, L. Rappaport, and H. K. Pratt (1970): *Planta*, *90*:333.

196. A. Musgrave, M. B. Jackson, and E. Ling (1972): *Nature New Biol.*, *238*:93.
197. B. A. Smit and M. L. Stachowiak (1990): *Plant Physiol.*, *92*:1021.
198. N. R. Bishnoi and H. N. Krishnamoorthy (1992): *J. Plant Physiol.*, *139*:503.
199. W. Wenkert, N. R. Fausey, and H. D. Watters (1981): *Plant Soil*, *62*:351.
200. M. C. T. Trought and M. C. Drew (1981): *J. Exp. Bot.*, *32*:509.
201. M. B. Jackson, M. C. Drew, and S. C. Giffard (1981): *Physiol. Plant*, *52*:23.
202. N. Nuridinov and B. B. Vartapetian (1980): *Fiziol. Rast. (Moscow)*, *27*:814.
203. D. M. Seliskar (1989): *Can. J. Bot.*, *68*:1780.
204. R. Jones and J. R. Etherington (1971): *J. Ecol.*, *59*:793.
205. D. M. Pearce and M. B. Jackson (1991): *Ann. Bot.*, *68*:201.
206. L. A. Donovan and J. L. Gallagher (1985): *Wetlands*, *5*:1.
207. I. D. J. Phillips (1964): *Ann. Bot.*, *28*:17.
208. W. Crocker, P. W. Zimmerman, and A. E. Hitchcock (1932): *Contrib. Boyce Thompson Inst.*, *4*:177.
209. K. J. Bradford, T. C. Hsiao, and S. F. Yang (1982): *Plant Physiol.*, *70*:1503.
210. M. B. Jackson and D. J. Campbell (1979): *New Phytol.*, *82*:331.
211. G. A. Kuznetsova, M. G. Kuznetsova, and G. M. Grineva (1981): *Fiziol. Rast. (Moscow)*, *28*:340.
212. M. Kawase (1974): *Physiol. Plant*, *31*:29.
213. R. L. Wample and D. M. Reid (1975): *Planta*, *127*:263.
214. M. Kawase (1979): *Am. J. Bot.*, *66*:183.
215. D. J. Osborne (1967): *Symp. Soc. Exp. Biol.*, *21*:305.
216. P. Whyte and L. C. Luckwill (1966): *Nature*, *210*:1380.
217. M. A. Hall (1977): *Ann. Appl. Biol.*, *85*:424.
218. H. Thomas and J. L. Stoddart (1980): *Annu. Rev. Plant Physiol.*, *31*:83.
219. R. S. Dhindsa, R. L. Plumb-Dhindsa, and D. M. Reid (1982): *Physiol. Plant*, *56*:453.
220. H. E. Jones and J. R. Etherington (1970): *J. Ecol.*, *58*:487.
221. R. H. Howeler (1973): *Soil Sci. Soc. Am. Proc.*, *37*:898.
222. S. H. F. Justin and W. Armstrong (1987): *New Phytol.*, *106*:465.
223. J. Armstrong, W. Armstrong, and P. M. Beckett (1988): *New Phytol.*, *110*:383.
224. J. Armstrong and W. Armstrong (1990): *Aqua. Bot.*
225. P. Laan, M. J. Berrevoets, S. Lythe, W. Armstrong, and C. W. P. M. Blom (1989): *J. Ecol.*, *77*:695.
226. T. V. Chirkova (1968): *Fiziol. Rast.*, *15*:565.
227. J. J. Phillipson and M. P. Contts (1978): *New Phytol.*, *80*:341.
228. D. D. Hook (1984): *Flooding and Plant Growth* (T. T. Kozlowski, ed.), Academic Press, Orlando, p. 265.
229. M. Kawase and R. E. Whitmoyer (1980): *Am. J. Bot.*, *67*:18.
230. W. Armstrong (1979): *Adv. Bot. Res.*, *7*:225.

231. M. Kawase (1981): *Hort. Sci.*, *16*:30.
232. R. Campbell and M. C. Drew (1983): *Planta*, *157*:350.
233. W. Armstrong (1971): *Physiol. Plant*, *25*:192.
234. D. M. Seliskar (1988): *J. Exp. Bot.*, *39*:1639.
235. M. B. Jackson, T. M. Fenning, and W. Jenkins (1985): *J. Exp. Bot.*, *36*;1566.
236. P. J. Kramer, W. S. Riley, and T. T. Bannister (1952): *Ecology*, *33*:117.
237. W. Armstrong (1978): *Plant Life in Anaerobic Environments* (D. D. Hook and R. M. M. Crawford, eds.), Ann Arbor Science, Ann Arbor, Michigan, p. 269.
238. D. A. Barber, M. Ebert, and N. T. S. Evans (1962): *J. Exp. Bot.*, *13*:397.
239. W. Armstrong (1975): *Environment and Plant Ecology* (J. R. Etherington, ed.), John Wiley and Sons, New York, p. 181.
240. M. S. Green and J. R. Etherington (1977): *J. Exp. Bot.*, *28*:679.
241. M. L. Otte, J. Rozema, L. Koster, M. S. Haarsma, and R. A. Broekman (1989): *New Phytol.*, *111*:309.
242. D. D. Hook, O. G. Langdon, J. Stubbs, and C. L. Brown (1970): *For. Sci.*, *16*:304.
243. D. D. Hook, C. L. Brown, and P. P. Kormanik (1971): *J. Exp. Bot.*, *22*:78.
244. R. L. Jat, M. S. Dravid, D. K. Das, and N. N. Goswami (1975): *J. Indian Soc. Soil Sci.*, *23*:291.
245. G. B. Sartoris and B. A. Belcher (1949): *Sugar*, *44*:36.
246. W. T. Jackson (1942): *Am. J. Bot.*, *42*:816.
247. R. L. Wample and D. M. Reid (1978): *Physiol. Plant*, *44*:351.
248. D. Fabijan, E. Yeung, I. Mukherjee, and D. M. Reid (1981): *Physiol. Plant*, *53*:578.
249. D. Fabijan, J. S. Taylor, and D. M. Reid (1981): *Physiol. Plant*, *53*:589.
250. C. J. Gill (1975): *Flora (Jena)*, *164*:85.
251. R. P. Tripeip and C. A. Mitchell (1984): *Physiol. Plant*, *60*:567.
252. T. T. Kozlowski, P. J. Kramer, and S. G. Pallardy (1991): *The Physiological Ecology of Woody Plants*, Academic Press, New York, p. 657.
253. L. P. Stolz and C. E. Hess (1966): *Proc. Amer. Soc. Hort. Sci.*, *89*:734.
254. G. R. Noggle and G. J. Fritz (1976): *Introductory Plant Physiology*, Prentice-Hall, Englewood Cliffs, New Jersey, p. 688.
255. B. Jacoby and B. Rudich (1980): *Ann. Bot.*, *46*:493.
256. J. K. M. Roberts, A. N. Lane, R. A. Clark, and R. H. Nieman (1985): *Arch. Biochem. Biophys.*, *240*:712.
257. R. M. M. Crawford (1966): *J. Ecol.*, *54*:403.
258. R. M. M. Crawford (1967): *J. Exp. Bot.*, *18*:458.
259. M. McManmon and R. M. M. Crawford (1971): *New Phytol.*, *70*:299.
260. D. D. Davies (1980): *The Biochemistry of Plants: A Comprehensive Treatise*, Vol. 2 (P. K. Stumpf and E. E. Conn, eds.), Academic Press, New York, p. 581.
261. M. E. Rumpho and R. A. Kennedy (1981): *Plant Physiol.*, *68*:165.
262. M. E. Rumpho and R. A. Kennedy (1983): *Plant Physiol.*, *72*:44.

263. A. Bertani, I. Brambilla, and F. Menegas (1980): *J. Exp. Bot.*, *31*:325.
264. A. M. Smith and T. ap Rees (1979): *Phytochemistry*, *18*:1453.
265. K. Wignarajah and H. Greenway (1976): *New Phytol.*, *77*:575.
266. S. Morrell and H. Greenway (1989): *Aust. J. Plant Physiol.*, *16*:469.
267. D. D. Hook and R. M. M. Crawford (1978): *Plant Life: Anaerobic Environments*, Ann Arbor Science, Ann Arbor, Michigan, p. 564.
268. R. A. Kennedy, M. E. Rumpho, and T. C. Fox (1992): *Plant Physiol.*, *100*:1.
269. J. K. M. Roberts (1989): *The Ecology and Management of Wetlands*, Vol. 1 (D. D. Hook, ed.), Croom-Helm Press, London, p. 392.
270. V. Saint-Ges, C. Roby, R. Bligny, A. Pradet, and R. Douce (1991): *Eur. J. Biochem.*, *200*:477.
271. F. Menegus, L. Cuttaruzza, A. Chersi, and G. Fronza (1989): *Plant Physiol.*, *90*:29.
272. J. K. M. Roberts, J. Callis, D. Jardetsky, V. Walbot, and M. Freeling (1984): *Proc. Natl. Acad. Sci. USA*, *81*:6029.
273. J. K. M. Roberts, J. Callis, D. Wemmer, V. Walbot, and O. Jardetsky (1984): *Proc. Nat. Acad. Sci. USA*, *81*:3379.
274. F. Menegus, L. Cattaruzza, M. Mattana, N. Beffagna, and E. Ragg (1991): *Plant Physiol.*, *95*:760.
275. C. A. Lemke-Keyes and M. M. Sachs (1989): *J. Herd.*, *80*:316.
276. R. A. Kennedy, M. E. Rumpho, and T. C. Fox (1987): *Plant Life in Aquatic and Amphibious Habitats* (R. M. M. Crawford, ed.), Blackwell Press, Oxford, p. 193.
277. B. G. Cobb and R. A. Kennedy (1978): *Plant Cell Environ.*, *10*:633.
278. K. D. Newman and T. T. van-Toai (1991): *Crop Sci.*, *31*:1252.
279. M. M. Sachs (1991): *Plant Life Under Oxygen Deprivation: Ecology, Physiology and Biochemistry* (M. B. Jackson, D. D. Davies, and H. Lambers, eds.), SPB Academic Publishing, The Hague, The Netherlands, p. 129.
280. I. Couce, S. Defontaine, J. P. Carde, and A. Pradet (1992): *Plant Physiol.*, *98*:411.
281. M. W. Smith and P. L. Ager (1988): *Hort. Sci.*, *23*:370.
282. S. R. Pezeshki (1992): *Ann. Sci. For.*, *49*:149.
283. D. L. Peterson and F. A. Bazzaz (1984): *Am. Midl. Natural*, *112*:261.
284. S. R. Pezeshki (1990): *For. Ecol. Manage.*, *33/34*:531.
285. S. R. Pezeshki, R. D. DeLaune, and W. H. Patrick, Jr. (1987): *Wetlands*, *7*:1.
286. S. R. Pezeshki, R. D. DeLaune, W. H. Patrick, Jr., and E. B. Moser (1989): *For. Ecol. Manage.*, *27*:41.
287. S. R. Pezeshki, R. D. DeLaune, and W. H. Patrick, Jr. (1992): *Proc. Soc. Wetl. and Sci.*,
288. S. R. Pezeshki, R. D. DeLaune, and S. Z. Pan (1991): *Vegetatio*, *95*:159.
289. S. R. Pezeshki (1991): *Trees*, *5*:180.

11

Stomata

Thomas M. Hinckley and Jeffrey H. Braatne

University of Washington
Seattle, Washington

INTRODUCTION

Plant growth and development are dependent on the uptake and transport of water, nutrients, and carbon; the conversion of light to chemical energy; and the fixation of carbon for structural and nonstructural (e.g., nectar, starch, and sugars) use. In these processes, water is necessary as a transport medium in the xylem and phloem, provides turgor pressure that maintains the shape and orientation of cells and organs, and is necessary for cell elongation and stomatal function. Water is also a constituent in many chemical reactions within plant cells. Therefore, the growth of plants is closely linked with the maintenance of a favorable water status.

Plants generally require a high water content, often greater than 75%, and a correspondingly high tissue water potential, mostly greater than -2 MPa. Given these high tissue water potentials and the relatively low water potential of the atmosphere (-95.6 MPa at 25°C and 50% RH), a very large potential energy gradient exists between the plant and the atmosphere. This driving force for water loss has been an important factor in the evolution of numerous plant characteristics and mechanisms to minimize and control water loss. However, the processes of carbon uptake and fixation require an exchange of gases, thus giving rise to mechanisms capable of balancing the uptake of carbon dioxide with the loss of water to the atmosphere.

In most plants, the exchange of water vapor and carbon dioxide in leaves largely occurs through specialized structures known as stomata. *Stoma* means "mouth" in greek, with *stomata* or *stomates* being the plural forms of *stoma*. The stomatal pore is formed between two *guard cells*, which are specialized cells of the epidermal tissue of leaves and other aerial plant organs. Most guard cells are either elliptical (kidney-shaped) or graminaceous (dumbbell-shaped, Fig. 1). The term stoma and stomata are used to refer not only to stomatal pores but also to the guard cells and adjacent cells forming the *stomatal complex.* If the adjacent cells are morphologically distinct, they are *subsidiary cells;* otherwise they are simply referred to as *neighboring cells.* Guard cells are capable of changes in shape that lead to either opening or closing of the pore or stoma. In this manner, when conditions are not favorable for CO_2 uptake, such as under conditions of insufficient light, or if desiccation of the leaf from water loss is imminent, stomata close. The functioning of stomata is thus critical in the maintenance of a balance between carbon uptake and water loss; stomatal function also affects the absorption of

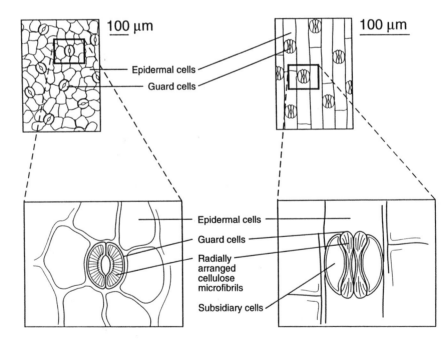

Figure 1 Stomata of vascular plants showing the radial alignment of cellulose microfibrils in the guard cells of (left) elliptical stoma (dicotyledon) and (right) graminaceous stoma (monocotyledon). (Modified from Refs. 15, 27, and 209.)

other gases such as toxic air pollutants and may in turn be affected by these compounds.

Mechanisms regulating stomatal function have been the subject of extensive research over the last century (1–11). There are also a number of reviews of methodology in stomatal studies (12–17). Rather than duplicate these reviews, we will summarize mechanisms that control stomatal function and describe how the integrative function of these mechanisms establishes a balance between carbon uptake and water loss. In the process, we will review mechanisms at the level of individual leaves, whole plants, and the plant canopy that control or affect stomatal function. We will also present two cases which illustrate how an understanding of stomatal function provides insights into the basic elements of plant structure and function. The first case involves an extensive series of studies on stomatal morphology and function ranging from individual tissues to closed canopies of native and hybrid species of *Populus*, while the second case documents the role of gender-specific stomatal function on water use and carbon fixation and, in turn, on the distribution of species across gradients of soil moisture. Much of our discussion will focus on woody plants.

STOMATAL MECHANISMS

The process of opening and closing of the stoma is regulated not only by biochemical and hydraulic factors but by the actual mechanics of movement in guard cells. In this section, our discussion will focus primarily on organelle and cellular functions.

Stomatal Mechanics

The aperture of a stoma is dependent on the internal turgor or pressure potential within guard cells as well as wall structure (4,18,19). Changes in stomatal aperture are governed by changes in guard cell turgor resulting from fluxes of water between guard cells and epidermal cells. In fact, epidermal cells appear to have a mechanical advantage over guard cells (19–22) such that an equal increase in the turgor pressure of both guard cells and epidermal cells can actually lead to a reduction in aperture of the stoma.

The relationship between turgor pressure and aperture is related to the physical characteristics guard cell walls (19,23). Nonuniform thickening of the guard cell wall clearly influences changes in cell shape as the stoma opens and closes (15). The radial orientation of cellulose microfibrils within guard cell walls (Fig. 1) also gives guard cells an elastic anisotropy (i.e., radial flexibility along specific axes as defined by the orientation of microfibrils). As a result, changes in turgor pressure lead to changes in the curvature and length of the guard cells (15,19,24–26). Opening of an elliptical stoma (Fig. 1, left) is mainly a result of swelling at the polar ends of the guard cells; the central portions of the dorsal and ventral walls also stretch and because guard cells are anchored at their polar ends, the cell walls curve outward to create an elliptical pore between the ventral walls (19). In a graminaceous guard cell (Fig. 1, right), the central portions do not stretch but are pushed apart when the polar ends swell, thus forming a slit-like rectangular pore.

Guard Cell Solutes and Biochemistry

Changes in the turgor pressure of guard cells result from fluxes of water into or out of the cell along a gradient of osmotic potential (3,6,11,23,27,28). The solutes involved in the establishment and maintenance of these osmotic gradients are organic acids, chiefly malate, and potassium, calcium, chloride, and hydrogen ions (Fig. 2). It is now accepted that upon illumination of the guard cell solute accumulation is jointly driven by (a) extrusion of protons across the plasma membrane,

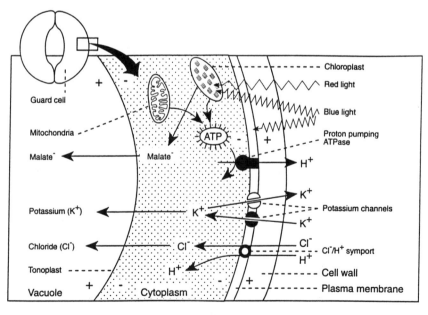

Figure 2   Solutes and biochemistry of ion fluxes in guard cells. (Modified from Ref. 209.)

powered by ATP, and (b) hyperpolarization of the plasma membrane
(11,29,30). Proton extrusion across the plasma membrane can occur by
either of two mechanisms: (a) activation of a proton-translocating
ATPase in the plasma membrane (31,32) or (b) activation of a plasma
membrane redox chain (33,34). The extrusion of protons generates the
electrical driving force for cation entry as well as a pH gradient across
the plasma membrane, conditions required for the maintenance of the
osmotic gradients underlying differentials in turgor pressure between
guard cells and epidermal cells.

The accumulation of K^+ in guard cells upon illumination has been
well documented (30,35–45) as fluxes of K^+ ions into and out of guard
cells strongly parallel changes in stomatal aperture. These fluxes probably
occur through at least two separate K^+ ion channels (Fig. 2). One channel is activated when the plasma membrane is hyperpolarized by light
(46–54) and appears to be the main channel by which K^+ enters guard
cells (46–54). This channel may be inactivated by increases in cytosolic
Ca^{2+} (55,56), which in turn acts not only to depolarize the plasma membrane but also to activate a second channel governing fluxes of K^+ out of
the guard cell (52,56–58). A recent review by Blatt (59) details the precise function of these ion channels within guard cells relative to ion channels located in other cells of the plant body.

Chloride and malate appear to be the primary counterions to K^+
uptake (Fig. 2). The accumulation of chloride and malate during stomatal opening has been widely documented (60–74). Collectively, chloride
and malate serve to buffer increases in cytosolic pH caused by proton
extrusion (75,76), maintain polarization and/or dissipate the pH gradient
across the plasma membrane (11,30). It is postulated that the uptake of
chloride occurs through a H^+/Cl^- symport (Fig. 2), whereby H^+ and
Cl^- are cotransported across the plasma membrane, thus buffering cytoplasmic pH without altering the charge potential across the plasma membrane (66,75–77). However, biosynthesis of malate alone may be
sufficient to balance K^+ uptake in many plants (11,30,78). Malate is produced by the carboxylation of phosphoenolpyruvate (PEP). The enzyme
catalyzing this reaction within the cytosol, phosphoenolpyruvate carboxylase (PEPC), has been found to occur in the guard cells of a variety
of plants (62,68–70,72,76,77,79–84). Carbon products resulting from the
degradation of starch (stored within the guard cell chloroplast) serve as
the skeletal precursors (PEP) for malate. The degradation of starch and
associated biosynthesis of malate have been observed to occur concomitantly with stomatal opening (85), while the closure of stomata is accompanied by the formation of starch within guard cell chloroplasts (dark)
and the probable release of malate from guard cells (78,86).

Glycolysis, oxidative phosphorylation, and/or photophosphorylation have all been documented as potential energy sources for driving the proton extrusion requisite for the ion fluxes in guard cells (11,86). Glycolysis appears to be driven by the hydrolytic or phosphorlytic breakdown of starch (87,88). Phosphorylated three-carbon intermediates of glycolysis are subsequently partitioned between two separate biochemical pathways: (a) PEP from glycolysis is diverted for carboxylation and reduction to malate and (b) pyruvate fuels the citric acid cycle, electron transport, and oxidative phosphorylation (11,89). Observations documenting the functional integrity of photosystems I and II within the chloroplasts of guard cells (90–93) have also been associated with indirect evidence that energy derived from photophosphorylation may be associated with stomatal opening (94). Yet the role of these photochemical reactions in stomatal function remains unclear (9,11), as does the potential involvement of photosynthetic reactions within guard cell chloroplasts (95–100).

FACTORS CONTROLLING STOMATAL MECHANISMS

The aperture of the stoma is governed by a variety of environmental signals, including light level and quality, concentration of carbon dioxide, air humidity, and leaf/soil water status. In addition, the activity of growth substances such as abscisic acid and cytokinin affects stomatal function. Our discussion deals with these signals separately, yet in reality guard cells integrate all of these signals into well-defined stomatal responses.

Light

The activation of stomatal opening by light has been studied extensively (3,101,102). Red and blue light have been commonly used to examine the action spectrum for stomatal opening (103,104) and to stimulate alternate pathways of carbon metabolism and energy production in guard cells (105–109). Both red and blue light induce stomatal opening, yet evidence varies as to the relative role of different photoreceptor systems in this process (102,106). When guard cells are illuminated, unidentified receptors of blue light are activated, resulting in the subsequent activation of proton extrusion across the plasma membrane and the hydrolytic and/or phosphorylytic breakdown of starch within guard cell chloroplasts (11,101,104). There is some evidence that the blue light receptor may be a flavin, a carotenoid (34,102,110), or possibly phytochrome (111–113). The red light receptor appears to be chlorophyll (11,102), whose activation is associated with metabolic pathways of guard cell chloroplasts. It is now generally accepted that the blue light response of stomata is important at low light and the red light (chlorophyll) response is important at high light intensity (3,101,104,114). This argument is based on observations of a threshold for stomatal activation in red light and the lack of a threshold for blue light (101,104). Much of this work was derived from studies of two closely related orchids (*Paphiopedilum* and *Phragmipedium*), in which one of the genera, *Paphiopedilum*, lacked guard cell chloroplasts (104,115,116). This achlorophyllous species had a low leaf conductance and only a slight stomatal response to changes in concentration of CO_2 (116). These studies revealed that relatively low levels of blue light ($10\,\mu mol/m^2/sec$) activated stomatal opening in the achlorophyllous species while red light had only a weak effect on stomatal aperture. Furthermore, the results were consistent with the hypothesis that blue light had a direct effect on stomatal aperture independent of possible chloroplast-driven changes in the concentration of carbon dioxide (101,102,104).

Carbon Dioxide

Stomatal apertures are inversely correlated with CO_2 concentrations in both the light and darkness (3,23,117,118). Elevated concentrations of CO_2 tend to close stomata, while stomata open in response to low CO_2 levels (118–120). Recent studies have confirmed that stomata are sensitive to intracellular carbon dioxide concentrations rather than concentrations of CO_2 at the leaf surface (9,121). These studies also confirm that the surfaces of guard cells facing the substomatal cavity are the primary sites of CO_2 perception (9,121). In a way, these results conform with widespread observations of nearly constant intercellular concentrations of CO_2 for C_3 and C_4 plants (118,122) and the general consensus that stomata optimize photosynthetic carbon fixation through the precise regulation of intercellular concentrations of CO_2 (9). Yet many questions remain as to whether or not intercellular CO_2 concentrations act as a controlling signal for stomatal function (9,118). These questions arise because stomatal responses to CO_2 vary with other factors, such as leaf or plant age and growth conditions as mediated by hormonal activity (i.e., auxin, abscissic acid, and cytokinins) (9,118). Additional confusion arises from the general approach of calculating intercellular concentrations of CO_2 on the basis of gas exchange measurements. Potential errors in this approach were recently noted in relation to the lack of uniformity in stomatal apertures across leaf surfaces (123–126), especially when leaves are water-stressed (123). As a result, it appears that the precise role of carbon dioxide in regulating stomatal aperture remains an open question.

The responses of plants and plant communities to rising atmospheric concentrations of carbon dioxide is currently a subject of extensive research and debate (9,127). It is generally assumed that the major consequences of elevated CO_2 (independent of those associated with global warming) will be (a) increased rates of CO_2 assimilation and (b) partial stomatal closure and decreased rates of transpiration per unit leaf area (9,118). The cumulative effect of these responses, if true, would be increases in the water use efficiency of C_3 plants. Yet such conclusions remain problematic as our understanding of the responses of photosynthesis to CO_2 greatly exceeds that associated with the effect of CO_2 on stomatal guard cells. We must overcome these and other deficiencies before we can effectively address the potential effects of elevated concentrations of CO_2 on the physiology and growth of C_3 plants.

Relative Humidity and Leaf/Soil Water Status

Studies have shown that exposure of a single leaf or whole plant to dry air accelerates transpiration, which in turn lowers leaf water status (128–130). In such instances, air humidity is an independent variable and leaf water potential a dependent variable. Leaf water potential may also directly affect stomatal conductance, which in turn will alter the transpiration rate. In this case, leaf water potential would be an independent variable (130). The existence of a transpiration/leaf water feedback loop is well documented, yet it also is clear that stomata may respond directly to air humidity (128,130–132). At the cellular level, water potential gradients within the leaf may cause epidermal cell turgor to be sensitive to changes in air humidity and thus be a potential trigger of stomatal closure. However, the general consensus remains that metabolic processes governing stomatal aperture appear to be involved in addition to hydraulic linkages between epidermal cells and guard cells as both contribute to stomatal closure (11,129,130). Recent micrometeorological measurements in plant canopies also support these conclusions (133). Yet the mechanism(s) governing stomatal response to humidity remain to be clarified. Meinzer and Grantz (134) propose that stomata possess a mechanism (perhaps related to the electrical potential of epidermal and guard cell membranes) capable of metering volumetric fluxes of water. Metering and subsequent regulation of differential fluxes of water via this mechanism would be dependent on both hydraulic and biochemical messages originating in other locations within the plant.

Several researchers have documented a strong relationship between stomatal function and soil water status (135–140). Masle and Passioura (141) demonstrated that leaf growth in wheat seedlings was more closely associated with both (a) bulk density of the soil and (b) soil water content than with leaf water potential. Ludlow (142) argued that the apparent sensitivity of many species, including important crop species, to the soil environment may be under intense natural selection. That is, a plant's ability to reduce water loss as soil water availability begins to decrease (due to drought, compaction, root exploration, etc.), but before substantial changes in soil or leaf water potential have occurred, has clear survival benefits and generally selects for improved water use efficiency, especially in wild plant populations. Unfortunately, such genetically programmed responses even under managed field conditions, would clearly lead to declines in growth and productivity.*

Plant Growth Substances

Abscisic acid (ABA) has long been implicated in a variety of plant responses to leaf and soil water status (9,148). ABA acid appears to have a major role in long-term adjustments of stomatal function to conditions of leaf and soil water deficits (4,148). Plants subjected to a period of water stress had elevated levels of ABA associated with closure of stomata (149–152). While this suggests the possibility that ABA is responsible for the changes in stomatal activity, it has also been noted that stomatal opening often returns to prestress levels well before ABA concentrations have declined. As a result, the potential transfer of ABA between active and nonactive pools within the epidermis has been seen as an important component of these stomatal responses to water stress (153).

Most studies have concentrated on ABA produced within the leaf where the trigger for its production may be changes in leaf water potential. [In contrast, Schildwacht (154) proposed that ABA might be produced in response to rapid hydraulic pulses originating from the root where no change in leaf water potential would have been easily detected.] Bates and Hall (155) initially hypothesized that a signal from the roots may be involved in the response of stomata to small changes in soil water potential that were insufficient to affect bulk leaf water status. Recent evidence places a strong emphasis on root ABA production and associated xylem transport to the leaf (140); however, there may be some qualifications. Munns and King (156) observed that root-stressed plants

---

* The currently accepted means of measuring leaf water potential is either with a Scholander pressure chamber or with a thermocouple psychrometer (143–145). From such measures the diurnal and seasonal curves of leaf water potential may be plotted and compared with other measures of physiological and growth processes. These techniques, especially the pressure chamber, have become so accepted that they are regarded as the standard. However, recent information using the pressure probe has called into question the validity of data measured with the pressure chamber (146,147). Until this issue is resolved, much of the current literature on water status and stomatal behavior has significant elements of uncertainty.

had altered stomatal behavior as others have shown. They extracted the sap coming from the roots in the xylem, treated unstressed plants, and observed a response similar to that noted in the stressed plant. However, when they removed all the ABA from the extracted sap, a response was still observed, though not as strong.

As is the case with many plant processes, the regulation of stomatal activity by growth regulators may involve the balance of two or more substances. While ABA can stimulate stomatal closure, cytokinin has been shown to have the opposite effect (119,157) and the effects of an increase in ABA following a stress period can sometimes be countered by application of cytokinin (158m). Thus, the root system may also act as an important source of additional growth substances, such as cytokinin, that function in a manner that is antagonistic to ABA.

Additional Factors

In contrast to metabolic messages, Teskey et al. (161) stressed the importance of a much more rapid message coming from a distant point in the conducting system to the foliage. He placed the base of a 1-mm long foliated shoot of *Abies amabilis* into water, allowed time for equilibration to occur, and then cut 50% of the way through the shoot. A rapid closure of stomata was observed (within 10 min) with no change in bulk needle water potential. A very different response was noted when the branch was completely cut through. In that case, water potential declined and stomata remained open until water potential approached the turgor loss point and then stomata gradually closed. Clearly how changes in water potential or changes in the flux of water are perceived by the stomata is important in how stomata respond (134).

Boyer (162), Chapin (163), and Schulze et al. (136) stressed that plants respond to both hydraulic and metabolic signals and that plants act as integrated systems (Fig. 3). In a sense, stomata function as multisensory hydraulic valves; a variety of different stimuli, including the level and quality of light, intercellular concentrations of carbon dioxide, humidity, cellular and tissue water potentials, and hormonal signals are sensed by guard cells and integrated into well-defined stomatal responses. In fact, based on data collected from root, stem, and leaf organs, whole plants, and canopies at various stages of leaf area development, Meinzer and Grantz (134) stated: "A striking coordination is observed between prevailing levels of stomatal opening and the hydraulic capacity of the soil, roots and stem to supply leaves with water." As a consequence of this coordination, transpiration approaches some maximum per unit ground area and does not vary appreciably from this maximum over a wide range of leaf areas. They concluded that stomata metered transpiration based on fluxes of root-derived materials in the xylem sap. Recent

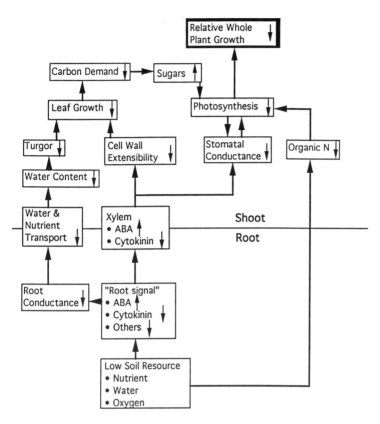

Figure 3 Diagram illustrating some of the linkages between low soil resources (temporally, spatially or continuously induced) and whole-plant growth (163). The low soil resource may be rapidly or slowly perceived depending on how the resource changes and the distribution of roots relative to the resource. Whether a change in root conductance and thus xylem transport will manifest itself as a change in plant water content will depend on how soon and to what degree stomata close and reduce water loss. Most of the relationships take between 12 and 48 hr to manifest themselves; however, changes in whole-plant growth rate may take longer. In addition, there are longer term feedbacks so that patterns of carbon allocation change favoring root growth.

evident indicates that stomata may also regulate transpiration in such a way as to avoid xylem tensions reaching the point of catastrophic xylem cavitation (164). In fact, stomatal conductance may track xylem hydraulic conductivity (165). Given these observations, one can easily picture how the entire system might be linked so as to avoid a critical collapse in any one component (162,163). Thus, the work of Meinzer and

Grantz (134) is a very elegant illustration of the importance of examining stomatal responses at several scales of biological organization simultaneously.

SCALING STOMATAL MECHANISMS: LEAF TO CANOPY

When discussing stomatal mechanisms in plants, one must keep in mind the scale or level of biological organization. For example, from the scale of the leaf or the whole plant, stomata represent the single location where transpiration may be regulated. (Changes in cuticular conductance, foliage morphology, and crown structure also play important roles in altering water loss, but they provide no means of short-term regulation). However, when one considers higher levels of biological organization (i.e., the whole stand), the relative importance of stomatal regulation may not be significant in regulating transpiration (166–169). It seems contradictory to think that the regulatory role of stomata decreases or even disappears as one's level of focus goes from the leaf to the canopy. Similarly, the generation of net assimilation (A) vs. carbon dioxide (Ci) curves in leaves has, until recently, assumed that stomata behave uniformly across a leaf. However, when individual stoma or small groups of stomata were examined, patchy as opposed to uniform stomatal apertures were noted (123). Terashima et al. (125) observed that "the high degree of patchiness is likely to represent an inefficiency in leaf function that is worthy of consideration in relation to crop productivity." Although nonuniform stomatal apertures have been observed under laboratory conditions, it may not be present in the same species under field conditions. However, the impact of patchy stomatal behavior on whole-plant productivity has not been evaluated to date. These examples indicate that scaling up or down is a critical element in our understanding and measurement of biological processes.

It is important to define both scaling and integration. When one examines the hierarchical structure of biological systems (Fig. 4), one finds that physiologists, whether working on agronomic or ecologically important species, tend to focus on biological processes at the organ or associated sublevels. In contrast, ecologists focus on whole organisms or groups of organisms or even landscapes. This is where socially relevant and evolutionarily important issues are played out. For example, issues of global climatic change, air pollution, plant yield and management all tend to be at canopy, ecosystem, or higher levels of biological structure. In addition, issues of evolutionary importance, such as how traits with high survival value (e.g., water use efficiency) become fixed and permit adaptation to variable environments, can only be addressed at popula-

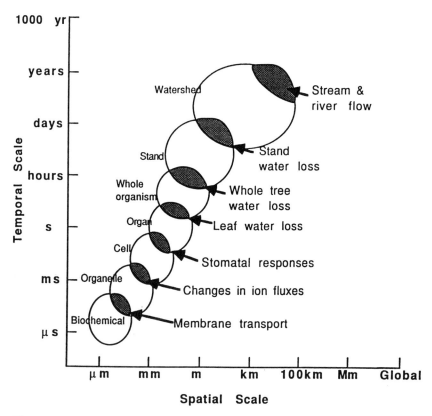

Figure 4 Temporal and spatial scales of processes associated with stomata. Bubbles represent hierarchical levels while shaded areas between bubbles indicate areas where exchange and feedbacks from one level to another may occur.

tion levels. Scientists tend to integrate or scale to these higher levels through models, the use of indicators (e.g., stable isotopes, remotely sensed parameters), by summing all the parts, or by actual measurement. Each of these steps has its own set of weaknesses.

Integration may be defined as the summation of a process of interest, which has been measured at a particular space–time range, to a greater time–space range. Scaling is defined as the mechanisms by which a change, whose effect occurs at a lesser time and space scale, is propagated up through scales of increased space and time (170,171). For example, when relative humidity decreases during the afternoon and stomata close in response, how does this change affect the loss of water from the plant canopy to the atmosphere? Integrating one would simply multiply the average leaf conductance by the leaf area of the stand, whereas in scaling one would also consider the temporal and spatial variation in leaf conductance, and the effects of canopy structure and wind speed on canopy boundary layer conductance. These and other elements of scaling which are critical to our understanding of biological processes have been reviewed in a recent book (172).

It has been a difficult process to scale information derived from small-scale physiological studies in order to predict larger scale phenomena for two reasons. The first problem is one of spatial and temporal heterogeneity. In a forest or crop stand, levels of physiological activity vary significantly within the canopy as a result of spatial variation in microclimate as well as genetic and environmental factors. Leverenz et al. (173) estimated the number of samples of leaf conductance required at any point in time to be within ±10% of the actual mean canopy conductance for a *Pinus sitchensis* stand. In spite of the relatively low heterogeneity in such a plantation forest, 35 samples would be necessary. The collection of such numbers presents a major physical barrier to extrapolating results from the organ (i.e., foliage) to the stand.

A second problem, which has only recently been fully appreciated by plant scientists, is that under field conditions the physiological responses governing exchange of mass and energy, which may be easily quantified on a small scale, may not control stand or regional level responses. The reason for this is that in many canopies the leaves are substantially *decoupled* with respect to properties of the bulk atmosphere above the canopy. In all plant canopies, leaves exchange water vapor, CO_2 and other gases with the air immediately adjacent to them (the boundary layer). In a poorly coupled (decoupled) canopy, the boundary layer is relatively thick and gases are not exchanged rapidly with the atmosphere above the stand. When this boundary layer becomes enriched with water vapor, water loss from the leaves slows as the vapor pressure difference between the leaf interior and the air at the leaf surface is reduced. Once this occurs, changes in stomatal opening or other plant parameters have little or no effect on water fluxes from the stand as a whole. In these cases, canopy transpiration will be determined largely by the input of solar energy rather than the humidity of the bulk atmosphere. In a well-coupled canopy, on the other hand, the boundary layer is thin and transpired water vapor is swept away rapidly. Under these circumstances, an increase in stomatal opening increases the rate of water loss from the leaves, which in turn increases the rate of water loss from the canopy as a whole. Thus, it seems clear that the degree of coupling between the canopy and the atmosphere has a strong effect on the canopy's response to changing environmental conditions and must be understood if we are to gauge accurately the role of stomatal changes at the leaf level in this process.

In an important paper entitled, "Stomatal Control of Transpiration: Scaling Up from Leaf to Region," Jarvis and McNaughton (167) introduced the *omega factor* as a way of reconciling the contrasting opinions held by physiologists and meteorologists concerning the role of stomata in controlling transpiration. Among physiologists it is widely believed that stomata are invariably dominant in controlling transpiration, whereas meteorologists maintain that estimates of evapotranspiration for well-watered vegetation need not consider stomatal activity. This apparent conflict is not one of scientific evidence but of interpretation of data gathered at incompatible scales. On a small scale, and especially under cuvette conditions where most physiologists measure transpiration, stomatal behavior does control water loss from the leaf. However, at a large scale, many canopies (especially the dense, uniform canopies studied by meteorologists) are so decoupled from the atmosphere that the main factors controlling water loss are those that stir and mix the canopy boundary layer (e.g., wind speed and canopy roughness). The omega factor, Ω_x (where x can be the leaf, whole plant, or canopy), a dimensionless coefficient that varies between 0 and 1, measures the degree to which conditions at the leaf or canopy surface and those in the free air stream are decoupled. It approaches 1 when conditions at the leaf or canopy surface are poorly coupled with those of the free air stream or atmosphere. Table 1 lists values of Ω from a number of different forest and crop types.

It would be highly desirable if a similar approach could be applied to describe coupling of leaves or canopies and the atmosphere with respect to concentrations and fluxes of CO_2. Unfortunately, the analysis of driving forces and conductances governing CO_2 assimilation is not entirely analogous to the situation with regard to water vapor. In addition to purely physical limitations on diffusion, CO_2 assimilation is regulated by biochemical reactions that do not respond linearly to changes in the CO_2 gradient between the leaf interior and the air. Furthermore, environmental variables such as light, which have no direct effect on transpiration, directly influence the biochemical capacity for CO_2 fixation. There is also evidence that long-term exposure to altered CO_2 levels results in species-specific adjustments in leaf morphology and photosynthetic biochemistry (174,175). This makes it much more difficult to apply an analysis of coupling to predict the impact of a rise in the bulk atmospheric CO_2 concentration on carbon fixation.

"Control" coefficients describing the sensitivity of photosynthetic CO_2 fixation to small changes in stomatal conductance, boundary layer conductance, and other variables have been derived recently for leaves

Table 1 Values of the Omega Factor (Ω) for Different Vegetation Types and Species

Vegetation type/species	Conditions	Ω
Temperate forest		
Picea sitchensis (Sitka spruce)		0.1
Pinus sylvestris (Scots pine)		0.1
Fagus sylvatica (European beech)		0.2
Acer negundo (boxelder)	Male	0.28
Acer negundo (boxelder)	Female	0.34
Hybrid *Populus* plantation	Moist	0.8
Tropical forest		
Anacardium excelsum	Dry	0.5
Anacardium excelsum	Wet	0.7
Gmelina arborea		0.9
Tectona grandis (teak)		0.9
Triplochiton sceroxylon		0.6
Horticultural crops		
Apple		0.3
Cherry		0.1
Citrus		0.3
Agricultural crops		
Alfalfa		0.9
Cotton		0.4
Sugarcane		0.9
Wheat		0.6

Source: Data synthesized from Refs. 167, 193, 197, 198, and Dawson, pers. commun.

held in gas exchange cuvettes (176) and for leaves under field conditions (176,177). The interactions and feedback among variables influencing control coefficients and CO_2 flux become particularly complex under field conditions. For example, the behavior of the stomatal control coefficient (C') for photosynthesis is considerably more complex than its partially analogous stomatal coupling coefficient ($1 - \Omega$) for transpiration. The control coefficient for photosynthesis can be negative if an increase in stomatal conductance enhances transpirational cooling, lowering leaf temperature sufficiently to reduce assimilation capacity despite the rise in intercellular concentrations of CO_2. Another example is the covariance between boundary layer conductance and vapor pressure at the leaf surface in the field. Although an increase in boundary layer conductance should enhance CO_2 assimilation, the boundary layer control coefficient will be diminished by the accompanying decrease in vapor pressure at the leaf surface, which will limit assimilation by reducing stomatal conductance.

CASE STUDY: *POPULUS* SPP. AND THEIR HYBRIDS

In 1978, a major cooperative program between the University of Washington and Washington State University was initiated, aimed at the genetic improvement of the native black cottonwood (*Populus trichocarpa* T. & G.) and its hybridization with other *Populus* species for short rotation culture (178–180). In the process of integrating traditional breeding, molecular genetics, bioculture, soils, ecology and physiology, morphology and anatomy, numerous insights have been gained into (a) the structure and pattern of genetic variation in *P. trichocarpa*; (b) how this variation combines with other *Populus* species to form heterotic hybrids; and (c) how a number of components contribute to the vigor and stability of these hybrids in a variety of environments. However, for the purpose of this chapter, we will restrict our presentation to stomatal function, where we have a considerable amount of information derived from studies ranging from leaf tissues to that of plant canopies.

In contrast to the documented observations of most researchers, the stomata of well-watered *P. trichocarpa* were unresponsive to vapor density gradient, leaf water potential, or light in initial field studies (181). An examination of leaves excised from shoots of *P. trichocarpa* showed that desiccation and subsequent wilting led to little or no change in leaf conductance. In contrast, the stomata of *P. trichocarpa x P. deltoides* hybrids closed rather rapidly as excised leaves were desiccated. The response of *P. deltoides* stomata was intermediate. Following these laboratory studies on excised leaves, the response of leaf conductance to gradual desiccation was studied using material grown in 5-L pots under greenhouse conditions. Under such conditions, complete stomatal closure was observed in *P. trichocarpa x deltoides* hybrids and *P. deltoides* while only partial closure was noted in *P. trichocarpa*. In *P. trichocarpa*, the greatest reductions in conductance (to less than 20% of fully open values) were noted in young, expanding foliage while the conductance of mature leaves was 50% or greater than unstressed leaves. These results suggested that our understanding of the stomatal behavior of *P. trichocarpa* was influenced by the nature of the experimental design (i.e., rapid vs. gradual desiccation, the use of incompletely vs. fully mature foliage, and excised vs. attached leaves). Under field conditions in which periods of drought develop slowly, stomata of *P. trichocarpa* were found to be responsive to vapor density gradient and soil moisture (182). This responsiveness increased as the soil dried. Leaf water potentials had no apparent affect on stomatal closure and actually increased with closure.

In order to explain these observations, stomatal activity was examined in epidermal strips taken from leaves grown in the greenhouse (152,183). When clones of *P. trichocarpa* were well watered, stomata remained open in spite of plasmolysis of the guard cells and their loss of turgor. Even osmotic potential of -7.2 MPa failed to close stomata of *P. trichocarpa* completely. While not entirely conclusive, these observations suggest that the maintenance of open stomata in spite of the loss of turgor in the guard cells was the result of an abnormal physical characteristic of the guard cell wall, which was normalized to some degree by a period of water stress, especially if the stress occurred while leaves were still expanding. In contrast to the response of *P. trichocarpa*, stomata from epidermal peels of both *P. deltoides* and *P. trichocarpa x deltoides* hybrids closed prior to plasmolysis of guard cells or the turgor loss point of intact leaves.

The inability of *P. trichocarpa* stomata to close when the guard cells were plasmolyzed was not due to an inability of this species to produce an increase in foliar ABA in response to water stress as is the case for "wilty" mutants of tomato and potato (183). Application of ABA to developing leaves of well-watered *P. trichocarpa* plants induced a normal state of stomatal function, though it was not clear if this condition would have persisted. On the other hand, stomata from completely mature leaves did not respond to exogenous applications of ABA. These results lead to the following conclusions:

1. The inability of *P. trichocarpa* stomata to respond when plants are maintained in a well-watered condition resulted from a low "resting" level of ABA in the foliage. Such conditions might be frequently observed with young material grown under well-watered conditions. Under unmanaged field conditions for trees greater than 3 or 4 years old, the population of leaves developed under such conditions is probably rather small.

2. ABA has a role in the development of stomatal function. A critical level appears to be necessary to induce changes in guard cell walls that are needed for normal stomatal function, i.e., stomata that close in response to low leaf water potentials. Interestingly, similar or partially dysfunctional stomatal responses have also been observed in micropropagated *Delphinium* (184). Exposure of immature *P. trichocarpa* leaves to periods of water stress apparently increases the level of ABA sufficiently to normalize stomatal function. The closer the leaves are to maturation, the less likely it is that stomatal function can be altered either by periods of water stress or by exogenous application of ABA.

3. As studies increase in spatial scale (epidermal strips to whole plants in the field) and their temporal scale (rapid aerial desiccation to longer term soil desiccation), the abnormal response of *P. trichocarpa* stomata appears to be replaced by more typical response patterns (182,185). However, studies at lower space—time scales were critical in elucidating genetic and developmentally unique aspects of stomatal function in this species.

4. Stomatal responses to light and vapor pressure deficit appear to be strongly regulated by the soil environment (137,157,182).

Perhaps the most interesting observation from this series of studies was the responsiveness of hybrid stomata to desiccation. Instead of being intermediate between the parents as in the case of many other hybrid features, stomatal response of hybrids either equaled or exceeded that of *P. deltoides*. This apparent superiority may in part be due to heterosis (186–188). The exception to this trend was in hybrids with the more whitish undersurface leaves characteristics of *P. trichocarpa*. Molecular analysis showed them to be triploids (or aneuploids) containing two genomes of *P. trichocarpa* and one genome of *P. deltoides*, i.e., TTD hybrids (189,190). Not surprisingly, their stomatal responsiveness was closer to that of *P. trichocarpa*, as were other leaf traits.

Similar to the superior response noted in hybrids between *P. trichocarpa* and *P. deltoides*, hybrids derived from other interspecific crosses (e.g., *P. trichocarpa x maximowiczii*) demonstrated superiority relative to either parent. In all cases, the stomatal responses of *P. trichocarpa* were significantly lower than hybrid genotypes (Braatne, unpublished data). Bassman and Zwier (191) observed similar results except that stomata of *P. trichocarpa* from eastern Washington (warm, dry continental climate) were more responsive than those of either a *P. trichocarpa x deltoides* hybrid or *P. deltoides*. Similar to our findings, they found little stomatal sensitivity to desiccation in *P. trichocarpa* from western Washington (cool, maritime climate). Interestingly, stomata of intraspecific hybrids of *P. trichocarpa* (western x eastern Washington genotypes) were very responsive to low leaf water potentials (Braatne, unpublished data). Subsequent studies have revealed further differences in growth, and in photosynthetic and morphological variables between eastern and western Washington clones of *P. trichocarpa* (179,191,192).

Most recently, Hinckley et al. (193) linked measurements of water loss from individual leaves to water transport of whole branches with water transport in the entire stem to stand water loss in *Populus* (Fig. 5). This study was performed during the fourth growing season in individual trees of an F_1 hybrid derived from a *P. trichocarpa x deltoides* cross.

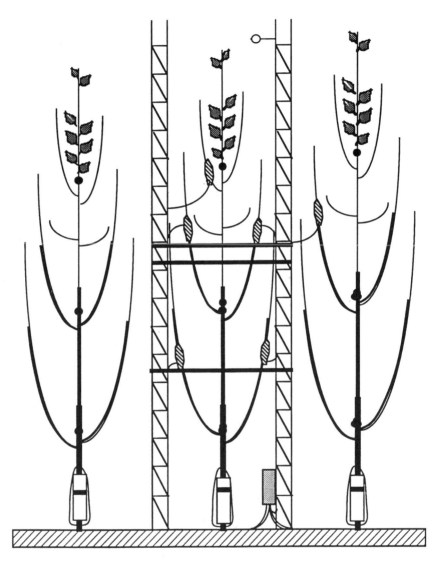

Figure 5 Diagram of *Populus* stand and associated scaffolding system. The relative positions of the Dynamax gauges (branches) and the whole-tree sap flux systems are indicated.

Towers and scaffolding were used to access the canopy; whole tree transpiration was measured at breast height using the methods of Cermak and Kucera (168); branch transpiration was measured using Dynamax gauges (194,195); and leaf conductances from leaves attached to measurement branches were recorded using a LiCor 1600 steady-state porometer.

Leaf conductances were highest from leaves in the upper canopy and lowest from those in the lower canopy (Fig. 6A). Maximum conductances occurred before solar noon (≈ 1300 hr) and declined thereafter. Highest rates of branch water loss occurred from upper canopy branches and lowest from lower canopy branches (Fig. 6B). Water loss from middle and upper canopy branches was relatively constant from 1130 to almost 1930 hr despite considerable declines in leaf conductance (e.g., from more than 400 to less than 300 mmol/m^2/sec). The average leaf transpiration rates, estimated from porometer measurements, were 50–100% higher than values obtained at the branch level (Dynamax gauges). In addition, there was tendency for the highest porometer-based transpiration rates to be observed between 1500 and 1730 hr. These differences reflect the inability of the porometer-based measurements to account for the boundary layer conductance surrounding the leaf and the degree to which intact leaves on branches are coupled to the driving gradient for branch transpiration. Patterns of whole-tree water loss demonstrated a plateau similar to that for the upper and middle crown branches; however, whole tree rates decreased after 1700 hr whereas they did not for these branches (Fig. 6C).

During the plateau of maximum transpiration from the middle and upper branches, average leaf conductance was calculated to be 409 mmol/m^2/sec (corresponds to an average from the porometer of approximately 312 mmol/m^2/sec) while crown conductance was 181 mmol/m^2/sec. These findings result in a boundary layer conductance of about 325 mmol/m^2/sec (i.e., less than leaf conductance). Based on a recent formulation from Jarvis and McNaughton (169), we calculated Ω to be about 0.8, which indicates that a 10% change in leaf conductance would cause transpiration to change by only 2% (at constant vapor pressure deficit, etc.). *Populus* is thus very poorly coupled when compared with a conifer (i.e., $\Omega = 0.1$).

Perhaps our greatest surprise involved the comparison of stand water loss for this fast-growing, deciduous hardwood located on an alluvial soil and with direct access to groundwater to that from a stand of evergreen conifers growing on a droughty, upland, gravelly, glacial outwash soil. Maximum stand water loss for *Populus* was approximately 4.50 mm/day. This is the same maximum as had been noted for *Pseudotsuga menziesii*

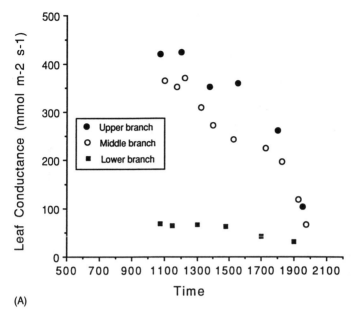

(A)

Figure 6 Results from the study of *Populus* hybrid 50-196 on August 3, 1992. (A) Average leaf conductance values from the upper (7–8 values per point), middle (12–15 values per point), and lower branches (5–7 values per point) are shown. (B) Daily course of branch sap flux for branches from the upper, middle, and lower crown. (C) Daily course of whole tree sap flux. (Data from Ref. 193.)

(196). The conifer stand had a slightly smaller leaf area index (10 vs. 12) and a much lower maximum foliar conductance (~ 60 mmol/m^2/sec) than did the *Populus* stand. However, the evergreen conifer is probably better coupled to the bulk atmosphere than the deciduous hardwood (see Table 1). As a result, very dramatic differences in stomatal morphology and conductance between these two species disappear when considering water loss from whole canopy.

Had we used only leaf conductance measurements in this study, we would have had a sampling nightmare (see Ref. 173) and we would have greatly overestimated canopy water loss. On the other hand, had we used only whole-tree measures of water loss, we would have not identified important components of spatial and temporal heterogeneity in transpiration nor any of the unique genetic and developmental controls over water loss. It is a long way from studies of epidermal peels to stand level measures of water loss, but these studies with *Populus* and the work of other researchers (134,197) demonstrate the need to link physiological

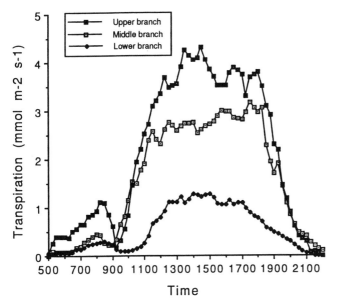

(B)

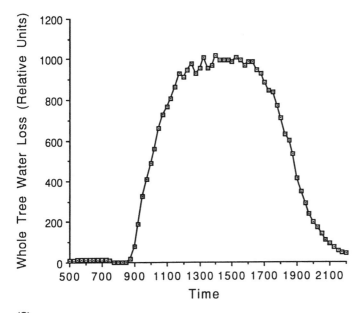

(C)

processes at different levels of structural organization. As stated by Meinzer (198), plant water status and ecosystem water balance are governed largely by actual fluxes of water vapor rather than stomatal conductance. Therefore, measurements of stomatal conductance alone often do not provide enough information to evaluate the physiological and ecological consequences of contrasting patterns of stomatal activity in different species.

CASE STUDY: HABITAT PARTITIONING BETWEEN THE SEXES OF DIOECIOUS PLANTS

A number of authors have observed that the ratio of male to female members of a dioecious species can be skewed under conditions of different resource availability. Such skewedness has been observed to occur with elevation (199), with soil pH (200), and with soil moisture (201–203). In both dioecious and monecious species, the reproductive effort, especially female flower and seed production, negatively impacts vegetative growth (204,205); therefore, habitat partitioning would be expected to occur between the sexes of dioecious species whereby females would occupy the most benign environments, which aids in allowing the costs associated with reproduction to be met (202,207). Such costs are not incurred by males. Herein we will concentrate on examples where water availability appears to cause this type of habitat partitioning between male and female plants, and where stomatal function in relation to water stress largely controls this partitioning.

Dawson and Bliss (202) observed that female plants of the arctic dwarf shrub *Salix arctica* were more common in wetter, higher nutrient environments than male plants. Small changes in microrelief resulted in large changes in soil moisture, soil temperature, and soil nutrition. As a consequence, female plants tended to be located in slight depressions whereas males were generally restricted to more xeric rises. Female plants had higher leaf conductances than male plants when soil temperatures were low, a common feature of these wet, arctic environments. In contrast, male plants had higher leaf conductances than female plants when soils were warm. In addition, leaf conductance of male plants was less sensitive to decreasing leaf water potential and increasing vapor pressure deficits than female plants. They speculated that these physiological differences represented specializations associated with the spatial segregation of males and females among habitats with differing resources. Females tended to be specialized in habitats where resources were more plentiful, thereby enabling them to outgrow male plants located in the same habitat.

In a more recent study, Dawson (206) and Dawson and Ehleringer (203,207) observed significant spatial segregation of male and female trees of the dioecious boxelder (*Acer negundo*) along canyons to the east and north of Salt Lake City (Fig. 7). Female trees were most common near the stream (i.e., riparian sites) whereas male trees were more common upslope from stream channels (i.e., drier sites). Under both laboratory and field conditions, female trees had higher leaf conductances, greater rates of photosynthesis, and poorer short- and long-term water use efficiencies than male trees. Stomata of male trees were more sensitive to declining water potentials and increasing leaf-to-air vapor pressure deficits. In other words, male trees had physiological attributes that allowed them to avoid water stress brought on by low water potentials and high evaporative demand. Female trees had faster growth rates, a greater allocation to reproductive effort than male trees, and poor stomatal control of water loss, restricting them for the most part to wet, streamside habitats. The differences observed between the sexes arise from differences in stomatal sensitivity to changes in the bulk leaf and whole-plant water status. Ongoing work suggests that slight differences in hydraulic architecture (after Ref. 164) and in differential sensitivity to root-borne hormone concentrations may underlie these clear gender-specific stomatal responses (Dawson, pers. commun.).

Dawson and Ehleringer (203,207) also used the relative abundance of the stable isotope of water, deuterium (D), to H, or the D/H ratio (δD), in order to elucidate the sources of water utilized by male and female trees. Smaller diameter trees were restricted to using water in the upper soil layers, presumably due to their shallow rooting depths, whereas

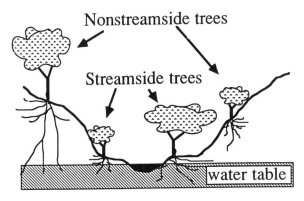

Figure 7 Diagram of a typical small canyon stream around Salt Lake City and the associated distribution of *Acer negundo*. (After Ref. 207.)

larger diameter trees used groundwater, regardless of whether they occurred in streamside or nonstreamside habitats (Fig. 7). During the growing season there are large changes in soil moisture but relatively minor changes in groundwater. Trees with access to groundwater would be able to grow longer and faster. Moreover, stream channels dry out every 4–7 years; trees that use groundwater would not be impacted by these drought years (Dawson, pers. commun.). The greater drought resistance capabilities of males favored survival on the upslope or drier sites whereas the more rapid growth characteristics of the females gave them a competitive advantage over the males close to stream channels.

Unlike the case study of *Populus, Acer negundo* shows only a 18–24% difference in leaf transpiration, based on a comparison of porometer to stem flow gauges. In addition, Ω calculated for male and female canopies was 0.28 and 0.34, respectively; this relatively high degree of canopy coupling has led to a very strong correlation between water use efficiency estimated on a short-term basis (instantaneous) and that estimated from carbon isotope discrimination (long-term; Dawson, pers. commun.).

In this particular example, Dawson and Bliss (202,208) and Dawson and Ehleringer (207) were able to develop an elegant story because they combined short-term measures of leaf function (i.e., leaf conductance, photosynthesis, water potential, etc.) with longer term measures of leaf function (i.e., stable isotopes of both carbon and water). This combination of time scales was then compared with the spatial distribution of individuals over a microlandscape in order to link physiological traits with performance across habitats.

Whether one deals with patchy stomatal behavior or microscale differences in habitat preferences by species or different sexes of the same species (208), it is important not only to document the phenomenon at the scale of immediate interest but to understand how lower and higher levels of biological organization interact and affect the phenomenon in question. Both the *Populus* and habitat partitioning case studies illustrate the importance of incorporating a number of observational or study scales when examining processes or patterns in nature.

ACKNOWLEDGMENTS

Populus research performed under subcontract no. 19X-43382C with Oak Ridge National Laboratory under Martin Marietta Energy Systems, Inc. contract DE-ACO5-840R21400 with the U.S. Department of Energy. Title of Project: "Genetic Improvement and Evaluation of Black Cottonwood for Short-Rotation Culture," R. F. Stettler, P. E. Heilman, and T. M. Hinckley, principal investigators.

The authors thank Drs. H. D. Bradshaw, Jr., R. Ceulemans, T. E. Dawson, E. D. Ford, F. C. Meinzer, D. S. Neuman, P. J. Schulte, D. G. Sprugel, R. F. Stettler, and E. Van Volkenburgh for extensive discussions and for sharing ideas and data which led directly or indirectly to many of the ideas and concepts presented herein. Critical reviews were provided by T. E. Dawson, R. F. Stettler, and E. Van Volkenburgh.

REFERENCES

1. Jarvis and T. A. Mansfield (1980): *Stomatal Physiology*, Cambridge University Press, Cambridge.
2. I. E. Teare and M. M. Peet (1983): *Crop–Water Relations*, John Wiley and Sons, New York.
3. E. Zeiger (1983): *Annu. Rev. Plant Physiol.*, 34:441.
4. T. A. Mansfield and W. J. Davies (1985): *Bioscience*, 35:158.
5. E. D. Schulze (1986): *Annu. Rev. Plant Phhysiol.*, 37:247.
6. E. Zeiger, G. D. Farquhar, and I. R. Cowan (1987): *Stomatal Function*, Stanford Univ. Press, Stanford, CA.
7. K. Raschke, R. Hedrich, U. Reckmann, and J. I. Schroeder (1988): *Bot. Acta*, 101:283.
8. C. M. Willmar (1988): *Biol. J. Linnean Soc.*, 34:205.
9. T. A. Mansfield, A. M. Hetherington, and C. J. Atkinson (1990): *Annu. Rev. Plant Physiol. Plant Mol. Biol.*, 41:55.
10. F. C. Meinzer (1993): *Trees* (in press).
11. G. Tallman (1993): *Crit. Rev. Plant Sci.* (in press).
12. Z. Sestak, J. Catsky, and P. G. Jarvis (eds.). (1971): *Plant Photosynthetic Production: Manual of Methods*, Dr. W. Junk, The Hague.
13. C. B. Field, T. J. Ball, and J. A. Berry (1989): *Plant Physiological Ecology: Field Methods and Instrumentation* (R. W. Pearcy, J. Ehleringer, H. A. Mooney, and P. W. Rundel, eds.), Chapman Hall, New York, p. 209.
14. R. W. Pearcy, E.-D. Schulze, and R. Zimmerman (1989): *Plant Physiological Ecology: Field Methods and Instrumentation* (R. W. Pearcy, J. Ehleringer, H. A. Mooney, and P. W. Rundel, eds.), Chapman Hall, New York, p. 137.
15. J. Weyers and H. Meidner (1990): *Methods in Stomatal Research*, Longman, Essex, England.
16. M. R. Kaufman and F. M. Kelliher (1991): *Techniques and Approaches in Forest Tree Ecophysiology* (J. P. Lassoie and T. M. Hinckley, eds.), CRC Press, Boca Raton, p. 117.
17. W. K. Smith and D. Y. Hollinger (1991): *Techniques and Approaches in Forest Tree Ecophysiology* (J. P. Lassoie and T. M. Hinckley, eds.), CRC Press, Boca Raton, p. 141.
18. R. A. Fischer (1973): *J. Exp. Bot.*, 24:387.
19. P. J. H. Sharpe, H.-I. Wu, and R. D. Spence (1987): *Stomatal Function*, Stanford Univ. Press, Stanford, CA, p. 91.

20. D. W. DeMichele and P. J. H. Sharpe (1973): *J. Theor. Biol.*, *41*:77.

21. M. Edwards, H. Meidner, and D. W. Sheriff (1976): *J. Exp. Bot.*, *27*:163.

22. H. Meidner and P. Bannister (1979): *J. Exp. Bot.*, *30*:255.

23. K. Raschke (1979): *Encycl. Plant Physiol.*, *7*:383.

24. P. J. H. Sharpe, and H.-I. Wu (1978): *Plant Cell Environ.*, *1*:259.

25. H.-I. Wu and P. J. H. Sharpe (1979): *Plant Cell Environ.*, *1*:259.

26. H.-I. Wu, P. J. H. Sharpe, and R. D. Spence (1985): *Plant Cell Environ.*, *8*:269.

27. H. Meidner and T. A. Mansfield (1968): *Physiology of Stomata*, McGraw-Hill, London.

28. T. C. Hsiao (1976): *Encyl. Plant Physiol.*, *2B*:195.

29. K. Raschke and G. D. Humble (1973): *Planta*, *115*:47.

30. E. A. C. MacRobbie (1987): *Stomatal Function*, Stanford Univ. Press, Stanford, CA, p. 126.

31. S. R. Gallagher and R. T. Leonard (1982): *Plant Physiol.*, *70*:1335.

32. A. Serrano (1989): *Annu. Rev. Plant Physiol. Plant Mol. Biol.*, *40*:61.

33. H. F. Bienfait (1985): *J. Bioener. Biomembr.*, *17*:73.

34. A. S. Raghavendra (1990): *Plant Cell Environ.*, *13*:105.

35. S.-I. Imamura (1943): *Jap. J. Bot.*, *12*:251.

36. T. Yamashita (1952): *Siebold. Acta Biol.*, *1*:51.

37. M. Fujino (1967): *Sci. Bull. Fac. Educ. Nagasaki Univ.*, *18*:1.

38. R. A. Fischer (1968): *Science*, *160*:784.

39. R. A. Fischer (1972): *Aust. J. Biol. Sci.*, *25*:1107.

40. G. D. Humble and T. C. Hsiao (1969): *Plant Physiol.*, *44*:82.

41. G. D. Humble and K. Raschke (1971): *Plant Physiol.*, *48*:447.

42. C. K. Pallaghy and R. A. Fischer (1974): *Z. Pflanzenphysiol.*, *71*:332.

43. E. A. C. MacRobbie and J. Lettau (1980): *J. Membr. Biol.*, *53*:199.

44. E. A. C. MacRobbie and J. Lettau (1980): *J. Membr. Biol.*, *56*:249.

45. W. H. Outlaw (1983): *Physiol. Plant*, *59*:302.

46. J. I. Schroeder, R. Hedrich, and J. M. Fernandez (1984): *Nature*, *312*:361.

47. J. I. Schroeder, K. Raschke, and E. Neher (1987): *Proc. Natl. Acad. Sci.*, *84*:4108.

48. M. R. Blatt (1987): *Planta*, *170*:272.

49. M. R. Blatt (1987): *J. Membr. Biol.*, *98*:257.

50. M. R. Blatt (1988): *J. Membr. Biol.*, *102*:235.

51. J. I. Schroeder (1988): *J. Gen. Physiol.*, *92*:667.

52. J. I. Schroeder and R. Hedrich (1989): *Trends Biochem. Sci.*, *14*:187.

53. G. Thiel and M. R. Blatt (1991): *J. Plant Physiol.*, *138*:326.

54. J. I. Schroeder and H. H. Fang (1991): *Proc. Natl. Acad. Sci. USA*, *88*:1158.

55. J. I. Schroeder and S. Hagiwara (1989): *Nature*, *338*:427.

56. K. Fairley-Grenot and S. M. Assmann (1991): *Plant Cell Physiol.*, *29*:907.

57. S. Hosoi, M. Iino, and K. Shimazaki (1988): *Plant Cell Physiol.*, *29*:907.

58. J. I. Schroeder (1989): *J. Membr. Biol.*, *107*:229.

59. M. R. Blatt (1991): *J. Membr. Biol.*, *124*:95.

60. G. D. Humble and K. Raschke (1971): *Plant Physiol.*, *48*:447.

61. C. K. Pallaghy and R. A. Fischer (1974): Z. *Pflanzenphysiol.*, *71*:332.
62. W. G. Allaway (1973): *Planta*, *110*:63.
63. M. G. Penny, L. S. Kelday, and D. J. F. Bowling (1976): *Planta*, *130*:291.
64. K. Raschke and H. Schnabl (1978): *Plant Physiol.*, *62*:84.
65. H. Schnabl (1978): *Planta*, *144*:95.
66. P. Dittrich, M. Mayer, and M. Meusel (1978): *Planta*, *144*:305.
67. A. S. Raghavendra (1980): *Aust. J. Plant Physiol.*, *7*:663.
68. H. Schnabl (1980): Z. *Naturforsch.*, *35*:621.
69. H. Schnabl and K. Raschke (1980): *Plant Physiol.*, *65*:88.
70. P. H. Brown and W. H. Outlaw (1982): *Plant Physiol.*, *70*:1700.
71. D. Laffray and P. Louguet (1982): *J. Exp. Bot.*, *33*:771.
72. H. Schnabl, C. Elbert, and G. Kramer (1982): *J. Exp. Bot.*, *33*:996.
73. D. J. F. Bowling (1987): *J. Exp. Bot.*, *38*:1351.
74. B. Michalke and H. Schnabl (1987): *J. Plant Physiol.*, *130*:243.
75. F. A. Smith and J. A. Raven (1979): *Annu. Rev. Plant Physiol.*, *30*:289.
76. A. Kurkdjian and J. Guern (1989): *Annu. Rev. Plant Physiol. Plant Mol. Biol.*, *40*:271.
77. D. Sanders, M. Hopgood, and I. R. Jenings (1989): *J. Membr. Biol.*, *108*:253.
78. C. A. Van Kirk and K. Raschke (1978): *Plant Physiol.*, *61*:361.
79. C. M. Willmer (1980): *New Phytol.*, *84*:593.
80. H. Schnabl (1983): *Physiol. Veg.*, *21*:955.
81. H. Schnabl and C. Kottmeier (1984): *Planta*, *162*:220.
82. B. Michalke and H. Schnabl (1990): *Planta*, *180*:188.
83. M. C. Tarczynski and W. H. Outlaw (1990): *Arch. Biochem. Biophys.*, *280*:153.
84. C. M. Willmer, Y. Petropoulou, and Y. Manetas (1990): *J. Exp. Bot.*, *41*:1103.
85. W. H. Outlaw and J. Manchester (1978): *Plant Physiol.*, *64*:79.
86. S. M. Assman and E. Zeiger (1987): *Stomatal Function*, Stanford Univ. Press, Stanford, CA, p. 163.
87. E. Beck and P. Ziegler (1989): *Annu. Rev. Plant Physiol. Plant Mol. Biol.*, *40*:95.
88. N. L. Robinson and J. Preiss (1987): *Plant Physiol.*, *85*:360.
89. A. S. Raghavendra and T. Vani (1989): *J. Plant Physiol.*, *135*:3.
90. E. Zeiger, P. Armond, and A. Melis (1980): *Plant Physiol.*, *67*:17.
91. D. A. Grantz, T. Graan, and J. S. Boyer (1985): *Plant Physiol.*, *77*:956.
92. K. Schimazaki and E. Zeiger (1985): *Plant Physiol.*, *78*:211.
93. E. Zemel, I. Leizerovich, and S. Gepstein (1988): *Plant Physiol.*, *88*:518.
94. A. Schwartz and E. Zeiger (1984): *Planta*, *161*:129.
95. S. Madhaven and B. N. Smith (1982): *Plant Physiol.*, *69*:273.
96. W. H. Outlaw, M. C. Tarczynski, and L. C. Anderson (1982): *Plant Physiol.*, *70*:1218.
97. E. Zemel and S. Gepstein (1985): *Plant Physiol.*, *78*:586.
98. K. Shimazaki, J. Terada, K. Tanaka, and N. Kondo (1989): *Plant Physiol.*, *90*:1057.

99. M. C. Tarczynski, W. H. Outlaw, N. Arnold, V. Neuhoff, and R. Hampp (1989): *Plant Physiol.*, *89*:1088.
100. U. Reckmann, R. Scheibe, and K. Raschke (1990): *Plant Physiol.*, *92*:246.
101. T. D. Sharkey and T. Ogawa (1987): *Stomatal Function*, Stanford Univ. Press, Stanford, CA, p. 195.
102. E. Zeiger (1990): *Plant Cell Environ.*, *13*:739.
103. T. C. Hsiao, W. G. Allaway, and L. T. Evans (1973): *Plant Physiol.*, *51*:82.
104. E. Zeiger, M. Iino, K. Shimazaki, and T. Ogawa (1987): *Stomatal Function*, Standford Univ. Press, Stanford, CA, p. 210.
105. T. Ogawa, H. Ishikawa, K. Shimada, and K. Shibata (1978): *Planta*, *142*:61.
106. A. Schwartz and E. Zeiger (1984): *Planta*, *161*:129.
107. G. Tallman and E. Zeiger (1988): *Plant Physiol.*, *88*:887.
108. B. T. Mawson and E. Zeiger (1991): *Plant Physiol.*, *96*:753.
109. M. Poffenroth, D. Green, and G. Tallman (1992): *Plant Physiol.*, *98*:1460.
110. M. A. Pemadasa (1982): *J. Exp. Bot.*, *33*:92.
111. M. G. Holmes and W. H. Klein (1985): *Planta*, *166*:348.
112. N. Roth-Beherano and C. Itai (1987): *Physiol. Plant*, *70*:85.
113. A. Nejidat, C. Itai, and N. Roth-Bejerano (1989): *Plant Cell Physiol.*, *30*:945.
114. E. Zeiger and C. Field (1982): *Plant Physiol.*, *70*:370.
115. W. E. Williams, C. Grivet, and E. Zeiger (1982): *Plant Physiol.*, *69*:84.
116. S. M. Assman and E. Zeiger (1985): *Plant Physiol.*, *77*:461.
117. T. A. Mansfield, A. J. Travis, and R. G. Jarvis (1980): *Stomatal Physiology*, Cambridge University Press, Cambridge, UK, p. 119.
118. J. I. L. Morison (1987): *Stomatal Function*, Stanford Univ. Press, Stanford, CA, p. 229.
119. K. Wardle and K. C. Short (1981): *J. Exp. Bot.*, *32*:303.
120. J. I. L. Morison and R. M. Gifford (1983): *Plant Physiol.*, *71*:789.
121. K. A. Mott (1988): *Plant Physiol.*, *86*:200.
122. J. I. L. Morison (1985): *Plant Cell Environ.*, *8*:467.
123. W. J. S. Downton, B. R. Loveys, and W. J. R. Grant (1988): *New Phytol.*, *110*:503.
124. G. D. Farquahr, K. T. Hubrick, I. Terashima, A. G. Condon, and R. A. Richards (1987): *Prog. Photosyn. Res.*, *4*:209.
125. I. Terashima, S. C. Wong, C. B. Osmond, and G. D. Farquhar (1989): *Plant Cell Physiol.*, *29*:385.
126. K. A. Mott, Z. G. Cardon, and J. A. Berry (1993): *Plant Cell Environ.*, *16*:25.
127. F. A. Bazzaz and E. D. Fajer (1992): *Sci. Am.*, *266*:68.
128. T. Gollan, N. C. Turner, and E.-D. Schulze (1985): *Oecologia*, *65*:356.
129. E.-D. Schulze (1986): *Annu. Rev. Plant Physiol.*, *37*:247.
130. E.-D. Schulze, N. C. Turner, T. Gollan, and K. A. Shackel (1987): *Stomatal Function*, Stanford Univ. Press, Stanford, CA, p. 311.
131. N. C. Turner, E.-D. Schulze, and T. Gollan (1984): *Oecologia*, *65*:338.

132. N. C. Turner, E.-D. Schulze, and T. Gollan (1985): *Oecologia*, 65:348.
133. D. A. Grantz and F. C. Meinzer (1991): *Agric. For. Meteorol.*, 53:169.
134. F. C. Meinzer and D. A. Grantz (1991): *Physiol. Plant.*, 83:324.
135. J. B. Passioura (1988): *Plant Cell Environ.*, 11:569.
136. E.-D. Schulze, E. Steudle, T. Gollan, and U. Schurr (1988): *Plant Cell Environ.*, 11:573.
137. B. A. Smit, D. S. Neuman, and M. Stachowiak (1990): *Plant Physiol.*, 92:1021.
138. J. Zhang and W. J. Davies (1989): *J. Exp. Bot.*, 37:1471.
139. J. Zhang and W. J. Davies (1989): *Plant Cell Environ.*, 12:73.
140. F. Tardieu, J. Zhang, N. Katerji, O. Bethenod, S. Palmer, and W. J. Davies (1992): *Plant Cell Environ.*, 15:193.
141. J. Masle and J. B. Passioura (1987): *Aust. J. Plant Physiol.*, 14:643.
142. M. M. Ludlow (1989): *Structural and Functional Responses to Environmental Stresses: Water Shortage*, SPB Academic, The Hague, p. 269.
143. P. F. Scholander, H. T. Hammel, E. D. Bradstreet, and E. A. Hemmingsen (1965): *Science*, 143:339.
144. G. A. Ritchie and T. M. Hinckley (1975): *Adv. Ecol. Res.*, 9:165.
145. N. C. Turner (1988): *Irrig. Sci.*, 9:289.
146. R. Benkert, A. Balling, and U. Zimmermann (1991): *Bot. Acta*, 104:423.
147. J. B. Passioura (1991): *Bot. Acta*, 104:405.
148. K. Raschke (1987): *Stomatal Function*, Stanford Univ. Press, Stanford, CA, p. 254.
149. M. G. Beardsell and D. Cohen (1975): *Plant Physiol.*, 56:207.
150. E. G. Newville and W. K. Ferrell (1980): *Can. J. Bot.*, 58:1370.
151. J. W. Radin and R. C. Ackerson (1981): *Plant Physiol.*, 67:115.
152. P. J. Schulte and T. M. Hinckley (1987): *Tree Physiol.*, 3:103.
153. D. A. Grantz, T. D. Ho, S. J. Uknes, and J. M. Cheeseman (1985): *Plant Physiol.*, 78:51.
154. P. M. Schildwacht (1989): *Planta*, 177:178.
155. L. M. Bates and A. E. Hall (1981): *Oecologia*, 50:62.
156. R. Munns and R. W. King (1988): *Plant Physiol.*, 88:703.
157. D. S. Neuman, S. B. Rood, and B. A. Smit (1990): *J. Exp. Bot.*, 41:1325.
158. J. W. Radin, L. L. Parker, and G. Guinn (1982): *Plant Physiol.*, 70:1066.
159. P. G. Blackman and W. J. Davies (1985): *J. Exp. Bot.*, 36:39.
160. L. D. Incoll and P. C. Jewer (1987): *Stomatal Function*, Stanford Univ. Press, Stanford, CA, p. 281.
161. R. O. Teskey, T. M. Hinckley, and C. C. Grier (1983): *J. Exp. Bot.*, 34:1251.
162. J. S. Boyer (1989): *Plant Cell Environ.*, 12:213.
163. F. S. Chapin, III (1991): *Bioscience*, 41:29.
164. J. S. Sperry and M. T. Tyree (1988): *Plant Physiol.*, 88:574.
165. D. L. Shumway, K. C. Steiner, and T. E. Kolb (1993): *Tree Physiol.*, 12:41.
166. J. B. Passioura (1979): *Search*, 10:347.
167. P. G. Jarvis and K. G. McNaughton (1986): *Adv. Ecol. Res.*, 15:1.

168. J. Cermak and J. Kucera (1990): *Silva Carelica, 15*:101.
169. K. G. McNaughton and P. G. Jarvis (1991): *Agric. Forest Meteorol., 54*:279.
170. E. J. Rykiel, Jr. (1985): *Aust. J. Ecol., 10*:361.
171. T. A. Pickett, J. Kolasa, J. J. Armesto, and S. L. Collins (1989): *Oikos, 54*:129.
172. J. R. Ehleringer and C. B. Field (1993): *Scaling Physiological Processes: Leaf to Globe*, Academic Press, San Diego.
173. J. Leverenz, J. D. Deans, E. D. Ford, P. G. Jarvis, R. Milne, and D. Whitehead (1982): *J. Appl. Ecol., 19*:835.
174. F. I. Woodward (1987): *Nature, 327*:617.
175. F. I. Woodward and F. A. Bazzaz (1988): *J. Exp. Bot., 39*:1771.
176. I. E. Woodrow, J. T. Ball, and J. A. Berry (1990): *Plant Cell Environ., 13*:339.
177. I. E. Woodrow, J. T. Ball, and J. A. Berry (1986): *Prog. Photosyn. Res., 4*:225.
178. T. M. Hinckley, R. Ceulemans, J. M. Dunlap, A. Figliola, P. E. Heilman, J. G. Isebrands, G. Scarascia-Mugnozza, P. J. Schulte, B. Smit, R. F. Stettler, E. Van Volkenburg, and B. M. Wiard (1989): *Structural and Functional Responses to Environmental Stress*, SPB Academic, The Hague, p. 199.
179. T. M. Hinckley, J. H. Braatne, R. Ceulemans, P. Clum, J. Dunlap, D. Neuman, B. Smit, G. Scarascia-Mugnozza, and E. Van Volkenburgh (1992): *Ecophysiology of Short Rotation Forest Crops*, Elseiver, London, p. 1.
180. R. Ceulemans, G. Scarascia-Mugnozza, B. M. Wiard, J. H. Braatne, T. M. Hinckley, R. F. Stettler, J. G. Isebrands, and P. E. Heilman (1992): *Can. J. For. Res., 22*:1937.
181. P. J. Schulte, T. M. Hinckley, and R. F. Stettler (1987): *Can. J. Bot., 65*:255.
182. S. R. Pezeshki and T. M. Hinckley (1988): *Can. J. For. Res., 18*:1159.
183. P. J. Schulte and T. M. Hinckley (1987): *Plant Cell Environ., 10*:313.
184. J. M. Santamaria, W. J. Davies, and C. J. Atkinson (1993): *J. Exp. Bot., 44*:99.
185. J. H. Braatne, T. M. Hinckley, and R. F. Stettler (1992): *Tree Physiol., 11*:325.
186. N. J. Bate, S. B. Rood, and T. J. Blake (1988): *Can. J. Bot., 66*:1148.
187. R. Matyssek and E.-D. Schulze (1987): *Trees, 1*:219.
188. R. F. Stettler and R. Ceulemans (1993): *Clonal Forestry I: Genetics and Biotechnology*, Springer-Verlag, Berlin, p. 68.
189. H. D. Bradshaw and R. F. Stettler (1993): *Theor. Appl. Genet., 86*:301.
190. H. D. Bradshaw and R. F. Stettler (in press): *Theor. Appl. Genet.*
191. J. H. Bassman and J. C. Zwier (1991): *Tree Physiol., 8*:145.
192. J. Dunlap, J. H. Braatne, T. M. Hinckley, and R. F. Stettler (1993): *Can. J. Bot., 71*:1304.

193. T. M. Hinckley, J. R. Brooks, J. Cermák, R. Ceulemans, J. Kucera, F. C. Meinzer, D. A. Roberts, and M. Wampler (in prep.): *Tree Physiol.*
194. J. L. Heilman and J. M. Ham (1990): *Hortscience, 25*:465.
195. M. Gutierrez, R. Harrington, F. C. Meinzer, and J. H. Fownes (in prep.): *Tree Physiol.*
196. L. J. Fritschen, L. Cox, and R. Kinerson (1973): *For. Science, 19*:256.
197. F. C. Meinzer, G. Goldstein, N. M. Holbrook, P. Jackson, and J. Cavelier (in press): *Plant Cell Environ.*
198. F. C. Meinzer (in press): *Trees.*
199. J. B. Mitton and M. C. Grant (1980): *Am. J. Bot., 67*:202.
200. P. A. Cox (1981): *Am. Nat., 117*:295.
201. N. M. Waser (1984): *Oikos, 42*:343.
202. T. E. Dawson and L. C. Bliss (1989): *Oecologia, 79*:332.
203. T. E. Dawson and J. R. Ehleringer (1991): *Nature, 350*:335.
204. S. Eis, E. H. Garman, and L. F. Ebell (1965): *Can. J. Bot., 43*:1553.
205. S. A. Vasiliauskas and L. W. Aarssen (1992): *Ecology, 73*:622.
206. T. E. Dawson (in press): *Stable Isotopes and Plant Carbon/Water Relations*, Academic Press, San Diego.
207. T. E. Dawson and J. R. Ehleringer (1993): *Ecology, 74*:798.
208. T. E. Dawson and L. C. Bliss (in press): *Func. Ecol., 7.*
209. L. Taiz and E. Zeiger (1991): *Plant Physiology*, Benjamin/Cummings, Menlo Park, CA.

12

Plant Response to Air Pollution

James A. Weber, David T. Tingey, and Christian P. Andersen

U.S. Environmental Protection Agency
Corvallis, Oregon

INTRODUCTION

Definitions of air pollution invoke the concept that compounds in the atmosphere must have an undesirable or adverse effect on a biological receptor(s) before they can be considered pollutants. One definition of an air pollutant is given by McCabe (1) which reads:

> The presence in the outdoor atmosphere of substances or contaminants, put there by man, in quantities or concentrations and of a duration as to cause any discomfort to a substantial number of inhabitants of a district or which are injurious to public health, or to human, plant or animal life or property . . .

When applying this concept of injury/adversity it is important to remember that the definition of injury/adversity is frequently in the eye of the observer, i.e., not all effects of air pollutants are equally important to all observers.

An example of this differential variation is the common usage in the literature on pollutant effects on vegetation of the terms *injury* and *damage* (2,3). Injury encompasses all measurable plant reactions that do not impair the intended use of the plant, e.g., agronomic yield. These injury responses can include changes in metabolism, reduced photosynthesis, leaf necrosis, leaf drop, and altered growth or quality if yield is not

357

impaired. In contrast, damage includes all effects that reduce the intended human use (economic production, genetic resources, or cultural values) or value of the plant to the ecosystem (ecological structure and function). Vegetation responses that are classified as damage are frequently considered to be more important than those classified as injury.

Statement of Goals

While specific pollutants induce plant responses peculiar to that pollutant, much of the process of uptake and many of the subsequent effects (especially indirect effects) are similar. The emphasis in this chapter will be on processes directly affected by pollution (direct effects) and on subsequent responses of processes not directly affected (indirect effects). A conceptual framework will be used to combine the known direct and indirect effects into an integrated picture of whole-plant response. Response of plants to O_3, the most widely spread pollutant in the United States, will be used as an example.

History

The first studies of the effects of air pollutants on vegetation began in Tharandt (Saxony), Germany in the mid-nineteenth century (4). Stöckhardt (5) published his pioneering studies on the effects of smoke from silver smelters on adjacent vegetation. Subsequent investigations have established the phytotoxic nature of various gases (primarily SO_2) from metallurgical and industrial sources (6–9). Observed effects on vegetation were not limited to foliar injury or to plant growth and yield, but also include more complex cases, such as hydrogen fluoride (HF) which can be incorporated into the food chain and induce fluorosis in animals eating the contaminated vegetation (10).

The types of air pollutants affecting vegetation have changed significantly during the last century. Initially, problems were related to specific industrial or manufacturing sources (point sources) whose emissions were frequently associated with particular vegetation responses. With the introduction of higher smokestacks and emission controls, the local concentrations of pollutants near industrial sources have decreased and the pollutants have been more widely distributed (albeit at lower concentrations) with the result that the association between individual sources and vegetation response has been reduced. Concurrently, the impact of air pollution from transportation and urban sources (area sources) has increased (11–13). Area sources are characterized by numerous small-to-medium emission sources, making it virtually impossible to associate impacts on vegetation with specific emission sources.

During the past 20–30 years exposure (concentration × time) to SO_2, HF, and other pollutants from point sources has decreased. Photochemical oxidants such as O_3 in the lower troposphere, produced by photochemical reactions of nitrogen oxides and hydrocarbons (13,14), have become the principle air pollutants in more industrialized countries. Currently, tropospheric O_3 at elevated concentrations is the most widely distributed air pollutant in the United States and causes more than 90% of the crop damage (15). Elevated O_3 levels have also been reported in Canada, central and western Europe, the Mediterranean area, and Asia (2,16). However, in less developed areas where many of the industrial sources have little or no controls on emissions, other pollutants dominate the impacts on vegetation.

Observations of vegetation injury can provide a means of identifying and monitoring pollutants emitted from a source and, in the absence of air monitoring equipment, provide a means of estimating the geographic distribution of the pollutant. When plants are injured by an air pollutant, visible foliar symptoms characteristic of the pollutant may develop. These visible foliar symptoms have been used to identify the causative compound and distinguish the pollutant effects from possible biotic or nutritional factors (17–21). However, foliar symptoms must be used with caution as symptoms are not always conclusive. Air monitoring for pollutants is also important for diagnosing the probable causative compound. Elevated atmospheric concentrations of the pollutant should occur prior to the development of foliar symptoms.

Pollutants can also have significant impacts on vegetation while inducing few or no visible symptoms (22–24). The only certain way to establish whether a pollutant is impacting plant growth and yield is by comparison with control plants grown under the same environmental conditions but not exposed to the pollutant (2,25). When such a comparison is made, premature senescence is frequently the only visible symptom associated with the pollutant-induced growth/yield loss. However, this symptom cannot be detected without reference to untreated control plants.

Over the last century many of journal articles, conference proceedings, and books have documented the effects of air pollution on vegetation. Some recent publications that provide more details of air pollution effects on vegetation are included in the references (2,9,26–37).

Within the United States, the Clean Air Act requires that the administrator of the U.S. Environmental Protection Agency (EPA) establish primary and secondary national ambient air quality standards (NAAQS) to protect public health and welfare from any known or anticipated adverse effects caused by criteria air pollutants (e.g., O_3, SO_2,

NO_x). The scientific bases for NAAQS's are contained in air quality criteria documents (e.g., 11,12,38) which describe all identifiable effects ascribed to a pollutant (associated with specific concentrations and exposure durations). These criteria documents are updated periodically and provide good summaries of the effects of specific air pollutants on vegetation, as well as information on pollutant sources, atmospheric reactions/ transformations, and impacts on humans.

Factors that Influence Pollutant Exposure

The effects of air pollutants on vegetation are controlled by factors that influence the nature of the exposure, particularly:

Type of compounds emitted or formed in the atmosphere
Their concentration, duration and frequency of occurrence
The meteorology and topography of the area

and by factors that influence plant response to the pollutant exposure, including:

The biological susceptibility of the organism
The prevailing climatic and edaphic factors

Historically, air pollutants have been divided into primary and secondary forms. A primary pollutant (e.g., SO_2, HF, or heavy metals) is one that is injurious in the same chemical form as when it was emitted into the atmosphere. A secondary pollutant is formed in the atmosphere through reactions among precursors emitted into the atmosphere. Ozone (O_3) is a typical secondary pollutant that is formed by action of sunlight on nitrogen oxides and hydrocarbons in the atmosphere (14,38). In addition to O_3, these photochemical reactions yield organic peroxides, organic nitrates, etc.

Air pollutants are not equally injurious to vegetation. It is difficult to develop generalizations but compounds such as SO_2 and NO_2, which can be metabolized by the plant and incorporated in the plant's sulfur or nitrogen pools, are typically less toxic than compounds such as O_3 or peroxyacetyl nitrate (PAN), which cannot be metabolized.

Ambient pollutant concentrations are influenced by a number of factors, including the amount of compound emitted per unit time from the source, the wind speed during dispersal, and atmospheric stability (Fig. 1). Pollutant exposures are characterized by the concentration, duration, and frequency of occurrence (39,40). Even though a compound can be toxic, the presence of the compound in the atmosphere does not assure that the vegetation will be injured since pollutant exposure may be at

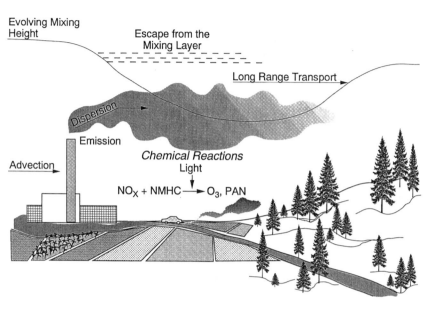

Figure 1 Examples of processes involved in emission, transport, and exposure of vegetation to atmospheric pollutants.

concentrations, durations, and/or frequencies that are too low to induce injury.

A few generalizations can be stated regarding the factors that contribute to vegetation injury (39–41):

Ozone phytotoxicity is the consequence of the amount of compound absorbed into the leaf, not the concentrations in the air around the plant (42,43).

High concentrations for a short time produce more foliar injury than a lower concentration for a longer time, even though the resultant external dose is similar for both cases.

Vegetation injury increases with the frequency of occurrence of elevated pollutant concentrations.

Analyses of a number of studies indicated that impacts on plant growth and yield are the result of cumulative exposures to higher pollutant concentrations (39,40).

Atmospheric factors that control the movement, rate of dilution, and removal of a pollutant from the atmosphere have a major influence on pollutant concentration (44–46). Because air pollutants are transported by wind, concentrations are typically higher downwind from a source than upwind, leading to pollutant exposures in areas that are not themselves sources of air pollution. When sources are located in valleys, wind transport of pollutants can produce high concentrations on adjacent slopes and ridges.

If compounds are emitted at a constant rate, atmospheric pollutant concentrations will be higher at low wind speed or during a temperature inversion since these factors retard dispersion and subsequent dilution of the compound. Also an inversion layer reduces the mixing volume and concentrates the precursors in the atmosphere below the inversion layer, yielding higher resultant concentrations. Since atmospheric pollutants are frequently concentrated near the base of the inversion layer, short-term high concentrations of pollutants can occur at ground level during the "break-up" of the layer through transport of parcels of air containing elevated pollutant concentrations to ground level.

An example of the spatial distribution of O_3 is presented (Fig. 2) to show how the meteorological factors and various emission sources are integrated and influence vegetation exposure. Because vegetation effects are the consequence of cumulative exposures to higher pollutant concentrations, the data are shown as the SUM06 exposure index (i.e., the sum of the average hourly O_3 concentration above 0.06 ppm). In general, the pollutant exposure increases from west to east as a consequence of prevailing wind direction and an increasing number of sources.

The highest O_3 exposures were located in the mountainous regions of the mid-Atlantic states. Elevated concentrations were measured in the major agricultural areas of the midwest and southeast and in much of the forested areas of the eastern United States. In the United States, terrain has a pronounced effect on pollution distribution. Air basins with large cities (e.g., Los Angeles) typically have high levels of air pollutants, whereas those without cities will likely have low levels. In addition to the spatial variability in exposure illustrated in Figure 2, year-to-year variability in exposure at specific locations does occur. For example, the coefficient of variation in O_3 exposure at rural sites was approximately 35% (47).

Factors that Influence the Plant Response to Pollutant Exposure

Plant species differ in their susceptibility to air pollutants. For example, legumes such as soybeans and clover are more susceptible to O_3 than grains such as corn or wheat (38,48). Correspondingly, different cultivars, ecotypes, and varieties also display different susceptibilities to air pollutants. Susceptibility to pollutants has been shown to be under genetic control (33,36,49). Although diverse mechanisms have been pro-

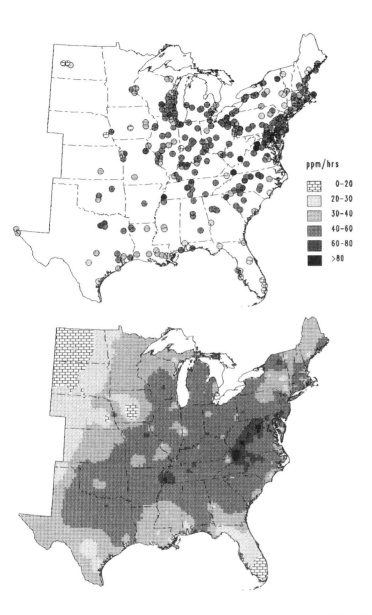

Figure 2 Spatial distribution of ambient O_3 exposure for 1988. Circles on the upper map illustrate the O_3 exposure (SUM06, which accumulates all concentrations $\geqslant 0.06$ ppm for the 3-month period during the summer) at rural monitoring sites. The gradient surface on the lower map was interpolated from the monitoring locations using an inverse distance weighting (second-power) method. The search window was limited to the nearest 10 monitored sites. Data are not shown for the western United States because there are few monitoring sites and the mountainous terrain makes special extrapolation more difficult.

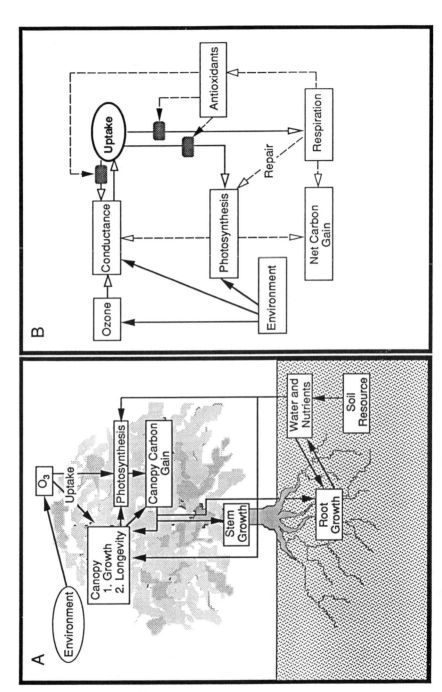

posed to explain this differential susceptibility, no generally accepted mechanism(s) can explain the differences (42,43,50).

Climatic and edaphic factors also alter the susceptibility of vegetation to a pollutant exposure. A range of climatic factors such as relative humidity, air temperature, light intensity, and photoperiod have been studied and shown to influence the response of vegetation to pollutants (2,32,38). Similarly, edaphic factors such as soil moisture, soil nutrients, and soil oxygen concentration also influence plant response (2,29,32,38). Although the response of various plants to individual climatic or edaphic factors varies widely, those factors that control stomatal conductance are also assumed to control in large part plant response to pollutants; however, differences in stomatal conductance are not the sole controlling factor (42,43,50).

CONCEPTUAL FRAMEWORK

Air pollutants, such as O_3, frequently have their major effects on the ability of a plant to amass and utilize carbon (Fig. 3). For those pollutants taken up directly from the air the major processes affected are those determining carbon gain. For air pollutants deposited on the soil the direct effects occur in the roots, for instance on nutrient uptake. Pollutant concentration is affected by environmental conditions (light, temperature, presence of precursors) at the time of formation and by atmospheric transport. The rate and total amount of pollutant taken up from the air can affect photosynthesis, respiration, leaf conductance, and leaf longevity. All of these factors in turn affect canopy carbon fixation and net accumulation of photosynthate which is allocated to various parts of the plant as well as to repair and resistance processes. Changes in the amount of photosynthate available for plant functions feed back to canopy carbon gain through reduction in root activity, canopy density (changes in

---

Figure 3 (A) Hypothesized interaction of O_3 with plant function emphasizing carbon allocation. The primary effect of O_3 is assumed to occur in the leaves by reducing leaf longevity and photosynthetic capacity. In turn, the amount of assimilate available for growth and respiration is reduced, leading to reductions in canopy size and root growth, both of which feed back on carbon assimilation. (B) Hypothesized effects of O_3 uptake within a leaf. The amount of O_3 taken up is controlled by stomatal conductance. Antioxidants and repair processes can compensate for the deleterious effects of O_3 uptake at the expense of assimilate export. After sufficient O_3 uptake and/or at sufficiently high rate of uptake photosynthesis, conductance, and respiration can be affected.

leaf production and senescence), and photosynthetic capacity of leaves. Ozone can also affect pollen viability, production of flower buds, and phloem function.

The response of plants to air pollutants often involves several processes, making predictions difficult. Models of plant growth have been developed which attempt to project the known physiological effects of air pollution stress over time on plant growth (51,52). These models are especially useful in assessing the pollutant effects on trees which will likely experience several stresses at different times over their life span. A conceptual model of plant response to O_3 uptake is presented in Figure 3. Models of plant response to air pollution must address issues of uptake, direct effects, indirect effects, and integrated response. A number of models have emphasized the pathway of uptake, using the electrical analog (53,58) either explicitly or implicitly.

Movement of an air pollutant such as O_3 involves bulk transport in the air, transport by convection within the canopy, and diffusion across the leaf boundary layer and through the stomates. Modeling of the consequences of O_3 uptake is not as well developed primarily because the mode of action of O_3 is complex, involving a number of interactions, and is not well described quantitatively. Leaving aside the initial events in O_3 interaction (e.g., membrane damage), the first measurable whole-plant effect frequently is a reduction in overall photosynthetic capacity through loss of leaves (accelerated senescence and/or decreased production) and/or through reduction in photosynthetic capacity of the leaves. In addition, the respiratory rate may increase as a result of repair respiration (59,60). These changes lead to a reduction in the amount of total nonstructural carbohydrate (TNC) available for growth of other organs, especially roots (61). Simulation models have recently been developed for describing the effects of O_3 on plant function. Both TREGRO (tree growth model; 51) and PGSM (plant growth simulation model; 52) were developed specifically to simulate the response of plants, especially trees, to O_3 and acid rain exposure.

DIRECT EFFECTS

The site of action of a pollutant varies as a function of the pollutant and of the plant organ, e.g., heavy metals and acid precipitation can affect the root whereas SO_2, NOx, and O_3 affect the shoot. Direct effects can elicit indirect responses in the plant, i.e., response in a nontarget organ. The magnitude of the direct response will be a function of external and internal factors that limit the amount of pollutant reaching the target.

A basic assumption in the analysis of plant response to air pollution is that the pollutant must enter the leaf and interact with the mesophyll cells and/or the guard cells. Effects on cuticle integrity have been studied in relation to acid rain; however, these effects do not appear to have primary importance in the reaction of leaves to O_3 (62). In the following discussion, the processes directly involved in pollutant response will be divided into two categories: (a) the relationship between exposure and uptake, and (b) metabolic effects.

Uptake

The response of vascular plants to O_3 may be viewed as the culmination of a sequence of physical, biochemical, and physiological events. Ozone in the ambient air does not impair plant processes or performance, only the O_3 that diffuses into the plant. An effect will occur only if a sufficient amount of O_3 reaches the sensitive cellular sites within the leaf. The O_3 diffuses from the ambient air into the leaf through the stomata, which can exert some control on O_3 uptake to the active sites within the leaf. Ozone injury will not occur if (1) the rate of O_3 uptake is sufficiently small that the plant is able to detoxify or metabolize O_3 or its metabolites; or (2) the plant is able to repair or compensate for the O_3 impacts (38).

Movement of O_3 from the air to the leaf is controlled by mixing in the bulk atmosphere and by boundary layer conductance of the canopy and around the leaf. Several papers have discussed the theory and measurement of these components of gas movement (e.g., 63). In general, limitations to O_3 uptake imposed by canopy and leaf boundary layer conductance decrease as wind speed increases and the leaf becomes more coupled to the bulk atmosphere (64). The density distribution of leaves within the canopy and roughness of the canopy also influence the reduction in wind speed with increasing depth into the canopy. Once an O_3 molecule has reached the leaf surface it can interact with the cuticle (62) or enter the leaf through an open stomate. Cuticular conductance is quite low (a few orders of magnitude smaller than stomatal conductance) for the major air pollutants and is generally neglected in the analysis of pollutant uptake (65). Stomatal conductance, which can vary between 0 and 0.4 mol/m^2/sec (0–0.01 m/sec), is the primary means by which a plant can regulate uptake (and loss) of gases. Conductances associated with the internal air space and the cell walls are frequently neglected because they are not usually limiting (i.e., they are large). Stomatal conductance also varies diurnally as a function of environmental conditions, plant water relations, etc., thereby affecting O_3 uptake. Fowler et al. (66; also see 46) argue that, given the low deposition velocity of O_3 to leaf surfaces, stomatal conductance is the dominant factor controlling O_3 uptake in forest canopies. However, the amount of O_3 taken up (effective or internal dose) is a complex function of the O_3 exposure profile, meteorological conditions, and plant stress level (67).

Movement of O_3 within the leaf is controlled by the geometry of the mesophyll air space (and its associated conductance) plus the thickness of the mesophyll cell wall (68). The initial reaction of O_3 is apparently quite rapid with the result that the kinetics of these reactions (or mesophyll conductance) can be neglected for practical purposes. This assumption is supported by recent evidence that indicates that internal O_3 concentration is at or near zero (69). A crucial component of this analysis is the ratio of the diffusivities of H_2O and O_3. In earlier work (e.g., 56) the ratio was assumed to be proportional to the square root of the ratio of the molecular weights ($\sqrt{48/18} = 1.63$); however, theoretical considerations indicate that this is not the case. Since no direct measure of this ratio has been made, Laisk et al. (69) used an equation of Chem and Othmer (70) to calculate the ratio as 1.68. Using conductance (boundary layer and stomatal) along with the appropriate correction for molecular diffusivity should provide a good means of simulating O_3 uptake.

Early studies of the effect of O_3 on leaf gas exchange found evidence for a direct effect of O_3 on stomatal conductance. At high concentrations (>250 ppb) leaf conductance in some species declined (71–76). Also, long-term exposure to lower concentrations of O_3 eventually produce a reduction in leaf conductance. These results led to the conclusion that the initial effect of O_3 was on the functioning of the stomates. More recent data, however, have shown that calculated internal CO_2 concentration does not decline with O_3-induced reduction in photosynthesis, an unexpected response if stomates were directly affected (77–79). This response to O_3 is consistent with an effect on mesophyll metabolism, especially photosynthesis, leading to a reduction in net CO_2 uptake which in turn reduces stomatal conductance through an increase in intercellular CO_2 (80). However, it must be pointed out that the relative importance of stomatal vs. mesophyll response is difficult to determine because the evidence is indirect, depending on calculation of internal CO_2 concentrations.

Concurrence of other stresses, such as water stress, will affect response to air pollutants, primarily through reduced stomatal conductance. With the rise in atmospheric CO_2 stomatal conductance will decrease, thereby reducing O_3 uptake. However, increasing temperature will likely induce increased stomatal conductance as well as increased production of tropospheric O_3 (81). The net result on O_3 uptake will depend on the relative strengths and interactions of the various stresses.

Metabolism

While O_3 uptake, and not the external exposure, determines the response of the plant to O_3, very few papers provide the necessary data on concurrent O_3 and leaf conductance for calculating uptake. The quantitative relationship between O_3 uptake and response has not been clearly established; however, it appears that those plants with higher leaf conductance, and therefore greater O_3 uptake, are more sensitive (57). Additional factors, e.g., detoxification mechanisms, are also likely to be important. Reports of thresholds (82) of O_3 exposure and/or uptake may be the result of either or both of these mechanisms. As noted by Cape and Unsworth (83), "If plants have detoxification mechanisms which are kinetically limited, the rate of uptake may be important, so that even an integrated absorbed dose may be insufficient to account for observed responses."

Being a strong oxidant, O_3 can and will react with a number of compounds within the cell, including antioxidants (84–86) and unsaturated lipids (87). Ozone can also enter into reactions that produce various reactive compounds such as peroxides and free radicals which in turn can affect cell function. Both in vivo and in vitro techniques have been used to study these interactions of O_3 with plant cells (87,88). While it is possible to show that O_3 affects a variety of molecular components of a cell, two factors make it difficult to develop a quantitative biochemical model of the response of plant metabolism to O_3. First, O_3 interacts with a range of compounds within the cell, i.e., it is nonspecific, although membranes appear to be a major site of action. Second, unlike some gaseous pollutants (e.g., NOx and SO_2), the fate of O_3 once in the cell cannot be followed directly because O_3 cannot be labeled in such a way as to unequivocally follow its fate.

The consequences of O_3 uptake at the biochemical level are potentially quite diverse and depend on the tissue involved. One frequently observed effect is an increase in leakiness of membranes, especially the plasmalemma (89). Changes in membrane function affect a wide range of cell processes, through loss of metabolites and breakdown in compartmentation, and can lead to an increase in tissue senescence. Exposure of leaves to O_3 leads to reduction in photosynthetic capacity (90), stimulation of respiration (59), and accelerated senescence (91,92). Ozone levels found in most environments do not appear to reduce photosynthetic rates immediately but require exposures of several days to weeks to elicit a response. It is not clear that O_3 can penetrate directly to the chloroplast; however, changes in chloroplast function are clearly a result of O_3 exposure. Ribulose-1,5-bisphosphate carboxylase, the enzyme responsible for the primary fixation of carbon in most plants, decreased after O_3 exposure (93). Changes such as these in biochemical functioning are reflected at higher levels of organization by decreased carbon fixation, increased respiration, and increased senescence.

Other Physiological Responses to Ozone

Production of mature fruit has been found to be disproportionately affected by air pollutants (see below). One probable reason for this response is the reduced availability of TNC through a reduction in integrated canopy photosynthesis and/or increased respiratory demand. However, more direct effects on reproductive processes are also possible (see Ref. 94). While it is not possible to simulate quantitatively these direct effects because a quantitative description is lacking, sufficient data exist to provide an idea of the effects of air pollutants on pollen viability, flowering, and seed production. The timing of pollutant exposure relative to pollination, flower bud initiation, fruit development, and so forth is likely to be crucial both for the vulnerability of the process to direct effect of O_3 and for the availability of TNC.

Pollen is probably the most vulnerable of all plant tissues to O_3 exposure, since it must be in direct contact with the ambient air during dispersal. Several in vitro studies have shown that pollen germination is sensitive to O_3 concentrations ranging from 0.06 to 1 ppm (95–98). Data on in vivo effects are more limited and to some extent contradictory to the in vitro studies (99–101).

Some work in the 1930s and 1950s indicates that air pollutants can affect the hormonal balance in plants. Went (102) reported on the effect of photochemical smog, of which O_3 is a major component, on a widely used growth regulator bioassay. In Utrecht, the Netherlands, Kögl et al. (103) found daily and seasonal fluctuations in the *Avena* coleoptile test for the growth regulator auxin. These fluctuations could not be ascribed to any known diurnal or seasonal periodicity. Similar data were obtained at the phytotron at the California Institute of Technology (Pasadena). Hull and Went (104) reported that these fluctuations were apparently connected with air pollution because the fluctuations disappeared not only in the *Avena* coleoptile test but also in the dwarf pea test for gibberellins, when charcoal-filtered air was used.

Recent work has shown an interaction between O_3 exposure and abscisic acid, primarily through variation in stomatal conductance (105–108). In addition, changes in ethylene production have been associated with varied O_3 exposure (109,110). While it is not yet possible to incorporate these responses into mathematical models, these results show that this area requires more study if we are to understand whole-plant response to air pollutants.

ORGANISMIC RESPONSE

Photosynthesis and allocation processes are sensitive indicators of physiological stress and reflect the relationship between pollutant sensitivity of plants and the availability of energy (carbon) for compensation and repair processes (43,111). Carbon is allocated throughout the plant on the basis of source sink patterns that vary seasonally and ontogenetically. Waring and Schlesinger (112) and Waring (113) provide a conceptual priority scheme for competing sinks within a tree, using relative strength (metabolic activity) of various carbon sinks. Canopy tissues have high priority for fixed carbon, whereas bole cambium, carbohydrate storage pools, and secondary metabolites (e.g., defense compounds) are lower in priority. Thus, bole growth receives carbohydrate after canopy tissue needs are met, suggesting that bole growth will be decreased before canopy growth under carbohydrate-limiting conditions. Frequently, reproductive structure are the lowest priority, leading to disproportionately large losses in fruit production (22–24). While these generalizations are fraught with exceptions, understanding "typical" allocation patterns and how these shift under stress provides a useful tool for studying whole-plant and ecosystem response to stress. Disruption of typical allocation processes by a pollutant can indicate the direction and magnitude of effects of that stress on other processes. In addition, studying carbon physiology offers the opportunity to observe how the plant integrates concurrent stresses in a whole-plant context.

Ozone causes both direct and indirect responses in plants, providing a good example of changing patterns of carbon distribution and utilization in plants under stress. As noted above, the magnitude of O_3 stress is determined by several factors, including (a) uptake, which is controlled by concentration and stomatal conductance, and (b) metabolic responses to O_3 uptake, including membrane repair, synthesis of antioxidant compounds (84), reduced photosynthesis, and premature senescence. Decreased photosynthate production (e.g., reduced photosynthesis or premature senescence) and increased carbon demand in the shoot (due to antioxidant synthesis, membrane repair and replacement) reduce carbon availability for allocation to other plant organs.

Disruption of growth or metabolism in nontarget organs, such as roots or stems in the case of O_3, is an indirect effect. If root growth is reduced, the plant may be predisposed to other stresses, such as moisture or nutrient stress. It is through this process that O_3 impacts growth and metabolism at the whole-plant level. Eventually the plant will establish a new balance between root and shoot function in response to the stress; however, homeostasis is probably never reached due to ever-changing environmental conditions (114). Although physiological factors besides carbohydrate allocation are involved in metabolic adjustments following O_3 exposure, carbon plays the central role in maintaining metabolic function.

Canopy Dynamics: Leaf/Needle Production and Senescence

Net carbon gain by a plant is an integration of (a) the range of photosynthetic capacities and conductance of the leaves, (b) the distribution of the leaves in space, (c) the dynamics of leaf production and senescence, and (d) environmental conditions. In species with indeterminate flushing, e.g., aspen and poplar, leaf production and loss can occur simultaneously for at least part of the growing season. In species with determinant flushing, e.g., pine, leaf production and loss are separated in time, at least within a given flush.

In aspen and poplar (*Populus* spp.), one of the early effects of O_3 exposure is an increase in the loss of older leaves (91,115). At low O_3 exposures, the increase in leaf loss can be at least partially compensated by increased leaf production; however, at high O_3 exposures leaf senescence rate increases relative to production, leading to a decline in canopy density. The combination of these responses produces a delayed reduction in the total canopy density but at the expense of resources used to produce the new leaves. Part of the stimulus for leaf production could be the increased availability of mineral nutrients through the export from senescing leaves (91). As O_3 exposure intensity increases, however, leaf production rate decreases and total canopy density decreases. Thus, O_3 increases canopy turnover at low concentrations but decreases canopy density through increased senescence and decreased production at higher O_3 exposures.

In pine, needle production and senescence are separated in time for a given flush. The number of flushes produced per year depends on the species and on the growing conditions. In ponderosa pine (*Pinus ponderosa* Laws.), one to two flushes are produced per year, generally with the second flush having fewer needles. Increased O_3 exposure leads to increased needle loss toward the end of the O_3 exposure (115), with needle loss continuing after exposure has ended. Stowe et al. (92) and Kress et al. (116) found a similar pattern in loblolly pine (*Pinus taeda* L.).

It is not clear as to whether acceleration of senescence is simply due to the induction of normal senescence, induction of a pathological response, or reduction in photosynthesis relative to respiration. Since one effect of O_3 is to produce leaky membranes, it is possible that the senescence response is not directly related to the response of photosynthesis. In any case, the effect of leaf loss will be a reduction in TNC available for the plant (61).

Below-Ground Processes: Roots and Mycorrhizae

Air pollution can affect roots and mycorrhizal associations by at least two mechanisms: (a) direct effects after deposition of contaminants to the soil and/or alteration of soil chemistry; (b) effects on the shoot, which indirectly alters root metabolism. It is difficult to identify the mechanisms involved in indirect responses, although carbohydrate availability for root metabolism is at least partially involved.

Heavy metals can be directly deposited to the soil or their availability can be altered through changes in soil chemistry resulting from pollutant deposition (117). For example, aluminum, a trivalent ion that becomes more available at low soil pH values, can impair cell division, increase cell wall rigidity, decrease root respiration, precipitate nucleic acids, and interfere with the uptake and transport of Ca, Mg, P, and Fe (118). However, fungal hyphae are capable of sequestering some heavy metals, so that mycorrhizal fungi may serve to reduce movement of heavy metals into host roots (119–121). Detoxification mechanisms in the fungus and plant root may be able to reduce the effects of toxic pollutants at the expense of carbohydrate from the shoot. Thus, stresses that reduce carbon transport to roots may impair the plant's ability to sequester toxic compounds at the site of uptake in the root.

Root growth and carbohydrate concentrations are often decreased as a result of O_3 exposure (61,122–132). Root starch was reduced in ponderosa pine by the end of one season of O_3 exposure (128). Coarse and fine root starch concentrations were also decreased by exposure to O_3 and were lower the following spring during shoot flush (61). Starch concentrations in O_3-treated seedlings were associated with suppressed growth of new roots in the spring following exposure. Decreases in root carbohydrates under O_3 stress result at least in part from decreased carbon allocation to roots (129,133,134).

The impact of pollutant stresses such as O_3 on allocation of carbon to roots and other plant organs suggests that carbon allocation patterns can serve as a diagnostic tool in the study of pollutant stress at the whole-plant level. The physiological mechanisms underlying altered patterns of allocation under pollutant stress are unknown in most cases. Some proposed pollutant stresses include decreased phloem loading (135), increased utilization of photosynthate in the leaves of affected plants for maintenance and repair processes (59,60), and decreased photosynthesis.

Climate Change

In future climate scenarios, elevated CO_2 concentrations may be associated with increased O_3 levels (81), with direct and interacting consequences in the shoot. However, the indirect effects of these stresses on roots are opposite, i.e., O_3 reduces and CO_2 increases carbon allocation to roots (133,136). Depending on the magnitude of each stress, O_3 stress may be partially ameliorated by increased CO_2. The degree to which increased CO_2 ameliorates the reduction in carbon allocation to roots under O_3 stress can be determined experimentally, which could provide information on plant integration of these stresses. Increased temperature is a third stress that is likely to co-occur with elevated CO_2 and O_3. In sum, an understanding of carbon allocation patterns provides a tool for studying and predicting the net effects of these and other stresses on plant growth.

ECOSYSTEM CONSEQUENCES

Effects of Air Pollutants on Agricultural Crops

The effects of air pollutants, as discussed above, can occur at several organizational levels ranging from molecular to organismal. If the molecular/cellular disturbances induced by the air pollutants are not repaired or compensated, effects are ultimately expressed as leaf injury, decreased growth and yield, changes in crop quality, alterations in susceptibility to abiotic and biotic stresses, and decreased reproduction.

Early studies that were conducted in Europe in the late 1800s and early 1990s were designed to assess the impacts of air pollutants (i.e., SO_2 or HF) that were emitted from various industrial sources and affected agricultural crops and forests. These studies established that air pollutants injured vegetation inducing foliar lesions and decreasing growth, developed criteria that could be used to identify foliar necrosis, and provided a tentative identification, based on symptomatology, of the probable causative pollutant. Subsequent studies developed quantitative relationships between pollutant exposure and crop yields, and established that growth of a wide range of crops could be reduced by air pollutants.

In the 1950s and 1960s researchers assessed the impact of O_3 on crop yield under field conditions by surveying foliar injury symptoms (137,138). However, injury symptoms and crop yield are usually not directly proportional due to the importance of various allocation processes and metabolic factors in determining plant yield. For example, significant yield loss can occur with little or no foliar injury (22–24). Also, foliar injury can be much greater than yield loss (139,140).

A number of studies in the United States have demonstrated that ambient O_3 concentrations can be sufficiently high to impair the yields and quality of a number of crops, including citrus, grape, tobacco, cotton, tomato, bean, soybean, peanut, onion, and potato (38). Controlled field studies have been conducted to determine the quantitative relationship between O_3 exposure and crop yields. These data have been used to estimate special crop loss and the economic losses from ambient O_3.

Yield loss is defined as a reduction in the marketable portion of the crop. For example, yield loss in corn is the consequence of reduced cob size, a decrease in the number of filled kernels, and reduced pollination, as indicated by darkened tips of the cobs in the higher O_3 treatments (Fig. 4A). The relationship between O_3 exposure and crop yield loss is nonlinear. Examples of this relationship between O_3 exposures and plant yield are shown in Figures 4B–D. These species and cultivars were selected to illustrate the response of several species/cultivars to O_3 as well as the type of year-to-year variation in plant response that may occur. Although all crops displayed a nonlinear relation, different crop species/ cultivars displayed differential susceptibilities to O_3. There is also year-to-year variability in crop susceptibility since the O_3 effect is also modified by climatic (e.g., temperature, relative humidity) and edaphic factors (e.g., soil water availability).

Estimated Crop Losses

The crop loss functions developed in controlled field exposures have been combined with information on the geographic distribution of ambient O_3 exposure and crop growing areas to estimate crop loss (43,141). The projections are based on (a) mathematical functions relating O_3 exposure to crop yield loss, (b) information on the geographic distribution of crops (142), and (c) distribution of monitored ambient O_3 exposures (Fig. 2).

An example of the spatial distribution of predicted soybean and wheat yield loss, using 1987 ambient air quality data, is shown as coded circles in the top maps of Figures 5 and 6. To estimate predicted relative yield loss values in nonmonitored areas, a distance weighting function was used to interpolate the predicted relative yield loss at the rural monitoring sites to all nonmonitored areas to produce a gradient surface over the study region. The interpolated gradient surface from these data are shown in the bottom maps of Figures 5 and 6. The gradient surface was constrained by the crop distribution data so that predicted losses were shown only for the principle areas where the crop was cultivated.

The predicted relative yield losses, using current (1982–1987) air quality, indicated that significant crop losses are expected to occur over the soybean and wheat growing regions (Figures 5 and 6). For both

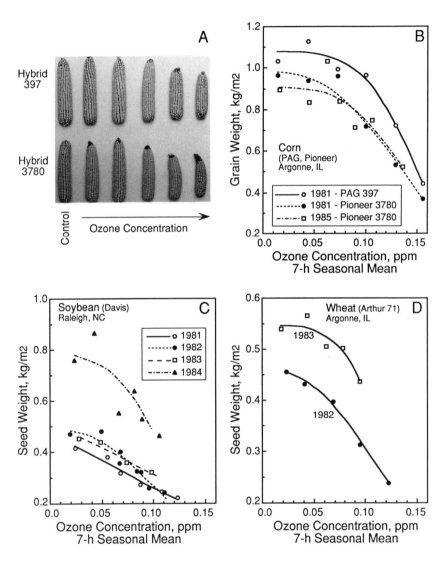

Figure 4 Examples of the effects of O_3 on the yield of several agricultural crops/cultivars. Part A illustrates how O_3 exposure changes the size and appearance of corn contributing to yield loss. Grain yield for corn (B), soybean (C), and wheat (D) as a function of ozone exposure. The O_3 concentrations are expressed as 7-hr seasonal means. The crops functions were selected to show the differences in response among species, years, and cultivars. The Weibull function was used to derive the nonlinear yield loss functions.

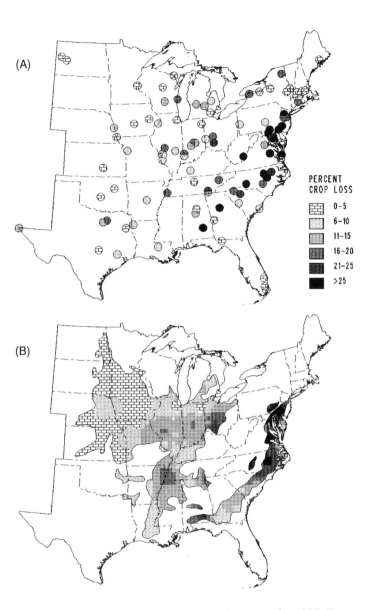

Figure 5 Predicted relative soybean yield loss (%) for 1987. Estimates are based on measured air quality and a composite soybean exposure–response function for O_3. The circles in top figure illustrate the estimated yield loss at each of the rural monitoring sites used to create the gradient surface. The gradient surface in bottom figure was interpolated from the monitoring locations using an inverse distance weighting method. The estimated crop losses are likely to be underestimated as the ambient air quality data were not adjusted for missing values. If the corrections had been made the measured air quality (SUM06) might have been larger at the various monitoring sites.

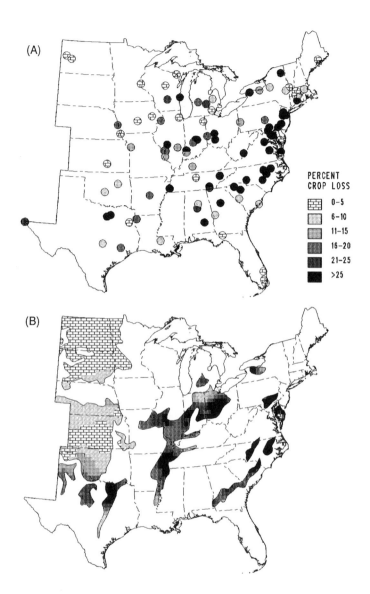

Figure 6 Predicted relative wheat yield loss (%) for 1987. Estimates are based on measured air quality and a composite wheat exposure–response function for O_3. The circles in top figure illustrate the estimated yield at each of the rural monitoring sites used to create the gradient surface. The gradient surface bottom figure was interpolated from the monitoring locations using an inverse distance weighting method. Although wheat is grown in other states, these states were omitted from the analysis because there were too few air monitoring sites to permit reliable interpolation. The estimated crop losses are likely to be underestimated as the ambient air quality data were not adjusted for missing values. If the corrections had been made the measured air quality (SUM06) might have been larger at the various monitoring sites.

crops, predicted relative yield losses ranged from 0 to >25%. The area-weighted yield losses for soybean and wheat (averaged over the 6 years) are 9.3% and 12.7%, respectively. Annual soybean yield losses for 1982–1987 ranged from 6.9% to 14.8% (Fig. 7). The annual yield losses of wheat for 1982–1987 ranged from 9% to 17% (Fig. 7). Largest annual yield losses, for the period 1982–1987, were observed in 1983. Differential changes in air quality across the crop growing areas are reflected in a trend of increasing soybean losses but decreasing wheat losses since 1983.

For both crops, localized areas of high loss were predicted for areas around the Chesapeake Bay and portions of the Ohio and Mississippi river valleys (Figures 5 and 6). For soybean, yield losses of 5–10% were forecast for 28% of the area while yield losses of 20% or greater were anticipated on less than 11% of the area. Similarly, for wheat, yield losses of 5–10% were forecast for 18% of the area while yield losses of 20% or greater were anticipated on 27% of the area.

Economic Impacts of O_3 on Agricultural Crops

Assessments of O_3-induced damage to agriculture found in the literature range from descriptions of procedures for simple monetary calculations of economic losses to more complex economic assessment methodologies (38,143–146). In the United States the annual economic costs of ambient

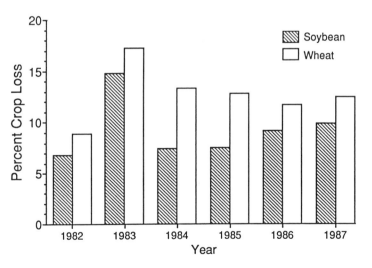

Figure 7 Annual area-weighted average yield loss for soybean and wheat over a 6-year period. Percentages were derived from spatially distributed data as shown in Figures 5 and 6.

O_3 on major agricultural crops in the United States range from \$1.3 to \$2.8 billion (Table 1). It is difficult to directly compare the results of the various studies as the crops, exposure–response information, environmental conditions, and assumed economic conditions differ considerably among studies. Although the conditions under which various studies were conducted may differ, results from these assessments illustrate that ambient O_3 levels impose significant costs on society (both producer and consumer).

Forests

Identifying the effects of a specific air pollutant on plants in a natural ecosystem is difficult due to the multitude of stresses affecting growth. Plants also have varying degrees of susceptibility to air pollution stress as they develop phenomenologically and ontogenetically. Perennial plants, in comparison to annuals, are repeatedly subjected to stresses year after year, and reductions in growth are often compounded over the life of the plant. Consequently, shifts in the competitive structure of ecosystems can result from chronic exposure to air pollution stress.

   Stresses that co-occur can have offsetting, additive, or multiplicative effects on plant response. It is probably rare to have a single, limiting stress that controls plant growth over the course of a season. The presence of concurrent stresses, each potentially affecting similar physiological plant processes, makes it difficult to separate the effects of one stress from another.

Table 1   Examples of Economic Assessments of the Impact of O_3 on U.S. Agriculture that Were Based on the NCLAN Database[a]

Region	Crops	Consumer benefit	Producer benefit	Total benefit
U.S.	Corn, cotton, soybeans	NR[b]	NR	2220
U.S.	Corn, cotton, soybeans, wheat	NR	NR	2400
U.S.	6 crops	1160	550	1770
U.S.	5 crops	NR	NR	1300
U.S.[c]	6 crops	1047	1590	2637
U.S.[d]	14 crops	683	2097	2780

Source: Derived from Refs. 38 and 144.
[a] Losses are expressed as millions of U.S. dollars (1980).
[b] NR, not reported.
[c] The assessment included an adjustment for moisture stress.
[d] The assessment included an adjustment for moisture stress and is expressed in 1982 dollars.

In natural ecosystems, the concept of predisposing factors and interacting stresses controlling growth is more realistic than the possibility that a single stress controls growth. While the debate continues on the role of air pollutants on forest growth and productivity, there is at least one example (southern California) where the combination of natural and anthropogenic stresses have acted in concert to reduce tree growth at the stand level. The result of the combined stresses was to change the competitive structure of the forest, including a change in dominant species in severely stressed areas.

In the San Bernardino Mountains of southern California trees have been exposed to high levels of smog containing O_3 from the Los Angeles basis since the 1950s (147,148). Visual damage on ponderosa pine was linked to the poor air quality in the region. The aspect of this work that makes it particularly noteworthy is the comprehensive experimental approach that was taken to evaluate the role of air pollution stress in these natural ecosystems. The study design addressed effects of the pollutant stress at several levels, including the physiological response at the plant level, and how growth reductions impacted stand and ecosystem structure and function.

Researchers found that ponderosa pine was particularly sensitive to O_3 concentrations in the smog. Injury included foliar injury, premature senescence, and lower nutrient retention in the foliage. Premature senescence reduced whole-plant photosynthetic capacity and radial growth. The trees weakened by O_3 stress were also more susceptible to root rot (*Fomes annosus*) and pine beetles (*Dendroctonus brevicomis*), which were the dominant stresses that resulted in tree mortality. As mortality increased, the understory composition changed to include O_3-tolerant shrub and oak species. The ecosystem structure was altered, including water, carbon, and nutrient fluxes, in response to elevated levels of O_3 and the predisposition of the dominant species to other stresses.

CONCLUDING REMARKS

Air pollutants have a negative impact on plant growth primarily through interfering with resource accumulation. Since leaves are in close contact with the atmosphere, many air pollutants, such as O_3 and NO_x, affect the metabolic function of the leaves and interfere with net carbon fixation by the plant canopy. Air pollutants that are first deposited on the soil, such as heavy metals, first affect the functioning of roots and interfere with soil resource capture by the plant. These reductions in resource capture (production of carbohydrate through photosynthesis, mineral nutrient and water uptake from the soil) will affect plant growth through changes in resource allocation to the various plant structures. When air pollution stress co-occurs with other stresses, e.g., water stress, the outcome on growth will depend on a complex interaction of processes within the plant. At the ecosystem level, air pollution can shift the competitive balance among the species present and may lead to changes in the composition of the plant community. In agroecosystems, these changes may be manifest in reduced economic yield.

REFERENCES

1. L. C. McCabe (1961): *Air Pollution*, Columbia University Press, New York, pp. 39–47.
2. R. Guderian, D. T. Tingey, and R. Rube (1985): *Air Pollution by Photochemical Oxidants: Formation, Transport, Control, and Effects on Plants* (R. Guderian, ed.), Springer-Verlag, Berlin, pp. 120–333.
3. D. T. Tingey, W. E. Hogsett, and S. Henderson (1990): *J. Environ. Qual.*, *19*:635–639.
4. K. Garber (1967): *Luftverunreinigung und ihre Wirkungen*, Gebrüder Borntraeger, Berlin.
5. J. A. Stöckhardt (1850): *Polyt. Centr. Bl.*, p. 257.
6. M. D. Thomas and R. H. Hendricks (1956): *Air Pollution Handbook* (P. L. Magill, F. R. Holden, and C. Ackley, eds.), New York.
7. M. D. Thomas (1960): *Air Pollution*, Columbia University Press, New York, pp. 233–278.
8. M. Katz and F. E. Lathe (eds.) (1939): *Effects of Sulfur Dioxide on Vegetation*, National Research Council of Canada, Ottawa.
9. R. Guderian (1977): *Air Pollution, Phytotoxicity of Acidic Gases and Its Significance in Air Pollution Control*, Springer-Verlag.
10. E. J. Catcott (1960): *Air Pollution*, Columbia University Press, New York, pp. 221–231.
11. U.S. Environmental Protection Agency (1982): *Air Quality Criteria for Oxides of Nitrogen*, EPA-600/8-82-026, Environmental Criteria and Assessment Office, Research Triangle Park, NC.
12. U.S. Environmental Protection Agency (1982): *Air Quality Criteria for Particulate Matter and Sulfur Oxides*, EPA-600/8-82-029, Evironmental Criteria and Assessment Office, Research Triangle Park, NC.
13. U.S. Environmental Protection Agency (1988): *Review of the National Ambient Air Quality Standards for Ozone. Assessment of Scientific and Technical Information*, OAQPS Draft Staff Paper, Air Quality Management Division, Office of Air Quality Planning and Standards, Research Triangle Park, NC.
14. K. H. Becker, W. Fricke, J. Löbel, and U. Schurath (1985): *Air Pollution by Photochemical Oxidants* (R. Guderian, ed.), Springer-Verlag, Berlin, pp. 1–125.
15. W. W. Heck, O. C. Taylor, R. Adams, G. Bingham, J. Miller, E. Preston, and L. Weinstein (1982): *J. Air Pollut. Cont. Assoc.*, *32*:353–361.

16. P. Greenfelt (ed.) (1984): *The Evaluation and Assessment of the Effects of Photochemical Oxidants on Human Health, Agricultural Crops, Forestry, Materials and Visibility*, Swedish Environmental Research Institute, Göteborg, Sweden (IVL-EM 1570).

17. J. J. Jacobson and A. C. Hill (eds.) (1970): *Recognition of Air Pollution Injury to Vegetation: A Pictorial Atlas*, Air Pollution Control Association, Pittsburgh, PA.

18. H. van Haut and H. Stratmann (1970): *Farbtafelatlas über Schwefeldioxid-Wirkungen an Pflanzen (Color-Plate Atlas of the Effects of Sulfur Dioxide on Plants)*, Verlag W. Giardet, Essen, p. 206.

19. C. R. Thompson, D. M. Olszyk, G. Katz, A. Bytnerowicz, P. J. Dawson, and J. Wolf (1984): *Air Pollution Injury on Plants of the Mojave Desert*, Statewide Air Pollution Research Center, University of California, Riverside.

20. S. S. Malhotra and R. A. Blauel (1990): *Diagnosis of Air Pollutant and Natural Stress Symptoms on Forest Vegetation in Western Canada*, Northern Forest Research Centre, Edmonton, Alberta, Canada, Information Report NOR-X-228.

21. H. J. M. Skelly, D. D. Davis, W. Merrill, E. A. Cameron, H. D. Brown, D. B. Drummond, and L. S. Dochinger (1990): *Diagnosing Injury to Eastern Forest Trees*, Agricultural Information Service, Pennsylvania State University, State College, PA.

22. W. A. Feder and F. J. Campbell (1968): *Phytopathology*, 58:1038–1039.

23. D. T. Tingey, W. W. Heck, and R. A. Reinert (1971): *J. Am. Soc. Hort. Sci.*, 96:369–371.

24. N. O. Adedipe, R. E. Barrett, and D. P. Ormrod (1972): *J. Am. Soc. Hort. Sci.*, 97:341–345.

25. D. T. Tingey (1989): *Biological Markers of Air Pollution Stress and Damage in Forests*, National Academy Press, Washington, DC, pp. 73–80.

26. J. C. Brandt (translator) (1987): *Acidic Precipitation*. VDI-Kommision Reinhaltung der Luft. Verein Deutscher Ingenieure, Düsseldorf.

27. W. W. Heck and C. S. Brandt (1977): *Air Pollution*, Vol. 3 (3rd ed.) (A. C. Stern, ed.), Academic Press, New York, pp. 157–229.

28. A. Wellburn (1988): *Air Pollution and Acid Rain*, Longman, Essex.

29. M. H. Unsworth and D. P. Ormrod (eds.) (1982): *Effects of Gaseous Air Pollution in Agriculture and Horticulture*, Butterworths, London.

30. S. Schulte-Hostede, N. M. Darrall, L. W. Blank, and A. R. Wellburn (eds.) (1987): *Air Pollution and Plant Metabolism*, Elsevier, London.

31. R. D. Nobel, J. L. Martin, and K. F. Jensen (eds.) (1989): *Proceedings of the Second US–USSR Symposium on Air Polution Effects on Vegetation Including Forest Ecosystems*, Northeastern Forest Exp. Sta.

32. W. W. Heck, O. C. Taylor, and D. T. Tingey (1988): *Assessment of Crop Loss from Air Pollutants*, Elsevier, London.

33. F. Scholz, H.-R. Gregorius, and D. Rudin (eds.) (1989): *Genetic Effects of Air Pollutants in Forest Tree Populations*, Proceedings of the Joint Meeting of the IUFRO Working Parties: Genetic Aspects of Air Pollution, Popula-

tion and Ecological Genetics, and Biochemical Genetics, Spring-Verlag, Berlin.

34. E.-D. Schulze, O. L. Lange, and R. Oren (1989): *Forest Decline and Air Pollution: A Study of Spruce* (Picea abies) *on Acid Soils*, Spring-Verlag, Berlin.

35. W. E. Winner, H. A. Mooney, and R. A. Goldstein (eds.) (1985): *Sulfur Dioxide and Vegetation: Physiology, Ecology, and Policy Issues*, Stanford University Press, Stanford, CA.

36. G. E. Taylor, Jr., L. F. Pitelka, and M. T. Clegg (1981): *Ecological Genetics and Air Pollution*, Springer-Verlag, New York.

37. J. R. Barker and D. T. Tingey (eds.) (1992): *Air Pollution Effects on Biodiversity*, Van Nostrand Reinhold, New York.

38. U.S. Environmental Protection Agency (1986): *Air Quality Criteria for Ozone and Other Photochemical Oxidants*, EPA-600/8-84-020, Environmental Criteria and Assessment Office, Research Triangle Park, NC.

39. W. E. Hogsett, D. T. Tingey, and E. H. Lee (1988): *Assessment of Crop Loss from Air Pollutants* (W. W. Heck, O. C. Taylor, and D. T. Tingey, eds.), Elsevier, London, pp. 107–138.

40. E. H. Lee, D. T. Tingey, and W. E. Hogsett (1988): *Environ. Pollut.*, 53:43–62.

41. E. H. Lee, W. E. Hogsett, and D. T. Tingey (1991): *Transactions: Tropospheric Ozone and the Environment* (R. L. Berglund, D. R. Lawson, and D. J. McKee, eds.), Air and Waste Management Association, Pittsburgh, PA, pp. 225–271.

42. D. T. Tingey and G. E. Taylor, Jr. (1982): *Effects of Gaseous Air Pollution in Agriculture and Horticulture* (M. H. Unsworth and D. P. Ormrod, eds.), Butterworths, London, pp. 113–138.

43. D. T. Tingey and C. P. Andersen (1991): *Ecological Genetics and Air Pollution* (G. E. Taylor, Jr., L. F. Pitelka, and M. T. Clegg, eds.), Springer-Verlag, New York, pp. 209–235.

44. R. W. Shaw and R. E. Munn (1971): *Introduction to the Scientific Study of Atmospheric Pollution* (B. M. McMormac, ed.), Reidel, Dordrecht, pp. 53–96.

45. A. C. Stern (ed.). (1977): *Air Pollution, 3rd ed., Vol. 1, Air Pollutants, Their Transformation and Transport*, Academic Press, New York.

46. D. Fowler (1992): *Air Pollution Effects on Biodiversity* (J. R. Barker and D. T. Tingey, eds.), Van Nostrand Reinhold, New York, pp. 31–51.

47. D. T. Tingey, W. E. Hogsett, E. H. Lee, A. A. Herstrom, and S. H. Azevedo (1991): *Transactions: Topospheric Ozone and the Environment* (R. L. Berglund, D. R. Lawson, and D. J. McKee, eds.), Air and Waste Management Association, Pittsburgh, PA, pp. 272–288.

48. A. S. Heagle, L. W. Kress, P. J. Temple, R. J. Kohut, J. E. Miller, and H. E. Heggestad (1988): *Assessment of Crop Loss from Air Pollutants* (W. W. Heck, O. C. Taylor, and D. T. Tingey, eds.), Elsevier, London, pp. 141–179.

49. M. L. Roose and A. D. Bradshaw (1982): *Effects of Gaseous Air Pollution*

in *Agriculture and Horticulture* (M. H. Unsworth and D. P. Ormrod, eds.), Butterworths, London, pp. 379-409.

50. D. T. Tingey and D. M. Olszyk (1985): *Sulfur Dioxide and Vegetation: Physiology, Ecology, and Policy Issues* (W. E. Winner, H. A. Mooney, and R. A. Goldstein, eds.), Stanford University Press, Stanford, CA, pp. 178-205.

51. D. A. Weinstein and R. Beloin (1990): *Process Modeling of Forest Growth Responses to Environmental Stress* (D. K. Dixon, R. S. Meldahl, G. A. Ruark, and W. G. Warren, eds.), Timber Press, Portland, OR, pp. 313-323.

52. C. Chen and L. E. Gomez (1990): *Process Modeling of Forest Growth Responses to Environmental Stress* (R. K. Dixon, R. S. Meldahl, G. A. Ruark, and W. G. Warren, eds.), Timber Press, Portland, OR, pp. 338-350.

53. J. H. Bennet, A. C. Hill, and D. M. Gates (1973): *J. Air Pollut. Cont. Assoc.*, 23:957-962.

54. R. A. O'Dell, M. Taheri, and R. L. Kabel (1977): *J. Air. Pollut. Cont. Assoc.*, 27:1104-1109.

55. V. J. Black and M. H. Unsworth (1979): *Nature*, 282:68-69.

56. G. E. Taylor, Jr. and D. T. Tingey (1982): *Effects of Gaseous Air Pollution in Agriculture and Horticulture* (M. H. Unsworth and D. P. Ormrod, eds.), Butterworths, London, pp. 113-138.

57. P. B. Reich (1987): *Tree Physiol.*, 3:63-91.

58. B. B. Hicks and D. R. Matt (1988): *J. Atmos. Chem.*, 6:117-131.

59. J. S. Amthor (1988): *New Phytol.*, 110:319-325.

60. J. S. Amthor and J. R. Cunningham (1988): *Can. J. Bot.*, 66:724-726.

61. C. P. Andersen, W. E. Hogsett, R. Wessling, and M. Plocher (1991): *Can. J. For. Res.*, 21:1288-1291.

62. G. Kerstiens and K. Lendzian (1989): *New Phytol.*, 112:13-19.

63. K. G. McNaughton and P. G. Jarvis (1991): *Agric. For. Meteorol.*, 54:279-302.

64. D. M. Gates (1980): *Biophysical Ecology*, Springer-Verlag, Berlin.

65. O. L. Lange, U. Heber, E.-D. Schulze, and H. Ziegler (1989): *Forest Decline and Air Pollution: A Study of Spruce* (Picea abies) *on Acid Soils* (E.-D. Schulze, O. L. Lange, and R. Oren, eds.), Springer-Verlag, Berlin, pp. 238-273.

66. D. Fowler, J. N. Cape, and M. H. Unsworth (1989): *Phil. Trans. Roy. Soc. London Series B*, 324:247-265.

67. G. E. Taylor, Jr. and P. J. Hanson (1992): *Agric. Ecosys. Environ.*, 42:255-273.

68. P. C. Harley, F. Loreto, G. Di Marco, and T. D. Sharkey (1992): *Plant Physiol.*, 98:1429-1236.

69. A. Laisk, O. Kull, and H. Moldau (1989): *Plant Physiol.*, 90:1163-1167.

70. N. H. Chem and D. F. Othmer (1962): *J. Chem. Eng. Data.*, 7:37-41.

71. A. C. Hill and N. Littlefield (1969): *Environ. Sci. Technol.*, 3:52-56.

72. D. B. Botkin, W. H. Smith, and R. W. Carlson (1971): *J. Air Pollut. Cont. Assoc.*, 21:778-780.

73. D. B. Botkin, W. H. Smith, R. W. Carlson, and T. L. Smith (1972): *Environ. Pollut.*, *3*:273–289.

74. J. H. Bennett and A. C. Hill (1973): *J. Environ. Qual.*, *2*:526–530.

75. J. H. Bennett and A. C. Hill (1974): *Air Pollution Effects of Plant Growth.* ACS Symposium Series (M. Dugger, ed.), American Chemical Society, Washington, DC, pp. 115–127.

76. V. J. Black, D. P. Ormrod, and M. H. Smith (1982): *J. Exp. Bot.*, *33*:1302–1311.

77. C. J. Atkinson, S. V. Robe, and W. E. Winner (1988): *New Phytol.*, *110*:173–184.

78. W. E. Winner, C. Gillespie, W.-S. Shen, and H. A. Monney (1988): *Air Pollution and Plant Metabolism* (S. Shulte-Hostede, N. M. Darrall, L. W. Blank, and A. R. Wellburn, eds.), Elsevier, London, pp. 255–271.

79. J. A. Weber, C. S. Scott, and W. E. Hogsett (1993): *Tree Physiol.*, *13*: 157–172.

80. G. D. Farquhar and T. D. Sharkey (1982): *Annu. Rev. Plant Physiol.*, *33*:317–345.

81. A. M. Hough and R. G. Derwent (1990): *Nature*, *344*:645–648.

82. B. D. Amiro, T. J. Gillespie, and G. W. Thurtell (1984): *Atmos. Environ.*, *18*:1207–1215.

83. J. N. Cape and M. H. Unsworth (1988): *Air Pollution and Plant Metabolism* (J. Schulte-Hostede, N. M. Darrall, L. W. Blank, and A. R. Wellburn, eds.), Elsevier, London, pp. 1–18.

84. H. Mehlhorn, G. Seufert, A. Schmidt, and K. J. Kunert (1986): *Plant Physiol.*, *82*:336–338.

85. A. S. Gutpa, R. G. Alscher, and C. McCune (1991): *Plant Physiol.*, *96*:650–655.

86. A. Hausladen, N. R. Madamanchi, S. Fellows, R. G. Alscher, and R. G. Amundson (1990): *New Phytol.*, *115*:447–458.

87. J. B. Mudd (1982): *Effects of Gaseous Air Pollution in Agriculture and Horticulture* (M. H. Unsworth and D. P. Ormrod, eds.), Butterworths, London, pp. 189–203.

88. R. L. Heath (1980): *Annu. Rev. Plant Physiol.*, *31*:395–431.

89. R. L. Heath (1987): *Advances in Phytochem.*, *21*:29–54.

90. M. M. Darrall (1989): *Plant. Cell Environ.*, *12*:1–30.

91. P. B. Reich and J. P. Lassoie (1984): *Plant. Cell Environ.*, *7*:661–668.

92. T. K. Stowe, H. L. Allen, and L. W. Kress (1992): *For. Sci.*, *38*:102–119.

93. M. S. Dann and E. J. Pell (1989): *Plant Physiol.*, *91*:427–432.

94. R. M. Cox (1992): *Air Pollution Effects on Biodiversity* (J. R. Barker and D. T. Tingey, eds.), Van Nostrand Reinhold, New York, pp. 131–158.

95. L. F. Benoit, J. M. Skelly, L. D. Moore, and L. S. Dochinger (1983): *Can. J. For. Res.*, *13*:184–187.

96. W. A. Feder (1968): *Science*, *160*:1122.

97. W. A. Feder and F. Sullivan (1969): *Phytopathology*, *59*:399.

98. R. A. Mumford, H. Lipke, D. A. Loufer, and W. A. Feder (1972): *Environ. Sci. Technol.*, *6*:427–430.

99. W. A. Feder, G. H. M. Krause, B. H. Harrison, and W. D. Riley (1982):

Effects of Gaseous Air Pollution in Agriculture and Horticulture (M. H. Unsworth and D. P. Ormrod, eds.), Butterworths, London, p. 482.

100. G. H. M. Krause, W. D. Riley, and W. A. Feder (1975): *Proc. Am. Phytopathol. Soc.*, 2:100.

101. W. J. Manning and W. A. Feder (1976): *Effects of Air Pollutants on Plants* (T. A. Mansfield, ed.), Cambridge University Press, Cambridge, pp. 47–60.

102. F. W. Went (1957): *The Experimental Control of Plant Growth*, Ronald Press, New York, pp. 309–310.

103. F. Kögl, A. J. Haagen-Smit, and C. J. van Hulssen (1936): *Z. physiol. Chemie*, 241:17–33.

104. H. M. Hull, F. W. Went, and N. Yamada (1954): *Plant Physiol.*, 29:182–187.

105. C. J. Atkinson, P. A. Wookey, and T. A. Mansfield (1991): *New Phytol.*, 117:535–541.

106. D. T. Tingey and W. E. Hogsett (1985): *Plant Physiol.*, 77:944–947.

107. G. Gupta, G. R. Sandhu, and C. Mulchi (1991): *J. Environ. Qual.*, 20:151–152.

108. R. A. Fletcher, N. O. Adepipe, and D. P. Ormrod (1972): *Can. J. Bot.*, 50:2389–2391.

109. H. Mehlhorn, J. M. O'Shea, and A. R. Wellburn (1991): *J. Exp. Bot.*, 42:17–24.

110. D. T. Tingey, E. H. Ross-Todd, and C. A. Gunderson (1988): *New Phytol.*, 110:301–307.

111. A. H. Johnson (1989): *Biologic Markers of Air Pollution Stress and Damage in Forests*, National Press, San Diego.

112. R. H. Waring and W. H. Schlesinger, (1985) *Forest Ecosystems: Concepts and Management*, Academic Press, San Diego.

113. R. H. Waring (1987): *Bioscience*, 37:569–574.

114. F. S. Chapin III, A. J. Bloom, C. B. Field, and R. H. Waring (1987): *Bioscience*, 37:49–57.

115. J. A. Weber, M. D. Plocher, and W. E. Hogsett (1991): *Bull. Ecol. Soc. Am.*, 72:282–283.

116. L. W. Kress, H. L. Allen, J. E. Mudano, and T. K. Stowe (1992): *Environ. Toxicol. Chem.*, 11:1115–1128.

117. J. L. Dawson and T. H. Nash III (1980): *Exp. Bot.*, 20:61–72.

118. C. D. Foy, R. L. Chaney, and M. C. White (1978): *Annu. Rev. Plant Physiol.*, 29:511–566.

119. M. D. Jones and T. C. Hutchinson (1988): *New Phytol.*, 108:461–470.

120. K. Cromack, Jr., R. L. Todd, and C. D. Monk (1975): *Soil Biol. Biochem.*, 7:265–268.

121. A. F. W. Morselt, W. T. M. Smits, and T. Linonard (1986): *Plant Soil*, 96:417–420.

122. W. E. Winner, J. S. Colman, C. Gillespie, H. A. Mooney, and E. J. Pell (1991): *Ecological Genetics and Air Pollution* (G. E. Taylor, Jr., L. F. Pitelka, and M. T. Clegg, eds.), Springer-Verlag, New York, pp. 177–202.

123. P. M. McCool and J. A. Menge (1984): *Soil Biol. Biochem.*, 16:425–427.

124. U. Blum and D. T. Tingey (1977): *Atmos. Environ.*, *11*:737–739.

125. W. J. Manning, W. A. Feder, P. N. Papia, and I. Perkins (1971): *Environ. Pollut.*, *1*:305–312.

126. T. D. Tingey and U. Blum (1973): *J. Environ. Qual.*, *2*:341–342.

127. W. E. Hogsett, M. Plocher, V. Wildman, D. T. Tingey, and J. P. Bennett (1985): *Can. J. Bot.*, *63*:2369–2376.

128. D. T. Tingey, R. G. Wilhour, and C. Standley (1976): *For. Sci.*, *22*:234–241.

129. R. D. Spence, E. J. Rykiel, Jr., and P. J. H. Sharpe (1990): *Environ. Pollut.*, *64*:93–106.

130. S. B. McLaughlin, R. K. McConathy, D. Duvick, and L. K. Mann (1982): *For. Sci.*, *28*:60–70.

131. K. F. Jensen (1981): *Environ. Pollut.*, *26*:246–250.

132. S. Meier, L. F. Grand, M. M. Schoeneberger, R. A. Reinert, and R. I. Bruck (1990): *Environ. Pollut.*, *64*:11–27.

133. P. M. McCool and J. A. Menge (1983): *New Phytol.*, *94*:241–217.

134. A. Gorissen and J. A. Van Veen (1988): *Plant Physiol.*, *88*:559–563.

135. R. P. Gould, P. E. H. Minchin, and P. C. Young (1988): *J. Exp. Bot.*, *39*:997–1007.

136. R. J. Norby, E. G. O'Neill, and R. J. Luxmoore (1986): *Plant Physiol.*, *82*:83–89.

137. A. A. Millecan (1971): *A Survey and Assessment of Air Pollution Damage to California vegetation in 1970*, Air Pollution Control Office, U.S. EPA Report APTD-0694.

138. A. A. Millecan (1976): *A Survey and Assessment of Air Pollution Damage to California Vegetation: 1970 through 1974*, California Department of Food and Agriculture.

139. A. S. Heagle, D. E. Body, and G. E. Neely (1974): *Phytopathology*, *64*:132–136.

140. R. J. Oshima, O. C. Taylor, P. K. Braegelmann, and D. W. Baldwin (1975): *J. Environ. Qual.*, *4*:463–464.

141. D. T. Tingey, D. M. Olszyk, A. A. Herstrom, and E. H. Lee (1993): *Ozone—Public Health and Welfare Effects* (D. McKee, ed.), Lewis Publishers, Inc., Chelsea, Michigan, pp. 175–206.

142. U.S. Dept. of Commerce (1987): *Census of Agriculture*, Bureau of the Census, Crops Harvested Maps, Vol. 2 Subject Series, Part 1.

143. R. M. Adams, S. A. Hamilton, and B. A. McCarl (1985): *J. Air Pollut. Contr. Assoc.*, *35*:938–943.

144. R. M. Adams, J. D. Glyer, and B. A. McCarl (1988): *Assessment of Crop Loss from Air Pollutants* (W. W. Heck, O. C. Taylor, and D. T. Tingey, eds.), Elsevier, London, pp. 473–504.

145. R. M. Adams, J. D. Glyer, S. L. Johnson, and B. A. McCarl (1989): *J. Air Pollut. Contr. Assoc.*, *39*:960–968.

146. B. C. Jordan, A. C. Basala, P. M. Johnson, M. H. Jones, and B. Madariaga (1988): *Assessment of Crop Loss from Air Pollutants* (W. W. Heck, O. C. Taylor, and D. T. Tingey, eds.), Elsevier, New York, pp. 521–535.

147. P. R. Miller, O. C. Taylor, and R. G. Wilhour (1982): *Oxidant Air Pollution Effects on a Western Coniferous Forest Ecosystem*, EPA Environ. Res. Brief (EPA-600/D-82-276).

148. P. R. Miller, J. R. McBride, and S. L. Schilling (1991): *Memorias del Primer Simposio Nacional: Agriculture Soctenible: Una opcion para el desarrollo sin deterioro ambiental*. Comision de Extudios Ambientales C.P. y. M.O.A. Intervational, pp. 161–172.

13

Environmental Effects of Cold on Plants

Maynard C. Bowers

Northern Michigan University
Marquette, Michigan

Popular belief places prolonged cold weather as the cause behind severe chilling injury to plants. However, most chill stress and injury is usually due to rapid and sudden temperature changes (often associated with a passing cold front) rather than extended extreme cold spells. Much, if not most, winter injury follows rapid, radical temperature drops to below-freezing levels following extended mild fall weather (1–3). This explains why extensive "winter" damage can occur after a warmer-than-average winter.

A plant's ability to endure low temperatures begins to develop as the late autumn days shorten and dormancy sets in. But even after leaf drop a plant may not be ready for temperatures far below freezing.

Temperature is an environmental condition which is continually changing, both during the day and with the season of the year. With these variations come changes in the rates of cellular processes, such as cell division, photosynthesis, respiration, lipid synthesis, and membrane alteration and repair, through the effects on the kinetics of biochemical reactions.

Each plant species has an optimum temperature range for growth and reproduction. Below this range low-temperature stress can result in reduced growth and other metabolic activities. Plant injury and death occur at still lower temperatures.

Plant temperatures closely parallel that of the environment at low temperatures even below the freezing point. A plant does not have the ability to control its own temperature. This is shown by the radiation cooling of plants below the air temperature at night and by radiation heating on the sunny side of trees to as high as 30°C above the shade side in winter. Plants are poikilotherms, i.e., they tend to assume the temperature of their environment. Low-temperature avoidance, therefore, cannot be developed. The few exceptions are the fleshy inflorescence of the Araceae and the so-called snow plants. These develop temperatures above that of their environment by rapid growth and the consequent rapid rate of release of respiratory heat. This occurs only at temperatures not low enough to cause injury. The conclusion is that low-temperature resistance can only be due to low-temperature tolerance.

Plants live year-round. They may not bear foliage, flowers, or fruit, but they still live to emerge the next year when the environment again becomes suitable for active growth. Often the suitability of the environment is determined by the maximum or minimum temperatures. Active plant growth is generally confined to a temperature range of about 10–40°C.

To survive low midwinter temperatures, most hardy plants need to be exposed to temperatures at or below freezing for some time before they become fully acclimated (4). If before such plants have had a change to acclimate a fall freeze occurs, they are exposed to a wide range of injury including death.

Periodically, extreme winter freezes occur which severely affect survival and production of perennial plants. These freezes are often advective freezes (involving the movement of air) which result from southern intrusion of cold arctic air masses. Advective freezes usually involve sudden drops in temperature which produce rapid cooling of plant organs and severe freezing injury. Extended advective freezes also result in low soil temperatures, particularly in the absence of a snow cover, the consequence of which is injury to the crown region and roots of both herbaceous and woody perennials. Midwinter occurrences of advective freezes are the most common; however, they may also occur in the spring and in the autumn.

Radiation freezes which involve little or no air movement result from the cooling of air near the ground due to radiation heat loss. Spring and fall are the usual times for this type of freeze and are commonly associated with maritime high-pressure air masses. Delayed acclimation during the autumn may allow frost injury to occur which in turn increases the susceptibility of the plant to winter injury.

The methods of adaptation to winter environments in temperate perennial herbaceous and woody plants have been studied from ecological, physiological, and morphological points of view. Raunkiaer's (5) life forms are well-known wintering forms that were categorized based on the amount and kinds of protection given to the resting buds and shoot apices of the plants during the nongrowing season. This life form system has been used to clarify the winter hardiness of resting buds to freezing stress in temperate zones (6–8).

The dormancy of temperate zone plants is considered to be another overwintering form, since inactive plant organs are in general more resistant than active organs to environmental stresses such as freezing and desiccation (9). Shoot apices and young foliage leaves are protected from desiccation and attacks by insects, fungi, and other disease-producing agents by the bud scales during the rest period. Accordingly, the differentiation of bud scales is also important for overwintering plants and has been used as a criterion categorizing the life forms of woody species (5).

ACCLIMATION

Acclimation is defined as the development of cold hardiness. The loss of their hardiness is termed *deacclimation*.

Perennial plants of the temperate zone undergo cyclic changes in hardiness each year. While actively growing, these plants display little if any cold hardiness, but with the cessation of growth in the autumn, they go into rest and harden. Rest is the physiological state in which the plant will not grow even though external conditions are favorable for growth. To break rest and allow growth to resume a period of chilling is required. Hardiness is lost when growth resumes in the spring.

Acclimation and deacclimation are active developmental processes initiated by certain environmental stimuli. In short, the beginning of acclimation and rest is precipitated by shortening days, changes in light quality, decreasing temperatures, and drought stress. Frost in the autumn brings on maximum hardiness. During midwinter, plants begin to deacclimate and reacclimate, to a limited extent, with exposure to changing environmental temperatures. This deacclimation becomes irreversible when spring growth temperatures become favorable. Considerable variation takes place in the rate and depth of acclimation throughout the many different organs and tissues of the plant. In addition, the rate and degree of acclimation is dependent on the age, vigor, and health of the plant.

Even once plants are properly acclimated, they may still be influenced by low-temperature stress in winter. To break their winter dormancy, plants need an extensive period of low temperatures. After this requirement is fulfilled, plants will rapidly lose their hardiness when exposed to mild temperatures (4).

Therefore, plants can lose hardiness, in midwinter, if abnormally warm weather continues for a long enough period of time. A subsequent rapid drop to below-freezing temperatures could bring about severe damage. Having several repeating cycles of freezing and thawing generally increases the damage. Such cycles do appear periodically, and indications of injury may not appear until bud break (10).

After deacclimation in the spring plants are still susceptible to low-temperature injury. Late frosts, particularly hard freezes following early spring thaws, are the cause of considerable injury to perennial plant species. The rate at which plants lose hardiness and at which temperature they do so varies.

All of these events (early fall and late spring frosts, unusually warm midwinters, and rapid fluctuations in temperature) can freeze the water inside plant tissues (11,12). This in turn can injure a plant either directly or indirectly.

CHILLING INJURY

Chilling injury has been defined as injury at temperatures low enough to cause damage but not cause the freezing of water (9). Common usage refers to temperatures between $0°$ and $15°C$, which causes irreversible damage to tropical and subtropical plants. Chilling injury frequently results in plants being "set back" so that maturation is delayed and yield reduced.

Many plants of tropical and subtropical origin sustain damage when exposed to low, nonfreezing temperatures. This chilling injury usually occurs below a critical threshold temperature of $10-12°C$ and evokes multiple and complex symptoms (13,14). For example, reproductive and photosynthetic functions can be impaired.

Chilling-sensitive plants exposed to low temperatures become injured (15,16). Characteristically, when kept under a constant level of chilling stress, injury is time-dependent and changes in physiological activities precede the development of visual symptoms of injury. Following a plant's return to nonchilling temperatures, other visual symptoms may develop. Leaves can take on a water-soaked appearance and necrotic lesions may become evident. Although the chilling injury may be reversed if the chilling period is sufficiently short, recovery in the postchilling period is not immediate and depressed rates of plant growth are commonly noted (17).

Chilling-induced changes in chlorophyll fluorescence in vivo have been utilized to rapidly assess plant response to low, nonfreezing temperatures (18–21). As plant tolerance to chilling temperatures is, at the vegetative level, described as the ability to withstand the development of chilling injury and to resume growth on return to nonchilling temperatures (15), it is evident that changes in chlorophyll fluorescence are monitoring a physiological characteristic that is intimately associated with vegetative chilling tolerance.

The return of chilled plants to nonchilling temperatures has been shown to cause an increase in leaf conductance in some plants and this has been cited as the cause of chilling-induced reductions in photosynthesis (22,23).

Visual symptoms of injury do not appear on plant leaves during the time that plants are subjected to the chilling stress, but rapidly appear and intensify following the return of the plant to 25/15°C (17). This delay in the materialization of injury symptoms suggests that these are temperature-dependent reactions which proceed very slowly at low temperatures but accelerate as the temperature is increased. The visual rating of chilling injury is correlated with the length of the chilling period.

Chilling injury is thought to be caused by loss of membrane fluidity, detected as membrane phase changes below the critical temperature in sensitive taxa. Biochemical consequences include altered reaction rate of the membrane-bound enzymes, ion leakage, and loss of compartmentalization. The resulting metabolic imbalances lead to cellular injury. Chilling may produce lesions in the thylakoid membranes of leaf chloroplasts and result in lowered photosynthesis. Most tropical plants are susceptible to chilling injury whereas temperate zone plants are less often affected (15).

No plant is consistently spared from low-temperature injury. Low-temperature injury is more serious when it occurs in the winter and affects the crown and root tissues.

FREEZING (OR FROST) INJURY

Various types of freezing injury result from untimely frosts in the spring and autumn or from extremely low temperatures in midwinter. The hardiness level varies with taxa, plant organ, and season. Freezing injury to the southwestern side of trees is attributed to thawing of the bark on cold, clear days as a result of solar radiation followed by rapid cooling as the sun sets or is covered by a cloud (24). The same type of stress may be involved in winter injury to evergreen foliage, but desiccation stress may also be involved (25).

Plant protoplasm can survive the lowest temperatures attainable (0°K approx.) if no ice forms on the tissues (26). Frost resistance is (with few exceptions) tolerance since the vast majority of plants cannot avoid freezing on exposure to extreme subfreezing temperatures. This tolerance exists only toward extracellular ice formation. No plant can survive the formation of microscopically visible crystals within their living cells, at least not if the ice is formed within the protoplasm. When plants are frozen extracellularly any injury that occurs is likely to be a dehydration injury.

Frost resistance varies seasonally. During the fall the plant "hardens," i.e., it slowly develops a greater and greater tolerance or hardiness, until the maximum is reached in midwinter and then it dehardens slowly until the minimum is reached in spring.

SUPERCOOLING

Most plants supercool below the freezing point of the water solutions within their tissues. Total avoidance of freezing by supercooling may provide some protection (to about $-4°C$) for tender plants. Most temperate zone plants which show freezing resistance freeze extracellularly. In these plants, ice first forms within the extracellular spaces and water is then withdrawn from the cell to the extracellular sites of freezing by the vapor pressure deficit (27). Ice does not penetrate the cell and intracellular freezing is avoided. When all thaws, the cells regain water and survive. However, injury can occur during extracellular freezing.

In nature, herbaceous plants experience slow cooling rates (1–2°C/hr) during the freezing event (9,28). This results in extracellular freezing when ice first forms in the dilute apoplastic solution. Such freezing, depending on the degree and duration of low temperature thawing, result in increased efflux of ions and organic solutes. The tissue also takes on a water-soaked appearance.

One aspect of the physiology of a freeze–thaw stress is the ability of the injured tissue to recover following moderate stress (29). During the course of recovery, water soaking completely disappears and the tissue returns to its prefrozen appearance (30).

Considerable evidence indicates that some hardy plants can attain their hardiness potential only if ice formation occurs at warm subzero temperatures before extensive supercooling occurs. Both hardy and nonhardy plant species can supercool extensively because most plants have no internal ice-nucleating agents active at temperatures higher than $-5°C$ (9). Nucleation at high subfreezing temperatures facilitates slow dehydration of cells in equilibrium with extracellular ice. Siminovitch and Scarth (31) reported that lethal intracellular freezing occurs in hardy cells that are allowed to supercool and that initiation of ice formation at temperatures close to $0°C$ favored cell survival. Olein (32) considered supercooling to be injurious to hardy plants because it promoted none-quilibrium freezing. Ice formation at relatively warm temperatures leads to controlled propagation of ice through the apoplast (9). Water is drawn from the cells by a gradient of water potential, leading to an increase in solute concentration and a depressed freezing point within the cells. Alternatively, supercooling followed by spontaneous ice nucleation at colder temperatures causes tissues to freeze too rapidly for cell dehydration to occur. In this situation, ice penetrates cells, ruptures membranes, and causes cell death (33).

Baertlein et al. (34) hypothesized that the efficient expression of ice nuclei in plant cells would induce the degree of supercooling and the presence of extracellular ice nuclei in plant cells would increase the freeze-tolerant plant's ability to tolerate the freezing process. Water in plant tissues will not supercool unless heterogeneous ice-nucleating substances are absent and the spread of ice from adjacent tissue can be prevented. The properties of a tissue which facilitate supercooling are undoubtedly a composite of several features and include physiological, structural, and morphological features (35). Presumably each component is critical and the loss of any one would prevent supercooling.

Temperate plants are not susceptible to chilling injury at temperatures above 0°C and tend to show signs of damage only after ice has formed within their tissues (36). As long as such periods of freezing are not prolonged and the rate of thawing is not too rapid, the formation of extracellular ice may not cause significant tissue damage in hardened plants. However, if extracellular ice persists, the gradient of water vapor pressure between the apoplast and the cell causes water to migrate from the cells to the apoplast, where it freezes, thereby increasing the amount of ice in the plant tissue. As well as causing mechanical damage to plant tissues, this process results in the progressive dehydration of the cell contents and an increase in the concentration of the cell sap. Consequently, the biochemistry of the cytoplasm is seriously disturbed. Rapid thawing may also have a lethal effect on frozen plants due to further disruption of cell metabolism and water relations.

ICE NUCLEATION

Ice nucleation–active bacteria can catalyze ice formation at temperatures as warm as $-2°C$ (37). These bacterial species were shown to be the predominant ice-nucleating agents on the surface of plants and the cause of frost injury in many plant species (37). The molecular basis for ice nucleation activity of *Pseudomonas syringae* has been extensively investigated (38–40).

The discovery that certain epiphetic bacteria have ice nucleating properties may be important to frost protection practices (41). *Pseudomonas syringae*, *Erwinia herbicola*, and *Pseudomonas fluorescens* possess special glycoproteins in their cell walls which have the property of nucleating ice above −5°C. In very tender plants the presence of these ice-nucleating bacteria has been found to increase with frost injury. Freezing injury in these plants is avoided by supercooling and the presence of ice-nucleating bacteria raises the supercooling point and thus the killing temperature. Furthermore, intrinsic nucleators are found in many plants capable of tolerating some freezing, which prevents appreciable supercooling (42).

DIRECT DAMAGE

Direct damage from low temperatures includes sunscald, winter burn on evergreens, blackheart, frost cracks, bud kill, dieback, death of blossoms and vegetative shoots, and outright death of species. Severe drought during the growing season usually increases winter injury to perennial plants. One effect of drought stress is to reduce photosynthesis. This depletes a plant's store of carbohydrates. Plants depleted in carbohydrates do not harden properly (4).

MORPHOLOGICAL DAMAGE

In general, chill-injured plants initially have water-soaked leaves or soft spots on fruits, which necrose or are invaded by secondary pathogens and begin to rot (43). The water soaking is due to a loss of membrane semipermeability and leakage of electrolytes into the free spaces. These symptoms can either occur rapidly (within several hours) or be delayed (several hours to several days) depending on the species, the severity of the chilling temperature, and the duration of the exposure.

Symptoms often depend on other environmental conditions that occur during the chilling event. For example, if leaf tissue is chilled in bright light, there is usually greater inhibition of photosynthesis (44) and greater chlorophyll bleaching (45) than when chilled in the dark. Similarly, plants chilled in low relative humidities often wilt due to decreased water absorption insufficient to match transpirational water loss. Plants chilled at high relative humidities do not experience the same high transpirational demand and consequently do not experience secondary water deficit (46).

MEMBRANE DAMAGE

The primary event in chilling injury is an alteration in the state of a cellular membrane from a relatively fluid liquid-crystalline state to a less fluid gel state (47). This change in state decreases membrane permeability and changes the activity of membrane-associated enzymes and enzyme systems, which result in major alterations in metabolism (48–50).

At the cellular level, the accumulation of cryoprotectants in the cytoplasm and organelles is important in the development of frost hardiness, but frost hardiness also frequently involves modifications in cell membranes themselves (51). In plants capable of acquiring strong frost hardiness, an increase in the content and unsaturation level of the membrane lipids is often observed, while in plants with a lower capacity for hardening, both of these changes in membrane lipids are less pronounced (52). Phospholipids seem to play a key role in frost resistance (53,54), but as far as photosynthesis is concerned the content and composition of the typical thylakoid lipids are also important.

It is not clear whether the low-temperature–induced increases in the amount of thylakoid lipids and their unsaturation level are primarily important in the acclimation of photosynthetic electron transport to low temperature (55) or whether they are involved in freezing tolerance too.

In field experiments on natural hardening, it is impossible to distinguish clearly between the effects on membrane lipids and CO_2 assimilation of low growth temperature, of hardening per se, and of freezing stress (56). The relationships are further complicated by fluctuations in the light intensity and possible photoinhibition. In fact, photosynthetic CO_2 fixation, which is often influenced by hardening, is also a most sensitive indicator of freezing injury of the leaves (57).

Hardening seems to increase the content of storage materials—starch in the chloroplasts and lipid globules in the cytoplasm (56). An increased content of tricylglycerols has frequently been observed in connection with hardening (58). This is probably an indication of the cessation of growth, which may be a prerequisite for hardening. A low growth temperature per se induced synthesis of cytoplasmic membranes, this being evident in both unhardened and hardened materials grown at 4°C (56). An endoplasmic reticulum is a locus of membrane biosynthesis, an abundance of trialkylglycerols, though evidently not a primary requirement for hardening, may be involved in hardened material in the avoidance of freezing damage, allowing rapid repair of stress-induced membrane rupture.

It seems unlikely that increases in the content of membrane lipids or in the unsaturation level of the fatty acid components are primary requirements for hardening. However, they seem to be involved in the ability to acquire pronounced frost hardiness, possibly allowing rapid cell membrane repair after frost stress.

If chilling tolerance is related to fatty acid unsaturation, one would expect that conditions that increase the plant's tolerance to chilling temperature would also increase the degree of unsaturation. Water-stressed *Phaseolus vulgaris* L. plants had a greater chilling tolerance than non-water-stressed plants (46). Markhart (14) pointed out that in his treatments of plants there was no difference in the fatty acid unsaturation. These results suggest that fatty acid unsaturation is related to chilling tolerance but other factors are also important.

ROOT APEX

The apical meristems of roots, which are composed of populations of dividing cells, exhibit different anatomical structures and rates of cell production in response to ambient temperatures (59). Included in this apical meristem is a cluster of cells known as the quiescent zone.

The quiescent zone contributes significantly to the repair of temperature-induced damage suffered by the meristem (60,61). The activation of the quiescent zone is characterized by many circumstances where proliferation in the surrounding meristem is impaired (62). The quiescent zone responds to cold-induced damage in the cap meristem by producing new cells at its acroscopic face which serve to regenerate this meristem (61). This response may be confined to a few cells, or it may involve the whole acroscopic face. After periods of cold treatment the quiescent zone reached a maximum of activity one day after the proximal meristem reached its minimum length (63).

RESPIRATION

Chilling temperatures have been shown to decrease (47) or increase (64) oxygen uptake. Intact plants or plant organs usually show a stimulation of oxygen uptake due to chilling, whereas isolated mitochondria show a decrease in activity (15). The stimulation of oxygen uptake has been thought to be due to uncoupling of the electron transport chain (65). More recently, however, stimulation of the alternative pathway of electron transport has been implicated as the pathway responsible for the increased oxygen consumption (64). It is not known if the increased activity of this pathway has a function in plant response to chilling stress or is just a symptom of injury.

PHOTOSYNTHESIS

Many plant species that are evolutionarily adapted to warm habitats are very susceptible to injury by low-temperature exposure. Brief exposures to low (0–12°C), but above-freezing, temperatures can have profound effects on season-long growth and productivity even though there may be no outward sign of damage. For many chill-sensitive species, an important element of the low-temperature–induced injury is an inhibition of photosynthesis.

Low temperature at night can also cause severe reductions in CO_2 fixation on the day following the chill. Like the inhibition caused by chilling in the light, it is clear that the primary loss of activity arises due to direct impairment of chloroplast function, but in the dark chilling it has not been possible to assign the cause to specific reaction (66).

Cold hardening of plants in natural environments is often accompanied by a decrease in the rate of photosynthesis. Complete inhibition of photosynthesis may occur during the winter in evergreen conifers, which can also acquire extreme frost hardiness (67,68). In plants which can achieve only moderate hardiness the effects of photosynthesis are usually less severe or nonexistent (69).

Thermal acclimation of photosynthetic activity, usually measured as changes in the optimum temperature for photosynthesis resulting from a change in growth temperature, has been demonstrated in a number of species from diverse environments (70,71).

From an ecological viewpoint, the potential or capacity for thermal acclimation is considered to be an important strategy for maximizing carbon fixation and seasonal net productivity, particularly by species native to areas which are subject to considerable temperature fluctuations during the growth season (72,73). Thermal acclimation would appear to be important for some plants indigenous to arctic and alpine regions (74).

Plants growing in these areas complete their life cycle over a relatively short growing season during which temperatures become progressively cooler shortly after germination and can fluctuate widely at any time during the season (75,76).

Photosynthesis in chilling-sensitive plants is markedly reduced by exposure to chilling temperatures. The damage is severely increased if the chilling occurs in the presence of light (77,78) or if the plant is water-stressed. Stomatal closure due to chilling-induced water stress is responsible for part of the decrease in photosynthesis (79), but direct chloroplast injury has also been reported (80). Decreased fluorescence from intact tissue or isolated chloroplasts suggests that the oxidative side of photosystem II is the site of injury (19,81,82). Consistent with this conclusion is the observed reduction in quantum yield of whole-plant photosynthesis (44).

An apparent decrease in chlorophyll content in leaves exposed to low temperatures is frequently a noted symptom of chilling stress (83) and has been suggested to be the result of photooxidation (84). However, studies with maize (*Zea mays* L. cv. Seneca Chief), a chilling-sensitive plant, have shown inhibitory effects of low temperatures on the synthesis of chlorophyll precursors (85). Chill-induced decreases in the accumulation of the porphyrin precursor 5-aminolevulinic acid were also evident (86).

Pine differs from maize in that some chlorophyll accumulation occurs in pine at all temperatures tested. Accumulation of carotenoids was also affected by low-temperature exposure, but to a lesser extent than chlorophyll (87). A decrease in chlorophyll b relative to chlorophyll a levels, found due to chilling, has been noted under some other conditions of stress such as winter stress (88).

Previous studies have suggested that deformation of the internal structure of the chloroplasts occurs in autumn- and winter-stressed pines (88,89). Chloroplasts may be damaged during freezing (90), and they may contribute to the freezing tolerance of other cell organelles, in particular the plasma membrane, by supply of metabolic energy, sugars, and lipids (91). The altered chlorophyll a/b ratio in child-stressed pine seedlings supports the suggestion that these seedlings may be deficient in the light-harvesting complex of photosystem II (87).

Reduction in the photosynthetic rate at low temperatures may be due to the decreased activities of photosynthetic enzymes (70,92,93), phosphate sequestration (94), and photoinhibition (95). Slow consumption of triose phosphate at low temperatures (96) and the accumulation of carbohydrate in the cold (97–99) suggests that photosynthesis is sink-limited in these conditions. Reduced export off carbon from leaves in the cold (99–102) may indicate low sink demand as well as a restriction of the transport process itself by low temperature (103).

Exposure of many plants to temperatures in the chilling range (0–12°C) increases the sensitivity of photosystem II to photoinhibition (44,104–107) because a given light level becomes increasing excessive when rates of photosynthesis decrease as a consequence of the low leaf temperatures. Furthermore, specific effects have been reported such as an inactivation of photophosphorylation (80,108). There is also a possibility that certain protective responses may not be fully effective at low leaf temperatures. Despite the fact that, initially, chilling damage appears to affect sites other than the photosystem II, the formation of large amounts of zeaxanthin in the xanthophyll cycle may be an important factor in the acclimation of plants to chilling temperatures (109).

Exposure of leaves to saturating light intensities, particularly in the presence of chilling temperatures, can cause long-term reductions in photosynthetic efficiency. Photoinhibition of photosynthesis can occur by two different mechanisms: (a) damage to the D1 protein of the photosystem II reaction center and (b) increases the rate of dissipation of excitation energy by nonphotochemical processes in the thylakoid membranes, which is considered to be a mechanism for reducing the rate of excitation of the photosystem II reaction center in order to minimize damage to the D1 protein (104). There is evidence that plants can suffer from photoinhibition in their natural field habitats (110–112). In temperate climates low air temperatures during the winter are often associated with clear skies and high light levels. Consequently, leaves of evergreen plants are often exposed to potentially photoinhibitory conditions in the field during winter (113).

PROTECTIVE RESPONSES

Cold treatment applied to petioles or stems of broad-leafed plants stops transport of photoassimilates in veins. Cold shock stops transport in the phloem whenever the temperature of a stem or petiole is suddenly lowered by 10°C. After 10–20 min at this temperature the movement of photoassimilates in the vein recommences, suggesting that the physical blockage of sieve tubes is not the sole cause of stoppage (114).

Abscisic acid has also been related to increased chilling tolerance in a number of species. Pretreatments with water stress, salt stress, nutrient stress, chilling stress, or exogenous application of abscisic acid all increased the endogenous level of abscisic acid and increased the tolerance to a subsequent chilling treatment (115,116). Markhart (117) demonstrated the protective effect of abscisic acid during chilling.

Root systems have considerable capacity to acclimate to chilling temperatures. An important aspect of this acclimation is the adjustment of the membrane that limits water flow through the root increasing its hydraulic conductance at the lower temperatures. The mechanism of this acclimation is suggested to involve compositional changes that alter membrane fluidity (117).

Cold acclimation in perennials is a seasonal process marked by an increased cold tolerance in fall reaching its maximum in winter. This is brought on by reduced day length and lower temperatures causing a progressive slowing of metabolism, the hardening of plant tissues, and the enforcement of dormancy in leaves and perennating buds. This is followed by a decrease in tolerance during the spring and reaching its maximum in the summer. Overwintering trees also enter into endodormancy during the same time that they develop cold hardiness.

Dormancy, whether innate or enforced, is clearly a widely adopted method of avoiding severe winter conditions. However, the survival of plant tissues, even in the form of dormant buds, depends ultimately on the ability of the cells to avoid freezing injuries.

Plant resistance to freezing injury can be seen as a series of successive "lines of resistance." The first line of defense in herbaceous and woody species is simply the depression of the freezing point of the water in the vacuole and cytoplasm due to the soluble solute content of the cell sap. The second line of defense—protecting critically important tissues of plants from freezing temperatures—can come into operation only after growth has ceased, dormancy has been established, and the plant tissues have been hardened by exposure to temperatures below 5°C for several days. The third line of defense is the slow cooling of hardened tissues to temperatures far below −40°C. This is required only in the most severe climates. A full explanation of these lines of resistance can be found in Fitter and Hay (36).

It is generally assumed that the influence of environmental factors on protective responses is mediated by plant hormones (118), proteins (119), and enzymes (120). Cytokinin levels in seedlings increased in response to chilling. Chilling has a promoting effect on the breakage of dormancy in Scots pine seedlings; the increase in the cytokinin content of buds and needles at this stage indicates the probable involvement of cytokinins in this process (121).

Early studies on the seasonal variation in protein content of cortical bark cells of black locust (*Robinia pseudoacacia*) demonstrated the accumulation of soluble proteins in the fall with a parallel increase in freezing tolerance. This was followed by a subsequent decline in protein content and the loss of cold hardiness in the spring when growth resumed (122). Several studies since then have provided evidence for quantitative and qualitative differences in protein content of bark between nonacclimated and cold-acclimated woody plants (123–126).

There are a number of reports of increases in the enzyme rubisco at lower temperatures, e.g., in the arctic-alpine species *Oxyria digyna* (127), in the C4 plant *Atriplex lentiformis* (128), and in the grass *Dactylis glomerata* (129). Gas exchange studies also support the view that rubisco increases after growth at lower temperatures. Acclimation to low temperatures involves increases in photosynthetic capacity at the lower temperatures and must require an increase in the activity of any enzyme, such as rubisco, which might potentially limit the rate of photosynthesis at the lower temperature (120).

After exposure to cold, the content of a range of enzymes and of leaf-soluble protein was increased. This agrees with a number of observations of increases in the amounts of a wide range of proteins at low temperatures (130). The increases in rubisco, and in soluble protein in general, have considerable implications for nitrogen use in plants at low temperature. This suggests that nitrogen-limited plants will be less able to acclimate to low temperature (120). By contrast, Treshow (131) had said that plants with moderate to low nitrogen are harder and more tolerant to temperature extremes.

Recent studies on mixed-grass prairie soils of southern Alberta have shown that over 50% of *Bouteloua gracilis* root biomass is lost during the course of the winter months, while at the same time marked increases in the levels of soil dehydrogenase and soil phosphatase activity are found (132,133). One hypothesis offered as a partial explanation for this degree of winter mass loss is that of the role that freeze–thaw cycles may play in commuting and degrading soil surface litter and below-ground biomass. This hypothesis was explored by Iverson and Sowden (134), who found that levels of soil enzyme activity rose substantially after initial freeze–thaw events, although their results were less conclusive regarding the effects that multiple freeze–thaw cycles might exert. This work was further expanded by Dormaar et al. (133), who suggested that for both *Festuca*- and *Bouteloua*-dominated grasslands of southern Alberta the chemical composition of below-ground root biomass was such that a rapid breakdown to lower molecular weight substrates occurred with repeated freeze–thaw events, ultimately leading to a substantial burst of mineralization upon spring warming.

Extraorgan freezing (135) as a protective response is defined as ice segregation outside an organ, by which the organ can survive freeze-induced dehydration, preventing ice penetration inside the organ. To clarify factors governing cold adaptation of conifers, it is necessary to explain the survival mechanism of shoot and flower primordia which are the least hardy tissues of conifers. Differential thermal analysis (136) of excised buds of boreal conifers showed that low-temperature exotherms derived from slowly frozen but lethally injured shoot primordia shifted markedly to a lower temperature when cooled very slowly, because water in the shoot primordia moves out through the crown and freezes (137). It was also demonstrated that the shoot primordia of Alaskan spruces survived freeze dehydration to $-70°C$ when cooled very slowly (138). Extraorgan freezing in primordial shoots reasonably explains why the hardiest conifers which tolerate extreme freeze dehydration can winter in Siberia and Alaska where the air temperatures cool down to $-60°C$ or below. It may be postulated that in both boreal and temperate conifers the shoot and flower primordia survive freezing-induced dehydration by extraorgan freezing under natural conditions (137).

SUMMARY

One of the more challenging problems in modern plant biology is the elucidation of the mechanism of plant cell freezing tolerance and the genetic elements that govern the acquisition of this tolerance. The understanding of this process is of fundamental scientific importance and economic value. Since the ability to develop tolerance to freezing is a genetically inherited trait in plants (9,139–142), recent advances in plant molecular biology and biotechnology make it a distinct possibility that plants could be engineered to increase their freezing tolerance. Excellent reviews of plant freezing injury and tolerance are currently available (9,11,14,143–145).

If plants are suitably preconditioned or hardened by slow cooling, they can withstand remarkably low temperature extremes. During hardening, the permeability of the outer plasma membrane increases, favoring the rapid resorption of water from the melting intercellular ice crystals. Also, due to the water binding, the free water content of the cell decreases so that the amount of soluble sugar and protein increases, lowering the freezing point of the protoplast.

The primary effect of cooling plants below their optimum temperature range is the reduction of rates of growth and of metabolic processes. Consequently, the length of time required for completion of the annual growth cycle increases as the climate becomes cooler and there may be a critical mean temperature below which the plants of a given species cannot reproduce successfully.

Low-temperature injury occurs when a plant gives off more heat than it absorbs. Heat loss may occur in one of two ways: (a) by conduction and (b) by radiation. Conductive heat loss occurs when the surrounding air is colder than the plant.

Radiation loss is due to heat passing into the atmosphere from warmer surfaces of the plant. Plants continuously radiate heat. They put more into the air than they can absorb and therefore become colder than the surrounding atmosphere. Most rapid radiation occurs when the sky is clear and quiet.

Needle-leaved coniferous trees and temperate deciduous broadleaved trees are characteristic of the generally continental, winter-cold climates of North America and Eurasia. These trees can survive freezing below $-30°C$ and thus are considered very hardy. Deciduous trees acclimate sooner and to a greater extent than evergreen trees. This may be associated with both response to photoperiod and the cessation of growth, which are apparently lacking in evergreen trees (146). The degree of winter hardiness of woody plants corresponds well with the minimum temperature at the distribution range of many plants

(147,148). In fact, the assessment of freezing resistance in midwinter has contributed to our understanding of plant cold adaptation and distribution.

The extent of injury is determined by how low, how fast, and for how long the temperature drops. The more temperature relations are studied, the clearer it becomes that every stage of plant development and reproduction is served best by a specific critical temperature, and any deviation from this will cause a stress reflected in poor growth, reproductive failure, and sometimes visible symptoms.

REFERENCES

1. D.F. Schoeneweiss (1975): *Annu. Rev. Phytopathol.*, *13*:193.
2. D.F. Schoeneweiss (1981): *Plant Dis.*, *65*:308.
3. D.F. Schoenewiss (1981): *J. Arboricult.*, 7:13.
4. P.H. Li and A. Sakai (1978): *Plant Cold Hardiness and Freezing Stress*, Academic Press, New York.
5. C. Raunkiaer (1934): *The Life Forms of Plants and Statistical Plant Geography*, Clarendon Press, Oxford, UK.
6. O. von Till (1956): *Flora (Jena)*, *143*:459.
7. W. Larcher (1980): *Physiological Plant Ecology*, 2nd Ed. (Translated by M.A. Biederman-Thorson), Springer-Verlag, Berlin.
8. F. Yoshie and A. Sakai (1981): *Jap. J. Ecol.*, *31*:395.
9. J. Levitt (1980): *Responses of Plants to Environmental Stresses, Vol. 1, Chilling, Freezing and High Temperature Stresses*, Academic Press, New York.
10. D.F. Schoeneweiss (1988): *Am. Nurseryman*, *168*:69.
11. M.J. Burke, L.V. Gusta, H.A. Quamme, C.J. Wieser, and P.H. Li (1976): *Annu. Rev. Plant Physiol.*, *27*:507.
12. A. Sakai and W. Larcher (1987): *Frost Survival of Plants*, Springer-Verlag, New York.
13. D. Graham and B.D. Paterson (1982): *Annu. Rev. Plant Physiol.*, *33*:347.
14. A.H. Markhart, III (1986): *HortScience*, *21*:1329.
15. J.M. Lyons (1973): *Annu. Rev. Plant Physiol.*, *24*:445.
16. G. Öquist and B. Martin (1986): *Photosynthesis in Contrasting Environments* (N.R. Baker and S.P. Long, eds.), Elsevier, Amsterdam, pp. 237–293.
17. S.E. Hetherington and G. Öquist (1988): *Physiol. Plant*, *72*:241.
18. R.M. Smillie (1979): *Temperature Stress in Crop Plants* (J.M. Lyons, D. Graham, and J.K. Raison, eds.), Academic Press, New York, pp. 187–202.
19. R.M. Smillie and S.E. Hetherington (1983): *Plant Physiol.*, *72*:1043.
20. M. Havaux and R. Lannoye (1984): *Photosynthetica*, *18*:117.
21. M. Havaux and B. Lannoye (1985): *Z. Pflanzenzüchtg.*, *95*:1.
22. B.G. Drake and F.B. Salisbury (1972): *Plant Physiol.*, *50*:572.
23. D. Pasternak and G.L. Wilson (1972); *New Phytol.*, *71*:683.

24. A. Sakai (1966): *Physiol. Plant.*, *19*:105.
25. W. White and C. Weiser (1964): *Proc. Am. Soc. Hort. Sci.*, *85*:554.
26. J. Levitt (1969): *Introduction to Plant Physiology*, C.V. Mosby, St. Louis.
27. H.A. Quamme (1987): *Can. J. Plant Sci.*, *67*:1135.
28. K. L. Steffen, R. Arora, and J.P. Palta (1989): *Plant Physiol.*, *89*:1372.
29. J.P. Palta, J. Levitt, and E.J. Stadelmann (1977): *Plant Physiol.*, *60*:398.
30. J.P. Palta, J. Levitt, and E. J. Stadelman (1977): *Cryobiology*, *14*:614.
31. D. Simonovitch and G.W. Scarth (1938): *Can. J. Res.*, *C16*:467.
32. C.R. Olein (1964): *Crop Sci.*, *4*:91.
33. P. Mazur (1977): *Cryobiology*, *14*:251.
34. D.A. Baertlein, S.E. Lindow, J.P. Nicholas, S.P. Lee, M.N. Mindrinos, and T.H.H. Chen (1982): *Plant Physiol.*, *100*:1730.
35. E.N. Ashworth (1984): *Plant Physiol.*, *74*:862.
36. A.H. Fitter and R.K.M. Hay (1981): *Environmental Physiology of Plants.* (Exp. Bot., Vol. 15), Academic Press, New York.
37. S.E. Lindow (1987): *Appl. Environ. Microbiol.*, *53*:2520.
38. C.A. Deininger, G.M. Mueller, and P.K. Wolber (1988): *J. Bacteriol.*, *170*:669.
39. S.E. Lindow, E. Lahue, A.G. Govindarajan, N.J. Panopoulos, and D. Geis (1989): *Mol. Plant–Microbe Interact.*, *2*:262.
40. P.K. Wolber, G.A. Deininger, M.W. Southworth, J. Vanderkerckhove, M. Van Montague, and C.J. Warren (1986): *Proc. Natl. Acad. Sci. USA*, *83*:7256.
41. S.E. Lindow, D.C. Arny, and C.R. Upper (1978): *Plant Cold Hardiness and Freezing Stress: Mechanisms and Crop Implications*, Vol. 1 (P.H. Li and A. Sakai, eds.), Academic Press, New York.
42. E.N. Ashworth and G.A. Davis (1984): *J. Am. Soc. Hort. Sci.*, *109*:198.
43. Y. Tatsumi and T. Murata (1981): *J. Jap. Soc. Hort. Sci.*, *50*:108.
44. S.B. Powles, J. Berry, and O. Björkman (1983): *Plant Cell Environ.*, *6*:117.
45. D. Bagnall (1979): *Low Temperature Stress in Crop Plants* (J.M. Lyons, D. Graham, and L.K. Raison, eds.), Academic Press, New York, pp. 67–80.
46. J.M. Wilson (1983). *Crop Reactions to Water and Temperature Stresses in Humid Temperature Climates* (C.D. Raper and P.J. Kramer, eds.), Westview Press, Boulder, CO, pp. 133–147.
47. J.M. Lyons and J.K. Raison (1970): *Plant Physiol.*, *45*:386.
48. G.J. Morris and A. Clarke (1981): *Effects of Low Temperature on Biological Membranes*, Academic Press, New York.
49. C.Y. Wang (1982): *HortScience*, *17*:173.
50. J. Wolfe (1978): *Plant Cell Environ.*, *1*:241.
51. M. Senser and E. Beck (1982): *Z. Pflanzephysiol.*, *108*:71.
52. M. Senser and E. Beck (1984): *J. Plant Physiol.*, *117*:41.
53. E. Sikorska and A. Kacperska-Palacz (1979): *Physiol. Plant.*, *47*:144.
54. E. Sikorska and A. Kacperska-Palacz (1980): *Physiol. Plant.*, *48*:21.
55. R.A.C. Mitchell and J. Barber (1986): *Planta*, *169*:429.

56. E. Aro and P. Karunen (1988): *Physiol. Plant.*, 74:45.

57. G.H. Krause, R.J. Klosson, A. Justenhoven, and V. Ahrer-Steller (1984): *Adv. Photosynth. Res.*, 4:349.

58. H.E. Nordby and G. Yelenovsky (1984): *Phytochemistry*, 23:41.

59. F.A.L. Clowes and R. Wadekar (1988): *New Phytol.*, 108:259.

60. F.A.L. Clowes and H.E. Stewart (1967): *New Phytol.*, 66:115.

61. P.W. Barlow and E.L. Rathfelder (1985): *Environ. Exp. Bot.*, 25:303.

62. F.A.L. Clowes (1976): *Cell Division in Higher Plants* (M.M. Yeoman, ed.), Academic Press, London, pp. 253–284.

63. P.W. Barlow and J.S. Adams (1989): *J. Exp. Bot.*, 40:81.

64. A.C. Leopold and M.E. Musgrave (1979): *Plant Physiol.*, 64:702.

65. R.P. Creencia and W.J. Bramlage (1971): *Plant Physiol.*, 47:389.

66. S. Martino-Catt and D.R. Ort (1992): *Proc. Natl. Acad. Sci. USA*, 89:3731.

67. M. Senser and E. Beck (1977): *Planta*, 137:195.

68. B. Martin, O. Mårtensson, and G. Öquist (1978): *Physiol. Plant.*, 44:102.

69. D. Rütten and R.L. Santarius (1988): *Physiol. Plant.*, 72:807.

70. J. Berry and O. Björkman (1980): *Annu. Rev. Plant Physiol.*, 31:491.

71. G. Öquist (1983): *Plant Cell Environ.*, 6:281.

72. B.F. Chabot (1979): *Comparative Mechanisms of Cold Adaptations* (L.S. Underwood, L.L. Tieszen, A.B. Callahan, and G.E. Folks, eds.), Academic Press, New York, pp. 283–301.

73. P. Grime (1979): *Population Dynamics* (R.M. Anderson, B.D. Turner, and L.R. Taylor, eds.), Blackwell, London, pp. 123–139.

74. W.D. Billings and H.A. Mooney (1968): *Biol. Rev.*, 43:481.

75. F.S. Chapin, III, and G.S. Shaver (1985): *Physiological Ecology of North American Plant Communities* (B.F. Chabot and H.A. Mooney, eds.), Chapman and Hall, New York, pp. 16–40.

76. B.T. Mawson, A. Franklin, W.G. Filion, and W.R. Cummins (1984): *Plant Physiol.*, 74:481.

77. B. Martin, D.R. Ort, and J.S. Boyer (1981): *Plant Physiol.*, 68:329.

78. A.O.N. Taylor, M. Jepsen, and J.T. Christeller (1972): *Plant Physiol.*, 49:798.

79. R.K. Crookston, J. O'Toole, R. Lee, J.L. Osbun, and D.H. Wallace (1974): *Crop Sci.*, 14:457.

80. M.P. Garber (1977): *Plant Physiol.*, 59:981.

81. C. Potvin (1985): *Plant Physiol.*, 78:833.

82. R.M. Smillie and R. Notts (1979): *Plant Physiol.*, 63:796.

83. R. Hodgins and R.B. van Huystee (1985): *Can. J. Bot.*, 63:711.

84. J.R. McWilliam and A.W. Naylor (1967): *Plant Physiol.*, 42:1711.

85. R. Hodgins and R.B. van Huystee (1986): *Plant Physiol.*, 125:325.

86. R. Hodgins and R.B. van Huystee (1986): *Plant Physiol.*, 126:257.

87. R. Hodgins and G. Öquist (1989): *Physiol. Plant.*, 77:620.

88. G. Öquist, O. Mårtensson, B. Martin, and G. Malmberg (1978): *Physiol. Plant.*, 44:187.

89. B. Martin and G. Öquist (1979): *Physiol. Plant.*, 46:42.

90. J.P. Palta and P.H. Li (1978): *Plant Cold Hardiness and Freezing Stress* (P.H. Li and A. Sakai, eds.), Academic Press, New York, pp. 49–71.

91. D.O. Ketchie, J.C.A.M. Bervaes, and P.J.C. Kuiper (1987): *Physiol. Plant.*, *71*:419.

92. M. Stitt (1987): *Progress in Photosynthesis Research*, Vol. 3 (J. Biggins, ed.), Nijhoff, Boston, pp. 685–692.

93. M. Stitt and H. Grosse (1988): *J. Plant Physiol.*, *113*:129.

94. C.A. Labate and R.C. Leegood (1988): *Planta*, *173*:519.

95. E. Ögren and M. Sjöstrom (1990): *Planta*, *181*:560.

96. R.F. Sage and T.D. Sharkey (1987): *Plant Physiol.*, *84*:658.

97. J. Azcon-Bieto (1983): *Plant Physiol.*, *73*:681.

98. C.J. Pollock (1986): *Current Topics Plant Physiol. Biochem.*, *5*:32.

99. M.J. Paul, D.W. Lawlor, and S.P. Driscoll (1990): *J. Exp. Bot.*, *41*:547.

100. G. Hofstra and C.D. Nelson (1969): *Can. J. Bot.*, *47*:1435.

101. C.J. Pearson and G. A. Derick (1977): *Aust. J. Plant Physiol.*, *4*:763.

102. J. Marowitch, C. Richter, and J. Hoddinott (1986): *Can. J. Bot.*, *64*:2337.

103. M.J. Paul, S.P. Driscoll, and D.W. Lawlor (1991): *J. Exp. Bot.*, *42*:845.

104. N.R. Baker (1991): *Physiol. Plant.*, *81*:563.

105. D. Greer (1988): *Aust. J. Plant Physiol.*, *15*:195.

106. E. Ögren and G. Öquist (1984): *Physiol. Plant.*, *62*:193.

107. S.B. Powles (1984): *Annu. Rev. Plant Physiol.*, *35*:15.

108. I.M. Kislyuk and M.D. Vaskovskii (1972): *Sov. Plant Physiol.*, *191*:688.

109. B. Demmig-Adams, K. Winter, A. Krüger, and F.C. Czygan (1989): *Plant Physiol.*, *90*:894.

110. O. Björkman and S.B. Powles (1982): *Carnegie Inst. Washington, Yearbook 80*:59.

111. W.W. Adams, I. Terashima, E. Brugnoli, and B. Demming (1988): *Plant Cell Environ.*, *11*:173.

112. E. Ögren (1988): *Planta*, *175*:229.

113. Q.J. Groom, N.R. Baker, and S.P. Long (1991): *Physiol. Plant.*, *83*:585.

114. S. McIntosh, D. Andrews, and J. Moorby (1990): *Can. J. Bot.*, *68*:266.

115. A. Rikin, A. Blumenfeld, and E. Richmond (1976): *Bot. Gaz.*, *137*:307.

116. A. Rikin, D. Atsmon, and C. Gitler (1979): *Plant Cell Physiol.*, *20*:1537.

117. A.H. Markhart, III (1984): *Plant Physiol.*, *74*:81.

118. L.E. Powell (1987): *HortScience*, *22*:845.

119. S. Sagisaka (1992): *Plant Physiol.*, *99*:1657.

120. A.S. Holaday, W. Martindale, R. Alred, A.L. Brooks, and R.C. Leegood (1992): *Plant Physiol.*, *98*:1105.

121. M. Qamaruddin, I. Dormling, and L. Eliasson (1990): *Physiol. Plant.*, *79*:236.

122. D. Simonovitch and D.R. Briggs (1949): *Arch. Biochem.*, *23*:8.

123. L.E. Craker, L.V. Gusta, and C.J. Weiser (1969): *Can. J. Plant Sci.*, *49*:279.

124. R.L. Hummel, T.M. Teets, and C.L. Guy (1990). *HortScience*, *25*:365.

125. S. Nakagawara and S. Sagisaka (1984): *Plant Cell Physiol.*, *25*:899.

126. M.K. Pomeroy, D. Siminovitch, and F. Wightman (1970): *Can. J. Bot.*, *48*:953.

127. B.F. Chabot, J.F. Chabot, and W.D. Billings (1972): *Photosynthetica*, 6:364.

128. R.W. Pearcy (1977): *Plant Physiol.*, 59:795.

129. K.J. Treharne and C.F. Eagles (1970): *Photosynthetica*, 4:107.

130. D.T. Patterson (1980): *Predicting Photosynthesis for Ecosystem Models I* (J.D. Hesketh and J.W. Jones, eds.), CRC Press, Boca Raton, pp. 205–235.

131. M. Treshow (1970): *Environment and Plant Response*, McGraw-Hill, New York.

132. J.F. Dormaar, S. Smoliak, and A. Johnston (1981): *J. Range Manage.*, 34:62.

133. J.F. Dormaar, A. Johnston, and S. Smoliak (1984): *J. Range Manage.*, 37:31.

134. K.C. Iverson and F.J. Sowden (1970): *Can. J. Soil Sci.*, 50:191.

135. A. Sakai (1979): *HortScience*, 14:69.

136. A. Sakai (1982): *Plant Cold Hardiness and Freezing Stress*, Vol. 2 (P.H. Li and A. Sakai, eds.), Academic Press, New York, pp. 199–209.

137. A. Sakai (1983): *Can. J. Bot.*, 61:2323.

138. A. Sakai (1979): *Plant Cell Physiol.*, 20:1381.

139. J.C. Bouwkamp and S. Honma (1969): *Euphytica*, 18:395.

140. V.W. Poysa (1984): *Cereal Res. Commun.*, 12:135.

141. G.E. Rehfeldt (1977): *Theor. Appl. Genet.*, 50:3.

142. J. Sutka and G. Kovacs (1985): *Euphytica*, 34:367.

143. C.J. Andrews (1987): *Can. J. Plant Sci.*, 67:1121.

144. P.L. Steponkus (1978): *Adv. Agron.*, 30:51.

145. P.L. Steponkus (1984): *Annu. Rev. Plant Physiol.*, 35:543.

146. R. Arora, M.E. Wisniewski, and R. Scorza (1992): *Plant Physiol.*, 99:1562.

147. H.L. Flint (1972): *Ecology*, 53:1163.

148. A. Sakai and C.J. Weiser (1973): *Ecology*, 54:118.

14

Photosynthetic Response Mechanisms to Environmental Change in C3 Plants

Rowan F. Sage and Chantal D. Reid

University of Georgia
Athens, Georgia

INTRODUCTION

The rate of photosynthesis is dependent on over 50 individual reactions, each of which potentially has a unique response to an environmental variable. The ability of plants to compensate for environmental effects on photosynthesis is critical to their performance and survival. Failure to adjust to environmental changes reduces competitive ability against species that have greater capacity to acclimate to the new conditions. Moreover, failure to adjust can be lethal in stressful environments. In an agricultural context, ineffective response of the photosynthetic apparatus depresses yield, with substantial economic cost. Understanding mechanisms controlling photosynthetic responses to environmental change is therefore important for understanding controls on plant productivity, species distribution, and the responses to climate change. This chapter summarizes recent concepts concerning response mechanisms of photosynthesis to environmental change in an attempt to bridge the gap between basic physiology found in textbooks and recent advances in the field. It is not intended as a detailed review of current literature since the field is too broad and dynamic to be fully covered here. Furthermore, while it is recognized that molecular processes ultimately control mechanisms of photosynthetic response to the environment, understanding of these mechanisms is uncertain and evolving rapidly. Therefore, this chapter will focus on responses occurring at the biochemical and organismal level of organization rather than at the molecular level.

GENERAL CONSIDERATIONS

To effectively describe mechanisms of photosynthetic response to the environment, physiological behavior should be considered at a range of levels within the leaf and throughout the plant. This is because the biochemistry of the photosynthetic apparatus is dependent on the whole plant for resources such as water and nutrients, while signal metabolites from the plant modulate photosynthetic activity through effects on stomatal function, resource allocation, and carbohydrate utilization. The time over which responses occur must also be considered. Three basic time scales can be delineated:

1. *Short-term responses,* which occur within minutes of an environmental change and typically involve preexisting components within the photosynthetic apparatus. Short-term responses are generally reversible.

2. *Long-term responses,* which can begin within 10 min but usually are pronounced in the days to weeks following an environmental change. These responses typically involve altered patterns of gene expression, reallocation of resources between the component processes of photosynthesis, and morphological change. The responses are not immediately reversible and often lead to the development of a visually different phenotype. Long-term responses represent acclimation if they improve performance in the altered environment.

3. *Adaptive responses.* At time scales covering multiple generations of a population, evolutionary changes in genotypes may occur, adapting a population to a modified environment.

Because construction and maintenance of the photosynthetic apparatus is expensive in terms of energy and mineral resources (1–3), it has been argued that species which efficiently use resources will be more successful (4,5). Efficient use of resources within the photosynthetic apparatus occurs when the capacities of the individual photosynthetic reactions equally limit the overall rate of photosynthesis (6,7). Environmental variation can perturb any balance within the photosynthetic apparatus, leading to a condition whereby one or a few reactions limit the rate of photosynthesis (1,8). Therefore, responses to environmental change may compensate for the differential effects of the environment, so that a balance is maintained between individual processes within the photosynthetic apparatus.

In the short term, responses to the environment involve the deactivation of catalytic sites of nonlimiting enzymes so that the regulated capacity of these enzymes in vivo is similar to the capacity of limiting enzymes (9). However, over the long term, these short-term adjustments do not improve the resource use efficiency of photosynthesis because a substantial fraction of the enzyme capacity remains deactivated, and no return is realized on the resources associated with deactivated enzyme (10). Long-term regulation could increase resource use efficiency by reducing the relative capacity of nonlimiting steps until a balance with limiting steps is reestablished. This may occur in one of three ways. First, protein contents can be modified so that excess enzyme capacity is degraded and released raw materials are reallocated into limiting enzymes. Second, if the environmental change caused a change in the cellular environment, such as pH, ionic concentration, or cell volume, acclimation may occur by reestablishing conditions within the cell which again favor high activity of the limiting enzymes. Third, classes of proteins termed molecular chaperones may be synthesized following an environmental change. Molecular chaperones are proposed to correct problems in protein structure and function in the new environment by facilitating proper transport, folding, and assembly of cellular proteins (11–13). Specific classes of molecular chaperones are commonly synthesized in response to stress, with the heat shock proteins being the best studied (13). Acclimation generally requires changes in gene expression and therefore requires an ability to sense the change in the environment and to coordinate gene transcription and translation so that appropriate genes are expressed while others are deactivated. Genes coding for the appropriate protein machinery must be present, yet this does not confer an ability to acclimate unless appropriate signal perception and transduction capacity is present in the leaf.

Patterns of Interaction Between the Photosynthetic Apparatus and the Environment

Substrate Effects

Photosynthesis is directly dependent on the availability of light and CO_2 except when these substrates are saturating. Any change in the level of light or CO_2 in the ambient environment rapidly affects photosynthesis, typically within a second of the change, because of the high surface-to-volume ratio of leaves and limited capacity to store light energy or CO_2 (14). Plants respond to changes in substrate deficiency in such a way as to enhance their acquisition of light and CO_2 from the environment. For example, the rate of CO_2 delivery to rubisco (ribulose-1,5-bisphosphate carboxylate/oxgenase) can be increased by controlling stomatal conductance, the thickness of the leaf boundary layer, and CO_2 formation from bicarbonate (15,16). Acquisition of light energy is dependent on leaf morphology, pubescence, orientation, and the number of light-harvesting complexes associated with the reaction centers of photosystem I or II (17,18). These responses may be rapid, occurring within minutes as in the case of stomatal closure or heliotropic leaf movement, or they may require days to complete, as in the development of a shade-acclimated plant from a sun-acclimated plant.

The supply of light and CO_2 is also dependent on either environmental parameters and processes in nonphotosynthetic organs. For example, water and nutrient deprivation within roots promotes the release of inhibitory signals such as abscisic acid (ABA) to the xylem, inducing stomatal closure (19). Low humidity also promotes stomatal closure (20), while temperature extremes may cause leaf and chloroplast movements which reduce light absorption (21,22).

Physical–Chemical Interactions

Each step in photosynthesis, photorespiration, and the reactions of starch and sucrose synthesis potentially has a unique response to conditions in the cellular environment, such as temperature, pH, ionic strength, membrane permeability, and metabolite concentration. If environmental factors affect the internal environment of the cell by altering membrane permeability, enzyme kinetics, pH, or the concentration of ions and metabolites, then they may affect photosynthesis by altering the physical and chemical properties of the photosynthetic constituents. For example, temperature affects photosynthesis directly through its effects on diffusion, membrane integrity, and the catalytic turnover rate of most photosynthetic enzymes (23). Salinity may affect photosynthesis by altering the ionic environment of the chloroplast (24). Direct effects of water stress on the photosynthetic apparatus include altered permeability of the thylakoids, and altered concentrations of metabolites and divalent cations (25,26). Compensation for these effects involves reestablishing a cellular environment that favors high activity of limiting enzymes. This is accomplished by modulating pH, ion and metabolite levels, altering membrane composition, and changes in the pattern of protein synthesis including the production of chaperone-type proteins better suited for the new environment.

Mineral Resource Availability

With few exceptions, the mineral elements used to construct the photosynthetic apparatus are acquired from the soil. Low soil availability of mineral nutrients and/or restrictions in uptake and delivery to developing leaves play a major role in determining the amount and activity of photosynthetic machinery present in a leaf. Nitrogen is the most important mineral nutrient because it is required in large amounts and is limiting in most soils (3,27). All of the macronutrients are directly involved in photosynthetic structure and function, and can limit photosynthetic activity should their availability in the soil decline (27,28). Many micronutrients (iron, copper) are also important, as they participate in electron transfer reactions in the chloroplast.

Sensory Perception of the Environment

Specific environmental receptors in plants sense changes in one or more environmental variable and trigger a cascade of responses, particularly at the level of gene expression (29,30). These responses may be immediate, such as stomatal opening by blue and red light, or delayed, such as developmental responses which lead to increased size of the light-harvesting complexes. Important receptors in the leaf include phytochrome, cryptochrome (a blue-light–absorbing flavoprotein), chlorophyll, and possibly CO_2 receptors in guard cells (31–33). When environmental perception occurs in nonphotosynthetic tissues, the signal can be transmitted to leaves by plant growth regulators such as ABA, auxins, and cytokinins (19). Sensory perception is important for regulating photosynthetic activity so that a balance exists between photosynthetic and nonphotosynthetic processes as well as between the resource demands of the plant and the availability of resources in the external environment (34).

Description of Environmental Responses

Environmental effects may best be described in terms of processes limiting photosynthesis. However, this can be problematic since the rate of photosynthesis is dependent on a multitude of processes spread throughout the plant. Biochemical processes in leaves have direct and immediate effects on photosynthetic CO_2 assimilation, while at the level of the whole plant a wide variety of processes have indirect effects. For example, desiccated roots release ABA to the transpiration stream, inducing stomatal closure (34,35). In its simplest sense, a process is limiting if any increase in the rate of that process increases the rate of photosynthesis. Regulatory activity can obscure "ultimate" limitations by reducing the activity of nonlimiting processes to balance the capacity of limiting processes. This leads to a condition whereby photosynthesis can be colimited by deactivated and fully activated processes. Ultimately, however, photosynthesis is limited by those processes which control the rate of photosynthesis if all steps were fully activated (36).

Control Analysis

Ideally, mechanistic studies of environmental responses should include a rigorous description of an environmental effect on the component processes controlling photosynthesis. This can be done using the control analysis of Kacser and coworkers (37), and was first adapted to photosynthetic systems by Ian Woodrow and coworkers (9,38–41) and Mark Stitt's group (42–45). In control analysis, the degree to which a process controls the rate of photosynthesis is described in terms of the control coefficient C_s, which is equal to

$$C_s = \frac{dA/A}{dS/S} \qquad [1]$$

where A is the net rate of CO_2 assimilation and S is the activity of the step in question (38). In other words, the control of a process over photosynthesis is equal to the relative change in the rate of photosynthesis resulting from a relative change in the rate of that process. If the control coefficient equals 1, the rate of photosynthesis is completely dependent on the process in question. If the control coefficient is 0, the process has no influence on the rate of photosynthesis. Control coefficients of all elements contributing to the rate of photosynthesis must sum to 1.

One of the strengths of control analysis is that a complete description of the activities of all photosynthetic processes is not required to effectively evaluate an environmental effect. Control coefficients can be described for single processes as well as multistep reaction sequences. Information is only required for the system of interest, the parameters of which are predefined, taking into account the degree to which processes can be effectively measured. For example, only the control coefficients of biochemical steps need be described if the system of interest is the chloroplast. If the system is an intact leaf, control coefficients for diffusional and biochemical components can be described. In leaves with shade and sun-acclimated layers of cells, the proportional contribution of each cell layer to whole-leaf photosynthesis can be described. For an entire plant, control coefficients for processes supporting photosynthesis (e.g., water and mineral uptake) can be estimated in addition to diffusional and biochemical processes associated with different layers in the canopy. Control analysis is a recent introduction in plant physiology, and relatively few labs have utilized the approach to study effects of specific environmental parameters on photosynthesis (e.g., 40,42,43,45–47). Introduction of mutants lacking activity in a specific photosynthetic step, and plants transformed with antisense DNA for individual photosynthetic proteins have allowed for control analyses of the photosynthetic biochemistry, and will likely prove useful in studying specific environmental effects (43–45,47–49). In addition, control analysis of electron transport has been performed whereby specific inhibitors have been used to selectively reduce the activity of electronic transport steps (46).

Methods

Environmental responses of photosynthesis are studied using five basic approaches (a) gas exchange of intact leaves; (b) fluorescence analysis; (c) biochemical analysis of enzyme activity and content; (d) biochemical analysis of metabolite pool sizes; and (e) molecular analysis of transcriptional, translational, and posttranslational regulation. Studies of responses in vivo have utilized a combination of gas exchange, fluorescence, and biochemical analysis. In higher plants, few studies have combined molecular analysis with more traditional approaches, but this is changing with the recent use of mutants and transgenic plants with antisense DNA for individual photosynthetic proteins (49). Molecular approaches will not be described in this chapter.

GAS EXCHANGE Rates of CO_2 uptake and water loss are used to determine the response of net CO_2 assimilation (A), stomatal conductance (g_s), and the intercellular partial pressure of CO_2 (C_i) to the environmental variable in question (50,51). Calculation of C_i factors out stomatal and boundary layer effects on photosynthesis, and allows assessment of the separate effects of the treatment on stomata and the photosynthetic biochemistry in intact leaves. A critical assumption in the calculation of leaf gas exchange is that the stomatal conductance is relatively uniform over the surface of the leaf (52). As a first approach, the gas exchange response to the environmental parameter of interest (e.g., nitrogen, light, humidity, or temperature) is determined. Except for a limited assessment of stomatal effects, which can be determined by comparing C_i, gas exchange responses to specific parameters are often difficult to interpret in terms of mechanisms because multiple components of the photosynthetic apparatus contribute to the response. To address mechanisms, the effect of the parameter on the response of A to C_i is determined. For example, if the parameter of interest is temperature, A/C_i responses would be determined at a range of temperatures. The A/C_i response is useful because specific components of the photosynthetic apparatus exhibit characteristic responses to C_i and are generally limiting at different C_i (1,50). At saturating light intensities, rubisco capacity typically limits photosynthesis at low C_i, while processes contributing to the regeneration of RuBP limit photosynthesis at high C_i (53).

FLUORESCENCE While fluorescence of chlorophyll a has long been used to evaluate the occurrence of photoinhibition (54), the introduction of the pulse amplitude–modulated (PAM) fluorimeter in the mid-1980s revolutionized fluorescence analysis of environmental effects on photosynthesis in intact tissues (55,56). The PAM fluorimeter provides a detailed assessment of the photochemical and nonphotochemical processes suppressing (quenching) chlorophyll a fluorescence. Important fluorescence parameters include minimal fluorescence, F_0, the amount of fluorescence occurring when all photosystem II (PSII) reaction centers are open and the primary electron acceptor, Q_a, is oxidized; F_m, the maximum fluorescence occurring when PSII reaction centers are closed (Q_a is reduced); steady-state fluorescence, F_s, the level of fluorescence occurring during steady-state photosynthesis; and variable fluorescence, F_v, which is the difference between F_m and F_0 (57). The ratio F_v/F_m is particularly useful because it is directly proportional to the quantum yield of photosynthesis and is a good indicator of damage to the PSII complex. Photochemical quenching (q_p) reflects the suppression of chlorophyll fluorescence occurring as a result of energy utilization by carbon fixation, photorespiration, and other processes which use ATP and photoreductant (58,59). Nonphotochemical quenching (q_n) reflects the suppression of chlorophyll a fluorescence due to thermal dissipation of absorbed energy either in the antennae complex of PSII or within the reaction center itself (58,60,61). By dissipating absorbed light energy as heat, the mechanisms of nonphotochemical quenching reduce the efficiency of light use by the photosynthetic apparatus and therefore reduce the quantum yield of photosynthesis (59). Increases in nonphotochemical quenching lead to a decrease in the ratio of variable to maximal fluorescence, F_v/F_m (59,61), which is detected with the PAM fluorimeter.

Photochemical quenching and F_v/F_m can be used to estimate the quantum yield of electron flow through PSII by

$$\Phi_{PSII} = \left(\frac{F_v}{F_m} \right) q_p \qquad [2]$$

In turn, the rate of electron transport through PSII, J_{PSII}, can be estimated by

$$J_{PSII} = \Phi_{PSII} I_a \qquad [3]$$

where I_a is absorbed irradiance (62). These procedures have been extensively utilized in recent years to further study the nature of photoinhibition (60,61), the extent of photorespiration in C3 and C4 plants (63–65), and the regulation of light harvesting and electron transport in response to a wide range of environmental variables (62,66–68).

BIOCHEMICAL ASSAYS Measuring the activity and, if possible, the amount of key photosynthetic constituents is a major technique for identifying photosynthetic response mechanisms, particularly over the long term where shifts in the content of key constituents can be important. Many studies have focused on rubisco, since this enzyme is the single largest nitrogen investment in leaves and is often considered to limit the rate of light-saturated photosynthesis in ambient air. Fortuitously, assay of rubisco activity and quantification of rubisco content are relatively easy. Rubisco activity is determined using either spectrophotometric or radioisotope assays, while rubisco content can be determined using immunological procedures or binding of [^{14}C]carboxyarabinitol bisphosphate (CABP), an analog of the intermediate in the rubisco carboxylation reaction which binds irreversibly to rubisco catalytic sites (69–72). A key criterion of the value of rubisco activity determinations is whether the activity measured in C3 plants is three to four times greater than the rate of light-saturated photosynthesis in air. Higher activity of rubisco in vitro is expected because rubisco activity is assayed at CO_2 saturation, while in leaves in ambient air the stromal CO_2 concentration is below the k_m of rubisco for CO_2 (1). In C4 plants, where rubisco operates near CO_2 saturation in vivo, rubisco activity and light-saturated photosynthesis are roughly equal (73).

Other major soluble proteins of photosynthesis frequently assayed include the highly regulated enzymes fructose 1,6-bisphosphate, sedoheptulose 1,7-bisphosphate, and sucrose phosphate synthase (74–77); a diagram of the reactions catalyzed by these enzymes is included in Appendix 1. Activity of these enzymes in vitro is commonly measured spectrophotometrically by coupling their activity to the oxidation or reduction of nicotinamide dinucleotides (75). For example, the activity of fructose 1,6-bisophophatase is determined by converting the product fructose 6-phosphate to glucose 6-phosphate, then to 6-phosphogluconate in a reaction which reduces NADP$^+$ to NADPH. NADPH has high absorbance at 340 nm, whereas NADP$^+$ does not (78). Therefore, the absorbance change at 340 nm is a direct indicator of fructose bisphosphatase activity as long as it catalyzes the limiting step in the reaction media. NADH also absorbs at 340 nm (and NAD$^+$ does not), allowing this compound to be used in reaction sequences where NADH is either a substrate or a product, as for example in the assay of ribulose 5-phosphate kinase (75). In all cases, care must be taken to ensure that the enzyme is not artificially deactivated by extraction and assay conditions, which can be problematic as numerous soluble enzymes are sensitive to effectors (protons, cations, substrates, products, regulatory metabolites; see Appendix 1 for specific examples) present in the stroma or cytosol (75). In addition, if the enzyme system under investigation includes isozymes in separate compartments, such as cytosol and chloroplast, fractionation of the isozymes prior to assay is often necessary to resolve an environmental effect (79).

Assay of thylakoid proteins is more problematic. Chlorophyll determination is a rough approximation of nitrogen investment in the thylakoid constituents. Approximately 65 mol thylakoid N per mol chlorophyll is present in leaves of terrestrial plants (3). PSII content can be assayed by the binding of radiolabeled atrazine. However, this procedure is problematic because two atrazine binding sites are present in leaves, one of high affinity on the acceptor side of PSII and one of lower affinity on the donor side of the complex (80,81). Functional PSII can be assayed in vivo using a technique where high-intensity flashes of light at a frequency of 4–5 Hz are used to drive O_2 evolution from leaf disks or chloroplasts (82–84). As long as there is no reduction in PSII efficiency, the rate of O_2 evolution is proportional to the number of functional PSII sites. This technique does not yield all PSII particles because a considerable fraction (up to 25%) may exist in a nonfunctional form (80,85,86). Spectrophotometric techniques for quantifying functional PSII include measuring absorbance changes brought about by the reduction of electron acceptors such as dichloroindophenol or the semiquinone Q_A (80,82,83,85,87,88).

Capacities of electron transport components are often assayed by determining partial-chain-electron-transport reactions using isolated thylakoids (e.g., 89–91). Specific donor and acceptor compounds are typically used to isolate the portion of the electron transport chain of interest (89,92,93). The key to these measurements lies in the ability to isolate thylakoid fragments with little loss in activity (94). This is feasible in spinach, peas, and similar mesophytes, but is problematic in leaves with an abundance of secondary compounds or fiber material that prevent effective isolation of functional thylakoids (85).

Specific electron transport proteins are assayed spectrophotometrically. Prominent techniques include flash-induced absorption changes at 518 nm to assay the activity of the chloroplast coupling factor for ATP formation; flash-induced absorbance changes at 702–705 nm to measure PSI content; fluorescence induction curves with and without the herbicide DCMU to estimate plastoquinone pool size; and spectral difference between hydroquinol-reduced minus ferricyanide-oxidized extracts to determine cytochrome f content (89). ATPase activity is also assayed by release of phosphate in the presence of octylglucopyranoside, with phosphate levels being assayed by absorbance changes or release of radioactive phosphate from labeled ATP (89,96).

Monoclonal antibodies are currently being developed to quantify individual photosynthetic proteins (e.g., sucrose phosphate synthase; 97,98) and may become the preferred technique in the years to come. Current approaches commonly separate the antibody–protein complex from the extract by electrophoresis, strain it, and quantify the protein by densitometry. A preferred procedure is to link the antibody to an easily measured compound, such as a dye or radioisotope, which can then be precisely quantified. The poor availability of affordable antibodies and a lack of suitable probes have limited the use of this technology in quantifying photosynthetic proteins, particularly in crude extracts. However, this will change as the technology spreads and individual groups adapt it for their own purposes. Finally, pigment concentrations, mainly chlorophyll, can be measured spectrophotometrically as a rough approximation of light-harvesting components. Chlorophyll a/b ratios are useful for assessing relative levels of antennae complexes associated with PSII vs. PSI (86). Nitrogen assays assess total nitrogen investment in the photosynthetic apparatus since 60–80% of all leaf nitrogen is associated with photosynthetic processes (3,72). Both leaf chlorophyll and nitrogen are useful reference values to assess relative changes in photosynthetic investment.

METABOLITE ANALYSIS   Analysis of photosynthetic metabolites is important because metabolites may build up in front of limiting reactions in photosynthetic carbon flow, whereas they are depleted immediately downstream from the limiting reactions. Ratios of major photosynthetic metabolites (e.g., PGA/triose phosphates; fructose 6-phosphate/glucose 6-phosphate) are particularly useful for identifying bottlenecks in carbon flow (48,99–102). However, analysis of steady-state metabolite levels can be limited because regulation of enzyme activity also regulates metabolite levels (36,103–105). Metabolites are usually assayed by coupled enzyme assays where oxidation or reduction of nicotinamide dinucleotides is assayed spectrophotometrically as described above (78).

In addition to analyzing metabolite and enzyme regulation in the steady state, limiting steps can be identified by studying the time course of photosynthetic responses to environmental change (36,104,105). Turnover rates of metabolites are generally much faster then enzyme regulation, so that moments after a step change in conditions, metabolites tend to pulse upward in front of limiting reactions and decline downstream from limiting reactions.

PHOTOSYNTHETIC LIMITATIONS

The net rate of CO_2 uptake by a leaf equals the rate at which rubisco carboxylates RuBP minus all forms of respiration or

$$A = v_c - (0.5v_0 + R_d) \qquad [4]$$

where A is the net CO_2 assimilation rate, v_c is the rate of carboxylation, v_0 is the rate of oxygenation, and R_d is the rate of mitochondrial respiration in the light (106). The rate of carboxylation is determined by (a) the availability of the substrates RuBP and CO_2, (b) the catalytic concentration of functional rubisco active sites, and (c) the competition for rubisco active sites between CO_2 and O_2 (106,107). Respiratory CO_2 loss is the sum of mitochondrial respiration in the light and photorespiration, where the rate of photorespiration is equal to half the oxygenation rate (106,108).

Describing limitations on photosynthesis in terms of these factors can be difficult because CO_2-dependent changes in the rate of photosynthesis reflect a CO_2–substrate effect as well as the competition between O_2 and CO_2. In addition, the effect of CO_2 on photosynthesis depends on whether rubisco capacity or RuBP availability is limiting (109). To rigorously describe the environmental response of photosynthesis, mechanistic models based on diffusional and biochemical processes have been developed. The diffusional component reflects the resistance to CO_2 diffusion from the outside atmosphere to the stroma. The biochemical component reflects the capacity of rubisco to carboxylate RuBP, the capacity of the photosynthetic system to regenerate RuBP, and the effects of CO_2 and oxygen on the rate of photorespiration. The following discussion begins with diffusion limitations and then addresses photorespiratory limitations before focusing on rubisco and RuBP regeneration limitations.

Diffusion Limitations

The rate of CO_2 movement from the bulk atmosphere to the chloroplast stroma is dependent on the gradient of CO_2 between the atmosphere and the stroma as well as path resistances to CO_2 diffusion. Major diffusional resistances occur at (a) the boundary layer of the leaf, (b) the stomatal pore, and (c) the liquid phase between the outer mesophyll wall and the site of carboxylation (110,111). Diffusive resistance of the intercellular air spaces may also be important. Boundary layer and stomatal resistances are air phase resistances. Mesophyll resistance is a liquid phase resistance that includes the solubilization of CO_2, and transfer of CO_2 across the wall, plasmalemma, and both chloroplast membranes (112). CO_2 may also diffuse across the cuticle and epidermis, but generally the resistance of this parallel pathway is so high that it is ignored.

The effect of path resistances on CO_2 diffusion is described using an equation derived from Fick's first law of diffusion (113) where A is dependent upon the above resistances and the gradient between the concentration of CO_2 in the atmosphere surrounding the leaf, C_a, and the concentration of CO_2 at the site of carboxylation, C_c, as follows:

$$A = \frac{C_a - C_c}{R_t} \qquad [5]$$

Since resistances in series are additive, the mesophyll, stomatal, and boundary layer resistances would sum to total resistance (R_t). Beginning in the late 1970s the emphasis shifted from describing the diffusional component in terms of resistance to describing diffusion in terms of conductances, g, where total conductance $g_t = 1/R_t$ (114). Conductances are advantageous because they are directly proportional to the CO_2 flux, and they can be expressed in the same units as photosynthesis when the CO_2 gradient is expressed in mole fraction (114–116).

All three major diffusional resistances can be controlled by the plant, though not necessarily to enhance CO_2 diffusion. Stomatal control is rapid, reversible, and, in stress-free conditions, closely coupled to photosynthetic activity and evaporative demand. As photosynthetic activity increases, stomatal conductance increases in the same proportion, maintaining a steady supply of CO_2 to the chloroplast. As a result, the intercellular partial pressure of CO_2 (C_i) is maintained at approximately 60–80% of the ambient CO_2 in nonstressed C3 plants (117,118). Stomatal conductance can be varied to compensate for changes in boundary layer conductance, so that the overall conductance to CO_2 diffusion from the ambient air to the intercellular spaces is relatively constant (119). During stress, stomatal conductance often declines to a greater extent than photosynthetic capacity, so that diffusional limitations on photosynthesis increase (120). In some cases, particularly following rapid stress, complete stomatal closure can almost eliminate photosynthesis over large portions of a leaf (121–123).

In contrast to stomatal resistance, boundary layer and mesophyll resistances are not modified by the plant over the short term as a means of controlling CO_2 flux, but over the long term they may modify CO_2 diffusion and can thus contribute to responses to the environment. Long-term modification of boundary layer and mesophyll resistances by the plant typically occur as a result of growth and maturation. Both resistances have frequently been overlooked as significant factors in the environmental response of photosynthesis. Boundary layer resistances are frequently ignored because gas exchange measurements of leaves are usually conducted in well-stirred cuvettes which reduce boundary layer resistance well below what may exist in a natural setting. Furthermore, estimating boundary layer resistances in natural environments is problematic for all but the simplest leaf shapes in stable environments (124–126). Mesophyll resistances were ignored until recently because techniques were unavailable for measuring their magnitude.

Stomatal Control

The greatest level of control over CO_2 diffusion is exerted by the stomata. Stomata are turgor-driven values whose aperture depends on the turgor pressure of the guard cells (127,128). Turgor pressure is controlled by solute regulation within the guard cell protoplast and the relative water content of epidermal tissues. Accumulation of potassium, chloride, and organic ions such as malate increase the osmotic activity within guard cell protoplasts, causing a reduction in water potential and an influx of water from surrounding cells (128). Proton efflux by an H-ATPase in the plasmalemma drives the import of ions from surrounding cells, and this ATPase may be a major site for the regulation of guard cell function (129,130). The control mechanism has not been completely described but appears to involve regulation of potassium ion channels by calcium ions within the guard cell (19,131,132). Increased concentration of Ca^{2+} in the cytosol inhibits stomatal opening, possibly by inhibiting inward K^+ channels and activating K^+ and anion efflux channels. This depolarizes guard cell membranes and allows the accumulation osmotica to leak out (132,133).

Stomatal conductance is sensitive to blue and red light, C_i, evaporative demand, temperature, and possibly photosynthetic activity within the guard cell itself (19,20,130,134,135). Blue and red light promote stomatal opening and greatly enhance the sensitivity of the guard cells to CO_2 (130,136). Blue light is required for maximum stomatal opening (136) and is postulated to activate the plasma membrane ATPase (130,137). Stomatal opening is highly sensitive to low fluences of blue light and saturates at relatively low light intensities (< 100 μmol/m^2/sec). The red light response is less sensitive to low light intensities than the blue light response, and saturates at higher intensities (136). Receptors for the light response of stomata have not been clearly identified, though a flavoprotein (cryptochrome) may be mediating blue light sensitivity while chlorophyll appears to mediate the red light response (130,138).

Mechanisms controlling the response to C_i have not been identified (33,139,140). Photosynthetic carbon metabolism in guard cells has been suggested as a possibility (33) although it remains unclear as to whether guard cells are able to photosynthetically fix CO_2 (134). CO_2 is necessary for maximal stomatal opening, largely because it is required for malate formation (19,141). This requirement appears to be met above a C_i of 100 μbar, as higher C_i promote stomatal closure (141). CO_2 may inhibit guard cell turgor by regulating photophosphorylation and ATP levels, or through direct regulation of ATPases in the guard cell plasma membrane (19,142).

Stomata also respond to growth regulators and ions transported in the transpiration stream. Abscisic acid (ABA) is the major inhibitory compound which is released from roots experiencing desiccation, salinity, and nutrient deprivation (35,143,144). ABA is an important mediator of stress effects throughout plants and is often referred to as the stress hormone (34). ABA produced in leaves also contributes to stomatal closure when leaves are stressed, but as yet no direct effects on the photosynthetic biochemistry have been clearly identified (120,145–147). Cytokinins and auxins promote stomatal opening and act as an antagonist to the action of ABA (148,149). However, ABA appears to override the effect of cytokinin and auxin at equivalent concentrations (19,149). ABA is proposed to close stomata by increasing calcium concentrations in the cytosol of the guard cell either by opening inward calcium channels (132) or by promoting release of internally stored calcium (133). ABA has also been proposed to increase sensitivity to preexisting calcium levels, possibly by activating calmodulin (133). In these models, calcium acts as a secondary messenger to ABA by transmitting the ABA signal into a biochemical response. Increased calcium concentration in the apoplast of the leaf also promotes stomatal closure, indicating that calcium may act as a long-distance messenger in addition to its role as a secondary messenger (133,150).

DESCRIBING STOMATAL LIMITATIONS The best description of stomatal limitation on photosynthesis is under debate (41,53,151). Early approaches utilized a resistance analogy whereby the limitation arising from stomatal resistance to CO_2 diffusion was described by the ratio of stomatal resistance divided by total resistance in the leaf, or

$$\text{Stomatal limitation} = \frac{R_s}{R_t} = \frac{R_s}{R_{bl} + R_s + R_m} \qquad [6]$$

where R_s is stomatal resistance, R_{bl} is boundary layer resistance, and R_m is a residual resistance often termed mesophyll resistance (151). This approach has serious drawbacks because it assumes that photosynthesis responds linearly to CO_2, which tends to be the case only at low C_i (53). Furthermore, it ineffectively deals with mesophyll transfer resistances and the photosynthetic biochemistry. In this procedure, effects of photosynthetic biochemistry and mesophyll transfer resistances are combined into a residual resistance, calculated as the inverse of the initial slope of the A versus C_i curve. The residual or mesophyll resistance, R_m, is inappropriate because the biochemical contribution to the residual resistance is not a diffusive process (112).

A straightforward approach to assessing stomatal limitations is the elimination technique described by Farquhar and Sharkey (53), where stomatal limitations are described in terms of the change in photosynthesis that would occur if the diffusive resistance of the epidermis was eliminated. In other words,

$$\text{Stomatal limitation} = \frac{A_a - A_i}{A_a} \qquad [7]$$

where A_a equals the rate of CO_2 assimilation at the ambient CO_2 partial pressure and A_i is the rate of CO_2 assimilation at the C_i corresponding to the ambient CO_2 partial pressure. This approach has been widely used to characterize the effect of the diffusive resistances on photosynthesis but is limited because it does not describe the sensitivity of photosynthesis to realistic changes in stomatal conductance (41,151).

Control analysis can be used to describe the sensitivity of photosynthesis to realistic changes in stomatal conductance. Here the fractional control of the stomata over photosynthesis is described by the control coefficient, C_g:

$$C_g = \frac{dA/A}{dg_s/g_s} \qquad [8]$$

where A is the net rate of CO_2 assimilation and g_s is the stomatal conductance. In practice, the control coefficient can be estimated as the slope of the A/C_i response at the operating C_i divided by the stomatal conductance at that C_i (41). This is similar to the sensitivity coefficient of Jones (151). The value of control analysis is that the sensitivity of photosynthesis to changes in stomatal conductance can be described at the operational C_i, and a complete description of the limitations on photosynthesis can be obtained when combined with an analysis of biochemical parameters. The value of control analysis relative to the elimination procedure of Farquhar and Sharkey (53) depends on the desired type of information. Control analyses describe the instantaneous control of stomata over photosynthesis, while the elimination procedure yields an integrated measure of control by describing the reduction in photosynthesis caused by the total stomatal resistance.

STOMATAL PATCHINESS   A critical assumption in estimating stomatal limitations is that the C_i can be accurately measured (52,53). This is difficult when stomatal conductance over the leaf loses uniformity. Under conditions of stress or following ABA treatment, groups of adjacent stomata can close completely, creating patches where gas exchange between leaf and atmosphere is essentially eliminated (52a,121–123,152,153). In leaves experiencing patchy stomatal closure, CO_2 assimilation rates fall, but estimated C_i may be constant because C_i calculations are (a) based on regions of the leaf where stomata are open and gas exchange occurs and (b) independent of leaf area. In cases of nonuniform stomatal closure, a drop in assimilation at relatively constant C_i may be erroneously interpreted as resulting from a lesion in the mesophyll biochemistry rather than stomatal closure. Identifying patchy stomatal closure can be straightforward if the researcher suspects its possibility (52a). Autoradiography of leaves fed [^{14}C]CO_2 (121,153), videoimaging fluorescent signals from leaves (152,154), comparing O_2 exchange of leaves at subsaturating (0.03%) and saturating CO_2 (10–15%) (155), and vacuum-infiltrating stomata with water (156) have been successfully used to document patchiness. Most studies documenting patchiness have either fed ABA through the petiole or imposed a rapidly developing (<7 days) drought stress on potted plants. In the field, where drought stress typically develops more slowly, patchiness appears to be less important (52a,157,157a).

Patchiness in drought-stressed leaves may also result from patchy reductions in the biochemical capacity of photosynthesis, with stomata closing in response to the reduction in photosynthesis (147). Further work is necessary to verify this possibility.

MIDDAY STOMATAL CLOSURE Diurnally, stomatal closure plays the leading role in reducing photosynthesis during midday periods. This behavior increases water use efficiency of photosynthesis by minimizing the rate of water loss under conditions whereby the cost of photosynthesis in terms of water is high (158). The degree of stomatal closure depends on the water status of the leaf and humidity gradient between leaf and air. If leaf water status declines or the humidity gradient increases, stomatal closure begins earlier in the day and persists longer into the afternoon, thereby inhibiting photosynthesis to a greater extent (158,159).

Boundary Layer Conductance

Conductance of the boundary layer is dependent on the thickness of the still air layer around the leaf. Decreased boundary layer thickness increases conductance and thus can affect photosynthesis directly by increasing CO_2 influx and indirectly by altering leaf temperature and transpiration rate (135). Thickness of the boundary layer is a function of leaf morphology: size, shape, orientation relative to the wind, and leaf pubescence (124,126). In addition, the structure of the leaf canopy substantially influences the boundary layer of leaves within the canopy. Dense canopies reduce air movement so that CO_2 diffusion to the leaf is lowered, reducing CO_2 availability by up to 30% (160,161). Wind is the major environmental parameter affecting boundary layer thickness (124). When wind directly affects photosynthesis, it is primarily through effects on boundary layer conductance to CO_2, H_2O, and convective heat transfer, with the changes arising from altered leaf temperature or transpiration (162,163).

Plants have little short-term control over boundary layer characteristics but can influence boundary layer properties through developmental modification of canopy size, leaf shape, orientation, and degree of pubescence. However, changes in these properties do not appear to occur in response to altered rate of CO_2 diffusion through the boundary layer, but in response to water and nutrient availability, or temperature stress. For example, water and mineral deficiency greatly reduce leaf size and number, leading to more open leaf canopies with thinner boundary layers for individual leaves (135). Water and high-temperature stress can induce pubescence, thus reducing boundary layer conductance (164). A prominent example of boundary layer modification occurs in xerophytes, where sunken stomata and leaf folding may reduce CO_2 diffusion. However, in some cases modifications of the boundary layer can have minor effects on photosynthesis, largely because stomata conductance may increase in response to increased humidity and reduced CO_2 in the boundary layer (119,160,165). This may offset reductions in the boundary layer conductance so that C_i is relatively unaffected.

Mesophyll Conductance

The conductance to CO_2 diffusion from the intercellular air spaces to the site of carboxylation has largely been ignored until recently, when a 30% drop in CO_2 concentration between intercellular air spaces and the chloroplast stroma was detected using steady-state measurements of carbon isotope discrimination by leaves (110–112,166). When all conductances between the stroma and the atmosphere are considered, chloroplast CO_2 concentrations in healthy, nonstressed leaves appear to be about 50–60% of atmospheric CO_2 levels for C3 plants (110). In most cases, the magnitude of the mesophyll transfer conductance increases proportionally with photosynthetic capacity and appears to be independent of the environment (110–112,167). Consequently, in plants of similar photosynthetic capacity, mesophyll transfer conductances tend to be equal, regardless of measurement or growth conditions (110,111). An interesting feature of the relationship between mesophyll conductance and photosynthesis is that the CO_2 concentration of the stroma would decline if mesophyll conductance did not increase proportionally with photosynthetic capacity. It is not clear as to how mesophyll conductance is increased in step with photosynthetic capacity. Possible mechanisms include increased mesophyll surface area in plants of higher photosynthetic capacity, increased chloroplast number, and increased activity of carbonic anhydrase (110). Carbonic anhydrase reversibly catalyzes hydration of CO_2 to bicarbonate (HCO_3^-) and speeds diffusion of CO_2 through the chloroplast in a process called facilitated transfer (16). The net direction of the reaction is determined by the difference between the ratio of CO_2 to HCO_3^- in the stroma and the equilibrium ratio under the same conditions. At the chloroplast boundary, CO_2/HCO_3^- is higher than equilibrium because spontaneous HCO_3^- formation occurs more slowly than CO_2 solubilization. Here carbonic anhydrase catalyzes net formation of HCO_3^-. Bicarbonate and CO_2 diffuse down their respective concentration gradients to the site of CO_2 fixation where removal of CO_2 by rubisco reduces the CO_2/HCO_3^- ratio below equilibrium. Carbonic anhydrase then converts HCO_3^- back to CO_2. At the stromal pH, the pool of HCO_3^- is much larger than CO_2 (nearly 10-fold higher at equilibrium) and the enhancement of CO_2 movement occurs because the much larger pool of HCO_3^- is incorporated into the diffusion pathway. Without carbonic anhydrase, the HCO_3^- pathway is ineffective because the spontaneous conversion between CO_2 and HCO_3^- is too slow (15,16). With high activities of carbonic anhydrase, the resistance to CO_2 transfer in the chloroplast is estimated to be two to three times lower than would occur if this enzyme were absent (16).

Despite changes in leaf structure, fiber content, and growth conditions, the uniformity of the decline in CO_2 between intercellular spaces and chloroplast stroma of various species indicates that conductance through the mesophyll is maximized. Thus, mesophyll conductance generally does not appear to be a major mechanism explaining the response of plants to environmental change. An important exception is found in wheat plants grown at high nitrogen nutrition. Here the ratio of mesophyll conductance to photosynthetic capacity is reduced, possibly because of increased chloroplast thickness, and contributes to a reduced response of photosynthesis to increases in leaf nitrogen content (110,168,169).

Rubisco Oxygenation and Photorespiration

The ratio of photorespiration to photosynthesis is dependent on three factors: (a) the concentration of CO_2 in the chloroplast stroma, (b) the concentration of O_2 in the stroma, and (c) the relative specificity of rubisco for CO_2 vs. O_2 (170–172). This is described as

$$\phi = \frac{\text{photorespiration}}{\text{photosynthesis}} = 0.5 \left[\frac{V_0}{V_c} \right] = \frac{0.5(O)}{(C)S_{rel}} \qquad [9]$$

where v_0 is the oxygenation rate, v_c is the carboxylation rate, O is O_2 concentration in the stroma, and C is CO_2 concentration in the stroma. S_{rel} is the relative specificity of rubisco for CO_2, where

$$S_{rel} = \frac{V_{cmax}K_o}{V_{omax}k_c} \qquad [10]$$

V_{cmax} is the V_{max} of rubisco carboxylase activity, V_{omax} is the V_{max} of rubisco oxygenase activity, k_c is the Michaelis constant for rubisco carboxylase activity, and k_o is the competitive inhibition constant of O_2 for carboxylation (107). Beside CO_2 and O_2, temperature is the only other major environmental parameter directly affecting photorespiration/

photosynthesis, and this is accomplished through effects on the relative specificity of rubisco for CO_2 and through effects on CO_2 and O_2 solubility (Fig. 1; 170,171). All other environmental parameters—water stress, humidity, wind, nutrients, and light—affect photorespiration/photosynthesis indirectly through effects on CO_2 diffusion or leaf temperature.

As temperatures increase, the specificity of rubisco for CO_2 relative to O_2 declines (171,172). This is largely because the affinity for CO_2 falls (k_c increases), whereas the affinity for O_2 is marginally affected (171). Solubility of both CO_2 and O_2 decline with increasing temperature, but to a greater extent for CO_2, so the ratio C/O declines (Fig. 1). The temperature response of photorespiration has a large dependence on C_i. At a C_i below the current operational C_i of most C3 plants (<200–300 μbar), photorespiration can become substantial, reducing photosynthesis by nearly 50% at 40°C (Fig. 2). Above ambient CO_2 levels, photorespiration is greatly reduced, rarely accounting for more than 20% of the rate of photosynthesis.

Limitation in the Capacity of Rubisco and RuBP Regeneration

A major advance in our understanding of environmental responses of photosynthesis comes from the development and application of mechanis-

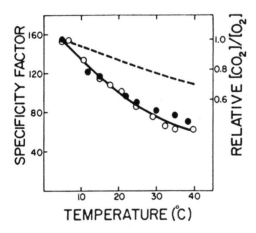

Figure 1   Effect of temperature on the CO_2/O_2 specificity factor of rubisco in spinach (*Spinacea oleracea*) (circles) and the equilibrium solubility ratio of CO_2 to O_2 relative to 5°C (dashed lines). Open circles represent determinations in atmospheric air mixtures at 21% O_2; filled circles represent determinations in air raised to 50% O_2. (From Ref. 171.)

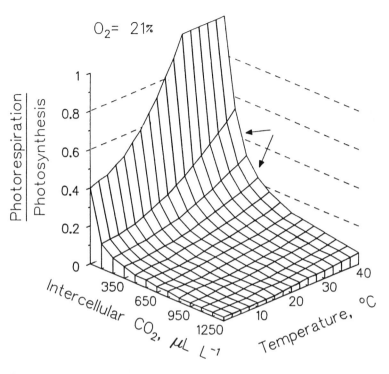

Figure 2   Modeled ratio of photosynthesis to photorespiration at 21% O_2 as a function of intercellular CO_2 and temperature in C3 plants. Arrows indicate the range of intercellular CO_2 typically observed in C3 plants. (From Ref. 173.)

tic models of the photosynthetic biochemistry. The most widely used model of photosynthesis in intact leaves is that developed by Farquhar and coworkers (106,107) to describe limitations on photosynthesis in terms of the capacity of rubisco to carboxylate RuBP and the capacity of thylakoid reactions to regenerate RuBP. This model was subsequently modified by Sharkey (50) and Harley and Sharkey (174) to account for effects of starch and sucrose synthesis on the RuBP regeneration capacity. Recent versions of the model describe photosynthesis in terms of three limiting processes: (a) the capacity of rubisco to consume RuBP, (b) the capacity of the thylakoid and Calvin cycle reactions to regenerate RuBP, and (c) the capacity of starch and sucrose synthesis to regenerate inorganic phosphate (P_i) from triose phosphates (Fig. 3; 50,174; see Appendixes 1–3 for review diagrams of thylakoid, stromal, and cytosolic reactions of photosynthesis). For simplicity, these three limitations will be referred to here as rubisco limitation, thylakoid limitation, and P_i regen-

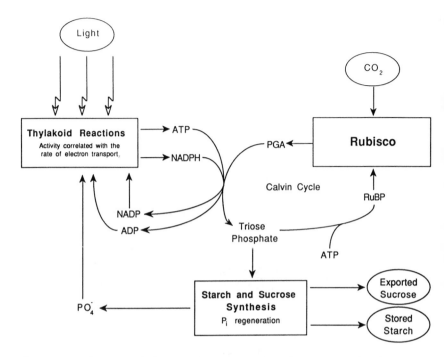

Figure 3 Schematic of photosynthesis in C3 plants highlighting the three major processes postulated to limit photosynthesis (in boxes), the point of entry of the major substrates (light and CO_2), and the exit point for the major products (sucrose and starch). See text for further discussion. Detailed metabolic pathways outlined in this figure are presented in the appendices.

eration limitation, respectively. Each of these three processes represents major points of direct interaction with the environment beyond the photosynthetic apparatus. Rubisco is the entry point for CO_2 and O_2; the thylakoid reactions are where light energy enters the photosynthetic system; and starch and sucrose synthesis are the exit points for photosynthetic products. These points of entry or exit are important because changes in CO_2, light, or utilization of photosynthetic products by the whole plant can force the photosynthetic apparatus into a condition whereby only one of the three general processes is the principle limitation on the rate of photosynthesis.

Rubisco-Limited Photosynthesis

Rubisco-limited photosynthesis may best be thought of as a limitation in the capacity of rubisco to consume RuBP and occurs when RuBP saturates rubisco catalytic sites (106,107,175). When rubisco capacity is limiting photosynthesis, the rate of carboxylation is as follows:

$$v_c = W_c = \frac{V_{cmax}(C)}{C + k_c(O/k_0)} \qquad [11]$$

where W_c is the term for rubisco-limited carboxylation (106). At 25°C, the k_c for rubisco is 11 μM (partial pressure equivalent about 270 μbar), while the k_0 is about 520 μM (the partial pressure equivalent is about 160 mbar; 50,171,176). In ambient air at 25°C, the CO_2 concentration in the stroma is typically near 6 μM, about 30–40% below the k_c of rubisco, so that rubisco operates well below CO_2 saturation. As Eq. [11] demonstrates, the carboxylation rate of rubisco-limited photosynthesis is directly dependent upon the V_{cmax} of rubisco as well as the CO_2 supply. V_{cmax} is a function of enzyme content and turnover rate per active site. Therefore, increasing CO_2 or the concentration of functional active sites of rubisco increase the potential rate of RuBP carboxylation.

Thylakoid-Limited Photosynthesis

As the stromal concentration of rubisco or CO_2 increases, the rate of RuBP consumption will increase until the concentration of RuBP is insufficient to saturate the catalytic sites of rubisco. At this point photosynthesis becomes limited by the rate of RuBP regeneration (107). Since some RuBP is bound to protein or in an nonreactive ionic state, RuBP becomes limiting below approximately 1.5–2 mol RuBP per mol rubisco active sites (177,178). RuBP regeneration is dependent on the rate of ATP and NADPH production, which in turn reflects light harvesting, photophosphorylation, and electron transport processes (see Appendix 3 for summary diagram of the electron transport in chloroplasts). Since many, if not most, leaves in a canopy are shaded at any one time (179), light harvesting and electron transport are common limitations on photosynthesis.

When thylakoid processes limit RuBP regeneration, the rate of carboxylation is described by

$$v_c = W_j = \frac{J}{a + b(\Phi)} \qquad [12]$$

where W_j is the rate of carboxylation when thylakoid reactions are limiting, J is the rate of electron transport, and Φ is the ratio of oxygenation to carboxylation (106). The value of constants a and b depends on ATP synthesis, NADPH synthesis, or other potentially limited aspects of the thylakoid system. If ATP availability is limiting, a = 4.5 and b = 5.25. If NADPH is limiting, a and b = 4 (106). The rate of electron transport is dependent on the limiting factor in thylakoids, which potentially include light harvesting, photophosphorylation, and any reaction in the electron transport chain (50,106). Since the rate of electron transport is closely coupled to the limiting thylakoid reaction, the overall rate of RuBP regeneration reflects the rate of electron transport. In addition, photorespiration reduces the rate of photosynthesis when thylakoid-dependent RuBP regeneration is limiting because O_2 competes with CO_2 for RuBP, and previously fixed CO_2 is released in photorespiratory metabolism (50). As a result, CO_2 and O_2 affect thylakoid-limited photosynthesis in addition to light intensity (50,108,180).

P_i Regeneration-Limited Photosynthesis

In addition to thylakoid processes and the Calvin cycle, starch and sucrose synthesis can influence the rate of RuBP regeneration by failing to metabolize triose phosphates as fast as they are produced (50,180–182). This leads to a sequestration of inorganic phosphate in organic forms. If P_i levels decline enough, photophosphorylation can become inhibited, reducing ATP synthesis and in turn RuBP regeneration (50,181). Theoretical treatment of the P_i regeneration limitation has been difficult since the behavior of starch and sucrose synthesis is insufficiently understood to be precisely modeled in an environmental context. Harley and Sharkey (174) used a simple empirical relationship to describe P_i regeneration–limited carboxylation as

$$v_c = W_p = 3T + 0.5v_o \qquad [13]$$

where W_p is the rate of carboxylation when P_i regeneration is limiting and T is the rate of triose phosphate use by starch and sucrose synthesis. The term $0.5v_0$ reflects the rate of triose phosphate use by the photorespiratory cycle. When P_i regeneration is limiting, the rate of photosynthesis is dependent on the rate of triose phosphate use by starch and sucrose synthesis as well as the rate of triose phosphate use by photorespiration. If photorespiration is reduced (by CO_2 increase or O_2 reduction) the rate of triose phosphate production increases because proportionally more RuBP is carboxylated. However, the rate of triose phosphate use by photorespiration is reduced by the same magnitude as the increase in the rate of triose phosphate production. Photosynthesis is unaffected by these changes because (a) starch and sucrose synthesis are maximized and cannot consume triose phosphates any faster, and (b) the increase in triose phosphate production is offset by the decrease in P_i regeneration in photorespiratory metabolism. This can be demonstrated by substituting Eq. [13] into Eq. [4]. In this way, changes in CO_2 and O_2 can have little effect on photosynthesis despite large effects on the rate of photorespiration (174).

If photorespiration results in a net return of P_i to the chloroplast, then a reduction in photorespiration by O_2 reduction or CO_2 increase can lower P_i levels in the chloroplast and cause a reduction in photosynthesis (174). A net return of P_i to the chloroplast occurs if some of the glycolate carbon leaving the chloroplast is not returned as glycerate, which must be phosphorylated in the chloroplast (174; Appendix 2). Export of glycolate carbon from the leaf is not unreasonable since some of the carbon exported to sinks may include photorespiratory intermediates, particularly amino acids. If export of glycolate carbon is significant, Eq. [13] must be modified to account for it (174).

Photosynthesis may also be inhibited if environmental conditions reduce the activity of one or more of the reactions of starch and sucrose synthesis. Leegood and coworkers (183,184) suggest that low temperature increases the P_i concentration required for an optimum rate of sucrose synthesis above that present in the cytosol. This might reduce the capacity for sucrose synthesis to the point where it becomes limiting. Alternatively, high rates of 3-phosphoglycerate (PGA) production which occur at high CO_2 may inhibit hexose monophosphate isomerase in the chloroplast, slowing starch synthesis (99).

Modeling the CO_2 Response of Photosynthesis

The "three-limitation" model describes the response of photosynthesis to an environmental factor in terms of the rate of CO_2 assimilation that would be realized if only one of the three potentially limiting processes was controlling CO_2 uptake. This is done by modeling the response of photosynthesis to a given variable, such as C_i, for each of the three limiting processes (Fig. 4). The actual rate of photosynthesis predicted at a given C_i would be the lowest of the three photosynthesis rates predicted by each limiting process. The distinct nature of the CO_2 response of each limiting process is described in Fig. 4. The CO_2 response of A when rubisco capacity is limiting is characterized by a substantial stimulation of A with increases in C_i to very high CO_2 levels (Fig. 4, curve A). The high degree of CO_2 sensitivity occurs because Rubisco activity is not CO_2-saturated until C_i rises above 1500 μbar and because of O_2 competition, particularly at low C_i. Therefore, increasing CO_2 stimulates rubisco-limited photosynthesis because more substrate is available to carboxylate RuBP and reduce the competitive inhibition of oxygen. Increasing the concentration of functional active sites or rubisco will increase the rate at which CO_2 can be scavenged from the stromal solution, increasing the rate of RuBP use. This leads to the name "rubisco-limited photosynthesis."

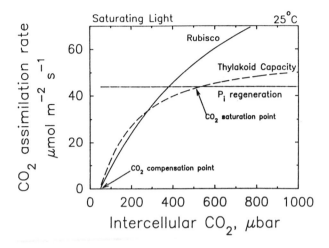

Figure 4 Characteristic responses to intercellular CO_2 of rubisco-limited, thylakoid-limited, or P_i regeneration–limited photosynthesis modeled according to Farquhar et al. (107) as modified by Sharkey (50) and Sage (175). Modeled parameters described in Sage (175).

If rubisco capacity was always limiting and the concentration of RuBP was always saturating for rubisco, the rate of photosynthesis at saturating CO_2 would equal the V_{cmax} for rubisco and be three to four times above what is observed at current atmospheric levels (1). Because RuBP regeneration capacity is finite, the rate of photosynthesis at high CO_2 is prevented from approaching the V_{cmax} activity of rubisco. If only thylakoid processes are limiting, a typical A/C_i response has high CO_2 sensitivity at low C_i and greatly reduced CO_2 sensitivity at elevated C_i (Fig. 4, dashed line). This CO_2 sensitivity of thylakoid-limited photosynthesis arises from the oxygenation of RuBP by rubisco and the release of previously fixed CO_2 by photorespiration. Oxygenase activity is high at low CO_2, but it declines as CO_2 levels increase (108). The subsequent rise in A with C_i is a reflection of increasing suppression of photorespiration. Thylakoid-limited photosynthesis will be CO_2-saturated and insensitive to reductions in O_2 concentration only when photorespiration is completely suppressed. This occurs at a C_i above 1000 μbar (50,185).

The initial slope of the A/C_i curve is largely dependent on rubisco capacity at light saturation (1,73,186), and a marked slope change between the linear response at low CO_2 and a nonlinear response of decreasing CO_2 sensitivity is indicative of a transition between rubisco and thylakoid limitations (1,169,176). In Fig. 4, the transition between rubisco and thylakoid limitations occurs at the C_i where the rubisco-limited curve intersects the thylakoid-limited curve. In practice, distinguishing the transition between rubisco and thylakoid limitations can be difficult because (a) both processes are highly CO_2-dependent at low C_i and (b) leaves are heterogeneous, with sun-adapted chloroplasts at the adaxial cell layers within a leaf and shade-adapted chloroplasts in abaxial cell layers (17,18,50,187,188). Therefore, changes in slope may be gradual, obscuring transitions between these two limitations. Two procedures help to determine whether rubisco or thylakoid processes are limiting at low C_i. First, when light harvesting limits photosynthesis, the slope of the A/C_i curve is light-dependent (189). Second, pulse amplitude-modulated (PAM) fluorescence can be used to estimate where thylakoid processes become limiting, since nonphotochemical quenching tends to be minimal under these conditions (67).

Under conditions where P_i regeneration is limiting, the rate of CO_2 assimilation is either unaffected or inhibited by rising C_i (Fig. 4; 174). This lack of a CO_2 response is used as a diagnosis for P_i regeneration limitations. P_1 regeneration–limited photosynthesis is also independent of, or inhibited by, O_2 reduction even though photorespiration is still occurring (50,108). Therefore, while lack of O_2 sensitivity is a useful indicator of a P_i-regeneration limitation on photosynthesis, it should not always be used to indicate reduced photorespiration (108).

The use of the three-limitation model to interpret A/C_i responses in plants enables one to diagnose the effect of an environmental factor on the photosynthetic biochemistry. Unless C_i reaches levels where photorespiration is completely suppressed (>1000 μbar), the following can be concluded: (a) the CO_2-saturated rate of photosynthesis is determined by P_i regeneration capacity; (b) at elevated CO_2, O_2- and CO_2-sensitive photosynthesis is a reflection of the RuBP regeneration capacity; and (c) rubisco is limiting at low C_i if the initial slope of the A/C_i response is independent of light intensity. If the initial slope is light-dependent, a thylakoid limitation is indicated at low C_i.

As described in Fig. 5, proportionally different changes in the capacity of the three limitations have characteristic effects on the A/C_i curve which help in the analysis of photosynthetic limitations. In panel A, reduction in the capacity of thylakoid-dependent RuBP regeneration reduces the rate of photosynthesis at high C_i but does not affect the rate of photosynthesis at low C_i until fairly low light intensities have been reached (compare curve A with curves B–D). The C_i at which a transition from a linear to markedly nonlinear response occurs declines with reduced thylakoid capacity relative to rubisco capacity. In panel B, a reduction in the P_i regeneration capacity reduces the CO_2-saturated rate of photosynthesis and, in turn, the CO_2 saturation point (compare curve A with curves E–G). If the P_i regeneration capacity is reduced low

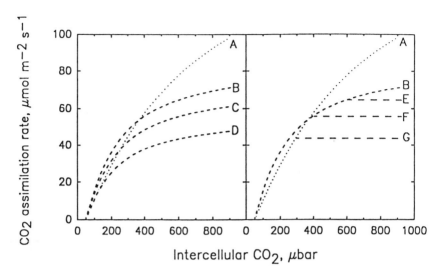

Figure 5   Modeled CO_2 response of photosynthesis at $27°C$ and 210 mbar O_2 for rubisco-limited photosynthesis (curve A, both panels), thylakoid-limited photosynthesis (curves B, C, D), and P_i regeneration–limited photosynthesis (curves E, F, G). Rubisco V_{cmax} is 180 $\mu mol/m^2/sec$. Curves B–D assume different electronic transport capacities where in B the maximum electron transport rate is 800 $\mu mol/m^2/sec$, in C; electron transport is 600 $\mu mol/m^2/sec$, and in D, electron transport is 400 $\mu mol/m^2/sec$. Curves E–G assume different P_i regeneration capacities where in E the P_i regeneration rate is 22 $\mu mol/m^2/sec$, in F it is 19 $\mu mol/$ m^2/sec, and in G it is 15 $\mu mol/m^2/sec$. (From Ref. 185.)

enough, thylakoid processes may never become limiting, since rubisco-limited photosynthesis may shift directly to P_i regeneration–limited photosynthesis at the CO_2 saturation point. In this instance, A/C_i responses of C3 plants may resemble those of C4 plants.

Single or Multiple Limitation

The three-limitation model has created controversy because it describes biochemical responses in terms of single limitations rather than in terms of shared control (e.g., see Refs. 9 and 190). These differences are not necessarily contradictory. First, P_i regeneration and thylakoid-mediated RuBP regeneration are each multistep processes whereby control may be spread over multiple enzymatic steps. Only rubisco limitation is a one-enzyme limitation. Second, photosynthesis is highly dependent on light and CO_2 availability, and the export of sucrose from the leaf. These external factors differentially affect the three limitations so that change in any one of these factors can quickly force the system to be limited by one of the three processes. This would be demonstrated by rapid changes in the control coefficients for the limiting processes as conditions change. Control analyses of environmental responses have documented dramatic shifts in the control of photosynthesis by Rubisco as light and CO_2 are altered (9,40,47,49). Interestingly, many plants appear to construct a photosynthetic apparatus which is approximately colimited by rubisco and thylakoid-dependent RuBP regeneration in atmospheric CO_2 and full sunlight (1,169,177,191). Changes in light, CO_2, and temperature disturb the balance, so that in the short term rubisco capacity becomes limiting at reduced CO_2 and thylakoid capacity becomes limiting at elevated CO_2 or reduced light intensity. An important question is whether short-term regulation and long-term acclimation to the new growth conditions reestablish the balance between these limitations. This will be addressed in the following sections.

RESPONSES TO ENVIRONMENTAL CHANGE: LIGHT, CO_2, TEMPERATURE, HUMIDITY, WATER, AND NITROGEN

Having reviewed mechanisms limiting photosynthesis, attention will now be focused on interpreting the responses of photosynthesis to changes in the environmental variables light, CO_2, temperature, humidity, water, and nitrogen.

Short-Term Responses to Light and CO_2

Although they are radically different substrates, light and CO_2 have similar short-term effects on photosynthesis since they both perturb the ratio of RuBP consumption to regeneration. Light provides the energy used in RuBP regeneration and is therefore a major determinant of RuBP supply, whereas CO_2 is the substrate used to carboxylate RuBP and is therefore a major determinant of the capacity of rubisco to consume RuBP.

Light Effects on Photosynthesis

Light has three major effects on photosynthesis. First, it provides the energy used in the production of ATP and NADPH. Second, light promotes the activation of key enzymes in the photosynthetic apparatus and stimulates stomatal opening. Third, light acts through photoreceptors such as phytochrome to modulate the development of leaves and the photosynthetic apparatus.

Light promotes enzyme activation largely through its effect on electron transport. Electron transport increases stromal pH and the concentration of divalent cations (mostly magnesium), altering the cellular environment to favor activation of photosynthetic enzymes in the stroma (192,193). Transported electrons are used to reduce ferredoxin and, in turn, thioredoxin, which activates enzymes by breaking disulfide bonds and forming sulfhydryls (193). Enzyme activation is also modulated by protein phosphorylation, with a protein kinase usually driving the phosphorylation reaction (49,194,195). Important light-activated enzymes in photosynthesis include rubisco, stromal fructose-1,6-bisphosphatase, NADP-triose phosphate dehydrogenase, ribulose-5-phosphate kinase, and sedoheptulose bisphosphate (14,193; Appendix 1). Generally, enzyme activation by pH, magnesium, and ferredoxin/thioredoxin acts as an on/ off switch which saturates at low irradiance (14,193,196). Fine control is often a result of enzyme–metabolic interaction (199,200). One exception is rubisco, whereby light acts as a fine controller of rubisco activity through ATP effects on rubisco activase (197,198).

As an energy source, light drives electron transport and proton movement for the production of NADPH and ATP. As light intensity increases, increased production of these compounds supports increased rates of RuBP regeneration and in turn increased photosynthesis. Eventually, the rate of RuBP regeneration may become large enough that RuBP effectively saturates all functional active sites of rubisco and photosynthesis becomes limited by rubisco capacity (1,50,175,189). At this point, photosynthesis becomes light-saturated and does not respond to further increases in light intensity, since rubisco capacity is not directly affected by light intensity (49,175,189). Decreasing the rubisco content or CO_2 availability reduces the capacity of rubisco to consume RuBP, and as a consequence rubisco becomes limiting at lower RuBP regeneration capacities and lower light intensities (175,189).

The light saturation point and light-saturated rate of photosynthesis is not always dependent on rubisco capacity. At elevated CO_2, electron transport capacity may become limited by electron flow through the cytochrome b_6-f complex (46,86,106,201). In such cases, electron transport rather than the rate of light harvesting can limit photosynthesis and determine the light saturation point. Alternatively, P_i regeneration may be low enough as to be limiting in atmospheric air, in which case the light saturation point is dependent on P_i regeneration relative to the thylakoid capacity to support RuBP regeneration. This often happens at cool temperatures (109,184).

LIGHT EFFECTS ON DIFFUSIONAL LIMITATIONS Beyond direct stimulatory effects on stomatal conductance at low to moderate light intensity, light does not appear to alter stomatal behavior in such a way as to affect diffusional limitations. As light levels increase, stomatal conductance increases with photosynthetic capacity. However, the increase in stomatal conductance is typically less than the increase in photosynthesis, so that C_i declines (140,202). Consequently, stomatal limitations may be higher at high light then at low light. Indirectly, higher light intensity may cause a reduction in stomatal conductance relative to photosynthesis by increasing leaf temperature and, in turn, leaf transpiration (see below).

CO_2 Effects

The CO_2 compensation point occurs at the C_i where photosynthesis equals the sum of photorespiration and dark respiration (106). At light saturation, rubisco capacity limits photosynthesis at the CO_2 compensation point because the rate of RuBP consumption by rubisco is depressed due to a deficiency of CO_2 (1,50,106,175,189). With increasing CO_2, rubisco consumption of RuBP increases until the C_i where RuBP is consumed as fast as it is produced. Below this C_i where RuBP regeneration becomes limiting, the stimulation of photosynthesis from increasing CO_2 results from both the increased availability of CO_2 and decreased oxygenase activity and photorespiration (50). Above the C_i where RuBP regeneration becomes limiting, the rise in photosynthesis with increase C_i results only from a reduction in photorespiration (106,108). At high C_i, the rate of RuBP carboxylation and triose phosphate production can become large enough to exceed the P_i regeneration capacity, and sequestration of P_i leads to a P_i limitation on photosynthesis (50,203).

Light intensity determines the thylakoid capacity to support RuBP regeneration and therefore it affects the C_i at which rubisco capacity and the capacity for RuBP regeneration are equal (175,189). Reductions in light intensity reduce the C_i at which an RuBP regeneration capacity becomes limiting and at which a rubisco limitation on photosynthesis shifts to an RuBP regeneration limitation. Eventually, at low light intensity (typically less than 100–200 μmol/m^2/sec), the capacity of the thylakoid reactions to support RuBP regeneration can become limiting at the CO_2 compensation point (189). When this occurs, RuBP regeneration capacity limits net CO_2 assimilation at all C_i and the initial slope of the A/C_i response exhibits light dependency (189). In contrast, when rubisco capacity is limiting, the initial slope of the A/C_i curve is independent of light intensity (189). Thus, the light dependency of the initial slope of the CO_2 response of photosynthesis is a useful probe of the biochemical limitations on photosynthetic activity (189).

DIFFUSIONAL EFFECTS OF CO_2   Stomatal conductance is directly controlled by C_i (204,205). As CO_2 increases, stomata close (19). In nonstressed leaves, the reduction in stomatal conductance with increasing CO_2 tends to maintain C_i/C_a constant, except at low C_i (between 100 and 200 μbar), where C_i/C_a generally declines as CO_2 increases (140,205;

Fig. 6). Water stress, low humidity, and ABA treatment increase stomatal sensitivity to rising CO_2, such that reductions in C_i/C_a at elevated C_i are more frequent and more pronounced following these treatments (19,140,140a). In stressed leaves, reduction in C_i/C_a with rising CO_2 is often accompanied by patchy stomatal closure (145; Mott, personal communication).

Short-Term Regulation of Photosynthesis in Response to Light and CO_2

Short-term regulation of photosynthesis has been proposed to compensate for environmental changes by reducing the in vivo capacity of nonlimiting components to balance the capacity of limiting components (67,104,175,189,206). For example, rubisco is deactivated when RuBP regeneration processes limit photosynthesis (104,175). Regulation spreads the control over photosynthesis from one or a few limiting steps to multiple steps and helps maintain constant the pools of photosynthetic metabolites (36,189,193). Stabilizing metabolite pools is important as many of the photosynthetic intermediates are powerful regulators of numerous enzymes in carbon metabolism (192,193,200). PGA and P_i in particular play a key role in regulating starch synthesis, sucrose synthesis, and the

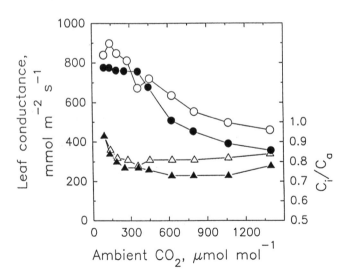

Figure 6   The CO_2 response of leaf conductance (O, $\bullet$) and the ratio of intercellular to ambient CO_2 (C_i/C_a; $\triangle$, $\blacktriangle$) in *Chenopodium album* grown at 34°C and either 350 μbar (O, $\triangle$) or 750 μbar CO_2 ($\bullet$, $\blacktriangle$). From Ref. 205a.

flow of carbon in the Calvin cycle (200,207). Extreme levels of metabolites would disrupt the coordination between RuBP production, RuBP consumption, and the utilization of triose phosphates.

Rubisco Regulation

Rubisco activity is primarily regulated by two mechanisms (197,208,209). The predominant mechanism is reversible carbamylation of a lysine residue in the active site (Fig. 7). The second mechanism is by reversible binding of phosphorylated sugars, principally carboxyarabinitol 1-phosphate (CA1P), an analog of the transition state intermediate in the carboxylation reaction. Regulation by carbamylation functions in response to changes in light intensity and/or CO_2 and O_2 in all C3 species examined (9,189,210). Decreasing light intensity or increasing CO_2 reduce the carbamylation state of rubisco when RuBP regeneration is limiting (9,36,189). When P_i regeneration is limiting, reducing O_2 causes a reduction in the carbamylation state of rubisco (211). Regulation by carbamylation appears to be most prevalent at intermediate light intensity (200–1000 μmol/m^2/sec or 10–50% of direct sunlight; 189,210,211a). In contrast, regulation by CA1P occurs in response to changes in light intensity only, and is prevalent at relatively low light levels (<10% of

1. Carbamylation Control

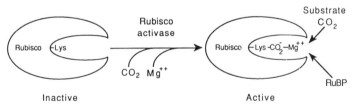

Inactive   Active

2. Inhibition by Carboxyarabinitol 1-phosphate

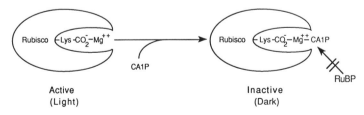

Active   Inactive
(Light)   (Dark)

Figure 7 Major mechanisms of rubisco regulation. Rubisco is regulated by either the reversible carbamylation of lysine residue (LYS) in the active site or binding of carboxyarabinitol 1-phosphate (CA1P) to carbamylated sites.

direct sunlight) or in darkness (189,209,210,211a). It appears that most species use CA1P to regulate rubisco (208,210,212).

These patterns of regulation are similar to predictions by models of rubisco regulation which assume rubisco activity is downregulated to balance limitations in RuBP regeneration (175). An example of the interaction between light and CO_2 on rubisco regulation is demonstrated in Figure 8. At high light, both modeled (Fig. 8A) and measured (Fig. 8B) responses of the activation state of rubisco demonstrate that rubisco deactivates in response to increasing C_i above approximately 300 μbar. This occurs because RuBP regeneration becomes limiting at about 300 μbar, and increases in C_i above this point increase the degree to which rubisco capacity is in excess. Reducing light intensity to intermediate levels (450 μmol/m^2/sec) also results in a RuBP regeneration limitation and rubisco deactivation. At intermediate light intensity, the carbamylation state of rubisco is highly CO_2-dependent because changes in CO_2 affect the extent to which rubisco capacity is in excess. Important deviations from the model occur at very low C_i (<100 μbar) where carbamylation may be limited for CO_2 and at low light where the CO_2 response of carbamylation is absent (Fig. 8B). These deviations may represent conditions where regulation by carbamylation becomes ineffective.

Rubisco carbamylation is dependent on rubisco activase, which catalyzes the removal of phosphorylated sugars from rubisco active sites (197,209). Phosphorylated sugars, primarily RuBP, bind tightly to decarbamylated sites preventing carbamylation, and their removal by activase

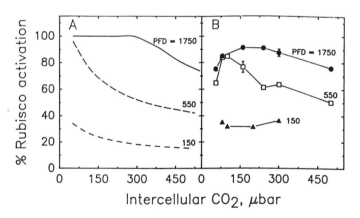

Figure 8 The CO_2 response in *Chenopodium album* of the activation state of rubisco at a photon flux density (PFD) of 1750, 550, or 150 μmol/m^2/sec. (A) Modeled response. (B) Measured response. (From Ref. 189.)

opens the sites up for carbamylation (209). It is not clear, however, if rubisco activase directly catalyzes carbamylation. Rubisco activase also removes CA1P and other inhibitory phosphorylated compounds, such as the "misfire" products 3-ketoarabinitol-1,5-bisphosphate and xylulose-1,5-bisphosphate, from rubisco catalytic sites (209,213). Misfire products are produced when rubisco isomerizes or epimerizes RuBP instead of carboxylating it (209).

The activity of rubisco activase and therefore rubisco activation is dependent on the rate of ATP production and electron transport (197,209,214,215). The rate of both ATP production and electron transport declines with reductions in light intensity, explaining the light dependence of rubisco activation. Reduction in rubisco activation with increases in C_i occurs because greater PGA reduction increases the rate of ATP consumption, which reduces the availability of ATP at constant photophosphorylation rate. When P_i regeneration capacity is limiting photosynthesis, increases in C_i or reduction in O_2 increase the production of triose phosphates, which reduces the pool size of P_i and reduces the rate of photophosphorylation. This reduces ATP availability and, in turn, the activation state of rubisco.

The time course of the change in rubisco activation following a change in conditions is presented in Figure 9A. In this experiment with bean (*Phaseolus vulgaris*), ambient CO_2 levels were suddenly increased to over 1000 μbar and O_2 levels were reduced from 21% to 2%. In response, photosynthesis rose slightly and then stabilized within a few minutes of the change. The activation state of rubisco began to deactivate within seconds of the change in conditions, stabilizing after 30 min or longer. Photosynthesis was unaffected by the change in the activation state of rubisco demonstrating the complete lack of control rubisco had over photosynthesis under these conditions. RuBP pool size rapidly declined to limiting levels following the switch and then recovered as rubisco deactivated. Similar responses in rubisco activation and RuBP pool size have been observed following changes in light intensity (103,105).

In the follow-up experiment, CO_2 was reduced to normal and O_2 was returned to 21% (Fig. 9B). This led to an increase in RuBP pools, but the rate of photosynthesis was initially depressed. This occurred because RuBP regeneration capacity became excessive in the instant after the reduction in CO_2 while the capacity of the deactivated rubisco was strongly limiting for photosynthesis. The activation state of rubisco recovered 5–6 min after the return to low CO_2. Photosynthesis increased in parallel with the rise in the activation state of rubisco, demonstrating the near-complete control rubisco had over photosynthesis under these conditions.

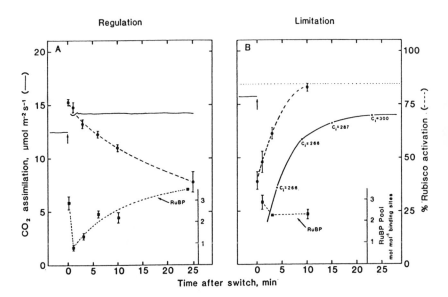

Figure 9 Time course for the response of photosynthesis (solid line), activation state of rubisco (filled circles), and RuBP pool size (filled squares) to either (A) a large-step increase in CO_2 partial pressure from 400 μbar at 180 mbar O_2, to 1500 μbar at 35 mbar O_2, or (B) a large step decrease in CO_2 from 1500 μbar at 35 mbar O_2 to 400 μbar at 180 mbar O_2 in *Phaseolus vulgaris*. C_i indicates the intercellular partial pressure of CO_2. Regulation refers to the condition where rubisco activity is regulated downward and is not limiting photosynthesis. Limitation refers to the condition where the activation state of rubisco directly limits photosynthesis. (From Ref. 216)

Regulation of Thylakoid Processes

Light energy becomes excessive whenever light harvesting exceeds the electron transport capacity. This occurs when electron transport is directly limiting or when the ability of carbon metabolism to utilize ATP and NADPH becomes limiting. When carbon metabolism is limiting, the pH gradient across the thylakoid membrane builds up and limits the rate of electron flow. In an immediate sense, PSII becomes overreduced and transfer of excitation energy from the pigment bed to PSII declines (59,61). Without a reduction in absorption of light, highly reactive species of oxygen can be generated in the antennae and core of PSII (60,217). These reactive species damage components of the light-harvesting apparatus and PSII complex, resulting in photoinhibition. To avoid photoinhibition, plants either reduce absorption of high light through chloroplast or leaf movements, or, more commonly, dissipate excess absorbed light as heat within the antennae or reaction center complexes (60,61,218). Mechanisms converting excess excitation energy to heat are part of a regulatory system which balances the production of excitation energy with the capacity of the carbon fixation and utilization reactions to consume that energy (61,66). Energy dissipation mechanisms are proposed to be activated by high pH gradients across the thylakoid membrane and the redox balance in the chloroplast (61,66,218). These mechanisms reduce the efficiency of light use and can reduce photosynthesis at low light if not rapidly reversed.

Two short-term mechanisms have been proposed for converting excess excitation energy to heat. One operates in the antennae of PSII; the other in the PSII core. In the antennae, the carotenoid zeaxanthin is proposed to either relax excited states of chlorophyll and/or reduce singlet oxygen (60,218,219). Zeaxanthin is formed from the carotenoid violaxanthin via antheraxanthin under conditions of high light and low pH in the thylakoid lumen (218). In low light and darkness, zeaxanthin can be converted back to violaxanthin, completing what is termed the xanthophyll cycle (60). In the core of PSII, excess energy may be dissipated by cycling excited electrons through a series of intermediates back to the P_{680} chlorophyll (86,217). In essence, electron flow is short-circuited within PSII with the liberation of heat. One intermediate in the loop may be a cytochrome b_{559} associated with the core but whose function is not definitely known (86).

In addition to the major processes regulating light harvesting, energization of the thylakoid membrane leads to phosphorylation of peripheral antennae complexes associated with PSII (18,220). When phosphorylated, peripheral antennae complexes of PSII (LHCII) dissociate from the PSII complex. Some of the dissociated LHCII complexes drift into the nonstacked regions of the thylakoid lamellae where they may associate with PSI complexes and pass excitation energy to PSI cores. Others remain undissociated and thermally dissipate excess energy (221). The condition whereby peripheral LHCII complexes are associated with PSII is termed state I; where they are phosphorylated and associated with PSI is termed state II; and where they are phosphorylated and not associated with either photosystem has been termed state III (221). Transitions between state I and state II balance energy flow between the photosystems and can partially compensate for differences in light quality which differentially stimulate electron flow through the reaction centers of the two photosystems (221). State transitions protect reaction centers from excess light energy by reducing antennae size of PSII and maintaining an even energy distribution between the photosystems so that electron transport is maximized (220,221). In higher plants, state transitions appear to contribute to the reduction of up to 20% of excess energy dissipation (222). Phosphorylation of PSII complexes also causes unstacking of grana by interfering with noncovalent bonds between LHCII complexes on adjacent thylakoids (223). This reduces light absorption and increases access of repair enzymes to damaged PSII particles (18).

Additional protection from photoinhibition occurs when chloroplasts turn the face of the grana stacks away from the light (21). This is reported to reduce light absorption by up to 20%. Chloroplast movement may be mediated by calcium–calmodulin interactions which affect the actin-myosin filaments in the cytoskeleton (21). Perception of light appears to involve both phytochrome and a flavoprotein blue light receptor (21). The blue light receptor appears to be responsible for minimizing light interception, whereas phytochrome may be more important in a low-light response which orients chloroplasts to maximize light interception (21).

Leaf reorientation may also minimize solar exposure in species exposed to high light, extreme temperature, or drought stress (22,224–226). This response is apparently mediated by a blue light receptor which affects turgor changes in cells of the petiole at the base of the lamina. In some species, such as soybean, favorable conditions promote orientations that maximize solar absorption, but under stress the response is reversed to minimize absorption (225,227). Development of a thick pubescence and highly reflective cuticles also serve to reduce light absorption under conditions of drought and heat stress, which promote photoinhibition (22,228).

THE NATURE OF PHOTOINHIBITION   In a strict sense, photoinhibition can be considered to result from damage to the light-harvesting apparatus in the thylakoids. In a broader view, photoinhibition is any process which results in extended reduction in quantum yield. In this sense, regulatory processes are photoinhibitory if the time for the relaxation of the regulation is minutes to hours (60,218,229). For example, activation of energy dissipation processes involving zeaxanthin or PSII cycling occur shortly after overenergization of the thylakoid membrane but may continue for hours after the excess light regimes have passed (60). This makes it difficult to distinguish between prolonged activity of protective mechanisms and high-light injury to the light-harvesting apparatus.

Damage resulting in photoinhibition appears to be concentrated in the reaction center of PSII, specifically the D1 protein which contains the plastoquinone binding site (86,217,230). Plants have a high capacity to repair damaged PSII particles though the timing of repair differs depending on species, environmental conditions, and the scope of the damage

(230). For example, repair is slowed by low temperature (231). Repair involves migration of the damaged PSII core to a stromal-exposed region of the thylakoid membrane, disassembly of the core, followed by insertion of newly synthesized D1 and reassembly (86,217,230). Because damage and repair are continuously occurring, turnover rates of D1 protein are among the highest in the chloroplast (86). Complete turnover of D1 can occur in as little as 60 min, and the mRNA transcript coding for D1 is more abundant than any other chloroplast transcript (217). As a result of these observations, the common view is that the capacity for repair is important in resistance to photoinhibition, and the rate of photoinhibition reflects the difference between rates of damage and repair (230). However, it has recently been suggested by Demmig-Adams and Adams (218) that in non-stressed plants adapted to high light, protection mechanisms such as a xanthophyll cycle predominate, and little actual injury occurs.

Photoinhibitory damage also results from the production of oxygen radicals by direct electron donation to O_2 at various points (reduced plastoquinone, PSI, ferredoxin) along the electron transport chain (232,233). Damage from this source is less localized than that occurring in the PSII complexes, affecting protein, nucleic acids, and lipids throughout the chloroplast. Plants have developed extensive antioxidation systems to minimize damage from oxygen radicals, the most prominent being superoxide dismutase, which converts superoxide radicals to O_2 and H_2O_2; tocopherol (vitamin E), which detoxifies free radicals within membranes; and peroxidase, which convert H_2O_2 to H_2O (232–234a). A major peroxidase detoxification system in chloroplasts involves reducing H_2O_2 to H_2O by channeling electrons from NADPH to H_2O_2 via glutathione and ascorbate (218,235,236). Ascorbate also appears to be important in the xanthophyll cycle, as it may facilitate the formation of zeaxanthin by transporting reducing power from NADPH to violaxanthin and antheraxanthin (218). Antioxidation systems in the chloroplast are inducible, and they are most pronounced under drought stress, low temperature, or following exposure to oxidizing pollutants (O_3, SO_2; 218,236,234a). Increased stress tolerance has been linked to the activity of these antioxidation systems (234,234a).

As a general rule, photoinhibition occurs following a dramatic change in conditions which either cause higher light exposure than previously experienced or reduce the capacity of the leaf to deal with high-light energy. This can result from sudden removal of shade, e.g., following tree fall in a forest, or the imposition of a stress which restricts carbon fixation and utilization, the capacity to protect the photosystems and antennae, antioxidant production, or the capacity for repair

(18,230,237). Plants in high nitrogen environments are far less likely to experience photoinhibition than plants on nitrogen-poor sites, owing to greater repair capacity and greater ability to develop a high carbon fixation capacity (238).

Regulation of Starch and Sucrose Synthesis

Starch and sucrose synthesis are tightly regulated in response to changes in carbon fixation because of the need to regenerate P_i and RuBP for continued photosynthesis and to store carbon in the cell for nonphotosynthetic periods. Enzymes of starch and sucrose synthesis are largely regulated by intermediates in their respective metabolic pathways, such as triose phosphates, hexose phosphates, and P_i (192,231; Appendix 1). Through its effect on photosynthetic activity, light regulates starch and sucrose synthesis by influencing the levels of these metabolites. However, the influence of light can be reversed under conditions whereby carbohydrate production is excessive. Reduced sink strength in nonphotosynthetic tissues can lead to elevated sucrose levels in leaves, causing sucrose or breakdown products of sucrose to feedback and inhibit further sucrose synthesis, and in turn induce a P_i regeneration limitation on photosynthesis (49,239,240). In this way, sinks may regulate photosynthesis (49,50,181,240). Starch synthesis is less able to respond to conditions of excess carbohydrate accumulation because starch has no known regulatory role in its own biosynthesis (49,207). Consequently, plants may be unable to deactivate photosynthesis to prevent excessive accumulation of starch, which is detrimental because large starch grains stretch and disrupt the integrity of the chloroplast, and may slow CO_2 diffusion in the chloroplast (49,241,242,243). This phenomenon is most pronounced in plants grown at high CO_2 or nutrient deficiency (49,241–243).

Sucrose synthesis is heavily regulated at cytosolic frustose-1,6-bisphosphatase, the gateway to sucrose synthesis (49,239; Appendix 1). This enzyme is primarily regulated by the powerful inhibitor fructose-2,6-bisphosphate (244). The concentration of fructose-2,6-bisphosphate is enhanced by high P_i, high fructose-6-phosphate, and low levels of triose phosphates. Sucrose synthesis is also regulated by the last two steps in the pathway, which are catalyzed by sucrose phosphate synthase (SPS) and sucrose phosphatase (49,98,239,245,246). Sucrose phosphate synthase is more important than sucrose phosphatase in regulation of sucrose synthesis and is controlled by reversible protein phosphorylation (49,195,246). Phosphorylation of SPS by a protein kinase deactivates sucrose phosphate synthase, whereas dephosphorylation by a phosphatase generally activates it (49,195,246). Sucrose and intermediates in sucrose synthesis modulate the activity of SPS (239; Appendix 1), but the extent to which they act directly on SPS as opposed to the phosphorylation/

dephosphorylation enzymes is not clear. Starch synthesis is regulated at ADP-glucose pyrophosphorylase, which catalyzes the next to last reaction in starch synthesis. Activity of ADP-glucose pyrophosphorylase is pro- moted by high PGA and inhibited by high P_i (207).

Long-Term Responses to Light

In constructing the photosynthetic apparatus, plants in natural environ- ments must balance the need to maximize light absorption in low-light environments with the danger of excessive light exposure and photoinhi- bition in high light (18). Mechanisms which maximize photosynthesis at one light intensity limit photosynthetic capacity at other light intensities (223,247). In the extreme, maladaptive resource allocation leads to chronic photoinhibition in high light and a net loss of carbon from the leaf in low light. Even in nonstressful environments, competition for light between neighboring plants tends to select for species whose allocation patterns allow them to more effectively utilize the light environment. Plants have thus developed a significant ability to modulate investment patterns to efficiently exploit the ambient light environment, though species adapted to relatively stable light environments exhibit the lowest potential for light acclimation (248,249). However, an ability to accli- mate to a wide array of environments has its cost in terms of reduced photosynthetic potential in predominately high- or low-light environ- ments.

Patterns of Long-Term Regulation in Response to Low Light
The most obvious feature in acclimation or adaptation to light is a change in leaf structure. Shade-grown leaves are thinner, with smaller, less numerous cells than sun leaves (18,30,247,250,251). This is advanta- geous because fewer cells are needed to utilize the available light. Developmentally, changes in leaf and chloroplast structure depend more on the total number of photons received in a day rather than maximum light intensity (252–254). Thus, plants continuously grown under inter- mediate light intensities may have leaves with higher photosynthetic capacity and sun-acclimated characteristics than low-light–grown plants exposed to high light for limited periods of the day. In addition to total quanta received, light quality plays an important role in shade acclima- tion under leaf canopies (a point often overlooked in low-light studies; 255). Below natural canopies, light is enriched in far-red wavelengths which are absorbed more strongly by PSI than PSII complexes (86). Growth in far-red–enriched light promotes development of extensive gra- nal stacks, reduces the amount of stroma lamellae, and increases the PSII-to-PSI ratio (86,255). Like state transitions, these adjustment com- pensate for differential light absorption by the photosystems and help to

balance energy flow between P_{680} and P_{700}. They can be considered long-term responses to changes in light quality, whereas state transitions are short-term responses.

At the chloroplast level, the following occur as plants acclimate or adapt to low light.

1. Chloroplast size and the amount of thylakoid membrane increase, while grana are increasingly stacked (90,223,255). In spinach, low-light–adapted chloroplasts have seven to eight thylakoids per granum in contrast to high-light–adapted chloroplasts which have about three thylakoids per granum (90). Species adapted to deep shade can have as many as 100 thylakoids per granum (18). In addition, granal stacks are broader in shade than sun plants. These changes increase the volume of the thylakoid lumen and reduce stroma lamellae and stromal volume relative to volume of the chloroplast.

2. The size of the light-harvesting apparatus increases leading to increased chlorophyll content per chloroplast (255). The size of the light-harvesting antennae associated with PSII increases due to attachment of additional LHCII to the periphery of the PSII antennae complex (18,86). This increases total chlorophyll content in the stacked region of the thylakoids. LHCII complexes are enriched in chlorophyll b compared to the pigment–protein complexes in the inner regions of both PSI and PSII antennae, and to the accessory complexes associated with PSI antennae (LHCI). The chlorophyll a/b ratio of LHCII is about 1.2, while that of LHCI is two to three times this (86). Consequently, the chlorophyll a/b ratio declines from above 2.8 in high-light–grown plans to below 2.5 in low-light–grown plants (18,90,247,255). However, changes in total chlorophyll content of leaves with changing growth irradiance are difficult to predict. At moderate light intensities during growth, leaf chlorophyll content may increase, but as growth irradiance further declines, total chlorophyll content may fall because leaves become thinner with fewer, smaller cells (247).

3. Relative to chlorophyll, the content of most electron transport components (plastoquinone, ATP synthase, cytochrome f, plastocyanin) declines as growth irradiance is reduced (3,84,90,223,256,257,258). The magnitude of the reduction is 25–50%, the more significant decline being observed for components restricted to the stroma lamellae. The ratio of PSII to PSI decreases with reductions in growth irradiance because PSI content tends to change in proportion with chlorophyll content, while PSII content tends to decline relative to chlorophyll (86,90). The exact response of the PSII to PSI ratio will be dependent on light quality as noted above.

4. The content of major soluble enzymes, particularly rubisco, declines as light intensity is reduced (3,18,248,259). Rubisco/CHL ratios

range from 0.5 mmol/rubisco/mol CHL in obligate shade plants grown in low light to over 12 mmol rubisco/mol CHL in obligate sun plants grown at high light (3,248). The potential range of rubisco/Chl is typically smaller within a single species. For example, in bean (*Phaseolus vulgaris*), rubisco/CHL ratios are as low as 2 mmol/mol in shade and approach 10 mmol/mol in sun; in the shade plant *Alocasia macrorrhiza*, rubisco/CHL ratios range between 0.5 and 3 mmol/mol (248).

The functional significance of these responses to low light is to reduce the synthetic and maintenance costs of the photosynthetic apparatus; reduce imbalances that may exist between light harvesting, electron transport, rubisco and the carbon utilization reactions; and improve the capacity to harvest low levels of light in the environment (3,260). These modifications enhance total carbon gain and resource use efficiency under low light, but also restrict the photosynthetic capacity and resource use efficiency under high light.

In shade-grown plants, acclimation to high light involves a reversal of these patterns (261,262). Furthermore, high-light acclimation involves an increased capacity to withstand photoinhibition. Photoinhibition is often important in the initial response of photosynthesis to high light environments, and is reduced or eliminated by either increased capacity for protection, repair, and carbon fixation, or by reduced light absorption capacity (18,237). In nutrient-rich habitats, an increase in the capacity for carbon fixation is a major component of the long-term response to high light, whereas in resource-poor environments, a permanent reduction in antennae size and light-harvesting capacity may instead occur (18,238,248,263).

The Time Course for Long-Term Acclimation

Following either an increase or a decrease in light intensity, changes in the content of the major photosynthetic constituents (PSII content, ATP synthase, rubisco, cytochrome f, chl *a/b* ratio) occur between 1 and 7 days after the change in light intensity, with the greatest rate of adjustment occurring 3–5 days after light levels change (261,262; Sage, unpublished). Most of the adjustments occur in developing leaves, as fully mature leaves are limited in their ability to respond (264). Short-term deactivation of excessive components should theoretically be reversed as excess enzyme capacity is degraded. This is observed with respect to rubisco as shown for kudzu (*Pueraria lobata;* Fig. 10). Following a transfer from high to low light, the activation state of rubisco in newly expanded leaves immediately declines. In the following week, rubisco/CHL declines, indicating a reduction of excess rubisco protein. During this period, the activation state of rubisco partially recovers, reflecting the removal of excess rubisco protein. The failure of the activation state

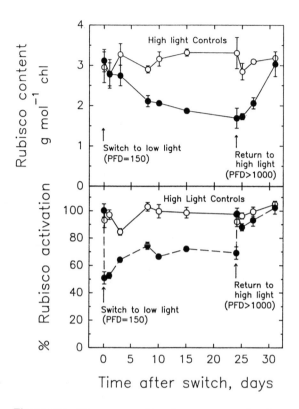

Figure 10 Time course for the response of rubisco content and the rubisco activation in kudzu (*Pueraria lobata*) to a reduction in photon flux density from high light (unshaded) to low light (shaded to provide a maximum PFD of 150 μmol/m^2/sec) at day 0, and following a return to high light on day 24 (Sage, unpublished).

of rubisco to fully recover indicates that a perfect balance between rubisco capacity and thylakoid limitations has not been reestablished. Rubisco is still in excess. Maintenance of excess rubisco is disadvantageous in stable light environments because nitrogen is diverted from limiting processes and carbohydrate is respired away in the maintenance process. Thus, nitrogen use efficiency is lower. However, in naturally shaded light environments with brief episodes of high light, failure to maintain excess rubisco and electron transport capacity would be detrimental since the plant would not be able to exploit high-light episodes. These patterns exhibited by kudzu appear to reflect a compromise between resource use efficiency and the probability of high-light exposure. Kudzu, like most plants, does not experience and is not adapted for constant light intensity during growth. Instead, an optimal allocation pattern at the predominant light intensity during growth appears to be sacrificed for flexibility in dynamic light environments.

Photosynthesis in Dynamic Light Environments

In a shaded environment, much of the incident light received during the day comes in a series of high-light episodes, or sunflecks, that range between 0.1 sec to over 30 min (265,266). The ability of plants to utilize sunflecks is a major determinant of photosynthetic performance in low-light environments. Important photosynthetic characteristics in variable light environments include (a) the magnitude of postillumination CO_2 fixation occurring after a sunfleck; (b) the rate of photosynthetic induction following an increase in PFD, which is a function of the rate of activation of photosynthetic enzymes and opening of stomata; and (c) the ability to maintain a high induction state between high-light episodes (265,267,268). Induction state is the rate of photosynthesis immediately after imposition of high light relative to the maximum steady-state rate of photosynthesis in a fully induced leaf. The value of mechanisms enhancing photosynthesis in dynamic light environments depends on the nature of the light variation. In environments with frequent sunflecks of short duration (<30 sec), maintenance of a high-induction state and postillumination CO_2 fixation are more valuable. During long-lasting sunflecks (>40 sec), postillumination CO_2 fixation loses importance (266).

POSTILLUMINATION CO_2 FIXATION   Postillumination CO_2 fixation can increase the utilization of a sunfleck by 160% compared to the amount of steady-state photosynthesis supported by an equivalent number of quanta (265). High capacity for postsunfleck CO_2 fixation is dependent on an ability to store protons in the thylakoid lumen and an ability to accumulate triose phosphates during the sunfleck (269). The increased size of the thylakoid lumen resulting from extensive granal stacking creates a large volume for storing protons, thereby enhancing the rate of photophosphorylation after the sunfleck (265,269). Accumulation of triose phosphates in the light is important because they can be made into RuBP using ATP derived from the stored protons in the thylakoid lumen. Shade-adapted plants have a greater capacity to accumulate triose phosphates during short sunflecks and this is correlated with higher capacities for postillumination CO_2 uptake than in sun plants (269). RuBP accumulation during sunflecks does not appear to be pronounced, possibly because of restricted carbon flux between triose phosphates and pentose phosphates in the Calvin cycle (269). Accumulation of RuBP during sunflecks could be detrimental because competition between ribulose 5-kinase and PGA kinase for ATP could reduce formation of 1,3-bisphosphoglyceric acid and, in turn, reduce the activity of triose phosphate dehydrogenase (see Appendix 1 for diagram of the Calvin cycle). Reduced activity of triose phosphate dehydrogenase reduces the rate of NADPH oxidation which would feed back on electron transport and lower the efficiency of light utilization during the sunfleck (269).

INDUCTION STATE IN A DYNAMIC LIGHT ENVIRONMENT   In leaves exposed to low light for extended periods (hours), induction of photosynthetic activity following transfer to high light is slow, requiring 10–30 min (14,270). Key phases of induction are a fast phase, involving activation of RuBP regeneration (completion time < 120 sec), and a slow phase, involving activation of rubisco and possibly stomatal opening (265,268). Photosynthetic induction also occurs during repeated bursts of high light if the period between bursts is short enough that significant relaxation of induction doesn't occur (266). Each high-light event promotes further induction until full induction is reached.

The length of time required for relaxation of induction varies between species, with shade-acclimated plants maintaining a high induction state for longer periods then high-light–acclimated plants (265,267). In shade-acclimated plants, relaxation of induction also exhibits a fast (half-time 2–5 min) and a slow phase (half-time 10–30 min). Relaxation of RuBP regeneration probably determines the fast response while the slow response depends on deactivation of rubisco (267,268). Thus, in environments where sunflecks occur frequently (within 10 sec of each other), the regeneration of RuBP is more limiting for photosynthesis, while in environments where sunflecks are further apart, rubisco activation or stomatal opening may be more limiting.

The slower decline in the induction state of shade plants may be related to their greater thylakoid lumen volume. A greater lumen volume in the thylakoids leads to greater proton storage per chloroplast in shade than sun plants, allowing for greater ATP synthesis after the sunfleck (265,269). Enhanced ATP production between sunflecks would promote higher activation of rubisco (by supporting greater activity of rubisco activase) and other enzymes whose activation state is dependent on ATP production.

Stomatal limitations in dynamic light environments vary with species and conditions (265,271). In moist, shaded environments, species may maintain high stomatal conductance despite substantial variation in the light regime, thereby minimizing stomatal limitations during induction (272). With increased evaporative demand and reduced water availability, stomatal conductance is more tightly controlled in response to light intensity and stomatal limitations increase (271).

Long-Term Response to CO_2

In the current atmosphere, short-term changes in CO_2 availability are significant only in situations where boundary layer resistances are large and/or following stresses which close stomata. Diurnal CO_2 gradients exist within the canopy because the lack of turbulent mixing precludes the resupply of CO_2 removed by photosynthesis during the day (161,273,274). In canopies with high photosynthetic rates, CO_2 levels may be depleted by 25% on wind-free days. Similarly, ambient CO_2 partial pressures in dense canopies, valley floors, or close to the ground may be up to 30% greater in the morning because of overnight respiration from soil and vegetation (161,275,276; H. Neufeld, pers. commun.). CO_2 availability is also depressed during drought and low-humidity stress due to stomatal closure. Stress-induced reductions in C_i generally appear to be less than 30%, but when stomatal patchiness occurs, CO_2 levels may drop to near the compensation point. With the possible exception of stress-induced reduction in C_i, changes in CO_2 availability are short-lived and do not appear to be of major consequence over the life of an average plant. In contrast, over evolutionary time, changes in atmospheric CO_2 may have been substantial. For much of Earth's history atmospheric CO_2 levels were well above today's level (277,278), whereas over at least the past 100,000 years atmospheric CO_2 levels have been 30–50% below the current level (279). Plant adaptation to changing CO_2 may explain some evolutionary trends observed over the past 100 million years, such as the evolution of C4 plants (173).

Concern is now focused on plant acclimation to rising atmospheric CO_2 caused by the high rate of fossil fuel consumption and forest destruction (280,281). Current projections are that atmospheric CO_2 levels will double within a century (282) and the CO_2 concentration is rising by an average of 1.5 μbar/year (283). More importantly, CO_2 levels in the atmosphere may triple from the preindustrial level in roughly 200 years. For plants adapted to a preindustrial CO_2 level of 250–280 μbar (284), this represents a substantial increase in CO_2 supply over a period of time in which adaptation is not possible for most species.

Biochemical Responses to Changes in Atmospheric CO_2

If acclimation generally leads to a realignment of limiting and nonlimiting processes, then acclimation to elevated CO_2 may involve a reduction in the degree to which rubisco capacity is excessive by investing less nitrogen in rubisco and/or by investing more of it into limiting processes contributing to RuBP regeneration. Furthermore, plants grown at elevated CO_2 commonly accumulate storage carbohydrates such as starch (49,242,285–287), indicating that the sink capacity of the plant does not increase with CO_2-induced increases in photosynthesis. Therefore, acclimation may also involve a reduction in both the rubisco content and the RuBP regeneration capacity, and an increase in N investment in nonphotosynthetic processes. Results from a wide range of high-CO_2 studies indicate that responses to elevated CO_2 vary between species and growth conditions to the degree that patterns of acclimation are difficult to generalize (10,49). Some trends are beginning to emerge which indicate that plants have a limited ability to acclimate to CO_2 enrichment.

EFFECTS ON THE A/C$_i$ RESPONSE   The range of variation in the response to long-term changes in CO_2 is demonstrated by comparison of A/C$_i$ responses of plants grown at ambient or elevated CO_2 (10,49,288–291). Four common patterns of response are observed:

1. Growth at high CO_2 results in a reduction in A at all C_i, with an increase in O_2 and CO_2 sensitivity of photosynthesis at high CO_2. This pattern is often seen in nutrient-stressed species and older leaves, and is often associated with extensive starch accumulation.
2. In plants grown at high CO_2, photosynthesis declines at low CO_2 while at high CO_2 the rate of photosynthesis and the O_2 sensitivity of photosynthesis increases. The rate of photosynthesis at the operational C_i is often unchanged. This pattern is suggested to be the most advantageous since it may result from reduced rubisco capacity and increased P_i regeneration capacity (10).
3. For plants grown at high CO_2, the rate of photosynthesis and the O_2 sensitivity of photosynthesis is increased at high CO_2 only, indicating a rise in the P_i regeneration capacity. This pattern is apparent in potato (*Solanum tuberosum*), a species in which the tuber can act as a large sink for carbohydrate, and in aspen (*Populus tremuloides*), a clonal species (10,288).
4. No effect is observed on the A/C$_i$ response after growth at elevated CO_2. This pattern is observed in many field-grown plants, and pot-grown plants where specific attempts are made to avoid root restrictions or nutrient limitations (289,292). The lack of any effect in the field indicates that plants may not acclimate to elevated CO_2. Instead, plants respond to perturbations in carbon/nutrient relations imposed by elevated CO_2.

EFFECTS ON RUBISCO   As with the A/C$_i$ responses, no consistent trend is observed in the response of rubisco to elevated CO_2, though responses in rubisco content or total activity reflect effects on photosynthesis (49,293). Species usually exhibit reduced rubisco activity following growth at high CO_2 (10,49,285,294–296). This reduction in rubisco activity may reflect indirect effects of high CO_2 such as faster growth rates and subsequent greater nutrient dilution, excessive carbohydrate accumulation, or restricted rooting volume (49,242,285,286,289,294, 297–299). Conditions generally promoting carbohydrate accumulation are known to promote a loss of rubisco content although the mechanism remains unclear (49).

One means of addressing whether rubisco remains excessive at elevated CO_2 is to examine its activation state. Sage et al. (10) observed that transferring plants of *Phaseolus vulgaris* to elevated CO_2 immediately reduced the activation site of rubisco by 20% or more, and no change in the activation state of rubisco occurred during a subsequent 4-week exposure to high CO_2 (Fig. 11). This change in the activation state of rubisco was not accompanied by any reduction in rubisco content on a CHL basis (Fig. 11), indicating that no shift in the capacity of rubisco relative to RuBP regeneration occurred and that rubisco capacity remained excessive. Four other annual crop and weed species grown at elevated CO_2 also exhibited prolonged reductions in the rubisco activa-

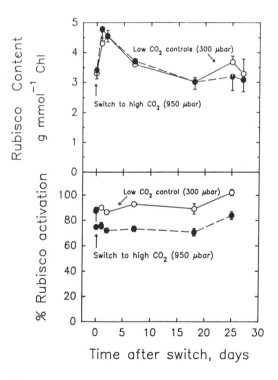

Figure 11 Time course for the response of rubisco content and the activation state of rubisco to a rise in ambient CO_2 from 300 μbar to 950 μbar on day 0 in *Phaseolus vulgaris*. (From Ref. 10.)

tion state (10), leading to the conclusion that acclimation to elevated CO_2 may not involve a proportionally greater reduction in rubisco content than the level of enzymes contributing to RuBP regeneration. By contrast, in *Pinus taeda* (loblolly pine), seedlings grown at elevated CO_2 (650 μbar) exhibited reduced rubisco content and an increased activation state of rubisco relative to seedlings grown at normal CO_2 (299a). This indicates that acclimation to elevated CO_2 in pines does involve a relatively greater reduction in rubisco allocation.

To further examine the photosynthetic response to growth CO_2, Sage and Reid (300) examined the long-term response of *Phaseolus vulgaris* transferred to low atmospheric CO_2(from 360 to 200 μbar). No long-term change in rubisco content or activation state was observed since plants at both low and ambient CO_2 level had the same amount of rubisco. In addition, the initial slope of the A/C_i response was unaffected by growth CO_2, indicating no change in rubisco capacity (Fig. 12). The CO_2-saturated rate of photosynthesis and the CO_2 saturation point

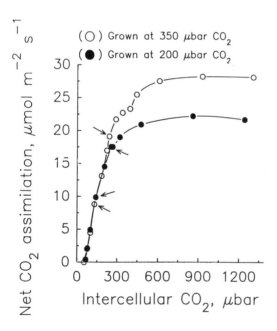

Figure 12   The CO_2 response of photosynthesis in *Phaseolus vulgaris* grown at a mean ambient CO_2 partial pressure of 200 or 350 μbar. Lower arrows indicate the operational C_i at a measurement CO_2 partial pressure of 200 μbar. Upper arrows indicate the operational C_i at a measurement CO_2 partial pressure of 350 μbar. (From Ref. 300.)

declined at low-growth CO_2, indicating that the P_i regeneration capacity declined in response to reduced carbohydrate production (Fig. 12). The A/C_i response of bean plants grown at low CO_2 indicates that acclimation to changing CO_2 does not involve allocation to rubisco capacity but does involve modulation of the capacity of starch and sucrose synthesis to match the rate of triose phosphate production.

Diffusion Limitations

Long-term exposure to CO_2 enrichment generally causes a reduction in stomatal conductance that is equivalent to or greater than that observed in species exposed to elevated CO_2 over the short term (Fig. 6; 33,291, 301,302). This reduction may maintain C_i/C_a constant, though often it is substantial enough to reduce C_i/C_a and increase the degree to which stomata limit photosynthesis (Fig. 6; 291,301–305). Plants grown at elevated CO_2 often have reduced C_i/C_a ratios at all C_i compared to plants grown at low CO_2 (33,291,306), indicating that the mechanism linking stomatal conductance to C_i has been affected during acclimation to elevated CO_2. The reduction in stomatal conductance with long-term exposure to higher CO_2 may result in part from decreased stomatal frequency over the leaf surface. Stomatal frequency is reported to have decreased linearly with increased atmospheric CO_2 since the preindustrial era (307,308), although these reports have been questioned (309). Experimental studies indicate that further reductions in stomatal density will accompany future CO_2 enrichment (310).

The transfer resistance from the intercellular air space to the stroma may also be sensitive to growth CO_2, although no direct measurements of this have been reported. Carbonic anhydrase activity increases as growth CO_2 decreases (311,312) and this may speed diffusion of CO_2 to the carboxylation site in the chloroplast stroma (15,16).

Acclimation to Elevated CO_2?

In contrast to light acclimation, which follows a characteristic pattern in most species, long-term responses to increasing CO_2 availability are highly variable except when nutrient supply and sink capacity remain high (49,289,291,313). With adequate nutrient and sink capacity there is often little difference between short- and long-term responses of photosynthesis to atmospheric CO_2 enrichment. Responses that are observed can frequently be attributed to excess carbohydrate accumulation rather than direct responses to changing CO_2 (10,49,243). This apparent lack of CO_2 acclimation is not particularly surprising, since variation in atmospheric CO_2 over the life of individual plants has been relatively minor until industrial times. Furthermore, light responses are mediated by known photoreceptors, whereas no CO_2 sensor in the mesophyll has been identified. Therefore, not only have plants lacked the need to acclimate to changing CO_2 in the atmosphere, they may also lack the ability to directly sense and respond to rising atmospheric CO_2.

Temperature

Temperature has a multitude of effects on the photosynthetic apparatus since each process potentially has a unique thermal dependence (23,314). Thermal extremes directly damage biochemical systems through protein denaturation, loss of membrane integrity, photoinhibition, and ion imbalance. Thermal stress will not be covered in depth here and the reader is referred to reviews on the subject by Kyle et al. (315), Long and Woodward (316), and Katterman (317). In this treatment, coverage will focus on response mechanisms over a nonstressful range of temperatures.

Temperature Response Curve in C3 Plants

In most C3 plants, photosynthesis increases with rising temperature up to between 20°C and 30°C, where a maximum CO_2 assimilation rate typically occurs (Fig. 13A). However, plants adapted to thermal extremes can exhibit thermal optima well above 30°C or below 20°C. For example, the arctic plant *Saxifraga cernua* can have a photosynthetic tempera-

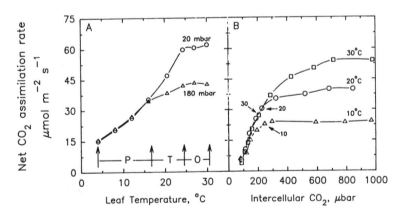

Figure 13 Temperature response of photosynthesis in *Populus fremontii* at either 180 mbar O_2 or 20 mbar O_2. (B) CO_2 response of photosynthesis at 10°, 20°, or 30°C in the same species at 180 mbar O_2. In A, the letter P indicates the region of the temperature response that is limited by P_i regeneration. T indicates the region where thylakoid processes such as electron transport may be limiting, and O indicates the region of optimal temperature. In B, the arrows indicate the operational C_i at an ambient CO_2 partial pressure of 300 μbar CO_2 for each of the three curves. (From Ref. 109.)

ture optima near $10°C$ (317a). The breadth of the thermal optimum of photosynthesis varies, but in most species the rate of photosynthesis declines above $30-35°C$. Temperature optima tend to narrow as photosynthetic capacity increases (318,319). Growth temperature has a pronounced effect on the photosynthetic temperature response with the optimum temperature of photosynthesis shifting toward the growth temperature. For every $5°C$ change in growth temperature between $10°C$ and $35°C$, the temperature optimum shifts approximately $2-3°C$ (135,314,317a,318). Temperatures promoting the highest rates of photosynthesis tend to correspond to thermal regimes to which the plant is naturally adapted. Acclimation to temperature regimes significantly different from optimal growth conditions reduces the maximum rate of photosynthesis but enhances photosynthesis at thermal extremes in the direction of the temperature change (23,314).

The temperature response of photosynthesis is also dependent on atmospheric humidity (318). As leaf temperature increases, the difference in water vapor concentration between the intercellular air spaces and the ambient atmosphere increases, unless supplemental humidification is provided. The increase in the leaf-to-air-vapor concentration difference promotes stomatal closure, further inhibiting photosynthesis at supraoptimal temperatures (320). Consequently, temperature response curves conducted at constant differences in vapor pressure difference (VPD) between leaf and air tend to have higher optima than those determined where the humidity gradient increases as temperature increases (321).

Oxygen sensitivity of photosynthesis increases with temperature, reflecting effects of two processes. First, at low temperature (below $10-20°C$, depending on species), photosynthesis is typically limited by P_i regeneration capacity and as a result is insensitive to changing O_2 and CO_2 (Fig. 13A; 109,183,184,322). As temperature increases, P_i regeneration becomes nonlimiting and in response photosynthesis becomes sensitive to O_2 and CO_2 (109,184). Second, as temperature increases, the oxygenation-to-carboxylation ratio increases (171,172). Therefore, if rubisco capacity or thylakoid processes are limiting photosynthesis, the inhibition of photosynthesis by photorespiration and, in turn, the sensitivity of photosynthesis to O_2 increases with the increase in temperature (109). Increases in photorespiration because of reduced CO_2 or elevated O_2 can reduce the temperature optimum in addition to reducing the maximum rate of photosynthesis (23,323).

Temperature Effects on the CO_2 Response of Photosynthesis

Temperature affects the A/C_i curve as follows:

1. The CO_2 compensation point increases with increasing temperature (185). This rise is due to stimulation of dark respiration and pho-

torespiration with increasing temperature, and is most pronounced above 20°C (23,170,172,191,324). Typical CO_2 compensation points range from near 35 μbar below 15°C to over 60 μbar above 35°C (172,185).

2. The initial slope is little affected by temperature, though it may decline slightly at low temperatures (<10°C) and above the temperature optimum (Fig. 13B; 1,109,185,191,325). The lack of a temperature effect on the initial slope reflects the thermal response of the kinetic properties of rubisco (23). Both the k_m of rubisco for CO_2 and the V_{max} of rubisco have similar thermal dependencies between 10°C and 30°C. Each has a Q_{10} near 2 (23). As demonstrated in Figure 14, the rate of the carboxylation reaction at CO_2 levels below the K_m are little affected by the rise in temperature when the V_{max} and the k_m of a reaction are equally stimulated by temperature.

3. Temperatures between 10°C and 30°C markedly stimulate the rate of photosynthesis in herbaceous plants at C_i above the initial slope region of the A/C_i response (Fig. 13b; 1,109,185). The CO_2-saturated rate of photosynthesis is often strongly stimulated, increasing with a Q_{10} between 1.5 and 2 (109,185). Above 30°C, increasing tempera-

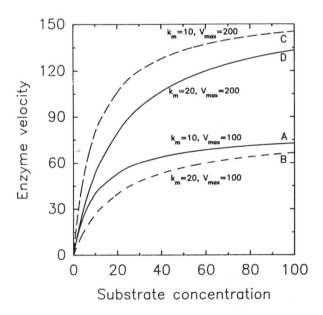

Figure 14 Effect of doubling k_m, V_{max}, or both k_m and V_{max} on the velocity of a hypothetical enzyme as a function of substrate concentration. Data calculated from the Michaelis–Menton equation.

tures may inhibit photosynthesis at elevated C_i in a pattern highly dependent on species and growth conditions (1,326).

4. The CO_2 saturation point increases with temperature, reflecting the lack of a thermal effect on the initial slope of photosynthesis and the marked stimulation of the CO_2-saturated rate of photosynthesis (Fig. 13B; 1,109,184,185,191). As a result, the range of C_i where the A/C_i response is linear increases with increasing temperature, indicating that rubisco limitations on photosynthesis are more likely at warmer temperatures.

At cooler temperatures, the CO_2 saturation point may fall below the operating C_i in ambient air. When this occurs, P_i regeneration is limiting photosynthesis in ambient air, and the temperature response of P_i regeneration determines the temperature response of photosynthesis in ambient air (109,184).

The temperature response of photosynthesis at elevated CO_2 reflects the temperature response of electron transport and starch and sucrose synthesis. Electron transport components have a Q_{10} between 1.5 and 2 up to the temperature optimum of photosynthesis at high CO_2 (23,191,314,326a). Above this optimum, electron transport can be inhibited because of high-temperature effects on thylakoids (318,324,326a). Photoinhibition also becomes prominent at high temperature (327). These effects reversibly inhibit photosynthesis at temperatures below which direct heat damage occurs (23,327–329).

Like electron transport, components of starch and sucrose synthesis have a Q_{10} of 2 or greater (239). This explains the strong temperature response of P_i regeneration–limited photosynthesis.

What Limits Photosynthesis at the Temperature Optimum?

P_i regeneration is limiting when temperature stimulates photosynthesis and O_2 has no effect. Thylakoid limitations are dominant when photosynthesis is stimulated by temperature and O_2 reduction enhances photosynthesis (Fig. 13). However, at the temperature optimum, limitations are less clear. In numerous cases, the operational C_i at the temperature optimum occurs on the initial slope of the A/C_i curve, indicating that rubisco is limiting (Fig. 13; 109,191). In other cases, the temperature optimum reflects the thylakoid-dependent capacity for RuBP regeneration because the operational C_i occurs above the initial slope region of the A/C_i curve. Reduced RuBP regeneration capacity at elevated temperature is associated with reduced rates of electron transfer and photophosphorylation (1,191,326,329). Rubisco also deactivates at elevated temperature, possibly in response to limitations in thylakoid processes (330,331).

Temperature Effects on Stomatal Limitations

In natural environments, increases in air temperature result in greater water vapor gradients between leaf and air and this promotes stomatal closure, increasing the stomatal limitation on photosynthesis (109,135,320). In addition, diffusional limitations will often increase with rising temperature because stomatal conductance has a greater effect on photosynthesis when the response of A to C_i at the operational C_i increases (109).

At a constant vapor concentration gradient between leaf and air, stomatal conductance tends to increase with increasing temperature, though the effect is often not pronounced (135,140a,320,332). In castor bean (*Ricinus communis*), increases in stomatal conductance with temperature are pronounced at low VPD only; at elevated VPD, stomatal conductance is insensitive to temperature (140a). In some species, such as sweet pepper (*Capsicum annuum*), stomata open with increasing temperature even though water vapor gradients become high (109,318). This response allows transpirational cooling to offset higher temperatures, but at a high cost in terms of water transpired.

At low temperature, stomatal limitations disappear completely when the CO_2 saturation point is less than the operational C_i (109,184). This is largely because the photosynthetic biochemistry loses CO_2 sensitivity at low temperature rather than stomatal opening (109). However, sudden low-temperature exposure can lead to high diffusional limitations as patchy stomatal closure may occur in response to chilling stress (333). This effect may be mediated by ABA (334).

Acclimation to Temperature Change

The ability to acclimate to temperature change is species-dependent with plants from thermally variable habitats typically exhibiting a greater and faster ability to acclimate (314,335). For example, short-lived desert annuals exhibit less shift in the temperature optimum of photosynthesis with changes in growth temperature than perennial forbs and shrubs which are photosynthetically active for longer periods of the year (335,336). The rate of acclimation is also highly variable, lasting anywhere from a few hours to weeks (314). While much of the variation is species-dependent, it is also dependent on the nature of the temperature change and the processes which become limiting in the new thermal environment.

Transfer of plants to low temperatures where P_i regeneration capacity becomes limiting is followed by an increase in the rate of photosynthesis and a relatively rapid appearance (within a day) of oxygen and CO_2 sensitivity of photosynthesis (101,183,184,337). Feeding P_i to leaves through the xylem stream also enhances photosynthesis in plants transferred to low temperature (183,184). These results indicate that rapid acclimation (hours to a few days) to low temperature involves an increase in the P_i regeneration capacity so that it no longer limits photosynthesis. The mechanism for increased P_i regeneration at low temperature is not clear, though it has been proposed that the P_i regeneration limitation is eliminated by an adjustment of P_i and photosynthetic metabolites in the cytosol and chloroplast to levels that favor higher activity of starch and sucrose biosynthetic enzymes (101,184,337). Increased activation of numerous enzymes (fructose-1,6-bisphosphate, sedoheptulose-1,6-bisphosphate, rubisco) occurs within minutes of a transfer to low temperatures, and may contribute to increases in photosynthesis often observed shortly after the drop in temperature (338). In addition, leaves with low carbohydrate levels are reported to have reduced photosynthesis at low temperatures compared to leaves of moderate carbohdyrate content, indicating that the carbohydrate status of the leaf influences acclimation potential (337).

A rapid phase of acclimation also occurs in response to a sudden increase in temperature of a few degrees or more (324). This acclimation is dependent on de novo protein synthesis and may be related to the synthesis of heat shock proteins (13,324). Heat shock proteins are proposed to compensate for deleterious effects of increased temperature on protein structure and assembly (13). It seems likely that optimal ion and metabolite levels may also be reestablished at warmer temperatures, though data on this is lacking.

Longer periods of acclimation (4–10 days) are associated with changes in protein levels, modification of the structure of major photosynthetic proteins, morphological change, and altered membrane composition (23,339–343). A reduction in temperature is associated with increased levels of soluble protein, notably rubisco, fructose-1,6-bisphosphatase, sucrose phosphate synthase, hexokinase, sedoheptulose bisphosphatase, and hexose phosphate isomerase (338). Leaves tend to be thicker at low temperature, having more cell layers and increased nitrogen content per leaf area (341,342). Thylakoids may also exhibit moderate adjustments. In rye (*Secale cereale*) at low temperature, the number of thylakoids per granum decreases, chlorophyll content may rise, and the maximum activity of PSI may increase (341). These changes in rye are correlated with an increase in photosynthesis and an increase in RuBP regeneration capacity, but it is not clear if they cause it (341). In *Saxifraga cernua*, plants acclimated to 10°C had greater electron transport activity and greater activity of PSI and PSII at a measurement temperature of 10°C than plants grown at 20°C (326a). These changes in S.

cernua were not accompanied by changes in levels of thylakoid proteins, indicating that acclimation to reduced temperature involves changes in the specific activity of electron transport components.

Structural modification of proteins may increase photosynthetic performance at thermal extremes (314). For example, the large subunit of rubisco from cold-acclimated plants is reported to have a different quaternary structure and exposure of sulfhydryl groups than preacclimated controls, and this may contribute to increased rubisco V_{cmax} and catalytic turnover rate at low temperatures (314,344).

Reductions in membrane integrity occur as a result of shifts in temperature beyond the range to which plants are acclimated (340,345). These interfere with protein–lipid interactions, ion permeability, and diffusion through the lipid bilayer. Proposed modifications in membrane structure as a result of temperature reduction include increased concentration of intrinsic proteins, increased fraction of polyunsaturated fatty acids, and altered ratios of the types of lipid incorporated into the membrane (340,341,345). These changes reduce the temperature at which the membrane undergoes a transition from a fluid to a gel-like state. As temperatures increase, the opposite trends occur, presumably increasing the stability of the membrane and lipophilic protein complexes at high temperature (345).

Plants adapted to low temperature exhibit similar responses as plant acclimated to cold climates. In an extensive comparison of species from high and low altitudes, Körner and coworkers observed that plants at high altitude had lower rates of leaf expansion, thicker leaves, higher leaf nitrogen content, and increased photosynthetic capacity relative to low-altitude species (342,346). The enhancement of photosynthesis occurred at all C_i (347), indicating that each photosynthetic component was stimulated by the growth conditions.

Atmospheric Humidity

Reductions in atmospheric humidity generally inhibit photosynthesis, particularly when the availability of soil water is low (158,159,347). These effects arise because reduced humidity increases the rate of transpiration, which in turn causes stomatal closure (348). The relationship between transpiration, E, and humidity is described as

$$E = g_l(W_{leaf} - W_{air}) \qquad [14]$$

where g_l is the leaf conductance to water vapor, W_{leaf} is the absolute humidity (water vapor concentration, usually expressed in terms of mole fraction) in the intercellular air spaces of a leaf, and W_{air} is the absolute humidity in the air (115). Differences in absolute humidity between leaf and air are also expressed in terms of VPD. A VPD of <10 mbar is mild; a VPD of >30 mbar represents a severe gradient (158). Since the intercellular air spaces within the leaf are assumed to be saturated with water vapor, and the saturation vapor pressure is an exponential function of temperature, absolute humidity inside the leaf depends on leaf temperature (116,124). In the atmosphere, absolute humidity is largely dependent on evaporation sources, precipitation, or movement of large air masses. Absolute humidity in the ambient air is often uniform throughout the day, though relative humidity (the absolute humidity divided by vapor content of the air at saturation) typically declines as air temperature increases. Since absolute humidity is relatively stable during the day, increases in leaf temperature (arising from increases in air temperature and higher radiation loads on the leaf) are primarily responsible for increases in VPD.

Stomatal conductance generally declines as VPD increases beyond 5 mbar (140a,158,159), though the response varies with species and water status of the leaf (20,158,348). Both water stress and ABA heighten stomatal sensitivity to VPD, initiating stomatal closure at lower VPD as well as promoting greater closure at any given VPD (349). At moderate VPD (10–20 mbar), the decline in stomatal conductance with increasing VPD reduces the degree to which transpiration increases but is not large enough to actually reduce the transpiration rate. Responses to moderate VPD are freely reversible, showing little hysteresis (348). In contrast at high VPD (>20 mbar), stomatal conductance can decline enough so that transpiration is actually reduced (348–350). High VPD also causes a depression of photosynthesis which can last for hours upon return to low VPD, and promotes oscillations in stomatal conductance (351; Mott, pers. commun.; Sage, unpublished).

Increases in VPD reduce photosynthesis by either (a) closing stomates and thus reducing C_i in a uniform pattern over the leaf surface; (b) closing stomates and reducing C_i in a nonuniform or patchy pattern; or (c) reducing the biochemical capacity of the photosynthetic apparatus. Reduction in biochemical capacity can cause a reduction in stomatal conductance since stomata are regulated to track the rate of photosynthesis (107,117,158). Reductions in photosynthesis occurring with mild to moderate VPD (<15 mbar) result largely from homogeneous reductions in stomatal conductance and C_i (140a,348,352; Mott, pers. commun.). As VPD increases above 20 mbar, however, further reductions in conductance often reflect patchy stomatal closure (52a,140a,352,352a; Mott, pers. commun.). The decline in transpiration observed at high VPD appears to be associated with nonuniform stomatal closure, and oscillations in stomatal conductance may reflect the opening and closing of stomatal patches (348,352a; Mott, pers. commun.). At high VPD, estimated C_i is often unaffected despite substantial reductions in photosynthesis (52a,140a,352,353,353a). This was initially interpreted to indicate that low humidity directly causes a reduction in the biochemical capacity for photosynthesis (353). However, no biochemical lesion has been identified in response to low humidity (348) and the recent reports of patchy stomatal closure at high VPD explain much of the reduction in photosynthesis at constant C_i (52a,140a,153,352). In summary, the effects of humidity on photosynthesis appear to result largely from stomatal closure, with a major exception being when high transporation causes severe drought stress in the leaf.

What are the mechanisms linking transpiration to stomatal conductance? Two hypotheses are favored. First, transpiration induces a reduction in leaf water content such that guard cells lose turgor. This is a hydropassive feedback mechanism in which stomata close in regions of localized water deficit (20,348,349). However, such feedback control cannot account for reductions in transpiration because any decline in transpiration will lead to a partial recovery of leaf water status and guard cell turgor (348,349). Alternatively, if transcuticular evaporation from the epidermis is significant enough to affect epidermal water content and, in turn, guard cell turgor independently of bulk leaf water status, then guard cells could continue to lose turgor despite reduced whole-leaf transpiration and improved leaf water status. This represents a hydropassive feedforward mechanism (354,355). Evidence is lacking for feedforward control based on cuticular transpiration, and the mechanism remains controversial (248,349,355).

The second hypothesis is a hydroactive control mechanism whereby a signal initiated in the bulk leaf tissue induces stomatal closure. This signal may be ABA, which is released from the mesophyll tissue when leaves become desiccated or transported from the roots in the transpiration stream (35,144). Release of ABA is well documented in water-stressed tissue, particularly wilted leaves, and is known to cause stomatal patchiness (121,123,144). Desiccated tissues have increased sensitivity to ABA (356) and high rates of transpiration can induce localized desiccation within leaves (357). Taken together, these results may explain why stomatal conductance becomes patchy at high VPD.

Drought Stress

Early studies recognized that desiccation reduces stomatal conductance, and for a many years there was a widespread view that stomatal closure was the primary cause of photosynthetic decline in drought-stressed leaves. Subsequent analysis of A/C_i responses from water-stressed leaves often demonstrated little reduction in C_i, indicating that stomata close in response to reductions in the capacity of the photosynthetic biochemistry (50,53). However, recent analyses reveal that in pot-grown plants, stomatal closure is actually the primary mechanism for reduction in photosynthesis in response to mild drought stress and no biochemical lesions have been identified (68,120,122,144,153,358,359). In many species, particularly those with bundle sheath extensions spanning the leaf mesophyll, stomata close in a patchy pattern, and this accounts for observed reductions in photosynthesis at constant C_i (52a,120,122,153,155,359). Evidence for this includes nonuniform $[^{14}C]CO_2$ fixation determined by autoradiography of stressed leaves and full recovery of photosynthesis by increasing CO_2 from 0.035% to saturating levels of CO_2 (10–15%; 52a). In mildly stressed leaves of *Phaseolus vulgaris*, patchiness fully explains photosynthetic reductions (359). In species lacking patchiness, reductions in C_i follow mild drought stress (90,359,360).

Mild drought stress is associated with less than a 10–20% reduction in leaf water content while severe drought stress is associated with loss of turgor and reductions in cell volume by over 20% (155,361). Reductions in cellular volume lead to increased ion concentration, which can reduce the biochemical capacity for photosynthesis by inhibiting the thylakoid capacity for RuBP regeneration (25,362–364). In severely drought-stressed leaves, RuBP levels, maximum quantum yield, the rate of CO_2-saturated photosynthesis, and dark variable fluorescence decline (25,26,68,155,363,365,366). The reduction in variable fluorescence in the dark indicates that photoinhibition has occurred (365). Rubisco capacity does not appear to account for reductions in the biochemical capacity of photosynthesis in drought-stressed leaves because rubisco activity and activation are unaffected by drought stress (153,366). Photophosphorylation may also be inhibited in severely desiccated leaves (365), though recent measurements of ATP synthase activity during drought stress discount this (358). Starch synthesis decreases in drought-stressed leaves, and sucrose synthesis may increase, leading to an increase in soluble sugars (26,102,367). However, species experiencing patchy stomatal closure may deactivate sucrose synthesis in response to low C_i (74,120), and this may promote P_i regeneration limitations (Sharkey, pers. commun.).

ABA, and possibly other signals such as calcium, plays a major role in mediating responses to drought stress throughout the plant (35,143,144). Many studies demonstrate that roots sense drying soil and release ABA to the xylem stream where it travels to the shoot and promotes stomatal closure, often in a patchy fashion (19,35,121,122). This release of ABA occurs with mild to moderate reduction in root water content and is interpreted as an early warning response which reduces the shoot demand for water well before severe desiccation sets in (34,35,159). ABA is not known to directly inhibit the photosynthetic biochemistry

(120) but does restrict shoot growth by inhibiting cell expansion and possibly meristematic activity (35,144). Reduction in shoot growth dramatically reduces canopy development, even when stomata and photosynthesis are marginally affected (368). This reduction in canopy size is more responsible for the loss in whole-plant carbon gain following drought stress than the loss of photosynthetic capacity of individual leaves (369,370). On the other hand, ABA stimulates root growth in drought-stressed plants which, combined with the reduction in shoot size, improves water status, productivity, and survival relative to plants where ABA-induced shifts in development do not occur (34).

Determining the role of ABA in the desiccation response is difficult because assays of leaf ABA content may be unable to resolve active ABA from larger pools of stored and inactive ABA. Also, ABA activity is influenced by leaf water content, with desiccated leaves being more responsive to ABA fed through the transpiration stream (356). This could occur because desiccated tissues have increased tissue sensitivity and/or because release of endogenous ABA from desiccated mesophyll cells enhances ABA concentrations at the receptor site. This helps explain stomatal patchiness because regions of the leaf with lower water content would be more sensitive to ABA arriving from the transpiration system.

Long-Term Responses to Drought Stress

A major criticism of many drought studies is that plants are grown in relatively small pots where stresses develop over the course of a few days. In the field, water stress typically develops over longer periods and is accompanied by modifications in osmotic potential, cell wall elasticity, root-to-shoot ratio, and hydraulic conductivity, all of which increase resistance to soil drought (34,35,371). Osmotic adjustment is particularly important during slowly developing drought stress because it minimizes reduction in chloroplast volume which is postulated to inhibit photosynthetic capacity during drought stress (362,363). In high-light environments, photoinhibition may also be common during long-term drought (371a) and may contribute to drought-induced reductions in the capacity of thylakoid-dependent RuBP regeneration.

Despite this information, a mechanistic understanding of photosynthetic responses to slowly developing drought is surprisingly vague given the large number of drought-response studies conducted. Patchy stomatal closure does not appear to be prevalent during slowly developing stress in potted or field-grown plants, in contrast to what is observed in the same species following rapid imposition of stress (52a,157,157a,359). Reductions in C_i and increased stomatal limitation often occur in slowly stressed plants (360,372,373), but this does not entirely account for reduced photosynthesis unless stomatal conductance become patchy

(157). Reductions in A at all C_i in slowly stressed leaves of the desert shrub *Encelia farinosa*, the desert forbs *Malvastrum rotundifolium* and *Lupinus arizonica*, and field-grown sunflower (*Helianthus annuus*) indicated that reductions in photosynthesis occur as a result of reduced rubisco capacity as well as reduced RuBP regeneration capacity (157,372,373). In *Encelia*, the reduction in photosynthesis with drought was accompanied by a proportional decline in leaf nitrogen content, indicating that adjustment to drought invovles reduced allocation of nitrogen to photosynthesis (372). By contrast, Matthews and Boyer (374) measured A/C_i responses of slowly stressed sunflower plants growing in pots; they reported little effect on the initial slope and a pronounced reduction in the rate of A at elevated C_i (374). This indicates that thylakoid-dependent RuBP regeneration rather than rubisco capacity is affected by long-term drought stress. Consistent with this, rubisco activity in field-grown sunflower is unaffected by slowly developing drought stress (360). At all stress levels in sunflower, *Encelia*, and *Malvastrum*, and at moderate stress in *Lupinus*, photosynthesis was sensitive to increasing CO_2 at high C_i (372–374), indicating that P_i regeneration is not an important limitation during long-term drought. Severely stressed *Lupinus arizonica* exhibited CO_2 insensitivity at elevated CO_2 (373), indicating a possible P_i regeneration limitation.

A common problem with studies of long-term drought conducted prior to 1988 is that the occurrence of stomatal patchiness was generally not considered and therefore cannot be ruled out as a major factor in the responses. Further studies are needed to definitely identify what role, if any, patchiness plays in long-term stress and, in the absence of patchiness, what mechanisms account for reductions in photosynthesis during moderate, and slowly developing, drought regimes.

Nitrogen

Nitrogen is the major mineral constituent invested in the photosynthetic apparatus, occurring in proteins, chlorophyll, and membrane constituents. Up to 80% of total leaf nitrogen is associated with photosynthesis (3). Consequently, any shortfall in N supply could affect N investment in photosynthesis. An important issue regarding N effects on photosynthesis is whether N allocation between photosynthetic and nonphotosynthetic processes declines as N deficiency increases. If allocation to photosynthetic tissues is constant, N deficiency affects photosynthesis simply by causing a reduction in the amount of photosynthetic machinery. An alteration of N allocation could either compensate for N deficiency by increasing N investment into leaves or aggravate N deficiency by allocating proportionally less N to leaves. In herbaceous plants, the latter possibility appears to be the case since N allocation to roots increases as N supply declines (375–377). Increased N allocation may be regulated by ABA, which is released from N-stressed roots and travels to the shoots where it inhibits leaf and stem expansion (3,143,378). While reduced photosynthetic capacity is significant, the reduction in leaf area is usually more responsible for reduced carbon gain by N-deficient plants (2,3,377).

The reduction in photosynthetic capacity per leaf area at low N is largely because of reduced biochemical machinery in the mesophyll cells (3,72,73,90,248,379). Increased stomatal limitations do not appear to be widespread despite the observation that ABA may be released by N-deficient roots (1,185,379). Based on A/C_i curves, the operational C_i occurring at ambient air is little affected by N nutrition except in very N-deficient leaves, where it may actually increase (185,379). The lack of an ABA effect on stomatal conductance may be related to an absence of water stress within leaves. Stomatal patchiness is not recognized as an important phenomenon following N stress, though Woo and Wong (380; discussed in Terashima et al., 123) may have witnessed it in low-N plants exposed to high CO_2.

Mechanistic studies of photosynthetic N responses recently focused on partitioning of N between the major components of the photosynthetic apparatus in order to determine whether N allocation within leaves changes with N availability. Studies of A/C_i responses do not generally detect a shift in balance between rubisco and RuBP regeneration as a function of N. Inflections in the slope of the A/C_i curve, which indicate a shift in limitation from rubisco to RuBP regeneration, occur at approximately the same C_i across a broad range of N (1,3,305,381). In some species, however, this is not the case. For example, in *Chenopodium album*, high-N plants exhibited reduced CO_2 saturation points, and decreased O_2 and CO_2 sensitivity when compared to moderate- and low-N plants (185). This indicates that high-N *C. album* plants have a reduced capacity for P_i regeneration relative to other photosynthetic processes.

In contrast to gas exchange studies, biochemical surveys of N allocation within the photosynthetic apparatus indicate that as leaf N declines, the fraction of N invested in rubisco is reduced (3,72,90,259). In C3 crop plants, rubisco as a function of leaf N ranges from 30% in well-fertilized leaves to below 15% in low-N leaves (3,72,73). By contrast, N investment into thylakoid components and Calvin cycle enzymes is relatively constant, composing approximately 25% of total leaf N across a range of N availability (3,72). Nonphotosynthetic components (mitochondria, nucleic acids) appear to be independent of leaf N, representing a constant 30–46 $mmol/m^2$ of N in mesophytic leaves (3,72). Consequently, as N decreases, the fraction of N invested in nonphotosynthetic processes increases from about 20% to 40% of leaf N (3). Overall, as leaf N decreases, the greatest shift in allocation is to produce less soluble protein, particularly rubisco. This should increase the C_i where a rubisco limitation gives way to an RuBP regeneration limitation. However, this is not observed. According to Evans and Terashima (381), high-N leaves may have reduced stromal CO_2 concentration because as N increases, mesophyll transfer resistance to CO_2 increases, possibly because chloroplast size increases (see also von Caemmerer and Evans, 110). Increased rubisco activity may compensate for reduced stromal CO_2 so that the balance between electron transport and rubisco capacity is conserved (381).

In most species, the response of A to leaf N per area is linear or nearly linear (319). Herbaceous annuals have the steepest slope and typically reach the highest leaf N contents (7). This combination of steep slope and high leaf N content explains the high photosynthetic capacity of many herbaceous annuals, particularly those adapted to open, highlight environments. The steep A-to-N response arises because as N investment increases above a minimum level most of the additional N is invested into the photosynthetic apparatus, particularly rubisco and other soluble enzymes (3). If the environment can support high rates of photosynthesis, then a high return on the N investment is realized. However, in most environments, light, water, and other nutrients are insufficient to support high rates of photosynthesis for extended periods, so that leaves with high photosynthetic capacities may actually have a reduced nitrogen use efficiency because a substantial fraction of N invested in the photosynthetic apparatus is not fully utilized (4,6). To avoid inefficient use of N, species from environments with reduced resource availability restrict their investment in photosynthesis and consequently have lower maximum leaf N contents than plants adapted to resource-rich habitats (2,4).

In plants adapted to low resource levels, such as shade plants, xerophytes, and low-nutrient–adapted schlerophylls, processes compensating for the low resource availability increase the cost of photosynthesis in terms of N (4,259,382). This reduces the response of A to N and lowers nitrogen use efficiency. For example, in shade-adapted plants, the investment of N in thylakoid components is substantial and somewhat fixed, while the capacity to synthesize substantial amounts of rubisco is limited (3). This enhances photosynthetic performance in low light but prevents the plant from taking advantage of high light levels where the maximum return of the N investment would be realized. Long-lived leaves are common in low-resource environments but require higher N expenditures in nonphotosynthetic functions, such as synthesis and storage of defense compounds or synthesis of wall material (250). Furthermore, in drought-stressed environments, photosynthetic capacity and nitrogen use efficiency may be reduced by increased diffusion limitations from stomatal closure and sunken stomata, or reduced light interception from leaf rolling, vertical leaf orientation, or increased reflectance (2,382). While these modifications reduce maximum nitrogen use efficiency in low-resource environments, performance and nitrogen use efficiency is greater than what would be observed if plants adapted to high-resource environments were transplanted to low-resource sites.

CONCLUSION: PHOTOSYNTHESIS IN A CHANGING GLOBAL ENVIRONMENT

This chapter has examined responses of plants to six major environmental parameters, all of which may change profoundly as global climate changes over the next few centuries. Acclimation and adaptation have been well documented for five of the parameters—light, temperature, nitrogen, humidity, and water availability—and appear to partly involve a readjustment of resource allocation so that no one physiological or environmental factor is completely limiting photosynthesis. However, in the case of the sixth parameter, CO_2, acclimation and adaptation have not been clearly demonstrated though theoretical patterns based on economic and physiological considerations have been proposed (10,33). If plants are indeed limited in their ability to acclimate to elevated CO_2, then the relatively rapid rise in atmospheric CO_2 over the next century may lead to a situation whereby photosynthesis in C3 species is maladapted for the atmospheric environment. In this situation, natural selection will tend to favor those genotypes which are better adapted to a high CO_2 environment. Successful traits may include reduced rubisco expression relative to RuBP regeneration capacity. However, rather than waiting for results of the ongoing global experiment in CO_2 enhancement, studies of fossilized remains from previous episodes of high atmospheric CO_2 may reveal traits that will adapt photosynthesis, and plants, to our future atmosphere.

APPENDIX 1: PHOTOSYNTHETIC CARBON REDUCTION CYCLE (CALVIN CYCLE), STARCH SYNTHESIS, AND SUCROSE SYNTHESIS (FIG. A1)

Small open circles are nonregulated enzymes, filled circles are moderately regulated enzymes, and thick-lined enlarged circles are extensively regulated enzymes.

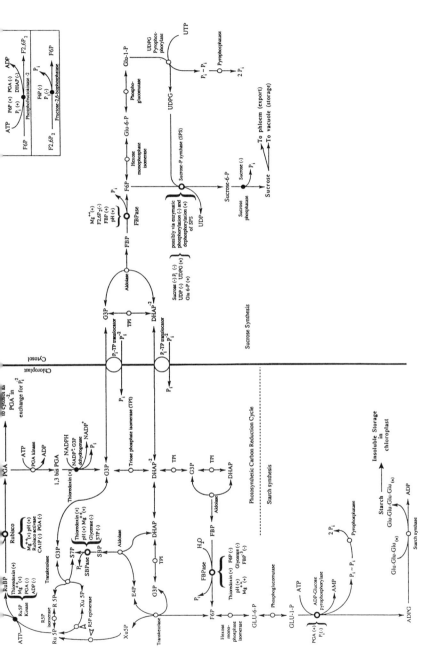

Figure A1 Photosynthetic carbon metabolism in C3 plants.

General Scheme

Two PGA molecules are formed when rubisco carboxylates RuBP. PGA is then reduced to the triose phosphates DHAP and G3P. Both are divalent anions and can be exported to the cytosol in exchange for a divalent P_i anion. In the cytosol, DHAP and G3P are used for sucrose synthesis. In the regeneration phase of the cycle, the reduced triose phosphates are combined and the carbons rearranged to yield Ru5P, which is then phosphorylated to RuBP. A major intermediate of this phase, F6P, can branch off into starch synthesis.

Energetics

ATP is required to phosphorylate each PGA to 1,3-bisPGA, and Ru5P to RuBP. NADPH is required to reduce each 1,3-bisPGA to glyceraldehyde-3-phosphate. Two NADPH and three ATP are thus required to reduce one CO_2 molecule to carbohydrate and regenerate the CO_2 acceptor RuBP. In addition, the equivalent of two ATP is needed to form F6P and Glu-6-P, and the equivalent of two ATP is required to attach an additional glucose molecule to a starch chain.

Phosphate Regeneration

Phosphate bound to carbohydrate is released by (a) the hydrolysis of FBP to F6P in the chloroplast, (b) the hydrolysis of FBP to F6P in the cytosol, (c) the hydrolysis of SBP to S7P in the chloroplast, (d) the hydrolysis of sucrose-phosphate to sucrose, (e) the formation of UDPG in sucrose synthesis, and (f) the formation of ADPG in starch synthesis.

Regulation

The key regulated steps are all largely irreversible and either (a) are entry points into a metabolic pathway (rubisco for the Calvin cycle, stromal FBPase and SBPase for the reduction phase of the Calvin cycle, and cytosolic FBPase for sucrose synthesis), (b) are near the end of a metabolic pathway (Ru5P kinase, ADPG pyrophosphorylase, sucrose phosphate synthase, and sucrose phosphatase), or (c) involve energy transduction (Ru5P kinase, NADPH-G3P dehydrogenase, ADPG pyrophosphorylase).
    Adapted from Refs. 49, 64, 193, 199, 200, 239, 245, 246.

Nonstandard Abbreviations

ADPG, ADP glucose; CA1P, carboxyarabinitol-1-phosphate; DHAP, dihydroxyacetone phosphate; E4P, erythrose-4-phosphate; FBP, fructose-

1,6-bisphosphate; FBPase, fructose-1,6-bisphosphate phosphatase; $F2,6P_2$, fructose-2,6-bisphosphate; G3P, glyceraldehyde-3-phosphate; Glu, glucose; Glu-1-P, glucose-1-phosphate; PGA, 3-phosphoglycerate; R5P, ribose-5-phosphate; Ru5P, ribulose-5-phosphate; RuBP, ribulose-1,5-bisphospate; SBP, sedoheptulose-1,7-bisphosphate; SBPase, sedoheptulose-1,7-bisphosphate phosphatase; SPS, sucrose phosphate synthase; S7P, sedoheptulose-7-phosphate; P, phosphate ester; P_i, inorganic phosphate; TP, triose phosphate; UDP, uridine diphosphate; UDPG, uridine-diphosphate glucose; UTP, uridine triphosphate; Xu5P, xylulose-5-phosphate.

APPENDIX 2: PHOTORESPIRATORY CYCLE IN HIGHER PLANTS (FIG. A2)

General Scheme

As a result of oxygenation of RuBP by rubisco, one PGA and one P-glycolate are formed. The P-glycolate is dephosphorylated in the chloroplast, and the glycolate produced is transported to the peroxisome. The released P_i is immediately available for photophosphorylation. Three fourths of the glycolate-carbon is rearranged and reduced to glycerate, which reenters the chloroplast and is phosphorylated by ATP to produce PGA. The remaining one fourth of the glycolate carbon is lost as CO_2. An assimilatory nitrogen cycle operates alongside the photorespiratory cycle in order to scavenge ammonium released during the decarboxylation of glycine and to regenerate amino donors for transamination of glyoxylate to glycine.

Energetics

The cost of photorespiration per oxygenation event is one half previously fixed CO_2, one half ATP for phosphorylation of glycerate, one half ATP for NH_4^+ reassimilation, and one half NADPH equivalent (as reduced ferredoxin) for nitrogen reassimilation. Total energy cost per oxygenation is one ATP and 1/2 NADPH equivalent. In addition, oxygenation consumes previously regenerated RuBP and competes with CO_2 for rubisco.

Adapted from Husic et al. (183).

Nonstandard Abbreviations

DHAP, dihydroxyacetone phosphate; Fd, ferredoxin; G3P, glyceraldehye-3-phosphate; NAD-MDH, NAD-malic dehydrogenase; C-THF, methyl–tetrahydrofolate complex; THF, tetrahydrofolate.

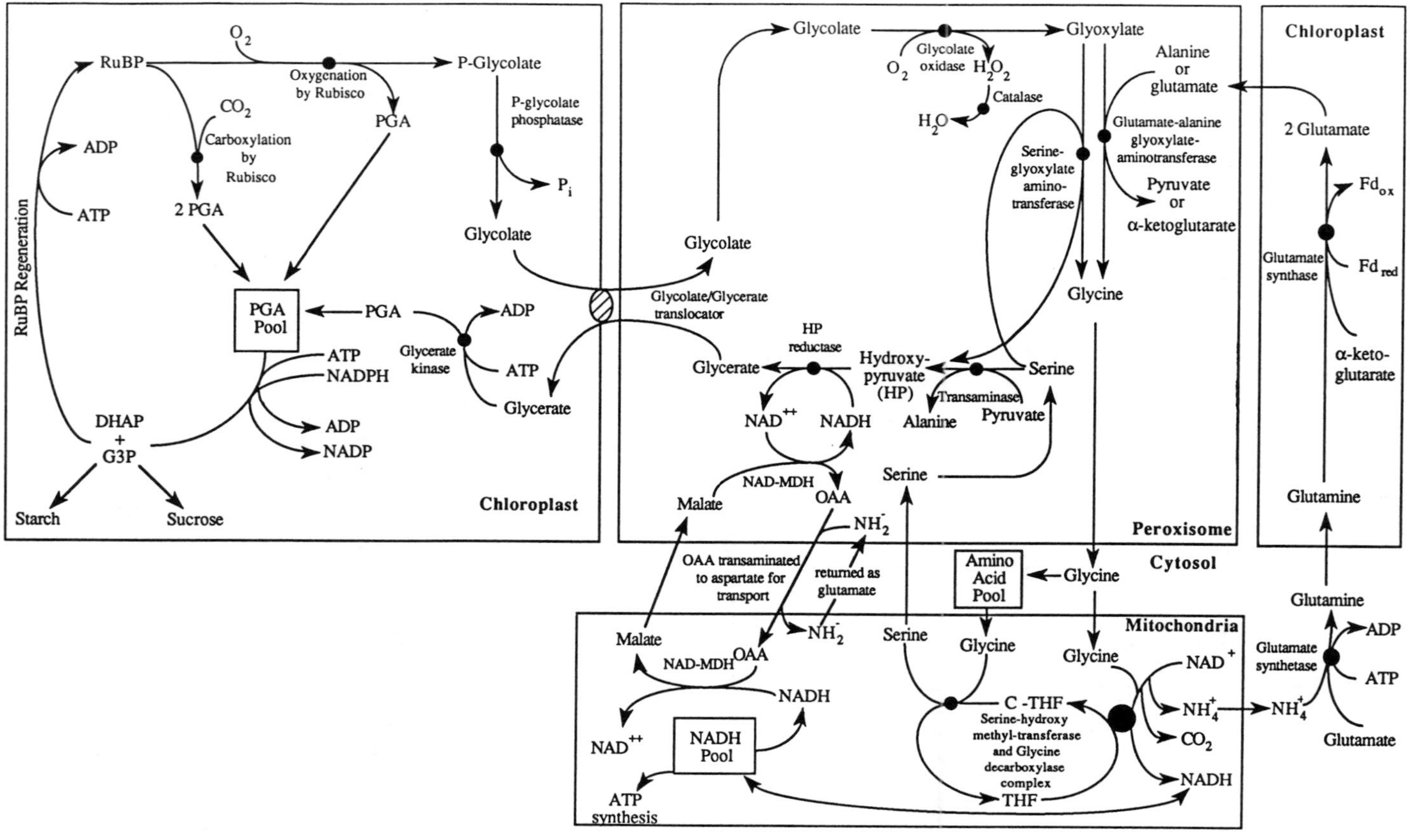

Figure 62. Photorespiratory cycle in higher plants

APPENDIX 3: ELECTRON TRANSPORT IN THE THYLAKOID MEMBRANE (FIG. A3)

General Scheme

Photosystem II complexes are largely localized in the appressed regions of the thylakoid membrane. Major components of PSII are (a) a water oxidizing complex (OEC); (b) two core polypeptides (D1 and D2) which bind the special pair chlorophyll a molecules that comprise P_{680}, pheophytin, and the quinones Q_A (a quinone permanently bound to the D2 subunit), and Q_B (a mobile plastoquinone which binds to D1 when oxidized); (c) an inner antennae complex of chlorophyll-binding proteins; and (d) an outer, or peripheral, antennae complex of chlorophyll-binding proteins generally termed LHCII, for light-harvesting complex. The inner complex has little if any chlorophyll b while LHCII has a chlorophyll a-to-b ratio of 1.2. Light energy absorbed by the PSII antennae complex is passsed to P_{680}, exciting an electron which moves via pheophytin and Q_A to Q_B. When reduced by two electrons, Q_B takes two protons from the stroma forming PQH_2, which separates from the PSII complex and diffuses into the matrix of the membrane. It then diffuses through the membrane to a cytochrome b_6-f complex. A second oxidized quinone (PQ) from the matrix of the membrane binds to the quinone binding site on D1, forming a new Q_B electron acceptor. Oxidized P_{680} replaces the lost electrons by oxidizing a tyrosine on the D1 subunit (Z), which in turn removes electron from manganese atoms in the oxygen evolving complex (OEC). When oxidized, the manganese atoms strip electrons from water, forming O_2 and releasing protons to the lumen.

The cytochrome complex is dispersed throughout appressed and nonappressed regions of the thylakoid membrane and contains three cytochromes (cytochrome f and two b-type cytochromes), and an iron-sulfur protein. Reduced plastoquinone binds to, and is oxidized by, the cytochrome complex releasing two protons to the lumen. One electron from the plastoquinone moves through an iron-sulfur subunit and cytochrome f to plastocyanin. A second electron of lower electrical potential than the first is passed from reduced plastoquinone to a low-potential cytochrome b (cyt b_L) and then to a cytochrome b of higher potential (cyt b_h). Cytochrome b_H then reduces an oxidized plastoquinone bound ot the cytochrome complex. Two more protons are taken from the stroma when this plastoquinone becomes fully reduced by the cytochrome complex. This particular plastoquinone will join the pool of reduced plastoquinone molecules in the matrix of the thylakoid membrane and eventually be oxidized by a cytochrome complex. The cycling of electrons through plastoquinone and the b-type cytochromes is termed the Q cycle, and it increases the efficiency of electron flow by pumping an additional

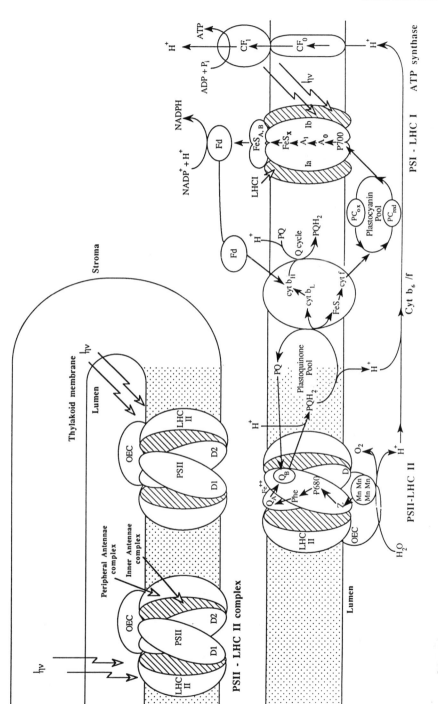

Figure A3 Electron transport in the Thylakoid membrane.

proton through the membrane for every electron passed through the cytochrome complex to plastocyanin.

Plastocyanin is a soluble, copper-containing protein in the thylakoid lumen which transports electrons from the cytochrome b_6-f complex to PSI. Photosystem I is localized on the stromal-exposed, nonappressed region of the thylakoid membrane. The core of PSI comprises two subunits, Ia and Ib, which bind the reaction center chlorophyll a molecules of P_{700}, and a series of electron transport intermediates—A_0 (a chlorophyll a molecule), A_1 (probably a phylloquinone), and iron–sulfur complexes. Like PSII, PSI has an inner antennae complex of chlorophyll a binding proteins, and an outer complex, termed LHCI, that has chlorophyll b in an a-to-b ratio of about 3.7. Upon absorption of light energy, excited electrons pass from P_{700} through A_0, A_1, and the iron–sulfur complexes to ferredoxin. Reduced ferredoxin can reduce NADP to NADPH or pass electrons to a cytochrome b in the cytochrome b_6-f complex, from which they will flow to plastoquinone and transport protons across the membrane. The cycling of electrons through ferredoxin and the cytochrome complex gives rise to the term cyclic photophosphorylation.

Energetics

Four electrons are transported to two NADPH for every oxygen molecule released by water oxidation. Four protons are released from the oxidation of two H_2O; four protons are transported across the membrane by the reduction of plastoquinone at PSII and its subsequent oxidation at the cytochrome complex; and four more protons are transported across the membrane in the Q cycle. In total (assuming complete Q-cycle function), 12 protons accumulate in the thylakoid lumen for every O_2 released. Assuming three protons are required to make one ATP from ADP and P_i, and all protons exit the lumen through ATP formation, up to four ATP can be produced per O_2 released.

Figure adapted from Anderson and Thomsen (384) and Refs. 86, 385–389.

ACKNOWLEDGMENTS

We thank Dr. Tom Sharkey and Dr. Howard Neufeld for valuable comments on this manuscript. Some of the unpublished research presented in this proposal was supported by National Science Foundation grant no. 8906390.

REFERENCES

1. S. von Caemmerer and G. D. Farquhar (1981): *Planta, 153*:376–387.
2. F. S. Chapin, A. J. Bloom, C. B. Field, and R. H. Waring (1987): *Bioscience, 37*:49–57.
3. J. R. Evans (1989): *Oecologia, 78*:9–19.
4. H. A. Mooney and S. L. Gulmon (1978): *Topics in Plant Population Ecology* (O. T. Solbrig, J. Cain, G. B. Johnson, and P. E. Raven, eds.), Columbia University Press, New York, pp. 316–337.
5. H. A. Mooney and S. L. Gulmon (1982): *Bioscience, 32*:198–206.
6. A. J. Bloom, F. S. Chapin III, and H. A. Mooney (1985): *Annu. Rev. Ecol. Syst.,16*:363–392.
7. C. Field and H. A. Mooney (1986): *On the Economy of Plant Form and Function* (T. J. Givnish, ed.), Cambridge University Press, Cambridge, pp. 25–55.
8. S. von Caemmerer and G. D. Farquhar (1984): *Planta, 160*:320–329.
9. I. E. Woodrow and J. A. Berry (1988): *Annu. Rev. Plant Physiol. Plant Mol. Biol., 39*:533–594.
10. R. F. Sage, T. D. Sharkey, and J. R. Seeman (1989): *Plant Physiol., 89*:590–596.
11. M. M. Sachs and T. D. Ho (1986): *Annu. Rev. Plant Physiol., 37*:363–376.
12. R. J. Ellis and S. M. Van der Vies (1990): *Annu. Rev. Biochem., 60*:321–347.
13. E. Vierling (1991): *Annu. Rev. Plant Phys. Plant Mol. Biol., 42*:579–620.
14. D. Edwards and D. Walker (1983): C_3, C_4: *Mechanisms, and Cellular and Environmental Regulation on Photosynthesis*, Univ. Calif. Press, Berkeley.
15. I. R. Cowan (1986): *On the Economy of Plant Form and Function* (T. J. Givnish, ed.), Cambridge University Press, Cambridge, pp. 133–170.
16. J. A. Raven and S. M. Glidewell (1981): *Physiological Processes Limiting Plant Productivity* (C. B. Johnson, ed.), Butterworths, London, pp. 109–136.
17. I. Terashima (1989): *Photosynthesis* (W. R. Briggs, ed.), Alan R. Liss, New York, pp. 207–227.
18. J. M. Anderson and C. B. Osmond (1987): *Photoinhibition* (D. J. Kyle, C. B. Osmond, and C. J. Arntzen, eds.), Elsevier, Amsterdam, pp. 1–38.
19. T. A. Mansfield, A. M. Hetherington, and C. J. Atkinson (1990): *Annu. Rev. Plant Physiol. Plant Mol. Biol., 41*:55–75.
20. D. A. Grantz (1990): *Plant Cell Environ., 13*:667–679.
21. W. Haupt and R. Scheuerlein (1990): *Plant Cell Environ., 13*:595–614.
22. D. Koller (1990): *Plant Cell Environ., 13*:615–632.
23. J. A. Berry and J. K. Raison (1981): *Physiological Plant Ecology I. Responses to Physical Environment, Encycl. Plant Physiol., New Ser.,* Vol. 12A (O. L. Lange, P. S. Nobel, C. B. Osmond, and H. Zeigler, eds.), Springer-Verlag, Berlin, pp. 277–338.
24. J. M. Cheeseman (1988): *Plant Physiol., 87*:547–550.
25. W. M. Kaiser (1987): *Physiol. Plant., 71*:142–149.

26. M. M. Chaves (1991): *J. Exp. Bot.*, *42*:1–16.
27. L. Natr (1972): *Photosynthetica*, *6*:80–99.
28. K. Mengel and E. A. Kirby (1982): *Principles of Plant Nutrition*, 3rd Ed., Int. Potash Inst., Bern.
29. J. M. Anderson, W. S. Chow, and D. J. Goodchild (1988): *Aust. J. Plant Physiol.*, *15*:11–26.
30. C. B. Osmond and W. S. Chow (1988): *Aust. J. Plant Physiol.*, *15*:1–10.
31. J. J. Casal and H. Smith (1989): *Plant Cell Environ.*, *12*:855–862.
32. H. Senger (1987): *Blue Light Response: Phenomena and Occurrence in Plants and Microorganisms*, Vol. 2 (H. Senger, ed.), CRC Press, Boca Raton, pp. 141–149.
33. K. A. Mott (1990): *Plant Cell Environ.*, *13*:731–737.
34. F. S. Chapin, III (1991): *Bioscience*, *41*:29–36.
35. W. J. Davies and J. Zhang (1991): *Annu. Rev. Plant Physiol. Plant Mol. Biol.*, *42*:55–76.
36. R. F. Sage, T. D. Sharkey, and J. R. Seeman (1988): *Planta*, *174*:407–416.
37. H. Kacser (1978): *Biochemistry of Metabolism, Biochemistry of Plants*, Vol. 11 (D. D. Davies, ed.), Academic Press, New York, pp. 39–67.
38. I. E. Woodrow (1986): *Biochim. Biophys. Acta*, *851*:181–192.
39. J. T. Ball, I. E. Woodrow, and J. A. Berry (1987): *Progress on Photosynthesis Research*, Vol. 4 (J. Biggins, ed.), Martinus Nijhoff, Dordrecht, pp. 221–224.
40. I. E. Woodrow and K. A. Mott (1988): *Aust. J. Plant Physiol.*, *15*:352–262.
41. I. E. Woodrow, J. T. Ball, and J. A. Berry (1990): *Plant Cell Environ.*, *13*:339–347.
42. M. Stitt (1989): *Phil Trans. Royal Soc. Lond.*, Series B, *323*:327–338.
43. A. L. Kruckeberg, H. E. Neuhaus, R. Feil, L. D. Gottieb, and M. Stitt (1989): *Biochem. J.*, *261*:457–467.
44. H. E. Neuhaus, A. L. Kruckeberg, R. Feil, and M. Stitt (1989): *Planta*, *178*:110–122.
45. H. E. Neuhaus and M. Stitt (1990): *Planta*, *182*:445–454.
46. U. Heber, S. Niemanis, and K.-J. Dietz (1988): *Planta*, *173*:267–274.
47. M. Stitt, W. P. Quick, U. Schurr, E.-D. Schulze, S. R. Rodermel, and L. Bogorad (1991): *Planta*, *183*:555–566.
48. W. P. Quick, U. Schurr, R. Scheibe, E.-D. Schulze, S. R. Rodermel, L. Bogorad, and M. Stitt (1991): *Planta*, *183*:542–554.
49. M. Stitt (1991): *Plant Cell Environ.*, *14*:741–762.
50. T. D. Sharkey (1985): *Bot Rev.*, *51*:53–105.
51. C. B. Field, J. T. Ball, and J. A. Berry (1989): *Plant Physiological Ecology. Field Methods and Instrumentation* (R. W. Pearcy, J. Ehleringer, H. A. Mooney, and P. W. Rundel, eds.), Chapman and Hall, New York, pp. 209–253.
52. G. D. Farquhar, K. T. Hubick, I. Terashima, A. G. Condon, and R. A. Richards (1987): *Progress in Photosynthesis Research*, Vol. 4 (J. Biggins, ed.), Martinus Nijhoff, Dordrecht, pp. 209–212.
52a. I. Terashima (1992); *Photosyn. Res.*, *31*:195–212.

53. G. D. Farquhar and T. D. Sharkey (1982): *Annu. Rev. Plant Physiol.*, *33*:317–345.
54. O. Björkman (1987): *Photoinhibition* (D. J. Kyle, C. B. Osmond, and C. J. Arntzen, eds.), Elsevier, Amsterdam, pp. 123–144.
55. U. Schreiber (1986): *Photosyn. Res.*, *9*:261–272.
56. H. P. Bolhár-Nordenkampf, S. P. Long, N. R. Baker, G. Öquist, U. Schreiber, and E. G. Lechner (1991): *Funct. Ecol.*, *3*:497–514.
57. O. Van Kooten and J. F. H. Snel (1990): *Photosyn. Res.*, *25*:147–150.
58. P. Horton and J. R. Bowyer (1990): *Methods in Plant Biochemistry*, Vol. 4 (J. L. Hardwood and J. R. Bowyer, eds.), Academic Press, New York, pp. 259–296.
59. N. R. Baker (1991): *Plant Physiol.*, *81*:563–570.
60. B. Demmig-Adams (1990): *Biochim. Biophys. Acta*, *1020*:1–24.
61. G. H. Krause and E. Weis (1991): *Annu. Rev. Plant Physiol. Plant Mol. Biol.*, *42*:313–349.
62. B. Genty, J.-M. Briantais, and N. R. Baker (1989): *Biochim. Biophys. Acta*, *900*:87–92.
63. R. B. Peterson (1990): *Plant Physiol.*, *92*:892–898.
64. R. B. Peterson (1991): *Plant Physiol.*, *96*:172–177.
65. J. P. Krall and G. E. Edwards (1991): *Aust. J. Plant Physiol.*, *18*:267–278.
66. E. Weis and J. A. Berry (1987): *Biochim. Biophys. Acta*, *894*:198–208.
67. T. D. Sharkey, J. A. Berry, and R. F. Sage (1988): *Planta*, *176*:415–424.
68. G. Cornic and J.-M. Briantais (1991): *Planta*, *183*:178–184.
69. G. J. Collatz, M. R. Badger, C. Smith, and J. A. Berry (1979): *Carnegie Inst. Wash. Yearbook*, *78*:171–175.
70. A. J. Keys and M. A. J. Parry (1990): *Methods in Plant Biochemistry*, Vol. 3 (P. J. Lea, ed.), Academic Press, New York, pp. 1–14.
71. A. Makino and C. B. Osmond (1991): *Plant Physiol.*, *96*:355–362.
72. T. D. Sharkey, L. V. Savitch, and N. D. Butz (1991): *Photosyn. Res.*, *28*:41–48.
73. R. F. Sage, R. W. Pearcy, and J. R. Seemann (1987): *Plant Physiol.*, *85*:355–359.
74. T. L. Vassey and T. D. Sharkey (1989): *Plant Physiol.*, *89*:1066–1070.
75. R. C. Leegood (1990): *Meth. Plant Biochem.*, *3*:15–37.
76. L. Copeland (1990): *Meth. Plant Biochem.*, *3*:73–85.
77. A. M. Smith (1990): *Meth. Plant Biochem.*, *3*:93–102.
78. O. H. Lowry and J. V. Passonneau (1972): *A Flexible System of Enzymatic Analysis*, Academic Press, New York.
79. T. D. Sharkey, J. Kozba, J. R. Seemann, and R. H. Brown (1988): *Plant Physiol.*, *86*:667–671.
80. P. Jursinic and R. Dennenberg (1989): *Photosyn. Res.*, *21*:197–200.
81. P. A. Jursinic, S. A. McCarthy, T. M. Brinker, and A. Stemler (1991): *Biochim. Biophys. Act.*, *1059*:312–322.
82. P. A. Jursinic and R. J. Dennenberg (1985): *Arch. Biochem. Biophys.*, *241*:540–549.
83. R. J. Dennenberg, P. A. Jursinic, and S. A. McCarthy (1986): *Biochim. Biophys. Acta*, *852*:222–233.

84. W. S. Chow, A. B. Hope, and J. M. Anderson (1991): *Aust. J. Plant Physiol.*, *18*:397–410.
85. D. R. Ort and J. Whitmarsh (1990): *Photosyn. Res.*, *23*:101–104.
86. A. Melis (1991): *Biochim. Biophys. Acta*, *1058*:87–106.
87. S. E. McCauley and A. Melis (1986): *Biochim. Biophys. Acta*, *849*:175–182.
88. S. W. McCauley and A. Melis (1987): *Photochem Photobiol.*, *4*:543–550.
89. W. S. Chow and A. B. Hope (1987): *Aust. J. Plant Physiol.*, *14*:21–28.
90. I. Terashima and J. R. Evans (1988): *Plant Cell Physiol.*, *29*:143–155.
91. W. R. D. Torre and K. O. Burkey (1990): *Photosyn. Res.*, *24*:127–136.
92. S. Izawa (1981): *Mech. Enzymol.*, *69*:413–433.
93. A. Trebst (1981): *Meth. Enzymol.*, *69*:675–715.
94. S. P. Robinson, G. E. Edwards, and D. A. Walker (1979): *Methodological Surveys in Biochemistry of Plant Organelles*, Vol. 9 (E. Reid, ed.), Ellis Horwood, Chichester, pp. 14–24.
95. D. A. Walker (1987): *The Use of the Oxygen Electrode and Fluorescence Probes in Simple Measurements of Photosynthesis*, Packard Publ. Ltd., Chichester, West Sussex.
96. U. Pick and S. Bassilian (1981): *Energy Coupling in Photosynthesis* (B. R. Selman and S. Sleman-Reiner, eds.), Elsevier, New York, pp. 251–260.
97. J. L. Walker and S. C. Huber (1989): *Plant Physiol.*, *89*:518–525.
98. J. L. Walker and S. C. Huber (1989): *Planta*, *177*:116–120.
99. K. J. Dietz (1985): *Biochim. Biophys. Acta*, *839*:260–263.
100. R. C. Leegood and S. von Caemmerer (1988): *Planta*, *174*:253–262.
101. C. A. Labate, M. D. Adcock, and R. C. Leegood (1990): *Planta*, *181*:547–554.
102. P. Quick, G. Siegl, E. Neuhaus, R. Feil, and M. Stitt (1989): *Planta*, *177*:535–546.
103. J. T. Perchorowicz, D. A. Raynes, and R. G. Jensen (1981): *Proc. Natl. Acad. Sci. USA*, *78*:2985–2989.
104. K. A. Mott, R. G. Jensen, J. W. O'Leary, and J. A. Berry (1984): *Plant Physiol.*, *76*:968–971.
105. R. T. Prinsley, K.-J. Dietz, and R. C. Leegood (1986): *Biochim. Biophys. Acta*, *849*:254–263.
106. G. D. Farquhar and S. von Caemmerer (1982): *Physiological Plant Ecology. II. Water Relations and Carbon Assimilation. Encycl. Plant Physiol.*, *New Ser.*, Vol. 12B (O. L. Lange, P. S. Nobel, C. B. Osmond, and H. Zeigler, eds.), Springer-Verlag, Berlin, pp. 549–588.
107. G. D. Farquhar, S. von Caemmerer, and J. A. Berry (1980): *Planta*, *149*:178–190.
108. T. D. Sharkey (1988): *Physiol. Plant*, *73*:147–152.
109. R. F. Sage and T. D. Sharkey (1987): *Plant Physiol.*, *84*:658–664.
110. S. von Caemmerer and J. R. Evans (1991): *Aust. J. Plant Physiol.*, *18*:387–305.
111. F. Loreto, P. C. Harley, G. D. Marco, and T. D. Sharkey (1992): *Plant Physiol.*, *98*:1437–1443.

112. J. R. Evans, T. D. Sharkey, J. A. Berry, and G. D. Farquhar (1986): *Aust. J. Plant Physiol.*, *13*:281–292.

113. W. Larcher (1980): *Physiological Plant Ecology*, 2nd Ed., Springer-Verlag, Berlin.

114. I. R. Cowan (1977): *Adv. Bot. Res.*, *4*:117–228.

115. J. T. Ball (1987): *Stomatal Function* (E. Zeiger, G. D. Farqhar, and I. R. Cowan, eds.), Stanford University Press, Stanford, pp. 445–476.

116. R. W. Pearcy, E.-D. Schulze, and R. Zimmermann (1989): *Plant Physiological Ecology. Field Methods and Instrumentation* (R. W. Pearcy, J. Ehleringer, H. A. Mooney, and P. W. Rundel, eds.), Chapman and Hall, New York, pp. 137–160.

117. S. C. Wong, I. R. Cowan, and G. D. Farquhar (1979): *Nature*, *282*:424–426.

118. J. A. Berry and W. J. S. Downton (1982): *Photosynthesis, Vol. 2. Development, Carbon Metabolism and Productivity* (Govindjee, ed.), Academic Press, New York, pp. 263–343.

119. F. C. Meinzer and D. A. Grantz (1991): *Physiol. Plant.*, *83*:324–329.

120. T. D. Sharkey, F. Loreto, and T. L. Vassey (1989): *Current Research in Photosynthesis, Vol. 4* (M. Baltscheffsky, ed.), Kluwer Academic, Dordrecht, pp. 549–556.

121. W. J. S. Downton, B. J. Loveys, and W. J. R. Grant (1988): *New Phytol.*, *108*:263–266.

122. W. J. S. Downton, B. J. Loveys, and W. J. R. Grant (1988): *New Phytol.*, *110*:503–509.

123. I. Terashima, S. C. Wong, C. B. Osmond, and G. D. Farquhar (1988): *Plant Cell Physiol.*, *29*:385–394.

124. P. S. Nobel (1991): *Physicochemical and Environmental Plant Physiology*, Academic Press, New York.

125. D. M. Gates (1980): *Biophysical Ecology*, Springer-Verlag, New York.

126. J. L. Monteith and M. H. Unsworth (1990): *Principles of Environmental Physics*, 2nd Ed., Edward Arnold, London.

127. P. G. Jarvis and T. A. Mansfield (1981): *Stomatal Physiology*, Cambridge University Press, Cambridge.

128. C. M. Willmer (1983): *Stomata*, Longman, New York.

129. R. Hedrich and J. I. Schroeder (1989): *Annu. Rev. Plant Physiol. Plant Mol. Biol.*, *40*:539–569.

130. E. Zeiger (1990): *Plant Cell Environ.*, *13*:739–747.

131. J. I. Schroeder and S. Hagiwara (1989): *Nature*, *338*:427–430.

132. J. I. Schroeder and S. Hagiwara (1990): *Proc. Natl. Acad. Sci. USA*, *87*:9305–9309.

133. S. Gilroy, M. D. Fricker, N. D. Read, and A. J. Trewavas (1991): *Plant Cell*, *3*:333–344.

134. W. H. Outlaw and L. M. Shen (1989): *Physiol. Plant.*, *77*:275–281.

135. N. C. Turner (1991): *Agric. For. Meteorol.*, *54*:137–154.

136. T. D. Sharkey and T. Ogawa (1987): *Stomatal Function* (E. Zeiger, G. D. Farghar, and I. R. Cowan, eds.), Stanford University Press, Stanford, pp. 195–208.

137. E. E. Serrano, E. Zeiger, and S. Hagiwara (1988): *Proc. Natl. Acad. Sci. USA*, 85:436–440.

138. T. D. Sharkey and K. Raschke (1981): *Plant Physiol.*, 68:1170–1174.

139. J. I. L. Morison (1985): *Plant Cell Environ.*, 8:467–474.

140. J. I. L. Morison (1987): *Stomatal Function* (E. Zeiger, G. D. Farqhar, and I. R. Cowan, eds.), Stanford University Press, Stanford, pp. 229–251.

140a. Z. Dai, G. E. Edwards, and M. S. B. Ku (1992): *Plant Physiol.*, 9:1426–1434.

141. K. Raschke (1979): *Physiology of Movements. Encycl. Plant Physiol.*, Vol. 7 (W. Haupt and M. E. Feinleib, eds.), Springer-Verlag, New York, pp. 383–441.

142. A. Shaish, N. Roth-Bejeramo, and C. Itai (1988): *Physiol. Plant.*, 76:107–111.

143. F. S. Chapin, III, D. T. Clarkson, J. R. Lenton, and C. H. S. Walter (1988): *Planta*, 173:340–351.

144. K. Cornish and J. W. Radin (1990): *Environmental Injury to Plants* (F. Katterman, ed.), Academic Press, New York, pp. 89–112.

145. S. P. Robinson, W. J. Grant, and B. R. Loveys (1988): *Aust. J. Plant Physiol.*, 15:495–503.

146. T. Graan and J. S. Boyer (1990): *Planta*, 181:378–384.

147. M. J. Lauer and J. S. Boyer (1992): *Plant Physiol.*, 98:1310–1316.

148. W. J. Davies and T. A. Mansfield (1987): *Stomatal Function* (E. Zeiger, G. D. Farqhar, and I. R. Cowan, eds.), Stanford University Press, Stanford, pp. 293–309.

149. L. D. Incoll and P. C. Jewer (1987): *Stomatal Function* (E. Zeiger, G. D. Farqhar, and I. R. Cowan, eds.), Stanford University Press, Stanford, pp. 281–292.

150. C. J. Atkinson, W. J. Davies, and T. A. Mansfield (1990): *New Phytol.*, 116:19–27.

151. H. G. Jones (1985): *Plant Cell Environ.*, 8:95–104.

152. P. F. Daley, K. Raschke, J. T. Ball, and J. A. Berry (1989): *Plant Physiol.*, 90:1233–1238.

153. T. D. Sharkey and J. R. Seemann (1989): *Plant Physiol.*, 89:1060–1065.

154. K. Raschke, J. Patzke, P. F. Daley, and J. A. Berry (1990): *Current Research in Photosynthesis*, Vol. 4 (M. Baltscheffsky, ed.), Kluwer Academic, Dordrecht, pp. 573–578.

155. G. Cornic, J.-L. Le Gouallec, J.-M. Briantais, and M. Hodges (1989): *Planta*, 177:84–90.

156. W. Beyschlag and H. Pfanz (1990): *Oecologia*, 82:52–55.

157. R. W. Wise, D. H. Sparrow, A. Ortiz-Lopez, and D. R. Ort (1991): *Plant Sci.*, 74:45–52.

157a. R. W. Wise, A. Ortiz-Lopez, and D. R. Ort *Plant Physiol.*, 100:26–32.

158. E.-D. Schulze and A. E. Hall (1982): *Physiological Plant Ecology II. Water Relations and Carbon Assimilation. Encycl. Plant Physiol.*, New Ser. 12B (O. L. Lange, P. S. Nobel, C. B. Osmond, and H. Zeigler, eds.), Springer-Verlag, Berlin, pp. 181–230.

159. E.-D. Schulze (1986): *Annu. Rev. Plant Physiol.*, *37*:147–174.

160. P. G. Jarvis and K. G. McNaughton (1986): *Adv. Ecol. Res.*, *15*:1–47.

161. F. A. Bazzaz and W. E. Williams (1991): *Ecology*, *72*:12–16.

162. J. A. Bunce (1985): *Plant Cell Environ.*, *8*:55–57.

163. P. R. Van Gardigan, J. Grace, and C. E. Jeffree (1991): *Plant Cell Environ.*, *14*:185–194.

164. J. R. Ehleringer and H. A. Mooney (1978): *Oecologia*, 37:183–200.

165. S. B. Idso and S. G. Allen (1988): *Agr. For. Meteorol.*, *43*:59–58.

166. P. C. Harley, F. Loreto, G. Di Marco, and T. D. Sharkey (1992): *Plant Physiol.*, *98*:1429–1436.

167. J.-L. Renou, A. Grebaud, D. Just, and M. André (1990): *Planta*, *182*:415–419.

168. J. R. Evans (1983): *Plant Physiol.*, *72*:297–302.

169. J. R. Evans and I. Terashima (1987): *Plant Cell Physiol.*, *29*:157–165.

170. S. B. Ku and G. E. Edwards (1977): *Plant Physiol.*, *59*:986–990.

171. D. B. Jordan and W. L. Ogren (1984): *Planta*, *161*:308–313.

172. A. Brooks and G. D. Farquhar (1985): *Planta*, *165*:397–406.

173. J. R. Ehleringer, R. F. Sage, L. B. Flanagan, and R. W. Pearcy (1991): *Trends Ecol. Evol.*, *6*:95–99.

174. P. C. Harley and T. D. Sharkey (1991): *Photosyn. Res.*, *27*:169–178.

175. R. F. Sage (1990): *Plant Physiol.*, *94*:1728–1734.

176. T. D. Sharkey (1986): *Biological Control of Photosynthesis* (R. Marcelle, H. Clijsters, and M. Van Poucke, eds.), Martinus Nijhoff, Dordrecht, pp. 115–126.

177. S. von Caemmerer and D. L. Edmonson (1986): *Aust. J. Plant Physiol.*, *13*:669–688.

178. J. R. Seeman and T. D. Sharkey (1986): *Plant Physiol.*, *82*:555–560.

179. D. R. Ort and N. R. Baker (1988): *Plant Physiol. Biochem.*, *26*:555–565.

180. T. D. Sharkey (1985): *Plant Physiol.*, *78*:71–75.

181. A. Herold (1980): *New Phytol.*, *86*:131–144.

182. M. N. Sivak and D. A. Walker (1986): *New Phytol.*, *102*:499–512.

183. R. C. Leegood and R. T. Furbank (1986): *Planta*, *168*:84–93.

184. C. A. Labate and R. C. Leegood (1988): *Planta*, *173*:519–527.

185. R. F. Sage, T. D. Sharkey, and R. W. Pearcy (1990): *Aust. J. Plant Physiol.*, *17*:135–148.

186. J. R. Seeman and J. A. Berry (1982): *Carnegie Inst. Wash. Yearbook*, *81*:78–83.

187. I. Terashima (1986): *J. Exp. Bot.*, *37*:399–405.

188. J. W. Leverenz (1988): *Plant Physiol.*, *74*:332–341.

189. R. F. Sage, T. D. Sharkey, and J. R. Seemann (1990): *Plant Physiol.*, *94*:1735–1742.

190. T. D. Sharkey (1989): *Phil. Trans. Royal Soc. Lond.*, *Series B.*, *323*:435–448.

191. M. U. F. Kirschbaum and G. D. Farquhar (1984): *Aust. J. Plant Physiol.*, *11*:519–538.

192. C. Cséke and B. B. Buchanan (1986): *Biochim. Biophys. Acta*, *853*:43–63.

193. F. D. Macdonald and B. B. Buchanan (1990): *Plant Physiology, Biochemistry, and Molecular Biology* (D. T. Dennis and D. H. Turpin, eds.), Longman, New York, pp. 239–252.
194. R. J. A. Budde and D. D. Randall (1990): *Plant Physiol.*, 94:1501–1504.
195. J. L .Huber, D. R. C. Hite, W. H. Outlaw, and S. C. Huber (1991): *Plant Physiol.*, 95:291–297.
196. R. C. Leegood (1985): *Photosyn. Res.*, 6:247–259.
197. A. R. Portis (1990): *Biochim. Biophys. Acta*, 1015:15–28.
198. Y. Lan, I. E. Woodrow, and K. A. Mott (1992): *Plant Physiol.*, 99:3061–3072.
199. A. Gardeman, D. Schimkat, and H. W. Heldt (1986): *Planta*, 168:536–545.
200. D. Schimkat, D. Heineke, and H. W. Heldt (1990): *Planta*, 181:97–103.
201. C. Wilhelm and A. Wild (1984): *J. Plant Physiol.*, 115:125–135.
202. J. T. Ball and J. A. Berry (1982): *Carnegie Inst. Wash. Yearbook*, 81:88–92.
203. T. D. Sharkey and P. J. Vanderveer (1989): *Plant Physiol.*, 91:679–684.
204. T. D. Sharkey, I. Imai, G. D. Farquhar, and I. R. Cowan (1982): *Plant Physiol.*, 69:657–659.
205. K. A. Mott (1988): *Plant Physiol.*, 86:200–203.
205a. R. F. Sage (1994): Photosyn. Res. (in press).
206. S. E. Taylor and N. Terry (1986): *Photosyn. Res.*, 8:249–256.
207. J. Preiss (1988): *Biochemistry of Plants*, Vol. 14 (J. Preiss, ed.), Academic Press, New York, pp. 181–254.
208. J. R. Seemann, J. Kobza, and B. D. Moore (1990): *Photosyn. Res.*, 23:119–130.
209. A. R. Portis (1992): *Annu. Rev. Plant Physiol. Mol. Biol.*, 43:415–437.
210. J. Kobza and J. R. Seemann (1988): *Proc. Natl. Acad. Sci. USA*, 85:3815–3819.
211. T. D. Sharkey, J. R. Seemann, and J. A. Berry (1986): *Plant Physiol.*, 81:788–791.
211a. R. F. Sage, C. D. Reid, B. D. Moore, and J. R. Seemann (1993): *Planta.*, 191:222–230.
212. J. C. Servaites, M. A. J. Parry, S. Gutteridge, and A. J. Keys (1986): *Plant Physiol.*, 82:1161–1163.
213. S. P. Robinson and A. R. Portis (1988): *FEBS Lett.*, 233:413–416.
214. S. P. Robinson and A. R. Portis (1988): *Plant Physiol.*, 86:293–298.
215. W. J. Campbell and W. L. Ogren (1990): *Plant Physiol.*, 92:110–115.
216. R. F. Sage, T. D. Sharkey, and J. R. Seeman (1987): *Progress in Photosynthesis Research*, Vol. 3 (J. Biggens, ed.), Martinus Nijhoff, Dordrecht, pp. 285–288.
217. J. Barber and B. Andersson (1992): *Trends Biochem.*, 17:61–66.
218. B. Demmig-Adams and W. W. Adams III (1992): *Annu. Rev. Plant Phys. Molec. Biol.*, 43:599–626.
219. A. J. Young (1991): *Physiol. Plant.*, 83:702–708.
220. W. P. Williams and J. F. Allen (1987): *Photosyn. Res.*, 13:19–45.
221. D. C. Fork and K. Satoh (1986): *Annu. Rev. Plant Physiol.*, 37:335–361.
222. P. Horton and A. Hague (1988): *Biochim. Biophys. Acta*, 932:107–115.

223. J. M. Anderson (1986): *Annu. Rev. Plant Physiol.*, *37*:93–136.

224. Y. Bao and E. T. Nilsen (1988): *Ecology*, *69*:1578–1587.

225. J. R. Ehleringer and I. N. Forseth (1989): *Plant Canopies: Their Growth, Form and Function* (G. Russell, B. Marshall, and P. G. Jarvis, eds.), Cambridge University Press, Cambridge, UK, pp. 129–142.

226. J. A. Gamon and R. W. Pearcy (1990): *Plant Cell Environ.*, *13*:267–275.

227. V. S. Berg and S. Heuchelin (1987): *Crop Sci.*, *30*:638–643.

228. J. R. Ehleringer (1980): *Adaptation of Plants to Water and High Temperature Stress* (N. C. Turner and P. J. Kramer, eds.), Wiley, New York, pp. 295–308.

229. G. Öquist, W. S. Chow, and J. M. Anderson (1992): *Planta*, *186*:450–460.

230. D. J. Kyle (1987): *Photoinhibition* (D. J. Kyle, C. B. Osmond, and C. J. Arntzen, eds.), Elsevier, Amsterdam, pp. 197–226.

231. G. Öquist, P. H. Greer, and E. Ögren (1987): *Photoinhibition* (D. J. Kyle, C. B. Osmond, and C. J. Arntzen, eds.), Elsevier, Amsterdam, pp. 67–87.

232. K. Asada and M. Takahashi (1987): *Photoinhibition* (D. J. Kyle, C. B. Osmond, and C. J. Arntzen, eds.), Elsevier, Amsterdam, pp. 227–287.

233. M. L. Salin (1987): *Physiol. Plant*, *72*:681–689.

234. L. S. Monk, K. V. Fagerstedt, and R. M. M. Crawford (1989): *Physiol. Plant*, *76*:456–459.

234a. M. J. Fryer (1992): *Plant Cell Environ.*, *15*:381–392.

235. R. G. Alscher (1989): *Physiol. Plant.*, *77*:457–464.

236. I. K. Smith, T. C. Vierhaller, and C. A. Thorne (1989): *Physiol. Plant.*, *77*:449–456.

237. J. A. Raven (1989): *Funct. Ecol.*, *3*:5–19.

238. P. J. Ferrar and C. B. Osmond (1986): *Planta*, *168*:563–570.

239. M. Stitt, S. C. Huber, and P. Kerr (1987): *The Biochemistry of Plants* (M. D. Hatch and N. R. Boardman, eds.), Academic Press, London, pp. 327–407.

240. C. H. Foyer (1897): *Plant Physiol. Biochem.*, *25*:649–657.

241. E. Medina (1971): *Carnegie Inst. Wash. Yearbook*, *70*:551–558.

242. G. Cave, L. C. Tolley, and B. R. Strain (1981): *Physiol. Plant.*, *51*:171–174.

243. W. J. Arp (1991): *Plant Cell Environ*, *14*:869–875.

244. M. Stitt (1990): *Annu. Rev. Plant Phys. Plant Mol. Biol.*, *41*:153–185.

245. M. Stitt and W. P. Quick (1989): *Physiol. Plant.*, *77*:633–641.

246. J. L. A. Huber, S. C. Huber, and T. H. Nielsen (1989): *Arch. Biochem. Biophys.*, *270*:681–690.

247. O. Björkman (1981): *Physiological Plant Ecology I. Responses to Physical Environment, Encycl. Plant Physiol., New Ser. 12A* (O. L. Lange, P. S. Nobel, C. B. Osmond, and H. Zeigler, eds.), Springer-Verlag, Berlin, pp. 57–107.

248. J. R. Seeman, T. D. Sharkey, J. L. Wang, and C. B. Osmond (1987): *Plant Physiol.*, *84*:796–802.

249. W. S. Chow, H. Y. Adamson, and J. M. Anderson (1991): *Physiol. Plant.*, *81*:175–182.

250. B. F. Chabot and D. J. Hicks (1982): *Annu. Rev. Ecol. Syst.*, *13*:229–259.
251. T. J. Givnish (1988): *Aust. J. Plant Physiol.*, *15*:63–92.
252. B. F. Chabot, T. W. Jurik, and J. F. Chabot (1979): *Am. J. Bot.*, *66*:940–945.
253. J. A. Bunce (1983): *Photosyn. Res.*, *4*:87–97.
254. J. P. Gaudillere, J. J. Devron, J. P. Bernoud, F. Jardinet, and M. Euvrard (1987): *Photosyn. Res.*, *13*:81–89.
255. J. Barber (1985): *Photosynthetic Mechanisms and the Environment* (J. Barber and N. R. Baker, eds.), Elsevier, Amsterdam, pp. 91–134.
256. J. R. Evans (1978): *Aust. J. Plant Physiol.*, *14*:157–170.
257. J. R. Evans (1988): *Aust. J. Plant Physiol.*, *15*:93–106.
258. W. S. Chow, L. Qian, D. J. Goodchild, and J. M. Anderson (1988): *Aust. J. Plant Physiol.*, *15*:107–122.
259. J. R. Evans and J. R. Seemann (1989): *Photosynthesis* (W. R. Briggs, ed.), Alan R. Liss, New York, pp. 407–424.
260. H. A. Mooney and N. R. Chiariello (1984): *Perspectives on Plant Population Ecology* (R. Dirzo and J. Sarukhan, eds.), Sinauer Assoc., Sunderland, MA., pp. 305–323.
261. W. S. Chow and J. M. Anderson (1987): *Aust. J. Plant Physiol.*, *14*:1–8.
262. W. S. Chow and J. M. Anderson (1987): *Aust. J. Plant Physiol.*, *14*:9–19.
263. C. B. Osmond (1983): *Oecologia*, 57:316–321.
264. D. A. Sims and R. W. Pearcy (1992): *Am. J. Botany*, 79:449–455.
265. R. W. Pearcy (1990): *Annu. Rev. Plant Physiol. Mol. Biol.*, 41:421–453.
266. R. L. Chazdon and R. W. Pearch (1991): *Bioscience*, 41:760–766.
267. T. L. Pons, R. W. Pearcy, and J. R. Seemann (1992): *Plant Cell Environ.*, *15*:569–577.
268. G. F. Sassenrath-Cole and R. W. Pearcy (1992): *Plant Physiol.*, 99:227–234.
269. T. D. Sharkey, J. R. Seeman, and R. W. Pearcy (1986): *Plant Physiol.*, 82:1063–1068.
270. J. Kobza and G. E. Edwards (1987): *Planta*, *171*:549–559.
271. A. K. Knapp and W. K. Smith (1990): *Physiol. Plant.*, 78:160–165.
272. R. W. Pearcy (1988): *Aust. J. Plant Physiol.*, *15*:223–238.
273. D. M. Schwartz and F. A. Bazzaz (1973): *Oecologia*, *12*:161–167.
274. D. D. Baldocchi, S. B. Verma, and N. J. Rosenberg (1981): *Ag. Meteorol.*, *24*:175–184.
275. N. T. Edwards and P. Sollins (1973): *Ecology*, 54:407–412.
276. D. D. Baldocchi, S. B. Verma, D. R. Matt, and D. E. Anderson (1986): *J. Appl. Ecol.*, 23:967–976.
277. A. C. Lasaga, R. A. Berner, and R. M. Garrels (1985): *The Carbon Cycle and Atmospheric CO_2: Natural Variations, Archean to Present* (E. T. Sundquist and W. S. Broecker, eds.), Geophysical Monograph 32, Washington, DC, pp. 397–411.
278. R. A. Berner (1990): *Science*, 249:1382–1386.
279. J. M. Barnola, D. Raynaud, Y. S. Korotkevich, and C. Lorius (1987): *Nature*, 329:408–414.

280. R. A. Houhgton (1991): *Global Climate Change and Life on Earth* (R. L. Wyman, ed.), Chapman and Hall, New York, pp. 43–55.

281. G. Marland (1990): *Trends '90* (T. A. Boden, P. Kanciruk, and M. P. Farrell, eds.), Oak Ridge National Laboratory, Report ORNL/CDIAC-36, Oak Ridge, TN, pp. 92–95.

282. M. C. MacCracken, M. I. Budyko, A. D. Hecht, and Y. A. Izrael (eds.). (1990): *Prospects for Future Climate: A Special US/USSR Report on Climate and Climate Change*, Lewis Publishers, Chelsea.

283. C. D. Keeling and T. P. Whorf (1990): *Trends '90* (T. A. Boden, P. Kanciruk, and M. P. Farrell, eds.), Oak Ridge National Laboratory, Report ORNL/CDIAC-36, Oak Ridge, TN, pp. 8–9.

284. A. Neftel, E. Moor, H. Oeschger, and B. Stauffer (1985): *Nature*, *315*:421–426.

285. M. M. Peet, S. C. Huber, and D. T. Patterson (1986): *Plant Physiol.*, *80*:63–67.

286. S. Yelle, R. C. Beeson, Jr., M. J. Trudel, and A. Gosselin (1989): *Plant Physiol.*, *90*:1465–1472.

287. S. C. Wong (1990): *Photosyn. Res.*, *23*:171–180.

288. T. D. Sharkey, F. Loreto, and C. E. Delwiche (1991): *Plant Cell Environ.*, *14*:333–338.

289. R. T. Thomas and B. R. Strain (1991): *Plant Physiol.*, *96*:627–634.

290. L. H. Ziska, K. P. Hogan, A. P. Smith, and B. G. Drake (1991): *Oecologia*, *86*:383–389.

291. K. P. Hogan, A. P. Smith, and L. H. Ziska (1991): *Plant Cell Environ.*, *14*:763–778.

292. L. H. Ziska, B. G. Drake, and S. Chamberlain (1990): *Oecologia*, *83*:469–472.

293. G. Bowes (1991): *Plant Cell Environ.*, *14*:795–806.

294. A. Von Schaewen, M. Stitt, R. Schmidt, U. Sonnewald, and L. Willmitzer (1990): *EMBO J.*, *9*:3033–3044.

295. R. T. Besford, L. J. Ludwig, and A. C. Withers (1990): *J. Exp. Bot.*, *41*:925–931.

296. A. J. Rowland-Bamford, L. H. Allen, Jr., J. T. Baker, and G. Bowes (1991): *Plant Cell Environ.*, *14*:577–583.

297. M. Mousseau and H. Z. Enoch (1989): *Plant Cell Environ.*, *12*:927–934.

298. J. Hoddinot and P. Joliffe (1987): *Can. J. Bot.*, *66*:2396–2401.

299. D. L. Ehret and P. A. Joliffe (1985): *Can. J. Bot.*, *63*:2026–2030.

299a. D. T. Tissue, R. B. Thomas, and B. R. Strain (1993): *Plant Cell Environ.*, *16*:859–865.

300. R. F. Sage and C. D. Reid (1993): *Photosynthetica*, *27*:605–617.

301. I. Nijs, I. Impens, and T. Behaeghe (1989): *Planta*, *177*:312–320.

302. D. Eamus (1991): *Plant Cell Environ.*, *14*:843–852.

303. N. Fetcher, C. H. Jaeger, B. R. Strain, and N. Sionit (1988): *Tree Physiol.*, *4*:255–262.

304. I. Nijs, I. Impens, and T. Behaeghe (1989): *J. Exp. Bot.*, *40*:353–359.

305. S. C. Wong (1980): *Carbon Dioxide and Climate: Australian Research* (G. I. Pearman, ed.), Aust. Acad. of Sci., pp. 159–166.

306. D. W. Lawlor and R. A. C. Mitchell (1991): *Plant Cell Environ.*, *14*:807–818.
307. F. I. Woodward (1987): *Nature, 327*:617–618.
308. J. Penuelas and R. Matamala (1990): *J. Exp. Bot., 41*:1119–1124.
309. C. H. Körner (1988): *Flora, 181*:253–257.
310. F. I. Woodward and F. A. Bazzaz (1988): *J. Exp. Bot., 39*:1771–1781.
311. C. W. Chang (1975): *Plant Physiol., 55*:515–519.
312. M. A. Porter and B. Grodzinski (1984): *Plant Physiol., 74*:413–416.
313. F. A. Bazzaz (1990): *Annu. Rev. Ecol. Syst., 21*:167–196.
314. G. Öquist and B. Martin (1986): *Photosynthesis in Contrasting Environments* (N. R. Baker and S. P. Long, eds.), Elsevier, Amsterdam, pp. 237–294.
315. D. J. Kyle, C. B. Osmond, and C. J. Arntzen (eds.) (1987): *Photoinhibition*, Elsevier, Amsterdam.
316. S. P. Long and F. I. Woodward (eds.) (1988): *Plants and Temperature*, Soc. Exp. Biol., Cambridge, UK.
317. F. Katterman (ed.) (1990): *Environmental Injury to Plants*, Academic Press, New York.
317a. B. T. Mawson, J. Svoboda, and R. W. Cummins (1986): *Can. J. Bot., 64*:71–76.
318. J. A. Berry and O. Björkman (1980): *Annu. Rev. Plant Physiol., 31*:491–543.
319. R. F. Sage and R. W. Pearcy (1987): *Plant Physiol., 84*:959–963.
320. R. K. Monson, M. A. Stidham, G. J. Williams III, G. E. Edwards, and E. G. Uribe (1982): *Plant Physiol., 69*:921–928.
321. J. Whiteside (1992): *Physiological Consequences of Resource Allocation in Two Annual Agricultural Species, Capsicum annuum and Lycopersion esculentum*, M. S. thesis, University of Georgia, Athens.
322. H. Schnyder, F. Mächler, and J. Nösberger (1984): *J. Exp. Bot., 35*:147–156.
323. M. R. Badger, O. Björkman, and P. A. Armond (1992): *Plant Cell Environ., 5*:85–99.
324. E. Weis and J. A. Berry (1988): *Plants and Temperature* (S. P. Long and F. I. Woodward, eds.), Soc. Exp. Biol., pp. 329–346.
325. S. B. Ku and G. E. Edwards (1977): *Plant Physiol., 59*:991–999.
326. P. J. Ferrar, R. O. Slatyer, and J. A. Vranjic (1989): *Aust. J. Plant Physiol., 16*:199–217.
326a. B. T. Mawson and W. R. Cummins (1989): *Plant Physiol., 89*:325–332.
327. M. M. Ludlow (1987): *Photoinhibition* (D. J. Kyle, C. B. Osmond, and C. J. Arntzen, eds.), Elsevier, Amsterdam, pp. 89–109.
328. E. Weis (1981): *Planta, 151*:33–39.
329. M. A. Stidham, E. G. Uribe, and G. J. Williams III (1982): *Plant Physiol., 69*:929–934.
330. E. Weis (1981): *FEBS Lett., 129*:197–200.
331. J. Kobza and G. E. Edwards (1987): *Plant Physiol., 83*:60–74.
332. P. J. Aphalo and P. G. Jarvis (1991): *J. Plant Physiol., 138*:12–16.
333. M. Peisker and I. Ticha (1991): *J. Plant Physiol., 138*:12–16.

334. D. A. Ward and D. W. Lawlor (1990): *J. Exp. Bot.*, *41*:309–314.
335. S. D. Smith and P. S. Nobel (1986): *Photosynthesis in Contrasting Environments* (N. R. Baker and S. P. Long, eds.), Elsevier, Amsterdam, pp. 13–62.
336. H. A. Mooney (1980): *Oecologia*, *45*:372–376.
337. C. A. Labate and R. C. Leegood (1990): *Planta*, *182*:492–500.
338. A. S. Holaday, W. Martindale, R. Alfred, A. L. Brooks, and R. C. Leegood (1992): *Plant Physiol.*, *98*:1105–1114.
339. H. J. Ougham and C. J. Howrath (1988): *Plants and Temperature* (S. P. Long and F. I. Woodward, eds.), Soc. Exp. Biol., pp. 259–280.
340. P. J. Quinn (1988): *Plants and Temperature* (S. P. Long and F. I. Woodward, eds.), Soc. Exp. Biol., pp. 237–258.
341. H. Huner, J. Williams, M. Krol, S., Boese, V. Henry, L. Lapointe, and T. Reynolds (1989): *Cur. Top. Plant Biochem.*, *8*:6–20.
342. Ch. Körner and W. Larcher (1988): *Plant and Temperature* (S. P. Long and F. I. Woodward, eds.), Soc. Exp. Biol., Cambridge, UK, pp. 25–58.
343. C. Guy (1990): *Environmental Injury to Plants* (F. Katterman, ed.), Academic Press, New York, pp. 35–61.
344. N. P. A. Hunter, J. P. Palta, P. E. Li, and J. V. Carter (1981): *Can. J. Biochem.*, *57*:1036–1041.
345. D. V. Lynch (1990): *Environmental Injury to Plants* (F. Katterman, ed.), Academic Press, New York, pp. 17–34.
346. Ch. Körner, P. Bannister, and A. F. Mark (1986): *Oecologia*, *69*:577–588.
347. Ch. Körner and M. Diemer (1987): *Funct. Ecol.*, *1*:179–194.
348. K. A. Mott and D. F. Parkhurst (1991): *Plant Cell Environ.*, *14*:509–515.
349. H. Nonami, E.-D. Schulze, and H. Zeigler (1990): *Planta*, *183*:57–64.
350. J. A. Bunce (1984): *J. Exp. Bot.*, *35*:1245–1251.
351. J. A. Bunce (1987): *J. Exp. Bot.*, *38*:1413–1420.
352. F. Loreto and T. D. Sharkey (1991): *Tree Physiol.*, *6*:409–415.
352a. K. A. Mott, Z. G. Cardon, and J. A. Berry (1993): *Plant Cell Environ.*, *16*:25–34.
353. T. D. Sharkey (1984): *Planta*, *160*:143–150.
353a. G. Bongi (1990): *Current Research in Photosynthesis*, Vol. 4 (M. Baltscheffsky, ed.), Kluwer Academic, Dordrecht, pp. 717–720.
354. G. D. Farquhar (1978): *Aust. J. Plant Physiol.*, *5*:787–800.
355. D. W. Sheriff (1984): *Plant Cell Environ.*, *7*:669–677.
356. F. Tardieu and W. J. Davies (1992): *Plant Physiol.*, *98*:540–545.
357. K. J. Bradford, T. D. Sharkey, and G. D. Farquhar (1983): *Plant Physiol.*, *72*:245–250.
358. A. Ortiz-Lopez, D. R. Ort, and J. S. Boyer (1991): *Plant Physiol.*, *96*:1018–1013.
359. D. Gunasekera and G. A. Berkowitz (1992): *Plant Physiol.*, *98*:660–665.
360. A. J. Fredeen, J. A. Gammon, and C. B. Field (1991): *Plant Cell Environ.*, *14*:963–970.
361. N. C. Turner (1987): *Proceeding of International Conference on Measurement of Soil and Plant Water Status*, Vol. 2 (R. J. Hanks and R. W. Brown, eds.), Utah State Univ., Logan, pp. 13–24.

362. M. Santakumari and G. A. Berkowitz (1990): *Plant Physiol.*, 92:733–739.
363. M. Santakumari and G. A. Berkowitz (1991): *Photosyn. Res.*, 28:9–20.
364. I. M. Rao, R. E. Sharp, and J. S. Boyer (1987): *Plant Physiol.*, 84:1214–1219.
365. G.-Y. Ben, C. B. Osmond, and T. D. Sharkey (1987): *Plant Physiol.*, 84:476–482.
366. C. Gimenez, V. J. Mitchell, and D. W. Lawlor (1992): *Plant Physiol.*, 98:516–524.
367. W. P. Quick, M. M. Chaves, R. Wendler, M. David, M. L. Rodrigues, J. A. Passaharinho, J. S. Pereira, M. D. Adcock, R. C. Leegood, and M. Stitt (1992): *Plant Cell Environ.*, 15:25–35.
368. I. N. Saab and R. E. Sharp (1989): *Planta*, 179:466–474.
369. T. C. Hsiao (1973): *Annu. Rev. Plant Physiol.*, 24:519–570.
370. K. C. Bradford and T. C. Hsiao (1982): *Physiological Plant Ecology. II. Water Relations and Carbon Assimilation. Encycl. Plant Physiol.*, New Ser., Vol. 12B (O. L. Lange, P. S. Nobel, C. B. Osmond, and H. Zeigler, eds.), Springer-Verlag, Berlin, pp. 263–324.
371a. S. B. Powles and O. Björkman (1982): *Carnegie Inst. Wash. Yearbook*, 81:76–77.
371. N. C. Turner and M. M. Jones (1980): *Adaptation of Plants to Water and High Temperature Stress* (N. C. Turner and P. J. Kramer, eds.), Wiley-Interscience, New York, pp. 87–104.
372. J. Comstock and J. Ehleringer (1984): *Oecologia*, 61:241–248.
373. I. N. Forseth and J. R. Ehleringer (1984): *Oecologia*, 57:344–351.
374. M. A. Matthews and J. S. Boyer (1984): *Plant Physiol.*, 74:161–166.
375. S. A. Levin and H. A. Mooney (1989): *Ann. Bot.*, 64:71–75.
376. M. Küppers, G. Koch, and H. A. Mooney (1988): *Aust. J. Plant Phys.*, 15:287–298.
377. R. F. Sage and R. W. Pearcy (1987): *Plant Physiol.*, 84:954–958.
378. F. S. Chapin III, C. H. S. Walter, and D. T. Clarkson (1988): *Planta*, 173:352–366.
379. A. Makino, T. Mae, and K. Ohira (1988): *Planta*, 174:30–38.
380. K. C. Woo and S. C. Wong (1983): *Aust. J. Plant Physiol.*, 10:75–85.
381. J. R. Evans and I. Terashima (1988): *Plant Cell Physiol.*, 29:157–165.
382. C. B. Field, J. Merino, and H. A. Mooney (1983): *Oecologia*, 60:384–389.
383. D. W. Husic, H. D. Husic, and N. E. Tolbert (1978): *Oecologia*, 60:384–389.
384. J. M. Anderson and W. W. Thomsen (1989): *Photosynthesis* (W. R. Briggs, ed.), Alan R. Liss, New York, pp. 161–182.
385. J. H. Golbeck (1992); *Annu. Rev. Plant Physiol. Mol. Biol.*, 43:293–324.
386. J. B. Marder and J. Barber (1989): *Plant, Cell Environ.*, 12:595–614.
387. L. Taiz and E. Zeiger (1991): *Plant Physiology*, Benjamin Cummings, New York.
388. O. Hansson and T. Wydrzynski (1990): *Photosyn. Res.*, 23:131–162.
389. D. P. O'Keefe (1988): *Photosyn. Res.*, 17:189–216.

15

Plant Respiratory Responses to the Environment and Their Effects on the Carbon Balance

Jeffrey S. Amthor

Woods Hole Research Center
Woods Hole, Massachusetts

As a result of the interdependence between respiration and other physiological processes it is very difficult to summarize briefly the effects of external conditions upon respiration of intact plants.

> P. Gaastra (1963), in *Environmental Control of Plant Growth* (L. T. Evans, ed.), Academic Press, New York, p. 136

INTRODUCTION

Plant growth[1] in general, and crop productivity and the success of natural vegetation in particular, are often positively related to the whole-plant *carbon balance*[2] (1). Carbon acquisition occurs during photosynthesis, i.e., the balance of photosynthetic carboxylations and photorespiratory decarboxylations (1–3). Other processes are responsible for the subsequent transformations and losses of photoassimilate (2), with respiration being of special significance. Indeed, it has been estimated that 30–70% of the carbon assimilated during photosynthesis may be lost as CO_2 during subsequent plant respiration (4–7). Respiration is not a simple loss of carbon, however, for it is respiration that supplies much of the usable energy (ATP), reductant (NAD[P]H), and carbon skeletons (intermediates) that are required for growth, maintenance, transport, uptake, and nutrient assimilation processes (8). It is useful to think of respiration as the primary link between carbohydrate generation during photosynthesis and the subsequent uses of carbon and energy in heterotrophic metabolism (Fig. 1) (9,10). Clearly, to understand the effects of the physical environment on plant growth, productivity, and success, the environmental controls of respiration as well as the controls of photosynthesis and morphogenesis must be known (2,7,11).

From a casual view of the above points that (a) the carbon balance is important to growth and success and (b) respiration is a quantitatively significant component of the carbon balance, it might be surmised that a change in the amount of respiration will lead to a change in the carbon balance and growth. Inasmuch as respiration is tightly coupled to the rest of metabolism (8,12,13), however, it is not likely that the respiration rate will change appreciably without a concomitant alteration of the rates of other metabolic processes. That is, in general, whole-plant respiration rate is positively and strongly related to whole-plant photosynthesis and growth[1] rates. But environmental *stress*[3] can influence respiration independently of effects on growth or photosynthesis by causing damage

---

[1] It is necessary to have a clear definition of *growth* in order to evaluate the quantitative role of respiration in plant carbon and energy use. Surprisingly, there is not a concise, generally accepted definition of plant growth that accounts for photoassimilate use in quantitative terms. A brief discussion is given in Ref. 83. For the present purposes, growth is the conversion of temporary storage pools, or "reserve material" in Ref. 41, of principally carbonaceous and nitrogenous compounds such as sugars and amides into new structural phytomass such as cell walls and membranes, organelles, and enzymes, and long-term storage such as seed endosperm reserves and tuber starch. Thus, I take a plant to be composed of (a) structure and long-term storage plus (b) temporary pools of the substrates of growth. At the cellular level, growth is made up of biosynthetic reactions, whereas at the whole-plant level processes such as translocation and nutrient assimilation are intimately linked to biosynthesis in growing cells and they may therefore be considered growth processes.

[2] Carbon balance is the net of carbon gains and losses without regard for the state of that carbon, i.e., structure and long-term storage or temporary pools of the precursors of growth. The whole-plant carbon balance is measured as CO_2 exchange and carbon losses to processes such as herbivory, leaching, exudation, and organ abscission (2). Losses of carbon may be from the temporary storage pools prior to growth or from structure and long-term storage following growth (11). I retract the statement "plant growth is the balance of photosynthetic gains and respiratory losses" (84), although this may be nearly the case over the long term.

[3] In the present context, *stress* is considered to be any factor in excess or limited supply that detrimentally affects growth or the carbon balance of a plant. Common notions of stress include low and high soil water contents, root zone salinity, air pollution episodes, and extremes in temperature. The primary concern is environmental conditions that are likely to affect plants in the field, so experimental treatments that are clearly unrealistic will be mostly ignored in order to avoid unnecessary complications in making generalizations about effects of the environment on a plant's metabolism and carbon balance.

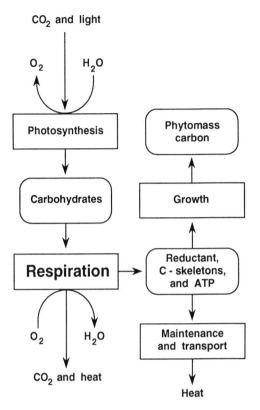

Figure 1 Summary scheme of the place of respiration in plant metabolism and growth. The flow of carbon and the exchange of energy (i.e., light and heat) and O_2 are emphasized. Carbohydrates formed during photosynthesis are the substrates of respiration, but the two processes may be temporally uncoupled. Here, "growth" includes the biosynthesis of new structure and long-term storage as well as the associated transport and nutrient assimilation processes.

or metabolic imbalances that require respiratory metabolism for the production of usable energy, reductant, or intermediates to reinstate plant health.

Examples of some specific respiratory responses to the environment are considered herein although the treatment is not meant to be encyclopedic. Instead, the approach is to outline general principles, with the aim of presenting widely applicable analyses and explanations of typical effects of the environment on respiration. The central tenet of this chapter, which contains a healthy dose of personal speculation, is that effects of the environment on respiration are generally mediated through effects on the processes that supply the substrates of respiration and growth, i.e., photosynthesis and translocation, or the processes that use the intermediates and end products of respiration, especially growth and repair. This is in contrast to direct effects of the environment on respiration. A corollary to this tenet is that it is not usually too important to ask first whether an environmental factor affects respiration, but rather to ask first what are the primary effects of a given environmental factor on plant metabolism and growth and then to ask what are the secondary and tertiary links to respiration.

Two related conceptual models will be considered as a framework for discussing effects of the environment on respiration. The first model ("pull-and-push") is in biochemical terms and is related to probable metabolic controls of respiration. The second model ("growth-and-maintenance") is generally applied at the whole-organ and whole-plant physiological levels, but it is readily reconciled with the underlying biochemistry and the pull-and-push model. I believe that these models are largely accurate and are useful in interpreting results of individual experiments, in making generalizations about metabolic responses to many environmental conditions, and in making predictions of plant response to conditions and circumstances for which experimental data do not exist. The principles underlying these models of the control of respiration rate will be considered in the next two sections. Then, after the fundamentals of the conceptual models have been outlined, some expected and observed responses by respiration to various environmental factors will be discussed.

BIOCHEMICAL CONTROL OF RESPIRATION AND THE PULL-AND-PUSH MODEL

At the biochemical level, respiration is the coordinated activity of glycolysis and the oxidative pentose phosphate network (14–17), the TCA cycle (14,18), and the mitochondrial electron transport and oxidative phosphorylation system (14,19–22). The most important *substrates* of higher plant respiration are transport and storage carbohydrates such as sucrose and starch, ADP, P_i, $NAD(P)^+$, and O_2. In fact, it is the uptake of O_2 that is the most common measure of respiration rate in excised tissue and cell and mitochondrial suspensions although nonrespiratory O_2 uptake can also occur. Respiratory enzymes and associated structures such as mitochondrial membranes are required for respiration, but except in some rapidly growing and in some senescing cells, the capacity of this respiratory machinery generally exceeds that required to support observed rates of substrate consumption (12). This implies that the respiratory machinery is often held in check, perhaps by substrate limitations, or perhaps by actual inhibitory mechanisms. This extra respiratory capacity—which incurs costs of synthesis and maintenance—might be important in metabolic responses to rapid changes in the environment to the extent that respiration rate can be increased immediately rather than only following the construction of new machinery.

The respiratory substrate that is supplied directly by the environment is O_2, and although environmental O_2 availability does not generally limit respiration, a water-saturated soil is a clear instance of a potential direct effect of the environment on root respiration (23,24). Low O_2 concentration inside bulky organs is here considered a property of such organs rather than the environment, but environmental temperature might affect the O_2 level inside bulky organs by influencing both the respiration rate and the solubility of O_2 in the cell sap (25). Carbohydrates, ADP, P_i, and $NAD(P)^+$ are here considered to be the substrates most likely to limit respiration.

The biochemical *products* of respiration are ATP, NAD(P)H, and carbon skeleton precursors of biosynthesis. These products are used in growth and by all other heterotrophic processes. A *byproduct* of respiration is CO_2—along with heat—and the most common measure of respiration rate in intact organs, plants, or plant communities is the rate of CO_2 efflux, in the dark for photosynthetic tissues. Even in the dark and in nonphotosynthetic tissues, however, CO_2 is exchanged in nonrespiratory reactions and CO_2 efflux (or O_2 uptake) is correctly called *apparent respiration*. Nonetheless, most nonphotosynthetic CO_2 (or O_2) exchange is associated with respiration and is called simply *respiration* herein.

Pull Mechanism of Respiratory Control

It is dogma that when the concentration of O_2 and the amount of respiratory machinery are both above limiting levels, as they usually are, the respiration rate is usually regulated by the rate of use of respiratory products and the concomitant regeneration of respiratory cosubstrates. Most notable is the regeneration of ADP and P_i during ATP use (8,12, 13,15,26,27). According to this paradigm, the rest of heterotrophic metabolism, i.e., sink activity, pulls respiration by using ATP and regenerating ADP and P_i (Fig. 2). Benefits to the plant of respiratory control by ADP regeneration are that respiration proceeds in concert with the needs for ATP (13), which is usually considered the "energy currency" of the cell (e.g., 28); carbohydrates are not consumed in the generation of unneeded ATP; and respiratory capacity does not limit ATP-requiring processes. Control of respiration rate by ADP regeneration suggests that an environmental factor eliciting a need for ATP will increase respiration

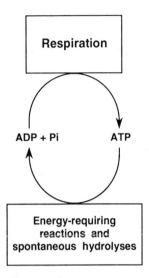

Figure 2 Simplified view of the regulation of respiration by ATP hydrolysis, i.e., regeneration of the respiratory cosubstrate ADP. This is generally considered to be significant in controlling respiration in vivo because the rate of respiration usually increases when plant tissue is presented with compounds that uncouple ATP synthesis from mitochondrial respiratory chain activity (13). Both glycolysis and the TCA cycle require ADP and generate ATP in substrate level phosphorylations, but most "heterotrophic" ATP is formed during oxidative phosphorylation by mitochondrial ATPases. In addition, the NAD^+ needed for activity of the TCA cycle is regenerated from NADH by the mitochondrial respiratory chain and respiratory chain activity can be slowed if ADP levels fall when ATP use slows. ATP can be used in heterotrophic metabolism or it can be hydrolyzed spontaneously or in futile cycles that are not coupled to obviously useful processes. At very low ADP concentrations, respiration rate is also controlled by the rate of proton leakage from the mitochondrial intermembrane space through the mitochondrial inner membrane into the matrix (e.g., 270).

rate. Conversely, when an environmental stress slows active processes, ATP use and respiration will tend to slow in turn. Rhodes (29) speculated that "membranes may play a central role in stress perception, responding to environmental . . . stimuli by altering the rate of pumping of ions." Because ATP is used for active membrane transport, respiration rate is sensitive to ion pumping rate, and this may form a link between stress and respiration. Increases in some active processes and decreases in others are probably common under stress, resulting in complicated changes in total ATP use and respiration (30).

The rate of respiration required to support processes consuming ATP is determined not only by the rate of ATP consuming processes but by the stoichiometry between substrate oxidation and ATP production in respiration. This stoichiometry is determined largely by the mitochondrial P/O. In healthy plant cells, 5–6 ATP molecules can be formed per O_2 molecule reduced (CO_2 released) during the oxidation of sugars (9,31,32). But all three sites of energy conservation, i.e., proton pumping, in the respiratory chain can be bypassed in plant cells (14,20–22,33,34), and when any of them are bypassed the P/O and the ATP/CO_2 (the ratio of ATP formed per CO_2 released in overall respiration when ATP is the sole product of respiration) are both reduced. In addition, futile cycling can hydrolyze ATP that is synthesized; when this occurs, it is without apparent benefit to the plant. A decrease in the *measured* P/O can occur if the activity of, say, lipoxygenase or polyphenol oxidase increases. These enzymes apparently can be induced by stress (35,36), although lipoxygenase activity may be primarily associated with pathogen infection or mechanical wounding (37). Such enzyme activity will not directly affect CO_2 efflux or the ATP/CO_2, only O_2 uptake and the measured P/O. Severe stress treatments can reduce the mitochondrial P/O (38,39) and in such cases more respiration is required to generate a given amount of ATP, i.e., the ATP/CO_2 declines.

Any factor that changes either the overall respiratory ATP/CO_2 or the mitochondrial P/O will alter the rate of respiration unless the demands for ATP are changed in parallel. Therefore, measures of the effects of environmental factors on the P/O of mitochondria are important in determining the reasons for any changes in respiration rate. The possibility of a change in the P/O *without* a change in respiration rate is also noteworthy and can arise when the rate of use of ATP decreases concomitantly with a decrease in the P/O. In the day-to-day life of most plants, however, environmental fluctuations and stresses probably do not greatly influence the P/O and respiration rate is governed by the interplay between substrate supply and product demand.

Whereas the amount of NAD^+ in intact plant cells may not usually limit respiration rate (13), the level of $NADP^+$ and the $NADP^+/NADPH$ are probably major regulators of the oxidative pentose phosphate network (17). In any case, $NAD(P)^+$ is required by respiration; reactions that oxidize $NAD(P)H$ have the potential to pull respiration and an environmental factor that increases the $NAD(P)H$ use rate might increase respiration rate. The generation of $NAD(P)^+$ is not invoked as often as ADP regeneration is when respiratory control is discussed.

The use of respiratory intermediates by nonrespiratory processes can also pull parts of respiration by virtue of the elimination of feedback inhi-

bition of several respiratory reactions. For example, phosphoenolpyruvate and glycerate-3-P are negative effectors of phosphofructokinase (17),
so an environmental condition that stimulates the use of these three-
carbon intermediates might enhance the rate of glycolysis. Conversely,
conditions that slow the use of later intermediates of glycolysis will slow
respiration. As with $NAD(P)^+$ regeneration, these relationships are not
usually thought to be as significant as ADP regeneration in regulating the
respiration rate in vivo.

Push Mechanism of Metabolic Control

Whereas the short-term control of respiration rate by ADP supply seems
to be a general phenomenon, the correlation between carbohydrate level
and respiration rate is often strongly positive as well (5,7,40–47) with the
relationship most striking in growing plants and organs (41). [In mature
or inactive organs and tissues, however, there may be a lack of correlation between carbohydrate level and respiration rate (5,42).] The relationship is often asymptotic, with a strong relationship at relatively low
carbohydrate levels and a much weaker or nonexistent relationship at
higher carbohydrate levels (e.g., Fig. 3). Two other sets of observations
are germane to this issue. One is the strong, positive, linear relationship

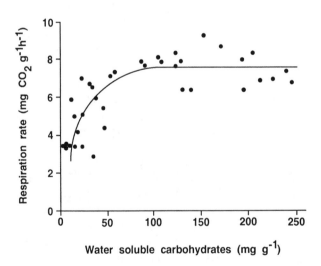

Figure 3 *Triticum aestivum* plant respiration rate as a function of water-
soluble carbohydrate content. Both respiration rate and carbohydrate content are
expressed per unit *structural* dry mass. The plotted points have been derived
from Figure 8 of (41) and the line has been fitted to them by eye.

between the rate of root respiration and the rate of translocation of carbon from the shoot to the root (48) and the other is the strong, positive, and often linear relationship between nighttime shoot respiration and previous daytime photosynthesis (49–56). Presumably, these two relationships are mediated through the level of carbohydrate in the respiring tissues. All these observations lead to the view that carbohydrate level, i.e., source activity, pushes growth, respiration, and the rest of metabolism. Based on this view, it is expected that environmental conditions that enhance photosynthesis will in turn increase respiration rate. I hasten to add, however, that in some cases carbohydrate level and respiration rate are poorly correlated (see 7), at least in the short term.

A reconciliation of the apparent contradiction between the control of respiration by the use of respiratory products such as ATP independently of carbohydrate supply and the often marked relationship between the level of carbohydrates and respiration rate, as well as the apparent inconsistencies in the relationship between respiration rate and soluble carbohydrate content of growing and mature tissues, has been most recently and elegantly pursued by a group of U.K. physiologists (7,57–60). They note that, except in starved tissue, an exogenous supply of carbohydrates usually increases respiration only after many hours, indicating that carbohydrates are not immediately limiting respiration. In many cases, however, an uncoupler does immediately increase respiration rate (12,13), even in starved cells (47,61). They (57,60) then cite evidence that over a period of hours to days carbohydrates *induce* processes such as growth, cell division, differentiation, and organ initiation, as well as increase respiratory capacity (see also 62) and alter protein quantity and quality (see also 63). These processes use ATP, reductant, and carbon skeletons, pulling respiration along and closing the circle of the simultaneous control of respiration by long-term average source and sink activities (Fig. 4) and the instantaneous rate of respiratory product use (Fig. 2). [Counter to induction of growth and respiratory capacity by carbohydrates, it has been reported that sucrose, glucose, and fructose can each repress the transcriptional activity of photosynthetic gene promoters (64; see also 65,285). In this case, the message from carbohydrates is to slow the production of additional carbohydrates, as opposed to increasing the rate of their use, in order to establish a balance between supply and demand.] The induction of growth processes by carbohydrates will be limited to immature organs and meristematic regions; it will not occur in mature organs. Thus the pushing of growth, and the consequent pulling of respiration, will be limited to cells capable of growth, which is consistent with the observations that a relationship between carbohydrate level and respiration rate is often absent in mature tissue. Mature leaf respiration rate can be positively related to carbohy-

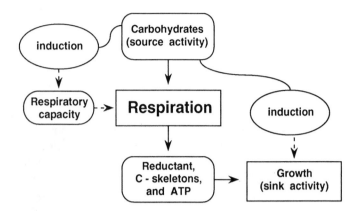

Figure 4 A proposed push mechanism of metabolic control in which a high carbohydrate level induces growth processes—as well as increasing respiratory capacity—which in turn use the products of respiration more quickly, thereby pulling respiration at a faster rate. But when sink activity is slow because of, say, an environmental stress, respiration can be held in check by a reduction in the rate of ADP regeneration irrespective of source activity. Also, when source activity is slow, i.e., when carbohydrate level is low, respiratory machinery may be broken down (47) and other heterotrophic processes may be repressed. These links provide mechanisms for the long-term balancing of source and sink activities.

drate level, however, because of the energetic requirements of translocating sugars from carbohydrate stores in leaves to stems, roots, or fruits (66). In addition, the rate of nitrate reduction in leaves in the dark, which consumes respiratory products, may be regulated by carbohydrate level (67), providing yet another link between carbohydrate level and a push of metabolism resulting in a pull of respiration (see also 9).

Another hypothesis concerning the basis of a positive relationship between carbohydrate level and respiration rate is that high sugar levels lead to engagement of the *alternative pathway*.[4] It has even been sug-

---

[4]*Alternative pathway* in the context of plant respiration refers to a cyanide-resistant transfer of electrons from the ubiquinone pool to a terminal oxidase bypassing mitochondrial complexes III and IV (14,20,33,34,70,267,268). No proton pumping across the inner mitochondrial membrane occurs along the alternative pathway (269) whereas two sites of proton pumping occur along the *cytochrome pathway* between the ubiquinone pool and complex IV (20,21). Thus, engagement of the alternative pathway reduces the mitochondrial P/O. The substrates of the alternative pathway are reduced ubiquinone (ubiquinol) and oxygen, the products are oxidized ubiquinone and water, and the byproduct is heat.

gested that "the alternative pathway is involved in the oxidation of sugars in excess of those required for the production of carbon skeletons for growth, ATP production, osmoregulation, and storage as carbohydrate reserves" (68). Although there is experimental evidence of a relationship between carbohydrate level and alternative pathway activity (e.g., 45,46), there is also counterevidence (e.g., 57). The alternative pathway is probably engaged when the ubiquinone pool of the respiratory chain is highly reduced (69), but it is not clear whether this is related to carbohydrate level. Moreover, because the role(s) of the alternative pathway is obscure (20,33,70), its relationship to overall respiration and the environment is unclear. We can be certain, however, that the notion that the alternative pathway acts to oxidize "excess" carbohydrates is controversial (16). It is also certain that I am not in a position to resolve this issue here.

Returning to the regulation of metabolism by the environment, it is common knowledge that stresses such as low water availability, low light, salinity, air pollution, and temperature extremes limit photosynthesis and carbohydrate production. To the extent that low carbohydrate levels reduce respiratory capacity and repress growth as outlined above, the common observation of slow growth and respiration as a result of chronic stress can be attributed to an inhibition of photosynthesis and subsequent downregulation of heterotrophic metabolism in response to carbohydrate depletion. On the other hand, several stresses reduce growth more than they limit photosynthesis and give rise to an increase in nonstructural carbohydrate levels. Clearly, a single chain of events cannot account for effects of all environmental conditions on growth and respiration. The key to understanding this dichotomy of limited respiration because of limited photoassimilate supply and limited respiration in spite of large pools of photoassimilate lies in a knowledge of the relative sensitivities of photosynthesis, translocation, and growth to various environmental conditions. Also, respiration rate cannot be pulled faster than the capacity of the machinery will allow, nor can its capacity be pushed to an infinite level. In any case, I take it as generally true that in the long term growth and respiration rates of plants are positively related to the supply of carbohydrates, but I recognize that various stresses can affect the stoichiometry of this relationship.

PHYSIOLOGICAL CONTROL OF RESPIRATION AND THE GROWTH-AND-MAINTENANCE PARADIGM

The most common and successful mathematical model of whole-plant or whole-organ respiration assigns metabolic processes to *growth* or *maintenance* categories (5,7,49,71–73). Although such "tidy-looking dichotomies can lead to severe oversimplifications of the complexities of our knowledge, beliefs and understandings" (74), this model is often useful in interpreting observed respiratory responses to the environment. This model can be written (here I express all terms in units of carbon or CO_2) as

$$R = g_{r,C} \, (dM_{G,C}/dt) + m_{r,C} M_{G,C} \qquad [1]$$

where R is the whole-organ or whole-plant respiration rate (mol CO_2/sec), $M_{G,C}$ is the amount of carbon in structure plus long-term storage (mol C), $dM_{G,C}/dt$ is the growth rate (mol C/sec), $g_{r,C}$ is the growth respiration coefficient (mol CO_2/mol C added to $M_{G,C}$), and $M_{r,C}$ is the maintenance respiration coefficient (mol CO_2/mol C in $m_{G,C}$/sec) (see also Table 1). Equation [1] is a concise restatement of the pull model of respiration in which the processes using respiratory products are classified as growth or maintenance. The growth and maintenance coefficients integrate an array of metabolic reactions and efficiencies. They are not constants, as has been assumed at times, but dynamic functions of many physiological and environmental factors (5). As suggested by Eq. [1], a knowledge of responses of both respiratory coefficients and growth rate to the environment is a first step in understanding respiratory responses to the environment.

Growth respiration rate, $R_{G,C}$ (mol CO_2/sec), is equal to $g_{r,C} \times$ the growth rate and maintenance respiration rate, $R_{M,C}$ (mol CO_2/sec), is equal to $m_{r,C} \times M_{G,C}$. Growth respiration occurs in growing tissue and plants whereas maintenance respiration occurs in all living cells (75). Note that in contrast to R, $R_{G,C}$, and $R_{M,C}$, the maintenance coefficient is a *specific*, or per-unit-carbon, respiration rate. It is common to consider the respiration supporting repair and acclimation processes as part of the maintenance term.

Another useful form of this model is obtained by dividing both sides of Eq. [1] by $M_{G,C}$:

$$r_C = m_{r,C} + g_{r,C} \, (dM_{G,C}/dt)/M_{G,C} \qquad [2]$$

where r_C is whole-organ or whole-plant specific respiration rate (mol CO_2/mol C in $M_{G,C}$/sec) and $(dM_{G,C}/dt)/M_{G,C}$ is the specific or so-called relative growth rate. Equations [1] and [2] can be applied to time periods of any length for which the average growth rate is known. Equation [2] shows a clear dependence of specific respiration rate on specific growth rate (Fig. 5), which has been observed consistently for some time (76–78). The model has been expanded elsewhere to include terms for processes such as N_2 fixation (79), ion uptake (80,81), mineral nitrogen assimilation (81), and translocation (9,66). In the growth-and-

Table 1 Symbols Used in This Chapter to Describe Higher Plant Respiration Within the Context of the Growth and Maintenance Model and the Carbon Balance[a]

Symbol	Name and definition
$g_{r,C}$	Growth respiration coefficient on a carbon basis. CO_2 efflux due to growth processes per unit carbon added to new structure and long-term storage (e.g., mol CO_2 mol^{-1} C). The coefficient is equal to $(1 - Y_{G,C})/Y_{G,C}$.
E_C	Growth efficiency on a carbon basis. Equal to $(dM_{G,C}/dt)/[(dM_{G,C}/dt) + R]$ and would be equal to $Y_{G,C}$ if $m_{r,C}$ were zero.
$E_{C,app}$	Apparent growth efficiency. Equal to $(P - R)/P$. This is a simple measure of the CO_2 balance of the plant (see footnote 2).
$M_{G,C}$	Moles of carbon in structure and long-term storage (see footnote 2).
$m_{r,C}$	Maintenance respiration coefficient on a carbon basis (e.g., mol CO_2 mol^{-1} C in $M_{G,C}$ sec^{-1}). It is convenient to express $m_{r,C}$ in units of per day.
P	Photosynthetic CO_2 uptake rate. Defined as the balance of photosynthetic carboxylations and photorespiration decarboxylations (e.g., mol CO_2 sec^{-1}).
R	Apparent respiration rate. CO_2 efflux rate from a whole plant, organ, or tissue (e.g., mol CO_2 sec^{-1}).
r	Specific respiration rate on a "structural" carbon basis. Equal to $R/M_{G,C}$ (e.g., mol CO_2 mol^{-1} C in $M_{G,C}$ sec^{-1}).
$R_{G,C}$	Growth respiration rate. Equal to $g_{r,C}$ times $dM_{G,C}/dt$ (e.g., mol CO_2 sec^{-1}).
$R_{M,C}$	Maintenance respiration rate. Equal to $m_{r,C}$ times $M_{G,C}$ (e.g., mol CO_2 sec^{-1}).
$Y_{G,C}$	True growth yield on a carbon basis. Carbon added to new structure and long-term storage per unit carbon used in growth processes (mol C mol^{-1} C). $Y_{G,C}$ is equal to $(dM_{G,C}/dt)/[(dM_{G,C}/dt) + R_{G,C}]$. This is also called the yield of the growth processes and the growth conversion efficiency.

[a] Symbols used vary in the literature. Rates and ratios are average values during a time interval of interest. All parameters are expressed in units of carbon because that is the most common means of expressing rates of photosynthesis and respiration and because it lends itself easily to a carbon balance approach to plant ecophysiology.

maintenance model, these additional processes are mostly included in the growth respiration term (72).

It is reiterated that growth respiration rate is related to growth rate in Eqs. [1] and [2]. Sometimes, however, the growth-and-maintenance paradigm is considered in terms of photosynthesis instead of growth per se with the empirical model of McCree (49):

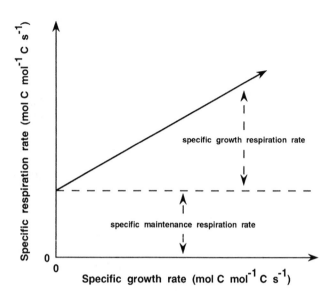

Figure 5 Relationship between specific respiration rate and specific growth rate as summarized in the growth and maintenance model of respiration. The slope of the line is the growth coefficient (mol CO_2 mol^{-1} C) and the ordinate intercept is the maintenance coefficient (mol CO_2 mol^{-1}C in $M_{G,C}$ sec^{-1}). This relationship describes well several sets of observations with whole plants and plant organs (e.g., 72,184,228,271–279). If the growth or maintenance coefficients change over time or with specific growth rate, nearly any relationship or degree of scatter might occur. The straight line relationship shown can even occur when both the growth and maintenance coefficients are nonlinear functions of growth rate. Chanter (280) has discussed statistical difficulties in fitting data to the model.

$$R = kP + cM_{G,C} \qquad\qquad [3]$$

where P is the average "gross" photosynthesis rate (mol CO_2/sec; see Table 1) during a period of interest and k and c are empirical constants. Equation [3] is related to Eqs. [1] and [2] by the relationships $g_{r,C} = k/(1 - k)$ and $m_{r,C} = c/(1 - k)$. An important parameter related to Eqs. [1] to [3] is $Y_{G,C}$, the "true growth yield" (71,82) [cf. PV of (83)]. It is equal to $1/(1 + g_{r,C})$.

Equation [3] is a source-limited push model of respiration in which substrate supply (P) implicitly drives growth and growth respiration. Note that in Eq. [3] respiration and growth are not strictly coupled, so that it can be applied only to time periods long enough for carbohydrate level induction processes to maintain a balance between photosynthesis and growth (84). Equation [3] emphasizes the often strong, positive relationship between the rates of photosynthesis and respiration and can be rearranged (71) into the form of a coupled relationship between photosynthesis and growth as mediated by respiration (Fig. 6).

The relative importance of the growth and maintenance components of respiration can be compared according to the amount of CO_2 released by each. For crops, season-long respiration is about evenly divided

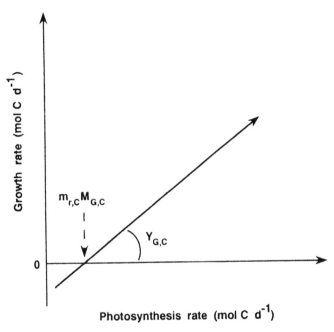

Figure 6 The coupled relationship between *daily* growth and *daily* photosynthesis derived from the growth and maintenance model of respiration (see 71). The equation of the line is:

$$dM_{G,C}/dt = Y_{G,C}(P - m_{r,C}M_{G,C})$$

and applies when no net change in the size of the temporary pools of photoassimilate occurs from day to day. An understanding of the effects of the environment on both the slope and the abscissa intercept can aid in the understanding of environmental effects on respiration and growth. A complication often arises under stress, however; the relative size of the temporary pool of the substrates of growth and respiration changes and the relationship between photosynthesis and growth is uncoupled. McCree (281) has outlined the use of this model, and its qualifications, from an experimental viewpoint.

between growth and maintenance processes (5). In tree stems also, half of respiration may be to support maintenance and the other half to support growth (85,86). Paembonan et al. (87) calculated that maintenance respiration accounted for 79% of above-ground respiration by a *Chamaecyparis obtusa* tree in the field over three annual cycles. Thus, growth and maintenance respiration are both of quantitative significance to the long-term carbon balance, but their ratio at any particular onto-genetic stage or in any particular organ can be far from unity; in mature cells and organs growth respiration is absent except for the translocation and nutrient assimilation processes contributing to growth elsewhere, whereas in actively growing cells respiration will be largely coupled to biosynthesis. Because growth and maintenance respiration are about equally important to a whole plant's carbon balance, it is necessary to understand the effects of the environment on both components of respiration.

Growth Respiration

The growth coefficient is a function of (a) the composition of the sub-strates of respiration and growth, (b) the composition of the end products of growth, (c) the pathways of biosynthesis, and (d) the energy efficiency of respiration, i.e., the ATP/CO_2 (83,93). Inasmuch as the environment influences any of these factors, $g_{r,C}$ and $Y_{G,C}$ will be affected. "Normal" values of the growth coefficient are in the range 0.15–0.65 mol C/mol C (94).

The environment is not likely to have a large impact on the composition of the carbon substrates of respiration and growth, although severe stress might result in the respiratory breakdown of proteins and lipids in addition to carbohydrates [but see (88)]. The source of nitrogen for growth, however, can vary among environments and affect the growth coefficient accordingly; theoretically, the cost of synthesizing protein from NO_3^-, including the heterotrophic assimilation of that NO_3^-, is greater than the cost of protein synthesis from NH_4^+ (Table 2). The respiratory costs incurred during N_2 fixation can be large (89). Pate and Layzell (89) present a theoretical analysis of the range of apparent respiratory costs of assimilating various forms of nitrogen by various pathways. In solution culture experiments, NH_4^+, compared to NO_3^-, induced a greater engagement of the alternative pathway, and thus a difference in the ATP/CO_2, in roots of *Triticum aestivum* (90) and *Plantago* spp. (91). This is presumably related to a large ratio of carbon skeletons to reductant used to assimilate NH_4^+ compared to NO_3^- and implies that the calculated differences shown in Table 2 for protein synthesis from NO_3^- vs. NH_4^+ are larger than they are in actual plants because a constant ATP/CO_2 was assumed in those calculations.

Table 2 Calculated True Growth Yields ($Y_{G,C}$) and Growth Respiration Coefficients ($g_{r,C}$) of Various Classes of Compounds Synthesized from Glucose and Inorganic Nitrogen via Least-Cost Biosynthetic Pathways and Including Related Transport Costs

Product of growth	Carbon content (g C in product per g product)	$Y_{G,C}$ (mol C in product per mol C in glucose consumed in growth)	$g_{r,C}$ (mol CO_2 released per mol C in product of growth)
Carbohydrates	0.451	0.944	0.059
Proteins (from NO_3^-)	0.532	0.520	0.925
Proteins (from NH_4^+)	0.532	0.779	0.284
Lignins	0.690	0.774	0.293
Lipids	0.774	0.607	0.648
Organic acids	0.375	0.982	0.018

Source: After Refs. 83, 93, 94.

Often the most dramatic effect of the environment on the composition of a plant is a change in the relative size of the temporary storage pools. Whereas such a change does not alter the growth coefficient as defined herein, it will generally alter the stoichiometry between photosynthesis and respiration. With a measure of CO_2 exchange only, i.e., without a measure of temporary pool sizes, growth may *appear* to be occurring at a faster (or slower) rate than it is. Moreover, it may appear that growth is proceeding at less (or more) cost because short-term storage is cheaper than growth (92). It is not known whether stress influences the pathways of biosynthesis for a given substrate–end product combination, however, so that the role of the environment in affecting $g_{r,C}$ via alternative pathways of biosynthesis is unclear.

Tissue grown in stressful environments may contain different lipids or proteins than tissue grown under more favorable conditions. The cost of synthesizing a unit mass of one type of lipid or protein can differ from the cost of synthesizing another type, but the effect of a change in lipid or protein *type* on the whole-plant growth coefficient will usually be small (93,94). The *amount* of protein in a plant, on the other hand, significantly affects whole-plant $g_{r,C}$. Plants growing in nitrogen-poor soils can be expected to have low nitrogen contents (95,96), and this should result in correspondingly low whole-plant protein contents and growth coefficients.

Stresses may alter whole-plant composition by affecting the partitioning of growth between root and shoot or between vegetative and reproductive organs (97) irrespective of the composition of a particular organ.

Since different organs on the same plant often have different growth coefficients (94,98), changes in partitioning among organs is likely to change the whole-plant growth coefficient (99). I emphasize that a small $g_{r,C}$ is not necessarily good for a plant and a reduction in $R_{G,C}$ need not lead to a more beneficial plant carbon balance, although a simple analysis of the type growth = photosynthesis − respiration implies that it will. Because lipids and proteins are necessarily more expensive to synthesize than are, for example, structural and long-term storage carbohydrates, a high lipid/protein seed or leaf must have a large $g_{r,C}$. An environmentally induced decrease in $g_{r,C}$ can reflect a decrease in tissue quality.

Because $R_{G,C}$ is the product of the growth coefficient and the growth rate, any factor that alters the rate of growth can change $R_{G,C}$ independently of a change in $g_{r,C}$. Thus, it is important to have good measures of growth to make sense of data within the context of either the pull-and-push or the growth-and-maintenance models. Environmental stress limits growth by definition (footnote 3) and it is likely to limit growth respiration as a consequence. I will even go as far as to say that reductions in growth respiration due to an environmental change or stress will usually be due to reduced growth rather than a change in the growth coefficient. This is not to say that an environmentally induced change in $g_{r,C}$ is unimportant but rather that the fundamental effect of the environment on growth respiration is often mediated through effects on growth rate. Indeed, limited $R_{G,C}$ is likely to be a reflection of limited growth and reduced productivity.

Maintenance and Repair Respiration

The maintenance coefficient reflects the metabolic costs of maintaining the status quo, including the repair of damaged cellular structures, or acclimating to a changing environment. Experimental estimates of $m_{r,C}$ have been quite variable, spanning the range 0.005–0.05/day at moderate temperatures (5,84,100).

About 40 years ago, James (25) speculated that the positive relationship between respiration rate and protein level may arise in part because the energy for maintenance of protein concentration against a continuous degradation is supplied by respiration. Based on the pioneering analysis of leaf maintenance by Penning de Vries (100) about 20 years ago, which followed earlier work in microbiology (82,101,102), it is now generally assumed that protein turnover is the maintenance process using the largest amount of respiratory products. Available data, however, indicate that less than half of maintenance respiration is to support protein turnover (10,103). Nonetheless, maintenance respiration is well related to protein content (5,6), implying that other maintenance processes, most notably active membrane transport, are linked to protein level and that maintenance respiration may be more appropriately related to protein content than to total or structural dry mass or carbon content (see 104,105). Price (78) came to a similar conclusion more than 30 years ago and wrote that "while protein analyses will not explain all respiratory alterations, reference to some such 'protoplasmic base' is essential for the intelligent appraisal of plant metabolism."

A presumed function of protein turnover in plants (106,107) [and animals (108)] is the replacement of one set of enzymes with another that is more beneficial at a particular ontogenetic stage or for a new set of environmental conditions. The primary function of intracellular ion transport, the other major maintenance process, is probably the maintenance of appropriate intercompartmental ion gradients against membrane leakage (100). Ion transport may constitute a large fraction of maintenance expenditures, but costs of ion gradient maintenance are ill defined, particularly with regard to whole, field-grown plants. Similarly, little is known of the *quantitative* nature of whole-plant protein turnover and its use of respiratory products in field-grown plants.

Because $R_{M,C}$ is the product of $m_{r,C}$ and $M_{G,C}$, any environmental factor that limits plant growth—the list is long—can limit whole-plant $R_{M,C}$ due to a small $M_{G,C}$ independent of affects on $m_{r,C}$. Environmental resource limitations are likely to elicit such a response. But if $m_{r,C}$ is increased by a toxic stress, $R_{M,C}$ need not be small despite a small $M_{G,C}$. Thus, although growth and growth respiration tend to be limited by stress, maintenance respiration may increase, decrease, or remain unchanged. As a generalization, respiration will be diminished by a stress to the extent that growth respiration is slowed because of growth limitations, but respiration will be increased by a stress to the extent that maintenance respiration is accelerated for repair and acclimation (11). This generalization leads to a coarse distinction between (a) environmental factors such as resource limitations that slow photosynthesis or growth without damaging cellular structures and (b) physical perturbations or toxins that damage cellular structures or elicit large increases in rates of ion pumping or protein turnover. Slowly developing water stress, low light, low nutrient availability, and low temperature are examples of factors that can limit photosynthesis or growth without otherwise damaging a plant. As a result of and following acclimation to these factors, overall metabolic activity will be slow, and the maintenance coefficient can be expected to be small as a result (84,100). Stress that causes cellular damage, on the other hand, can be expected to elicit an increase in maintenance expenditures, and also probably limit growth, because of the meta-

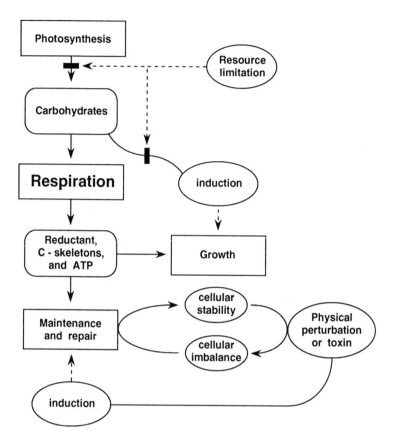

Figure 7 Possible interactions between respiration and either a resource limitation affecting photosynthesis and the supply of carbohydrates to heterotrophic metabolism or a physical perturbation or toxin. A resource limitation, by limiting the supply of carbohydrate, might repress growth processes over and above simple reductions due to substrate limitation. A physical perturbation or toxin, on the other hand, may induce various maintenance and repair processes as well as upset intracellular metabolite gradients. These might accelerate maintenance respiration and divert respiratory products away from growth.

bolic costs of repair and detoxification (Fig. 7). For example, environmental stresses that damage membranes might increase ATP use for the repair of those membranes and the reestablishment of intracellular ion gradients. Such accelerated ATP use can pull respiration along at a faster pace.

RESPIRATORY RESPONSES TO TEMPERATURE

Respiration rate in general (109) and maintenance respiration rate in particular (53,84) respond strongly to short-term changes in tissue temperature over a broad thermal range (Fig. 8). Beginning at, say, several degrees below 0°C—at which respiration is very slow but often measurable—respiration rate may increase exponentially with an increase in temperature up to at least the prevailing temperature of the plant's environment. Respiration rate then typically increases more or less linearly, passes through an inflection point, and reaches a maximum rate within a shoulder between perhaps 40 and 50°C. With still further increases in temperature, respiration rate declines sharply (5,109). The responses of respiration to temperature in and above the shoulder are time-dependent, i.e., the duration of exposure to high temperature determines in part the respiration rate at that temperature with longer exposures resulting in slower respiration (25). In nature, the high-temperature shoulder is seldom reached, so that respiration rate increases with any increase in temperature that is likely to be experienced by a plant. Diurnal changes in temperature may elicit twofold changes in the rate of respiration while seasonal changes in temperature can affect respiration to an even greater extent. In accordance with the pattern of respiratory response to temperature (Fig. 8), the Q_{10} or temperature coefficient of

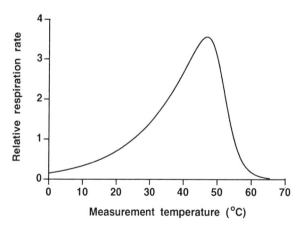

Figure 8   An idealized relationship between respiration rate and tissue temperature. This curve applies to short-term changes in temperature; long-term changes in temperature can elicit acclimatory responses by metabolism and respiration (see text and Fig. 9). Thornley and Johnson (282), for example, theoretically derive a biological temperature response with this shape.

respiration changes with temperature; it is not a constant. The Q_{10} is typically about 3 at low temperature and decreases as temperature increases (25,109), becoming less than unity above the high-temperature shoulder. The "temperature optimum" for respiration should not be derived from a relationship between respiration rate and plant temperature (e.g., Fig. 8) as might be done for leaf photosynthesis. To the extent that it is even useful to consider an optimal temperature for respiration, that optimum should be related to growth respiration rather than total respiration because growth and maintenance processes are likely to respond differentially to temperature. Simply maximizing the ratio $R_{G,C}/R_{M,C}$ with respect to temperature is not enough, however. The optimum temperature for respiration is the temperature that results in the greatest amount of growth in the long term. This indeed entails a favorable $R_{G,C}/R_{M,C}$ ratio but also requires rapid growth and ample substrate. In addition to direct effects of temperature on respiration, temperature also affects photosynthesis, translocation, and development, and these in turn are strong regulators of growth and respiration.

Acclimation and Adaptation

Rates of many physiological processes at any given temperature are quantitatively linked to temperature history (e.g., 110,111) as a result of acclimation processes (phenotypic adjustments to a change in temperature) and respiration is not an exception (112–123). These observations do not contradict the *short-term* response to temperature depicted in Fig. 8. Instead, a plant acclimated to a relatively high temperature can still show a response to temperature similar to that in Fig. 8, but with the entire curve shifted to the right, whereas a plant acclimated to a relatively low temperature will display a respiratory response to temperature that is shifted to the left. Respiratory acclimation can occur within a few hours to days following a change in temperature in controlled environments (114) and is presumably occurring at nearly all times in the natural environment. Different plants growing at and acclimated to different temperatures may have similar rates of respiration at their respective growth temperatures [e.g., Fig. 9 and (117)] although the extent of acclimation possible for a plant can depend on the environment to which the plant is adapted (genetically adjusted to the prevailing temperature) (116,117).

Weger and Guy (124) noted that many reports of respiratory acclimation to temperature have been for shoots and leaves rather than for roots. They grew *Picea glauca* plants with different root temperatures for 4 weeks and then measured the rate of growing-root respiration over a range of temperatures. No acclimation to temperature was observed and

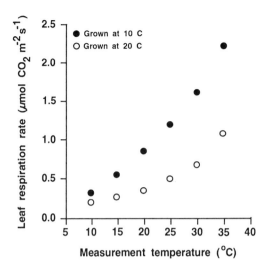

Figure 9 Leaf respiration rate on an area basis measured at six temperatures for *Festuca arundinaceae* grown at 10°C or 20°C (118). Note that the rate of respiration at 10°C by the leaf from the plant grown at 10°C is about the same as the rate of respiration at 20°C by the leaf grown at 20°C. This degree of acclimation, approaching "the homeostatic ideal" (117; and see 283), is apparently common in leaves and shoots.

an uncoupler (CCCP) did not increase respiration rate. Roots are strongly affected by shoots, and shoot temperature was held constant across their root temperature treatments, so that their observations may have resulted from an overriding regulation of root metabolism by shoot metabolism. It is not clear, therefore, whether *Picea glauca* roots will acclimate to a whole-plant temperature change. Smakman and Hofstra (125) had earlier noted for *Plantago lanceolata* with a shoot temperature of 21/16°C (day/night) but with root temperatures of either 13 or 21°C that roots did acclimate and root respiration rate was the same when measured at 13 and 21°C for roots cultured at those respective temperatures. The cause of the discrepancy between these two studies of root respiration is not evident and may reflect species differences. In any case, more studies of long-term root respiration responses to changing temperature are needed, especially for plants in the field (see 126). Indeed, there are still many important questions concerning the acclimation of shoot as well as root respiration—and growth—to temperature and I suggest that many of these questions can be usefully addressed within the context of the pull-and-push and growth-and-maintenance models.

Inasmuch as the pull-and-push and growth-and-maintenance models reflect reality, there is no need to propose a change in respiration per se during acclimation to a change in temperature. This is not to say that biophysical and biochemical changes to the respiratory machinery do not occur upon a change in temperature (127), but if growth or maintenance processes acclimate to a temperature change, observed changes in respiration can easily be reflecting an underlying change in those other processes. This begs the question of how and why those other processes acclimate to temperature, but the point I wish to make here is that our understanding of respiratory acclimation to temperature will increase most rapidly if we are examining the primary effects of temperature on growth and maintenance. Let us remember what respiration is for and how it can be regulated when considering environmental effects on its rate.

As contrasted with a given plant acclimated to a change in temperature, plants native to cool habitats often have faster respiration at a given temperature than do plants native to warm habitats (109,113,128–131). This is probably the result of a combination of adaptation and acclimation to temperature, and may reflect the conservation of similar rates of metabolism and growth in environments with different prevailing temperatures. In plants growing at high elevation (low temperature) mitochondria are more abundant than in low-elevation plants (132; and see 133), which presumably facilitates more rapid respiration at any particular temperature. Fifty years ago, Wager (128) alluded to the possibility that respiration is faster at low temperature for plants from cool habitats compared to warm habitats because growth is more rapid at low temperature in the former. Maintenance processes are also probably adapted to different habitats.

As suggested by both experimental data and the models, respiration is rapid when growth is rapid and respiration should be faster in growing organs or plants than in less active plants at any particular temperature. Such a growth rate–temperature interaction has been observed in trees growing in the field, where respiration is faster at a given temperature in spring (134,135), when growth is rapid, compared to other seasons, when respiration is due mostly to maintenance processes. Forest herbs show the same respiratory response to temperature as a function of growth rate in the field (136). In fact, whole-ecosystem respiration can be faster at a given temperature in the summer than in the winter (137,288), which is probably related to the timing of growth and other energy-requiring processes. In contrast, Marks (138) reported faster specific respiration rates at a given temperature during winter in *Rubus chamaemorus*.

In addition to ontogeny and temperature history, other factors also influence the respiratory response to temperature. For example, there is a strong positive effect of whole-plant carbohydrate level on respiration rate at a given temperature (139,140) which may be related to the induction by high carbohydrate levels of processes consuming respiratory products, most notably growth. A previous water stress (141) or frost (142) can also affect the relationship between temperature and respiration rate.

Low- and High-Temperature Extremes

Any temperature above or below the optimum, nebulously defined above, is stressful. The impact of that stress increases as the temperature diverges from the optimum. Raison (143) discussed low-temperature damage to the respiratory machinery in chilling sensitive plants within the context of changes in the activation energy of enzymes as a result of changes in the molecular ordering of membrane lipids [see also (144)]. This differs from a simple slowing of respiration at low temperature due to a general loss of cellular kinetic energy. The hypothesis of membrane phase transitions at low temperature has been criticized (145), but in any case low temperature can cause apparent damage to respiratory capacity in some plants, especially tropical species. The rate of cooling in a low-temperature experimental treatment is critical to the damage that ensues (146).

A general observation is that carbohydrates accumulate in plants at low temperature (147,148). This is most simply explained by a greater limitation on growth than on photosynthesis by low temperature, but low temperature might also limit translocation of sugars to sinks (148,149). Even though it is usually assumed that the growth respiration coefficient is independent of temperature (94,282), many experiments show a positive relationship between $g_{r,C}$ and temperature (5). This may be the result of alterations of any of the factors determining $g_{r,C}$ (93), but it seems most probable that low temperature leads to a small apparent $g_{r,C}$ via a decrease in growth relative to photosynthesis manifested by an accumulation of carbohydrate reserves. Thus, even if the actual $g_{r,C}$ is independent of temperature, the apparent $g_{r,C}$ may not be because storage is cheaper than growth. Whatever the underlying mechanisms are, low temperature tends to decrease the ratio of whole-plant respiration to photosynthesis.

A lack of an effect on, or even the stimulation of, the alternative pathway by low temperature has been noted (references listed in 124). For example, Zea mays seedlings grown at 14°C compared to 30°C displayed an enhanced alternative pathway capacity and oxidase protein amount (150). When seedlings grown at 14°C were transferred to 30°C, the amount of alternative oxidase declined over a period of 2–3 days (151). In leaves of the arctic species Saxifraga cernua, activity of the alternative pathway was insensitive to short-term changes in temperature between 5 and 25°C (121), so that the fraction of total mitochondrial electron transport occurring along the alternative pathway increased as temperature decreased and the mitochondrial P/O and the overall ATP/ CO_2 apparently decreased as temperature decreased. Contrary results have been reported for *Picea glauca* roots in which activity of the alternative pathway was markedly stimulated by a short-term increase in temperature (124). Moreover, the fraction of total O_2 uptake ascribed to the alternative oxidase increased with an increase in temperature (124). With regard to plants growing in situ, however, very little is known of the quantitative effects of low temperature on pathways of respiratory electron transport.

Semikhatova (152,153) claimed that the efficiency of oxidative phosphorylation is not reduced during high-temperature stress (41–42°C) but that increased energy needs for cellular repair pulls respiration more quickly. The decline in respiration rate with an increase in temperature above the shoulder (Fig. 8) is probably the result of metabolic damage. Even if the mitochondrial P/O is unaffected by high temperature (153), respiration will decline at very high temperature because the processes using ATP will slow upon their damage. To prevent long-term metabolic damage during a high-temperature period, rapid respiration in support of repair processes may be needed (154). As a matter of practical significance, most plants in nature do not experience temperatures high enough to cause metabolic damage.

The temperature of a plant is largely a function of environmental temperature and radiation fluxes, although leaves do cool themselves by opening stomata and evaporating water (155,156). In a few instances, however, plant organs may elevate their temperature significantly above that normally allowed by the environment through rapid, i.e., thermogenic, respiration. The most noted cases concern the inflorescences of several Araceae species (157). Thermogenic respiration in these inflorescences proceeds for up to several hours and may occur on two to five occasions (157). The rate of respiration may be inversely related to air temperature, resulting in a more or less constant inflorescence temperature across a range of air temperatures (158). The high inflorescence temperature results in the volatilization of attractants of insect pollinators (157) and as such may be important to reproduction, the inflorescence's *raison d'être*. Fast respiration in Araceae inflorescences is mediated by rapid activity of the alternative pathway, but activity of complex I or its bypass (21,22) must also be enhanced. Increased activity of glycolysis and the TCA cycle are clearly required as well (159), and glycolytic and TCA cycle capacities increase prior to the thermogenic period (160). There is thus a coordinated increase in the capacity and activity of the whole of respiration. It is unfortunate, therefore, that several papers seem to imply that the alternative pathway produces heat and the cytochrome pathway—and indeed, the rest of respiration—does not. Heat generation is the result of the oxidation of substrate by respiration; it is not limited to the alternative pathway. The cytochrome pathway is not completely efficient in terms of energy transfer during proton pumping, and even if it were, heat is generated during the subsequent hydrolysis of ATP formed during oxidative phosphorylation. Thus, unless growth is occurring and energy is being retained in newly formed phytomass, all respiratory electron transport, be it associated with the cytochrome pathway or the alternative pathway, generates heat directly or indirect (9,161). The alternative pathway *allows* rapid respiration, and therefore heat production, because it circumvents much of the respiratory control exerted by ATP use and ADP regeneration.

RESPIRATORY RESPONSES TO CARBON DIOXIDE

Earth's atmospheric CO_2 concentration is increasing on an annual basis, due mainly to human activities, at the rate of about 0.5%/year (162). This increase has several consequences for the carbon balance of plants because atmospheric CO_2 (a) is the carbon substrate of photosynthesis, (b) inhibits photorespiration, and (c) is a strong contributor to the so-called greenhouse effect and the consequent effects on global climate and weather. In addition, terrestrial plant respiration may be affected by long-term (hours to years) exposure to elevated daytime CO_2 ("indirect effects") and is apparently also influenced by short-term (immediate) changes in nighttime CO_2 level independent of the CO_2 history of the plant ("direct effects") [163].

Effects of CO_2 in the range 0–1000 ppm on apparent respiration have been assessed in three ways (286).

Procedure 1. Plants are grown under arbitrary conditions. The respiration rate of a given plant or organ is then measured at two or more CO_2 concentrations during the course of minutes to a few hours. Results are a gauge of the short-term *direct* effects of CO_2 on respiration and the processes using respiratory products, i.e., processes consuming carbon skeletons, reductant, and ATP.

Procedure 2. Plants are grown in two or more CO_2 concentrations for periods of weeks to months or years. (In some experiments, the CO_2 concentrations across treatments differ during the day but are equal at night.) Respiration rates of organs or whole plants are then measured at a common (single) CO_2 concentration. Results are a gauge of the *indirect* effects of long-term daytime CO_2 concentration on respiration and the processes using respiratory products.

Procedure 3. Plants are grown in two or more CO_2 concentrations for periods of weeks to months or years. (Again, CO_2 treatments may apply during the day but not the night.) Organ or whole-plant respiration rate is then measured at the respective growth (daytime) CO_2 concentrations. Results are a gauge of the combined *direct* and *indirect* effects of CO_2 and should reflect physiological acclimation to CO_2.

A single study may include more than one procedure; for example, plants can be grown in different CO_2 levels with respiration then measured at the respective growth CO_2 levels but also measured at one or more other CO_2 levels. The relative importance of direct and indirect effects of CO_2 on respiration can be assessed in this manner.

Indirect Effects

When other environmental factors are constant, moderate increases in daytime CO_2 levels can result in increased whole-plant photosynthesis, carbon accumulation, and growth (164–168 and Sage and Reid, this volume). A decrease in a plant's N/C (mass/mass) ratio may accompany the increase in carbon accumulation (169–171) although not in all cases (e.g., 172). Based on these responses to increased daytime CO_2 level, it was predicted that specific respiration rate might decline, that whole-plant respiration might increase, and that respiration per unit nitrogen might be relatively unaffected by long-term exposure to elevated CO_2 (163).

A common response to long-term CO_2 enrichment is a reduction in *specific* respiration rate (173–184). Poorter et al. (185) tabulated results from procedures 2 and 3 experiments, but in a lumped manner without procedure distinction. Therefore, a combination of direct and indirect effects of CO_2 on respiration rate is inherent in their summary. That summary shows that whole-plant specific respiration rate was slowed by long-term CO_2 enrichment in 13 of 16 studies (their table 1; see also 286). In those 13 studies, however, the mean respiration rates from control and elevated CO_2 treatments were not always significantly different. When expressed on a tissue nitrogen or protein basis rather than a dry mass basis, CO_2 history may have little effect on respiration rate. For example, Baker et al. (177) plotted respiration rate of plants grown across a wide range of daytime CO_2 levels vs. their nitrogen content along with a single straight line. The constancy of respiration rate at a given plant nitrogen content was implied by the straight line, although the individual data points from the low-CO_2–grown plants generally fell below the fitted line whereas the points from the high-CO_2–grown plants tended to be above the line, suggesting that the single line may have been an inadequate summary of the data. On a per-plant or ground area basis, respiration rate can be increased following long-term exposure to elevated CO_2 (e.g., 177). In another case, Wullschleger et al. (184) exposed field-grown *Liriodendron tulipifera* trees to several levels of CO_2 in open-top chambers. Specific leaf respiration rate was slower in elevated CO_2, but when expressed on a leaf nitrogen basis, respiration rate was not affected by CO_2, indicating that the cause of the observed CO_2 effect on respiration was a change in the composition of the leaf. Wullschleger et al. (184) determined leaf growth and maintenance respiration coefficients and found that elevated CO_2 invoked a decrease in the maintenance coefficient. If the maintenance coefficient was reduced because of a reduction in leaf protein content rather than slower turnover with the same amount of protein, it is expected that the growth coefficient would also have been reduced. Although the growth coefficient did tend toward smaller values in the elevated CO_2-grown leaves, the values were not different according to a regression analysis (184). In an earlier related experiment, Silsbury and Stevens (186) had found that elevated CO_2 (1000 vs. 320 ppm) reduced whole-plant $m_{r,C}$ (21%) and $g_{r,C}$ (15%; derived from their table 2) for pot-grown *Trifolium subterraneum*. Baker et al. (177) reported an increase in shoot $m_{r,C}$ but no change in $g_{r,C}$ with below-ambient daytime CO_2 level (see also 286).

An integrative measure of the overall place of respiration in growth and the carbon balance is the growth efficiency (GE_C, dimensionless) of Tanaka and Yamaguchi (187,188) which I write in terms of carbon:

$$GE_C = \frac{dM_{G,C}/dt}{(dM_{G,C}/dt) + R} \tag{4}$$

where growth and respiration rates are average values during a time interval of interest (Table 1; see also 5). The apparent growth efficiency ($GE_{C,app}$) is simply

$$GE_{C,app} = \frac{P - R}{P} \tag{5}$$

where P and R are average rates of photosynthesis and respiration during a time period of interest. This is the *apparent* GE_C because $P - R$ is not necessarily equal to growth (see footnotes 1 and 2). It can be anticipated that elevated CO_2 might increase $GE_{C,app}$ by virtue of an increase in carbon accumulation in nonstructural carbohydrates and increase in the C/N. Evidence does indeed indicate an increase in the $GE_{C,app}$ when daytime CO_2 is elevated (186,189–191,286). Thus, the ratio of carbon retained in the plant to that assimilated can increase in elevated CO_2, but this can be accompanied by a decrease in the quality of the matter accumulated. In the long term, and when active sinks are plentiful, however, growth efficiency may be unaffected by elevated CO_2 (192). Experimental results to date are, on the whole, easily interpreted in terms of the models used here: (a) higher CO_2 leads to more plant material per unit ground area and this requires relatively more growth respiration during its formation and subsequently more maintenance respiration; (b) the higher nitrogen content of low-CO_2-grown plants results in greater specific respiration rate because of larger growth or maintenance coefficients; and (c) for a number of reasons, respiration rate is coupled to nitrogen content irrespective of daytime CO_2 level.

As mentioned above, alternative pathway activity may be positively related to carbohydrate level. Because elevated CO_2 can result in increased carbohydrate levels (e.g., 193), engagement of the alternative pathway might be linked to daytime CO_2 concentration (44). The results of experiments designed to ascertain the effects of long-term CO_2 enrichment on engagement of the alternative pathway and its effects on growth and the carbon balance appear to me to be inconclusive (163,286). Moreover, some of the existing data may require reevaluation in light of the use of rather small rooting volumes and the potential consequent effects on source–sink relations (194). In any case, if the alternative pathway is markedly increased as a result of elevated daytime CO_2, increases in the growth and maintenance coefficients should be expected; these have not been observed [but see (45,286)]. It seems at this time that other factors have overridden any effects of elevated CO_2 on the alternative pathway. An exception to this, however, might be evident in the reports of increased leaf respiration on a leaf area basis following CO_2 enrichment (reviewed in 185). Increased respiration rate on a leaf area basis might also be a manifestation of increased energy use for increased translocation following enhanced photosynthesis (66,195,286).

Direct Effects

Over and above the indirect effects of daytime CO_2 level on respiration, CO_2 has a more direct, immediate effect on respiration rate. Increased CO_2 *in the dark*, in the range 0–1000 ppm, often causes an instantaneous and reversible reduction in leaf apparent respiration rate (e.g., 154,174,196–202). Although results have varied, a doubling of ambient CO_2 has slowed leaf CO_2 efflux rate by 10–30% in several cases (Fig. 10 and Table 3). Similarly, shoot and whole-plant apparent respiration rates can be negatively related to short-term changes in CO_2 (e.g., 203–205). The many observations of slowing of apparent respiration by short-term increases in CO_2 (Tables 3 and 4) suggest a feedback inhibition of plant heterotrophic metabolism by CO_2—a byproduct of that metabolism—at physiological concentrations. They also imply direct effects of increasing

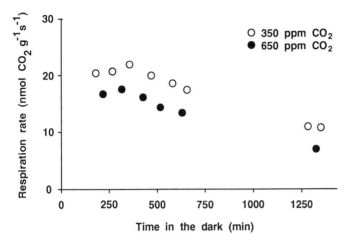

Figure 10 Effects of ambient CO_2 level on apparent respiration rate in leaves of *Rumex crispus* at 15°C (202). Data are from a single leaf attached to a plant grown in a glasshouse at a daytime CO_2 concentration of ≈ 350 ppm. Measurements were made in the dark in a leaf cuvette in an open gas exchange system in which CO_2 level was cycled between 350 and 650 ppm. Sufficient time was allowed at each CO_2 concentration to attain a steady CO_2 efflux rate. In all cases, apparent respiration rate increased when CO_2 level was decreased from 650 to 350 ppm and apparent respiration rate decreased when CO_2 was increased from 350 to 650 ppm. Similar results were obtained with other leaves from other plants.

atmospheric CO_2 on plant metabolism, and hence the global carbon cycle, over and above stimulated photosynthesis, inhibited photorespiration, altered patterns of photosynthate partitioning, and altered tissue composition. The direct inhibition of respiration by CO_2 is most prevalent in leaves, whereas whole-plant respiration is a more significant measure of respiration in terms of carbon balance. Leaf respiration, however, may constitute more than half of plant respiration in forests (reviewed in 207,208), though leaf respiration is less important in grain crops (5).

It is difficult to assign direct effects of CO_2 on respiration to either growth or maintenance components of respiration. In mature tissues, effects of CO_2 will be on maintenance or translocation, but in growing tissues the effects may be on any of the growth or maintenance processes. The mechanism(s) underlying a direct inhibition of respiration by CO_2 is unknown and several fundamental questions concerning direct effects of CO_2 on respiration should be addressed in future experiments with a range of economically and ecologically important species: (a) To what

Table 3   Summary of Published Effects of Ambient CO_2 on *Leaf* Apparent Respiration Rate (R) as a Function of Short-Term (minutes) Changes in Ambient CO_2 Concentration (C_a)[a]

Species	R_{700}/R_{350}	Ref.
Amaranthus hypochondriacus	0.86[b]	174
	0.96[c]	
Commelina cyanea	0.71[d]	201
Eucalyptus sp.	0.76	196
Glycine max	0.28[e]	197
	0.78[b]	174
	0.99[c]	
Lolium perenne	1.10[f]	178
Lycopersicon esculentum	0.88[b]	174
	0.79[c]	
Picea sitchensis	0.86	199
Rumex acetosa	0.59[g]	198
Rumex crispus	0.75[h]	202
	0.37[i]	
Xanthium strumarium	0.74	154
Zea mays	0.46	199

[a] Data were derived from text, tables, and figures of the cited papers and fitted to the equation $R = \alpha \exp(-\beta C_a)$ (see 202). Some of the cited experiments involved two CO_2 concentrations only. Parameter α encompasses factors such as units of measurement, species, growth conditions, and ontogenetic state. Parameter β describes the short-term effect of C_a on R. The calculated ratio of R at 700 ppm CO_2 to R at 350 ppm CO_2 (R_{700}/R_{350}) is shown.
[b] Grown at 350 ppm CO_2.
[c] Grown at 700 ppm CO_2.
[d] Excised whole leaf.
[e] Detached leaves at 21% O_2.
[f] Grown at 680 ppm CO_2.
[g] R_{700}/R_{350} was near zero after a 16-hr dark period.
[h] "Alternating C_a experiments."
[i] "Multilevel C_a experiments" based on C_a, not C_i as in Amthor et al. (202).

extent does CO_2 concentration in the air surrounding a leaf in the dark affect the rate of CO_2 efflux? (b) Does increased activity of phosphoenolpyruvate carboxylase (PEPcase) following short-term increases in CO_2 in the dark contribute to the observed slowing of apparent respiration rate? (c) Is some aspect of respiration per se affected by short-term increases in CO_2, or does CO_2 primarily affect another component of dark (i.e., heterotrophic) metabolism which in turn alters respiration rate through classic respiratory control mechanisms, i.e., the use of ATP, reductant, or carbon skeletons generated by respiration? (d) Does a short-term

Table 4 Summary of Published *Shoot* and *Whole-Plant* Apparent Respiration Rate as a Function of Short-Term Changes in Ambient CO_2 Concentration[a]

Species	R_{700}/R_{350}	Ref.
Amaranthus hypochondriacus (plant)	0.86[b]	174
	0.91[c]	
Castanea sativa (shoot)	0.21[d]	205
Glycine max (plant)	0.45[b]	174
	0.92[c]	
Lolium perenne (plant)	1.00[e]	178
Lycopersicon esculentum (plant)	0.66[b]	174
	0.70[c]	
Medicago sativa (shoot)	0.85[f]	204
	0.96[g]	
	0.77[h]	
	0.94[i]	
Pisum sativum (shoot)	1.16[j]	206
	0.82[k]	
Stylosanthes humilis (shoot)	0.26[l]	203
	0.42[m]	
	0.53[n]	

[a] Other comments are as in Table 3. See also (286).
[b] Grown at 350 ppm CO_2.
[c] Grown at 700 ppm CO_2.
[d] June 1 measurements on ambient CO_2-grown plants.
[e] Grown at 680 ppm CO_2.
[f] "Low-assimilate status" plants at 19°C.
[g] "High-assimilate status" plants at 19°C.
[h] "Low-assimilate status" plants at 19°C excluding data at 50 ppm CO_2.
[i] "High-assimilate status" plants at 19°C excluding data at 50 ppm CO_2.
[j] At 28°C.
[k] At 38°C.
[l] At 30°C.
[m] At 35°C.
[n] At 40°C.

change in ambient CO_2 level affect the partitioning of electrons between the cytochrome pathway and the alternative pathway in mitochondria? (e) Are any of the processes mentioned in b–d altered by long-term exposure to elevated CO_2; that is, does acclimation with respect to the direct effects of CO_2 on dark metabolism occur and, if so, to what extent?

It is significant that the role of PEPcase activity in the apparent inhibition of respiration by CO_2 in the dark is unknown. PEPcase is found in C3 as well as CAM and C4 species. It catalyzes the reaction $PEP + HCO_3^- \rightarrow$ oxaloacetate $+ P_i$, which results in the net uptake of

CO_2. The principle question concerning PEPcase with respect to direct effects of CO_2 on apparent respiration is, does dark CO_2 fixation account for any, or all, of the inhibition of net CO_2 efflux in the dark in elevated CO_2?

The activity of PEPcase is regulated by numerous metabolic effectors. Ting and Osmond (209) reported PEPcase $K_m(HCO_3^-)$ of 0.018 mM (i.e., 0.018 mol/m^3) in the C_3 *Atriplex hastata*. For PEPcase isolated from the CAM species *Kalanchoë daigremontiana*, $K_m(HCO_3^-)$ was 0.01–0.02 mM in the absence of effectors (210). Free PEPcase isolated from *Euglena*, however, had a $K_m(HCO_3^-)$ of 7.3–9.8 mM whereas it functioned in a multienzyme complex with a $K_m(HCO_3^-)$ of 0.7–1.3 mM (211). PEPcase can exist in several isoforms within a leaf (212), and these may have different $K_m(HCO_3^-)$ values. Van Oosten et al. (213) reported a decrease in specific activity of PEPcase in needles from *Picea abies* exposed to elevated CO_2 for 2 years in open-top chambers.

Water at 20°C and pH 7 equilibrated with air containing 350 ppm CO_2 has an HCO_3^- concentration of ~0.061 mM. The HCO_3^- concentration is lower at lower pH. According to Amthor (163), leaf intercellular CO_2 concentration in the dark need not rise above twice the ambient CO_2 concentration. Although it seems likely that PEPcase is saturated with respect to CO_2 under normal metabolic circumstances, this is not known to be the case for all important C3 species and an increase in CO_2 might enhance the rate of dark CO_2 fixation (286).

Two previous observations suggest that CO_2 does affect respiration per se and that stimulated dark CO_2 fixation does not account for all of the frequently observed slowing of apparent respiration rate. First, Gale's (154) experiments showing leaf injury from high temperature in the dark when CO_2 was increased from 320 to 840 ppm indicate that respiration, or the processes supported by respiratory products, is slowed by CO_2 so that normal repair processes were slowed in turn. Second, apparent respiration rate is highest at zero or near-zero CO_2 (e.g., 202,203), suggesting that the respiratory quotient should be unity at approximately zero CO_2 for tissues oxidizing primarily carbohydrates—as is assumed to be the case for mature leaves under most conditions—whereas respiratory quotients of about unity have been measured (214) at CO_2 levels that slow apparent respiration significantly (see also 287). In any case, a priority for future research is the determination of the rate of dark CO_2 fixation as a function of nighttime ambient CO_2.

To the extent that CO_2 slows respiration per se rather than only stimulating dark CO_2 fixation, it is meaningful to know whether the primary action is on a respiratory reaction(s) or on a process(es) consuming a respiratory product such as ATP or NADH. Thus, two underlying causes of respiratory inhibition by CO_2 can be considered. First, CO_2 might inactivate a respiratory enzyme or intermediate transporter through, e.g., carbamate formation (215), resulting in a true direct inhibition of respiration. If CO_2 inhibits respiration irrespective of the demands for respiratory products (i.e., carbon skeletons, ATP, and reductant), increased CO_2 has the capacity to slow overall metabolism regardless of metabolic needs. For plants requiring active repair and detoxification in the face of a stress such as air pollution or high temperature, unencumbered respiration may be essential for the maintenance of plant health (30,154). Thus, an inhibition of respiration by CO_2 may decrease the ability of a plant to cope with a harsh or changing environment. Second, CO_2 might slow a process consuming a respiratory product, again perhaps via carbamate formation or any of a number of other mechanisms (216). That is, CO_2 might slow a nonrespiratory energy- or carbon skeleton–requiring process with the result that respiration is slowed in turn through classic respiratory control mechanisms (8,12,27). In this case, respiration could fulfill its roles in cellular metabolism without hindrance by CO_2, except with respect to those nonrespiratory reactions that *are* affected by CO_2.

A reasonable appraisal of these two possibilities, i.e., inhibition by CO_2 of respiration itself or of a process using respiratory products, can be obtained with the use of uncouplers. If an increase in ambient CO_2 slows the uncoupled rate of respiration, a direct effect of CO_2 on the respiratory machinery is indicated. If, however, the uncoupled rate of respiration is not affected by ambient CO_2, it would seem that the primary effects of CO_2 take place outside the respiratory system itself. Palet et al. (217) suggest that CO_2 affects cytochrome oxidase activity whether respiration is coupled or not.

One of the most intriguing observations made to date regarding direct effects of CO_2 on respiration is that by Reuveni and Gale (204) of an increase in dry mass accumulation in *Medicago sativum* when nighttime, but not daytime, CO_2 concentration was increased. Similar findings have since been obtained with other species (J. Gale, pers. commun.). These results indicate changes in composition (e.g., more nonstructural carbon as observed under higher daytime CO_2) or increased coupling of respiration to growth under elevated nighttime CO_2. The first possibility could result from an inhibition of respiration and other processes converting photoassimilate to plant structure, producing a higher ratio of nonstructural to structural matter in higher nighttime CO_2. The second possibility might arise, for example, if engagement of the alternative pathway is "wasting" sugars and is inhibited to a greater extent by CO_2 than is the cytochrome pathway. The second possibility could also be due to the elimination by CO_2 of some general, and significant, inefficiency in respiration not related to the alternative pathway. This can only be the case, however, if respiration under ambient CO_2 levels entails such an inefficiency or uncoupling from growth and maintenance processes, and this flies in the face of the notion that respiration might be tightly coupled to useful processes under normal conditions!

There is evidence that respiration is less sensitive to CO_2 in tissues normally exposed to high CO_2 levels (218,219). For example, more than 2000 ppm CO_2 was required to inhibit CO_2 efflux from *Agave deserti* roots whereas such roots are normally exposed to ~1000 ppm CO_2 (24). Oxygen uptake by wheat apices, which can be tightly enclosed by several leaves that trap CO_2 released by respiration, was not affected by 500 ppm CO_2 (220). Wheat grain respiration was not different in atmospheres of 0 or 425 ppm CO_2 (table 6 in 221). Respiration in larger fruits and other bulky organs that normally contain high CO_2 concentrations is little affected by small changes in ambient CO_2. To the extent that this notion that the CO_2 sensitivity of respiration is related to the CO_2 concentration normally experienced by a tissue is generally true for plant cells, it raises the question, will plant respiration, or the processes pulling respiration, become adapted to higher CO_2 in the future and thus become relatively insensitive to direct inhibition by that higher CO_2? We do not know, but we should endeavor to find out.

RESPIRATORY RESPONSES TO STRESS

In addition to extremes in temperature, other environmental factors are stressful to plants and there are a plethora of reports in the literature of the effects of this or that stress on respiration rate. In many cases, these reports are about individual organs of low-light growth chamber–grown plants without concomitant measures of photosynthesis or growth. Although such studies can be important, they do not lend themselves well to the development of broad theory with respect to whole-plant carbon balances in the field.

The effects of resource limitations (see above) on respiration are generally not interesting. Resource limitations usually slow photosynthesis and growth (Fig. 7). Metabolism becomes lethargic. The need for respiration is reduced and respiration is pulled along slowly, but it is not otherwise much affected. On the other hand, physical perturbations and toxins are more exciting. Perturbations can elicit active responses from the plant and these responses in turn pull respiration for repair and maintenance along at a faster pace. Growth may be slowed and with it growth respiration, but other realms of metabolism get going!

Stresses in the Aerial Environment

Air pollutants such as ozone may have their primary impact on leaves by damaging membranes (222) and it can be expected that there will be a respiratory cost for cellular repair and reorganization (i.e., reestablishment of intra- and intercellular metabolite gradients) during and after an air pollution episode. This need for repair respiration will exist independently of the relationships between respiration and photosynthesis or growth, and while photosynthesis and growth are typically reduced by air pollutants (223,224), respiration may increase. For example, Barnes (225) observed that photosynthesis was usually depressed whereas respiration was stimulated in seedlings of four species of *Pinus* exposed to concentrations of ozone commonly observed in the field. In other instances too leaf and shoot respiration have been increased during and after exposure to air pollutants in general (223) and ozone in particular (30,226–234). The growth coefficient of *Phaseolus vulgaris* leaves was unaffected by low levels of ozone—but growth was reduced—whereas the maintenance coefficient was increased (30,228). Similarly, maintenance respiration was increased in *Raphanus* shoots exposed to sulfur dioxide (235). Increased maintenance respiration can apparently lead to reinstatement of normal dark metabolism within a few days following ozone exposure (Fig. 11).

An air pollution–induced increase in maintenance and repair respiration in leaves reduces the amount of photoassimilate, which is usually already limited by inhibited photosynthesis, that is available for translocation to other organs. (It also reduces the amount of ATP, NAD(P)H, and carbon skeletons available from a given amount of photoassimilate for growth processes in that leaf.) This might be expected to limit growth in those other organs so that a secondary respiratory response to air pollution will be slow respiration in nonleaf tissues as a result of low carbohydrate supply and slow growth. Circumstantial evidence exists for such a response to ozone in roots (233,236,237).

There have also been reports of slow respiration in leaves and shoots following exposure to ozone (e.g., 238). Such results can be interpreted in at least two ways. First, in the short term and generally being the case for individual leaves, high concentrations of ozone will cause large-scale damage and cell death. This halts all metabolic activity in the killed cells, including respiration and the processes using respiratory products. Thus, slowing of respiration by previous ozone exposure is often correlated with visible necrosis (personal observations). Such high-ozone con-

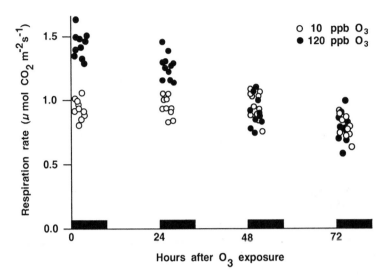

Figure 11 . The rate of respiration in expanded *Phaseolus vulgaris* leaves during the first few hours of each of the four nights following the third of three days of daytime exposure to 10 (O) or 120 (●) ppb ozone (J. S. Amthor, unpublished data). Multiple measurements were made on four to six plants from each ozone treatment. Plants were grown and treated in high-light growth chambers (228). Temperature was 25°C at night. Solid bars show dark (night) periods.

centration experiments are of interest when studying mechanisms of action but are limited in their applicability to plants growing in the field. Second, decreased respiration can result from ozone exposure because of the reduction in photosynthesis and carbohydrate levels and therefore growth limitations. Stated simply, it is expected that respiration will increase to the extent that maintenance and repair processes are enhanced as a result of cellular injury from an air pollutant, but respiration will decrease inasmuch as growth is slowed or tissue is killed.

NAD(P)H use may be of particular significance with regard to oxidative stresses because of the need for reductant for detoxification. A consideration of the stoichemistry between respiration and, for example, ozone reduction is instructive. Reduction of the products of one ozone molecule [i.e., one H_2O_2, one superoxide radical, and one hydroxyl radical (36)] may require about three reductant molecules, e.g., via ascorbate and GSH. If three NADPH are oxidized for reduction of each ozone molecule absorbed by a leaf, and ozone is taken up by a leaf at a rate of 0.03 μmol/m²/sec [found for an ambient ozone concentration of 100 ppb and a leaf conductance of 0.5 mol H_2O/m^2/sec (see 239)], the rate of use of NADPH for detoxification is 0.09 μmol/m^2/sec. Slightly more than one CO_2 is evolved for every two $NADP^+$ reduced in the oxidative pentose phosphate network [accounting for the ATP used to phosphorylate glucose in the formation of glucose 6-P to initiate activity of the oxidative pentose phosphate network (8)], so the rate of respiration, i.e., CO_2 release, to support the rate of NADPH use for detoxification as calculated above is about 0.045 μmol CO_2/m^2/sec. This reflects perhaps 2–10% of the respiration of leaves under normal conditions, so that respiration to support ozone reduction per se may be minor. The more substantial increases in leaf respiration observed following ozone exposure (226–234), when primary detoxification processes have presumably been completed, are attributable to repair processes. These repair processes can pull respiration by using ATP or NAD(P)H.

Many plants and plant organs exist in low-light environments. For example, plants growing in the understory of a forest are shaded by the overstory and leaves low in a crop canopy are shaded by leaves in the top of the canopy. The effects of low light on the carbon balance are most evident with respect to being a limitation on carbon gain in photosynthesis, but respiration plays a role here too.

Until the late 1960s and early 1970s, it was assumed that leaves low in a canopy would become parasitic after a large canopy leaf area was attained because their daily respiration would exceed their photosynthesis allowed by the shaded conditions. Thus, an optimum leaf area index was thought to occur when leaves at the bottom of a canopy no longer could maintain a positive daily carbon balance. It was envisioned that at larger leaf area indexes, daily canopy CO_2 uptake would decline because of an increase in parasitic leaves and because all light available for photosynthesis was being absorbed. This notion was supported by some data and several models, but the experimental procedures generating that data were generally inappropriate for their intended purposes. In any case, there have been a number of reports documenting an asymptotic relationship between canopy respiration rate and leaf area index (240–244) and that leaf respiration is positively related to light environment (240,245–247). Moreover, the growth coefficient of leaves may be positively related to light levels of the leaf environment (248).

The mechanism(s) controlling leaf respiratory acclimation to low light is unknown (246,247). In general, and as with other resource limitations, low light slows things down, and respiration is pulled along more slowly as a consequence. It can be anticipated that the slow respiration rates of leaves in low-light environments is due in part to a history of slow photosynthesis and a consequent repression of processes using respiratory products that are associated with rapid carbohydrate processing and translocation. The ability to acclimate to a change in light is important for early leaves on plants with rapid growth rates (such as many crops) where the early, lower leaves are shaded as time progresses. It can also be important to deciduous forest understory plants that leaf out prior to the overstory; metabolism can be rapid during the spring understory light phase and then slow after overstory closure results in light (i.e., photosynthesis) limitations. Acclimation to low light is reversible so that leaves have the capacity to increase their rate of heterotrophic metabolism in response to a transition from low to high light. This is beneficial, say, to understory plants that find themselves under a newly formed tree-fall gap.

Stresses in the Soil Environment

Soil temperature and other soil physical and chemical properties influence root metabolism, but the soil environment also affects shoot physiology and growth. Soil-related stresses can be categorized as either resource limitations or toxins. Lack of water (i.e., low soil water potential) is one of the two soil resource limitations most often studied by plant ecologists. The other is a low level of available nitrogen. Two toxins of concern are sodium (salinity) and aluminum (acidity).

McGree (92) reviewed effects of water stress and salinity on plants from the perspective of respiration and carbon balance. The salient points are that a slowly developing water stress, as occurs in the field during drought, reduces photosynthesis and growth, and that growth can be slowed more markedly than photosynthesis resulting in an accumulation of nonstructural carbohydrates (249). [Recall that low temperature may also inhibit growth to a greater extent than photosynthesis, giving rise to an accumulation of nonstructural carbohydrates; the same is true for nutrient deficiencies and salinity (250).] The decrease in growth can be expected to lessen the demand for respiration, which has been observed for *Sorghum bicolor* (56,251,252). Inasmuch as nonstructural carbohydrates accumulate under water stress and carbohydrate storage is cheaper than growth, it appears from measurements of CO_2 exchange that photosynthate is being used more efficiently, i.e., GE_{app} increases (see 252). Wilson et al. (251) observed a 5% increase in *S. bicolor* whole-plant apparent $Y_{G,C}$ (a 15% decrease in apparent $g_{r,C}$) over the nighttime leaf water potential range of -0.1 to -1.2 MPa. Irrigating a water-stressed plant can elicit an increase in respiration and a decline in GE_{app} (252). This can be taken as an indication of rapid growth and growth respiration following stress alleviation.

Many rapidly imposed stresses inhibit protein synthesis (turnover) (29,253) and because ATP is required for protein synthesis, respiration supporting protein synthesis may be reduced during a rapidly imposed stress. In contrast, protein turnover rate increased in field-grown *Hordeum vulgare* leaves during normal (i.e., slow) water stress development (254). The calculated costs of leaf protein turnover accounted for only a small fraction of measured leaf respiration, however, and the costs did not contribute greatly to the carbon balance. The general decline in overall metabolic activity during water stress can be expected to result in a decline in specific maintenance respiration, and indeed, the whole-plant maintenance coefficient was reduced by water stress in *S. bicolor* (251).

An additional aspect of water stress that can be important to whole-plant respiration and carbon use is the commonly observed increase in root/shoot ratio. (This increase is due not to increased root growth but to a greater proportional limitation to shoot growth.) Because the growth and maintenance respiration coefficients can differ in root and shoot (99), the whole-plant growth and maintenance coefficients can change accordingly under water stress. These changes will alter, if only slightly, the stoichiometry between growth and growth respiration, and whole-plant size and whole-plant maintenance respiration.

This consideration of water stress effects on respiration has been brief. The main points are simple: Water stress reduces growth which affects growth respiration. The stoichiometries between photosynthesis, respiration, and growth are altered and whole-plant maintenance respiration will generally decline due to an overall slowing of metabolic activity and the limited size of the stressed plant. Respiratory responses vary when plants are grown in small pots and subjected to rapidly developing water stress, as is common in laboratory experiments. On the whole, respiration is less sensitive to water stress than is growth (e.g., 41) or photosynthesis (e.g., 255). This must be due in part to the fact that growth and photosynthesis can be suspended for a time during stress, but respiration must always proceed, albeit often at a reduced rate.

From the perspective of whole-plant respiration and carbon balance, nitrogen limitations (a) limit photosynthesis and growth, (b) can lead to the accumulation of nonstructural carbohydrates, and (c) can affect the root/shoot ratio. The expected respiratory responses to these factors will often be the same as those just outlined for a water stress.

Plants in a saline environment function much as they do when growing in a drying soil, except that there are likely to be additional metabolic costs associated with more extensive ion pumping (i.e., active transport) for salt exclusion, secretion, or compartmentation. Because salinity can accelerate ion pumping, and because ion pumps are presumably driven by ATP, it is of interest to relate measures of respiration to measures of ATP production. In pea seedlings, NaCl treatment (77 mM) increased the rate of respiration but did not alter the P/O ratio (256). This indicates an increase in the rate of ATP turnover in NaCl-treated plants, as expected. The increase in respiration rate was accompanied by an increase in the *capacity* of respiration.

The apparent costs of salinity can be high. For example, the relationship between the rate of Na^+ secretion from leaves of *Distichlis spicata* growing in saline soil and the rate of leaf respiration (257) suggests that 3.8 mol of CO_2 was released for each mole of Na^+ secreted. If, on average, 5.5 ATP was formed in respiration for each CO_2 released (8,29), then 21 ATP was expended per Na^+ ion secreted! Clearly, processes in addition to Na^+ secretion were associated with the increased respiration. Salinity has enhanced, but also inhibited, respiration in other studies (e.g., 258–261).

A few studies have investigated the effects of salinity on respiration and growth within the context of the growth-and-maintenance model. For instance, whole-plant $Y_{G,C}$ was apparently decreased (i.e., $g_{r,C}$ was increased) in *Medicago sativa* as salinity increased (262). It has also been reported that salinity increased whole-plant $m_{r,C}$ in *Phaseolus vulgaris*, *Xanthium strumarium*, and *Atriplex halimus* (54) and in *Medicago sativa* (262). It is simplest to attribute this increase in the maintenance coefficient to the additional metabolic costs of active transport for ion exclusion, compartmentation, or osmotic adjustment under saline conditions. That is, such increases in the maintenance coefficient are not unexpected, but corroborate the notion that a stress eliciting active responses for repair, avoidance, or detoxification increases energy use, and therefore respiration, for those processes.

It was mentioned above that several stresses give rise to an accumulation of nonstructural carbohydrates and it is a longstanding observation that in nonhalophytes salinity leads to such an increase in nonstructural carbohydrate concentration while at the same time limiting growth (261). Some of the carbohydrate accumulated under salinity (and water stress) may contribute to osmotic adjustment (but see 250), but the starch accumulated under salinity (261) will not. In any case, carbon used for osmotic adjustment is not immediately available for growth. When growth was slowed and carbohydrates accumulated in *Hibiscus cannabinus* because of salinity, mature leaf respiration rate increased (261), which is contrary to the respiratory decline during carbohydrate accumulation observed with resource limitations.

Experimental data are limited, but a brief outline of possible respiratory responses to various toxic metals such as aluminum in the soil can be made. The mechanisms of tolerance or detoxification in situ are poorly understood, but the major hypotheses (i.e., exclusion from the cytosol or external chelation) each involve expenditures of respiratory products. A plant that cannot exclude or detoxify a metal is subject to damage and perhaps death whereas a plant that excludes or detoxifies a metal does so at a cost, but with the potential benefit of survival. One expectation then is that root respiration will be increased for various exclusion or detoxification processes when a root is challenged by toxic metals and perhaps more so in relatively tolerant plants. Respiratory products such as ATP and carbon skeletons may be consumed for pumping and chelation, and these respiratory products are then no longer available for growth or other processes. Respiration may continue to be pulled along by the exclusion/detoxification mechanisms until metal levels are diminished, and the result of this is an increase in $R_{M,C}/R_{G,C}$. The overall effect is a decrease in the relative use of photoassimilate for growth and reproduction, but again, the plant that can respond with rapid respiration when ATP and carbon skeletons are required may be better suited for survival (263,289).

SYNOPSIS

Respiration is an essential, and large, component of the carbon balance of plants; it is the primary metabolic link between the assimilation of carbon and energy in photosynthesis and the subsequent use of that carbon and energy in (a) growth, (b) the storage of assimilate in reproductive and perennial tissue, and (c) the support of the processes necessary for the maintenance and repair of existing phytomass. Thus, it is important to understand the relationships between environmental conditions and the efficiency and rate of respiration and of carbon use and loss. The premise underlying this chapter is that the environment affects respiration primarily by influencing (a) the processes supplying the carbohydrates used in growth and respiration and (b) the processes using the products of respiration, most notably ATP. These are not new ideas, but they are ideas worth emphasizing.

Two general classes of environmental factors can be considered with respect to effects on respiration or the processes regulating respiration. The first, *resource limitations*, can be subdivided into limitations of the substrates of photosynthesis (i.e., light and CO_2) and limitations of other resources such as water, nutrients, and heat. Low light or CO_2 limits photosynthesis without an increase in nonstructural carbohydrate pool size. Thus, the ratio of respiration to photosynthesis may be conserved or even increased. Growth and plant size, however, will be limited, and if the growth and maintenance coefficients are relatively unchanged both whole-plant growth and maintenance respiration will be reduced. This reduced respiration is a result of slower plant growth and smaller plant size. Water, nutrient, or heat limitations can slow growth more extensively than photosynthesis and result in an accumulation of nonstructural carbohydrates. Whole-plant growth and maintenance respiration will then each probably be slow, perhaps more so than as a result of a light or CO_2 limitation. In such a case, however, the ratio of respiration to photosynthesis may decrease because storage is generally cheaper than growth. The overall effect can be an apparent (see footnote 1) increase in the *efficiency* of photoassimilate use in growth.

The second class of environmental limitation on growth, carbon balance, and respiration is a *physical perturbation* or *toxin* such as ozone or salinity. With respect to maintenance respiration, this class of stress can call for an increase in metabolic activity for repair or detoxification processes and can lead to an increase in the maintenance coefficient. Whole-plant maintenance respiration can still be diminished, however, because limited growth will limit plant size. Perturbations and toxins in general probably cause a shift in photoassimilate use from growth to maintenance processes, thus exacerbating the effects of what is likely to be an already limited supply of photoassimilate. A stress that results in cellular death is a limiting case of the perturbation and toxin class. In the dead cells, respiration will cease—except perhaps within microbial decomposers—but respiration might increase or decrease in the cells surrounding those killed (264), depending in part on the nature and extent of the stress. Multiple stresses may elicit multiple respiratory responses.

In many respects, research to date has been feebly chipping away at the edges of the important questions about the effects of the environment on respiration. The difficulties of quantitatively studying respiration in situ are numerous. A major factor contributing to these difficulties is the ubiquitous nature of the process. Respiration is linked to nearly all other areas of metabolism. Respiration occurs in all living cells, above and below ground, at all times and under all conditions. If anything about a plant is affected by a change in the environment, respiration will probably be changed as a consequence. Determining how and when to get at things for a measurement in such a complex area of plant biology is no easy task, especially if one is to leave the plant relatively undisturbed. Moreover, it is necessary to have concomitant measures of processes such as photosynthesis, growth, protein turnover, nutrient uptake and assimilation, and intra- and intercellular transport in order to evaluate the *causes* of environmental impacts on respiration. The composition of growing tissue must also be known to evaluate the efficiency of growth and growth respiration (83,93,94,265,266). The notion that the rate of respiration is controlled by the rate of use of respiratory intermediates and end products begs the question of overall metabolic control. Herein the control of respiration has been neatly delegated to processes using respiratory products, while the mechanistic control of those processes has not been considered. The whole business is quite complicated, but the point is not that relevant experiments are impossible because of the large number of variables that must be considered but rather that a more holistic perspective than that usually evident from simple two-variable cause and effect experiments is needed for progress. It is not appropriate to, for example, measure only the water potential and respiration rate of a tissue.

It would be presumptuous to suggest that the models described herein are necessarily correct or that respiratory response to the environment can always be predicted. It is not presumptuous to state that the models are reasonable and provide a robust framework with which to study respiration and that it has been possible to quite successfully predict respiratory responses to many aspects of the physical environment. Future research needs to test these models, not just contribute to the collection of data. If the models fail seriously, they must be discarded. If the models fail marginally, they must be modified. If the models succeed, they must be tested further. Thus far, the models have performed well where they have been applied, but I add that we know too little about respiration in situ to be overly confident in their predictions. Perhaps most importantly, these models reinforce the need to measure more than just the rate of respiratory gas exchange because respiration is not an isolated process; respiration is an integral part of the whole of metabolism and an understanding of respiration requires an understanding of that whole.

If the rate of respiration was measured over all environmental conditions of interest to plant ecologists, the answer to the question, "What is the respiratory response to this and that?" could be "So and so found that respiration responded in such and such a way to this and that." Such an extensive survey of respiratory responses to the environment, however, is not presently possible. In the meantime, an answer to that question will often involve first a question of its own. For example, "What happens or can be expected to happen to photosynthesis, translocation, growth, and cellular integrity under conditions of this and that?" With this knowledge, the answer to the original question becomes, "Because these environmental conditions do this and that to these other processes and states, it can be expected that respiration will respond in such and such a way to this and that." The response of the carbon balance to the environment can be ascertained only with answers to both questions.

My point is that in the study of environment–respiration interactions the goal is not to develop a summary in the form of a parallel list of environmental factors and respiration rates, although such a summary could be useful. Instead, the goal is to uncover the underlying links between environmental conditions, respiration, and the rest of metabolism and physiology. The regulation of photosynthesis by environmental resource availability, the regulation of photoassimilate partitioning by environmental cues, the regulation of growth by substrate supply from photosynthesis and translocation, and the regulation of respiration by respiratory end product use in growth and maintenance processes are the bases of those links.

REFERENCES

1. R. W. Pearcy, O. Björkman, M. M. Caldwell, J. E. Keeley, R. K. Monson, and B. R. Strain (1987): *Bioscience*, 37:21.
2. H. A. Mooney (1972): *Annu. Rev. Ecol. Syst.*, 3:315.
3. C. B. Osmond, K. Winter, and H. Ziegler (1982): *Encycl. Plant Physiol. NS*, 12B:479.
4. T. Kira (1975): *Photosynthesis and Productivity in Different Environments* (J. P. Cooper, ed.), Cambridge Univ. Press, Cambridge, UK, p. 5.
5. J. S. Amthor (1989): *Respiration and Crop Productivity*, Springer-Verlag, New York.
6. M. G. Ryan (1991): *Ecol. Appl.*, 1:157.
7. J. F. Farrar (1985): *Plant Cell Environ.*, 8:427.
8. H. Beevers (1961): *Respiratory Metabolism in Plants*, Row Peterson, Evanston, IL.
9. J. S. Amthor (1993): *Ecophysiology of Photosynthesis* (E.-D. Schulze and M. M. Caldwell, eds.), Springer-Verlag, Berlin, p. 71.
10. J. S. Amthor (1994): *Physiology and Determination of Crop Yield* (K. J. Boote, ed.), American Society of Agronomy, Madison, WI, in press.
11. J. S. Amthor and K. J. McCree (1990): *Stress Responses in Plants: Adaptation and Acclimation Mechanisms* (R. G. Alscher and J. R. Cumming, eds.), Wiley-Liss, New York, p. 1.
12. H. Beevers (1970): *Prediction and Measurement of Photosynthetic Productivity* (I. Setlik, ed.), Pudoc, Wageningen, The Netherlands, p. 209.
13. H. Beevers (1974): *Plant Physiol.*, 54:437.
14. R. Douce (1985): *Mitochondria in Higher Plants*, Academic Press, Orlando.
15. T. ap Rees (1985): *Encycl. Plant Physiol. NS*, 18:391.
16. T. ap Rees (1988): *Biochem. Plants*, 14:1.
17. L. Copeland and J. F. Turner (1987): *Biochem. Plants*, 11:107.
18. J. T. Wiskich and I. B. Dry (1985): *Encycl. Plant Physiol. NS*, 18:281.
19. Y. Yatefi (1985): *Annu. Rev. Biochem.*, 54:1015.
20. R. Douce and M. Neuburger (1989): *Annu. Rev. Plant Physiol. Plant Mol. Biol.*, 40:371.
21. A. L. Moore and P. R. Rich (1985): *Encycl. Plant Physiol. NS*, 18:134.
22. J. M. Palmer and J. A. Ward (1985): *Encycl. Plant Physiol. NS*, 18:173.

23. B. J. Good and W. H. Patrick Jr. (1987): *Plant Soil*, 97:419.
24. J. A. Palta and P. S. Nobel (1989): *Physiol. Plant.*, 76:187.
25. W. O. James (1953): *Plant Respiration*, Oxford Univ. Press, London.
26. C. R. French and H. Beevers (1953): *Am. J. Bot.*, 40:660.
27. I. B. Dry, J. H. Bryce, and J. T. Wiskich (1987): *Biochem. Plants*, 11:213.
28. P. M. Ray, T. A. Steeves, and S. A. Fultz (1983): *Botany*, Saunders College, Philadelphia.
29. D. Rhodes (1987): *Biochem. Plants*, 12:201.
30. J. S. Amthor (1988): *New Phytol.*, 110:319.
31. L. Taiz and E. Zeiger (1991): *Plant Physiology*, Benjamin/Cummings, Redwood City, CA.
32. D. G. Nicholls and S. J. Ferguson (1992): *Bioenergetics 2*, Academic Press, London.
33. C. Lance, M. Chauveau, and P. Dizengremel (1985): *Encycl. Plant Physiol. NS*, 18:202.
34. R. Douce, R. Brouquisse, and E.-P. Journet (1978): *Biochem. Plants*, 11:177.
35. D. T. Tingey, R. C. Fites, and C. Wickliff (1976): *Physiol. Plant.*, 37:69.
36. E. F. Elstner (1987): *Biochem. Plants*, 11:253.
37. J. N. Siedow (1991): *Annu. Rev. Plant Physiol. Plant Mol. Biol.*, 42:145.
38. F. D. H. MacDowall (1985): *Can. J. Bot.*, 43:419.
39. T. T. Lee (1967): *Plant Physiol.*, 42:691.
40. G. L. Cunningham and J. P. Syvertsen (1977): *Photosynthetica*, 11:291.
41. F. W. T. Penning de Vries, J. M. Witlage, and D. Kremer (1979): *Ann. Bot.*, 44:595.
42. B. M. Coggeshall and H. F. Hodges (1980): *Crop Sci.*, 20:86.
43. L. E. Moser, J. J. Volenec, and C. J. Nelson (1982): *Crop Sci.*, 22:781.
44. J. Azcón-Bieto and C. B. Osmond (1983): *Plant Physiol.*, 71:574.
45. J. Azcón-Bieto, H. Lambers, and D. A. Day (1983): *Plant Physiol.*, 72:598.
46. J. Azcón-Bieto, D. A. Day, and H. Lambers (1983): *Plant Sci. Lett.*, 32:313.
47. E.-P. Journet, B. Bligny, and R. Douce (1986): *J. Biol. Chem.*, 261:3193.
48. A. A. Hatrick and D. J. F. Bowling (1973): *J. Exp. Bot.*, 24:607.
49. K. J. McCree (1970): *Prediction and Measurement of Photosynthetic Productivity* (I. Setlik, ed.), Pudoc, Wageningen, The Netherlands, p. 221.
50. P. J. M. Sale (1974): *Aust. J. Plant Physiol.*, 1:283.
51. G. K. Hansen and C. R. Jensen (1977): *Physiol. Plant.*, 39:155.
52. J. L. Heilman, E. T. Kanemasu, and G. M. Paulsen (1977): *Can. J. Bot.*, 55:2196.
53. K. J. McCree and J. H. Silsbury (1978): *Crop Sci.*, 18:13.
54. M. Schwarz and J. Gale (1981): *J. Exp. Bot.*, 32:933.
55. M. André, J. Massimino, A. Daguenet, D. Massimino, and T. Thiery (1982): *Physiol. Plant.*, 54:283.
56. K. J. McCree, C. E. Kallsen, and S. G. Richardson (1984): *Plant Physiol.*, 76:898.

57. I. J. Bingham and J. F. Farrar (1988): *Physiol. Plant.*, *73*:278.
58. I. J. Bingham and J. F. Farrar (1989): *Plant Physiol. Biochem.*, *27*:847.
59. J. H. H. Williams and J. F. Farrar (1990): *Physiol. Plant.*, *79*:259.
60. J. F. Farrar and J. H. H. Williams (1991): *Soc. Exp. Biol. Sem. Ser.*, *42*:167.
61. R. Douce, R. Bligny, D. Brown, A.-J. Dorne, P. Genix, and C. Roby (1991): *Soc. Exp. Biol. Sem. Ser.*, *42*:127.
62. M.-H. Avelange, F. Sarrey, and F. Rébillé (1990): *Plant Physiol.*, *94*:1157.
63. C. Baysdorfer and W. J. VanDerWoude (1988): *Plant Physiol.*, *87*:566.
64. J. Sheen (1990): *Plant Cell*, *2*:1027.
65. M. Stitt (1991): *Plant Cell Environ.*, *14*:741.
66. L. C. Ho and J. H. M. Thornley (1978): *Ann. Bot.*, *42*:481.
67. M. Aslam and R. C. Huffaker (1984): *Plant Physiol.*, *75*:623.
68. H. Lambers (1985): *Encycl. Plant Physiol. NS*, *18*:418.
69. I. B. Dry, A. L. Moore, D. A. Day, and J. T. Wiskich (1989): *Arch. Biochem. Biophys.*, *273*:248.
70. J. N. Siedow and D. A. Berthold (1986): *Physiol. Plant.*, *66*:569.
71. J. H. M. Thornley (1970): *Nature*, *227*:304.
72. J. D. Hesketh, D. N. Baker, and W. G. Duncan (1971): *Crop Sci.*, *11*:394.
73. W. F. Hunt and R. S. Loomis (1979): *Ann. Bot.*, *44*:5.
74. M. Warner (1992): *Religion and Philosophy* (M. Warner, ed.), Cambridge Univ. Press, Cambridge, UK, p. 1.
75. S. Brody (1945): *Bioenergetics and Growth*, Reinhold, New York.
76. F. Kidd, C. West, and G. E. Briggs (1921): *Proc. Roy. Soc. London*, *B92*:368.
77. R. S. Inamdar, S. B. Singh, and T. D. Pande (1925): *Ann. Bot.*, *39*:281.
78. C. A. Price (1960): *Encycl. Plant Physiol.*, *XII/2*:493.
79. J. D. Mahon (1977): *Plant Physiol.*, *60*:812.
80. I. R. Johnson (1983): *Physiol. Plant.*, *58*:145.
81. I. R. Johnson (1990): *Plant Cell Environ.*, *13*:319.
82. S. J. Pirt (1965): *Proc. Roy. Soc.*, *B163*:224.
83. F. W. T. Penning de Vries, A. H. M. Brunsting, and H. H. van Laar (1974): *J. Theor. Biol.*, *45*:339.
84. J. S. Amthor (1984): *Plant Cell Environ.*, *7*:561.
85. D. G. Sprugel (1990): *Trees*, *4*:88.
86. M. G. Ryan (1990): *Can. J. For. Res.*, *20*:48.
87. S. A. Paembonan, A. Hagihara, and K. Hozumi (1992): *Tree Physiol.*, *10*:101.
88. T. ap Rees (1990): *Biochem. Plants*, *2*:1.
89. J. S. Pate and D. B. Layzell (1990): *Biochem. Plants*, *16*:1.
90. A. J. Barneix, H. Breteler, and S. C. van de Geijn (1984): *Physiol. Plant.*, *61*:357.
91. T. Blacquière, R. Hofstra, and I. Stulen (1987): *Plant Soil*, *104*:129.
92. K. J. McCree (1986): *Aust. J. Plant Physiol.*, *13*:33.
93. F. W. T. Penning de Vries, H. H. van Laar, and M. C. M. Chardon (1983): *Potential Productivity of Field Crops Under Different Environments*, Int. Rice Res. Inst., Los Baños, Laguna, Philippines, p. 37.

94. F. W. T. Penning de Vries, D. M. Jansen, H. F. M. ten Berge, and A. Bakema (1989): *Simulation of Ecophysiological Processes of Growth in Several Annual Crops*, Pudoc, Wageningen, The Netherlands.
95. E. M. Birk and P. M. Vitousek (1986): *Ecology*, 67:69.
96. J. R. Evans (1983): *Plant Physiol.*, 72:297.
97. A. J. Bloom, F. S. Chapin, and H. A. Mooney (1985): *Annu. Rev. Ecol. Syst.*, 16:363.
98. H.-H. Chung and R. L. Barnes (1977): *Can. J. For. Res.*, 7:106.
99. R. S. Stahl and K. J. McCree (1988): *Crop Sci.*, 28:111.
100. F. W. T. Penning de Vries (1975): *Ann. Bot.*, 39:77.
101. S. B. McGrew and M. F. Mallette (1962): *J. Bacteriol.*, 83:844.
102. A. G. Marr, E. H. Nilson, and D. J. Clark (1963): *Ann. N.Y. Acad. Sci.*, 102:536.
103. T. J. Bouma, J. H. J. A. Janssen, M. J. de Kock, P. H. van Leeuwen, and R. de Visser (1992): *Physiol. Plant.*, 85:A51.
104. A. Barnes and C. C. Hole (1978): *Ann. Bot.*, 42:1217.
105. C. C. Hole and A. Barnes (1980): *Ann. Bot.*, 45:295.
106. R. C. Huffaker and L. W. Peterson (1974): *Annu. Rev. Plant Physiol.*, 25:363.
107. R. C. Huffaker (1990): *New Phytol.*, 116:199.
108. A. J. S. Hawkins (1991): *Funct. Ecol.*, 5:222.
109. D. F. Forward (1960): *Encycl. Plant Ecology*, XII/2, Springer-Verlag, Berlin, p. 234.
110. W. Larcher (1980): *Physiological Plant Ecology*, 2nd ed., Springer-Verlag, Berlin.
111. J. Berry and O. Björkman (1980): *Annu. Rev. Plant Physiol.*, 31:491.
112. B. R. Strain and V. C. Chase (1966): *Ecology*, 47:1043.
113. W. D. Billings and H. A. Mooney (1968): *Biol. Rev.*, 43:481.
114. D. A. Rook (1969): *N. Z. J. Bot.*, 7:43.
115. J. Woledge and O. R. Jewiss (1969): *Ann. Bot.*, 33:897.
116. N. J. Chatterton, C. M. McKell, and B. R. Strain (1970): *Ecology*, 51:545.
117. W. D. Billings, P. J. Godfrey, B. F. Chabot, and D. P. Bourque (1971): *Arctic Alpine Res.*, 3:277.
118. K. J. Treharne and C. J. Nelson (1975): *Environmental and Biological Control of Photosynthesis* (R. Marcelle, ed.), W. Junk, The Hague, p. 61.
119. R. W. Pearcy (1977): *Plant Physiol.*, 59:795.
120. B. T. Mawson, J. Svoboda, and W. R. Cummins (1986): *Can. J. Bot.*, 64:71.
121. A. K. McNulty and W. R. Cummins (1987): *Plant Cell Environ.*, 10:319.
122. Ch. Körner and W. Larcher (1988): *Symp. Soc. Exp. Biol.*, 42:25.
123. D. E. Collier and W. R. Cummins (1990): *Ann. Bot.*, 65:533.
124. H. G. Weger and R. D. Guy (1991): *Physiol. Plant.*, 83:675.
125. G. Smakman and R. Hofstra (1982): *Physiol. Plant.*, 56:33.
126. R. S. Holthausen and M. M. Caldwell (1980): *Plant Soil*, 55:307.
127. D. T. Patterson (1980): *Predicting Photosynthesis for Ecosystem Models* (J. D. Hesketh and J. W. Jones, eds.), CRC Press, Boca Raton, p. 205.
128. H. G. Wager (1941): *New Phytol.*, 40:1.

129. H. A. Mooney (1963): *Ecology*, 44:812.
130. L. G. Klikoff (1966): *Nature*, 212:529.
131. R. W. Pearcy and A. T. Harrison (1974): *Ecology*, 55:1104.
132. E. A. Miroslavov and I. M. Kravkina (1991): *Ann. Bot.*, 68:195.
133. F. T. Ledig and D. R. Korbobo (1983): *Am. J. Bot.*, 70:256.
134. R. S. Kinerson (1975): *J. Appl. Ecol.*, 12:965.
135. S. A. Paembonan, A. Hagihara, and K. Hozumi (1991): *Tree Physiol.*, 8:399.
136. Y. Yamamura (1984): *Bot. Mag* (Tokyo), 97:179.
137. G. M. Woodwell and W. R. Dykeman (1966): *Science*, 154:1031.
138. T. C. Marks (1978): *Ann. Bot.*, 42:181.
139. V. Breeze and J. Elston (1978): *Ann. Bot.*, 42:863.
140. M. J. Robson (1981): *Ann. Bot.*, 48:269.
141. K. W. Brown and J. C. Thomas (1980): *Physiol. Plant.*, 49:205.
142. W. Larcher (1969): *Photosynthetica*, 3:167.
143. J. K. Raison (1980): *Biochem. Plants*, 2:613.
144. J. A. Berry and J. K. Raison (1981): *Encycl. Plant Physiol.*, 12A:277.
145. J. Wolfe and D. J. Bagnall (1980): *Ann. Bot.*, 45:485.
146. K. L. Steffen, R. Arora, and J. P. Palta (1989): *Plant Physiol.*, 89:1372.
147. J.-E. Hällgren and G. Öquist (1990): *Stress Responses in Plants: Adaptation and Acclimation Mechanisms* (R. G. Alscher and J. R. Cumming, eds.), Wiley-Liss, New York, p. 265.
148. J. F. Farrar (1988): *Symp. Soc. Exp. Biol.*, 42:203.
149. R. M. M. Crawford and T. J. Huxter (1977): *J. Exp. Bot.*, 28:917.
150. C. R. Stewart, B. A. Martin, L. Reding, and S. Cerwick (1990): *Plant Physiol.*, 92:761.
151. C. R. Stewart, B. A. Martin, L. Reding, and S. Cerwick (1990): *Plant Physiol.*, 92:755.
152. O. A. Semikhatova (1970): *Prediction and Measurement of Photosynthetic Productivity* (I. Setlik, ed.), Pudoc, Wageningen, The Netherlands, p. 247.
153. O. A. Semikhatova and I. A. Daletskaya (1974): *Soviet Plant Physiol.*, 21:97.
154. J. Gale (1982): *J. Exp. Bot.*, 33:471.
155. K. Raschke (1956): *Planta*, 48:200.
156. D. M. Gates (1968): *Annu. Rev. Plant Physiol.*, 19:211.
157. B. J. D. Meeuse (1975): *Annu. Rev. Plant Physiol.*, 26:117.
158. K. A. Nagy, D. K. Odell, and R. S. Seymour (1972): *Science*, 178:1195.
159. T. ap Rees (1977): *Symp. Soc. Exp. Biol.*, 31:7.
160. A. J. MacDougall and T. ap Rees (1991): *J. Plant Physiol.*, 137:683.
161. K. Wohl and W. O. James (1942): *New Phytol.*, 41:230.
162. R. T. Watson, H. Rodie, H. Oeschger, and U. Siegenthaler (1990): *Climate Change: The IPCC Scientific Assessment* (J. T. Houghton, G. J. Jenkins, and J. J. Ephraums, eds.), Cambridge Univ. Press, Cambridge, UK, p. 1.
163. J. S. Amthor (1991): *Plant Cell Environ.*, 14:13.
164. W. E. Loomis (1960): *Encycl. Plant Physiol.*, V:85.

165. M. Stitt (1991): *Plant Cell Environ.*, *14*:741.
166. S. P. Long (1991): *Plant Cell Environ.*, *14*:729.
167. B. G. Drake and P. W. Leadley (1991): *Plant Cell Environ.*, *14*:853.
168. M. A. Ford and G. N. Thorne (1967): *Ann. Bot.*, *31*:629.
169. R. J. Norby, E. G. O'Neill, and R. J. Luxmoore (1986): *Plant Physiol.*, 82:83.
170. P. S. Curtis, B. G. Drake, and D. F. Whigham (1989): *Oecologia*, 78:297.
171. A. J. Rowland-Bamford, J. T. Baker, L. H. Allen Jr., and G. Bowes (1991): *Plant Cell Environ.*, *14*:577.
172. R. S. Loomis and H. R. Lafitte (1987): *Field Crops Res.*, *17*:63.
173. W. Spencer and G. Bowes (1986): *Plant Physiol.*, *82*:528.
174. J. A. Bunce (1990): *Ann. Bot.*, *65*:637.
175. J. A. Bunce and F. Caulfield (1991): *Ann. Bot.*, *67*:325.
176. K. P. Hogan, A. P. Smith, and L. H. Ziska (1991): *Plant Cell Environ.*, *14*:763.
177. J. T. Baker, F. Laugel, K. J. Boote, and L. H. Allen, Jr. (1992): *Plant Cell Environ.*, *15*:231.
178. G. J. A. Ryle, C. E. Powell, and V. Tewson (1992): *J. Exp. Bot.*, *43*:811.
179. J. A. Bunce (1992): *Plant Cell Environ.*, *15*:541.
180. B. G. Drake (1992): *Aust. J. Bot.*, *40*:579.
181. S. D. Wullschleger and R. J. Norby (1992): *Can. J. For. Res.*, *22*:1717.
182. S. D. Wullschleger, R. J. Norby, and D. L. Hendrix (1992): *Tree Physiol.*, *10*:21.
183. L. H. Ziska and A. H. Teramura (1992): *Physiol. Plant.*, *84*:269.
184. S. D. Wullschleger, R. J. Norby, and C. A. Gunderson (1992): *New Phytol.*, *121*:515.
185. H. Poorter, R. M. Gifford, P. E. Kriedemann, and S. C. Wong (1992): *Aust. J. Bot.*, *40*:501.
186. J. H. Silbury and R. Stevens (1984): *Advances in Photosynthesis Research*, Vol. 4 (C. Sybesma, ed.), Martinus Nijhoff/Dr. W. Junk, The Hague, p. 133.
187. A. Tanaka and J. Yamaguchi (1968): *Soil Sci. Plant Nutrition*, *14*:110.
188. J. Yamaguchi (1978): *J. Fac. Agr. Hokkaido Univ.*, *59*:59.
189. R. G. Dutton, J. Jiao, M. J. Tsujita, and B. Grodzinski (1988): *Plant Physiol.*, *86*:355.
190. H. Poorter, S. Pot, and H. Lambers (1988): *Physiol. Plant.*, *73*:553.
191. J.-P. Gaudillère and M. Mousseau (1989): *Acta Oecologica/Oecologia Plant.*, *10*:95.
192. B. Grodzinski (1992): *Bioscience*, *42*:517.
193. A. J. Rowland-Bamford, L. H. Allen Jr., J. T. Baker, and K. J. Boote (1990): *J. Exp. Bot.*, *41*:1601.
194. W. J. Arp (1991): *Plant Cell Environ.*, *14*:869.
195. G. T. Byrd, R. F. Sage, and R. H. Brown (1992): *Plant Physiol.*, *100*:191.
196. J. P. Decker and J. D. Wein (1958): *J. Solar Energy Sci. Eng.*, *2*:39.
197. M. L. Forrester, G. Krotkov, and C. D. Nelson (1966): *Plant Physiol.*, *41*:422.
198. P. Holmgren and P. G. Jarvis (1967): *Physiol. Plant.*, *20*:1045.

199. G. Cornic and P. G. Jarvis (1972): *Photosynthetica, 6*:225.
200. A. Kaplan, J. Gale, and A. Poljakoff-Mayber (1977): *Aust. J. Plant Physiol., 4*:745.
201. N. Thorpe and F. L. Milthorpe (1977): *Aust. J. Plant Physiol., 4*:611.
202. J. S. Amthor, G. W. Koch, and A. J. Bloom (1992): *Plant Physiol., 98*:757.
203. J. E. Begg and P. G. Jarvis (1968): *Agr. Meteorol., 5*:91.
204. J. Reuveni and J. Gale (1985): *Plant Cell Environ., 8*:623.
205. A. El Kohen, J.-Y. Pontailler, and M. Mousseau (1991): *C. R. Acad. Sci. Paris, t. 312, Série III*:477.
206. E. O. Hellmuth (1971): *Photosynthetica, 5*:190.
207. L. H. Allen, Jr. and E. R. Lemon (1976): *Vegetation and the Atmosphere, Vol. 2, Case Studies* (J. L. Monteith, ed.), Academic Press, London, p. 265.
208. A. Hagihara and K. Hozumi (1991): *Physiology of Trees* (A. S. Raghavendra, ed.), John Wiley and Sons, New York, p. 87.
209. I. P. Ting and C. B. Osmond (1973): *Plant Physiol., 51*:439.
210. D. L. Nott and C. B. Osmond (1982): *Aust. J. Plant Physiol., 9*:409.
211. J. S. Wolpert and M. L. Ernst-Fonberg (1975): *Biochemistry, 14*:1103.
212. M. Schulz, C. Hunte, and H. Schnabl (1992): *Physiol. Plant., 86*:315.
213. J.-J. Van Oosten, D. Afif, and P. Dizengremel (1992): *Plant Physiol. Biochem., 30*:541.
214. A. J. Bloom, R. M. Caldwell, J. Finazzo, R. L. Warner, and J. Weissbart (1989): *Plant Physiol., 91*:352.
215. G. H. Lorimer (1983): *Trends Biochem. Sci., 8*:65.
216. M. A. Mitz (1979): *J. Theor. Biol., 80*:537.
217. A. Palet, M. Ribas-Carbó, J. M. Argilés, and J. Azcón-Bieto (1991): *Plant Physiol., 96*:467.
218. A. Bown, D. Boulter, and D. A. Coult (1968): *Physiol. Plant., 21*:271.
219. C. C. Hole (1977): *Ann. Bot., 41*:1367.
220. P. Pheloung and E. W. R. Barlow (1981): *J. Exp. Bot., 32*:921.
221. P. Kriedemann (1966): *Ann. Bot., 30*:349.
222. R. L. Heath (1980): *Annu. Rev. Plant Physiol., 31*:395.
223. N. M. Darrall (1989): *Plant Cell Environ., 12*;1.
224. J. M. Pye (1988): *J. Environ. Qual., 17*:347.
225. R. L. Barnes (1972): *Environ. Pollut., 3*:133.
226. P. B. Reich (1983): *Plant Physiol., 73*:291.
227. L. Skärby, E. Troeng, and C.-Å Boström (1987): *For. Sci., 33*:801.
228. J. S. Amthor and J. R. Cumming (1988): *Can. J. Bot., 66*:724.
229. J. M. M. Aben, M. Janssen-Jurkovičová, and E. H. Adema (1990): *Plant Cell Environ., 13*:463.
230. M. B. Adams, N. T. Edwards, G. E. Taylor, Jr., and B. L. Skaggs (1990): *Can. J. For. Res., 20*:152.
231. J. D. Barnes, D. Eamus, and K. A. Brown (1990): *New Phytol., 115*:149.
232. C. A. Ennis, A. L. Lazrus, P. R. Zimmerman, and R. K. Monson (1990): *Tellus, 42B*:183.

233. H. Moldau, J. Söber, A. Karolin, and A. Kallis (1991): *Photosynthetica*, 25:341.
234. G. Wallin, L. Skärby, and G. Selldén (1990): *New Phytol.*, 115:335.
235. J. S. Coleman, H. A. Mooney, and J. N. Gorham (1989): *Oecologia*, 81:124.
236. G. Hofstra, A. Ali, R. T. Wukasch, and R. A. Fletcher (1981): *Atm. Environ.*, 15:483.
237. N. T. Edwards (1991): *New Phytol.*, 118:315.
238. Y.-S. Yang, J. M. Skelly, B. I. Chevone, and J. B. Birch (1983): *Environ. Sci. Technol.*, 17:371.
239. A. Laisk, O. Kull, and H. Moldau (1989): *Plant Physiol.*, 90:1163.
240. L. J. Ludwig, T. Saeki, and L. T. Evans (1965): *Aust. J. Biol. Sci.*, 18:1103.
241. K. J. McCree and J. H. Troughton (1966): *Plant Physiol.*, 41:1615.
242. R. W. King and L. T. Evans (1967): *Aust. J. Biol. Sci.*, 20:623.
243. J. H. Cock and S. Yoshida (1973): *Soil Sci. Plant Nutr.*, 19:53.
244. D. Joggi, U. Hofer, J. Nösberger (1983): *Plant Cell Environ.*, 6:611.
245. O. Björkman (1981): *Encycl. Plant Physiol. NS*, 12A:57.
246. A. L. Fredeen and C. B. Field (1991): *Physiol. Plant.*, 82:85.
247. D. A. Sims and R. W. Pearcy (1991): *Oecologia*, 86:447.
248. K. Williams, C. B. Field, and H. A. Mooney (1989): *Am. Naturalist*, 133:198.
249. R. Munns, C. J. Brady, and E. W. R. Barlow (1979): *Aust. J. Plant Physiol.*, 6:379.
250. R. Munns (1988): *Aust. J. Plant Physiol.*, 15:717.
251. D. R. Wilson, C. H. M. van Bavel, and K. J. McCree (1980): *Crop Sci.*, 20:153.
252. S. G. Richardson and K. J. McCree (1985): *Plant Physiol.*, 79:1015.
253. E. Vierling (1991): *Annu. Rev. Plant Physiol. Plant Mol. Biol.*, 42:579.
254. W. D. Hitz, J. A. R. Ladyman, and A. D. Hanson (1982): *Crop Sci.*, 22:47.
255. H. Brix (1962): *Physiol. Plant.*, 15:10.
256. A. Livne and N. Levin (1967): *Plant Physiol.*, 42:407.
257. R. S. Warren and P. M. Brockelman (1989): *Bot. Gaz.*, 150:346.
258. A. Kalir and A. Poljakoff-Mayber (1976): *Plant Physiol.*, 57:167.
259. H. M. Rawson (1986): *Aust. J. Plant Physiol.*, 13:475.
260. E. L. Taleisnik (1987): *Physiol. Plant.*, 71:213.
261. P. S. Curtis, H. L. Zhong, A. Läuchli, and R. W. Pearcy (1988): *Am. J. Bot.*, 75:1293.
262. M. G. T. Shone and J. Gale (1983): *J. Exp. Bot.*, 34:1117.
263. J. R. Cumming, A. Buckelew Cumming, and G. J. Taylor (1992): *J. Exp. Bot.*, 43:1075.
264. I. Uritani and T. Asahi (1980): *Biochem. Plants*, 2:463.
265. D. K. McDermitt and R. S. Loomis (1981): *Ann. Bot.*, 48:275.
266. K. Williams, F. Percival, J. Merino, and H. A. Mooney (1987): *Plant Cell Environ.*, 10:725.

267. T. Solomos (1977): *Annu. Rev. Plant Physiol.*, 28:279.
268. G. G. Laties (1982): *Annu. Rev. Plant Physiol.*, 33:519.
269. B. K. Ishida and J. M. Palmer (1988): *Plant Physiol.*, 88:987.
270. A. Kesseler, P. Diolez, K. Brinkmann, and M. D. Brand (1992): *Eur. J. Biochem.*, 210:775.
271. M. Kimura, Y. Yokoi, and K. Hogetsu (1978): *Bot. Mag.* Tokyo, 91:43.
272. R. Szaniawski (1981): *Z. Pflanzenphysiol.*, 101:391.
273. R. Szaniawski and M. Kielkiewicz (1982): *Physiol. Plant.*, 54:500.
274. J. Merino, C. Field, and H. A. Mooney (1982): *Oecologia*, 53:208.
275. J. Kallarackal and J. A. Milburn (1985): *Ann. Bot.*, 56:211.
276. E. G. Reekie and R. E. Redmann (1987): *New Phytol.*, 105:595.
277. A. Kallis and T. Golovko (1988): *Acta Physiologiae Plantarum*, 10:123.
278. S. Mariko (1988): *Bot. Mag.* Tokyo, 101:73.
279. R. Matyssek and E.-D. Schulze (1988): *Trees*, 2:233.
280. D. O. Chanter (1977): *J. Appl. Ecol.*, 14:269.
281. K. J. McCree (1986): *Photosynthetica*, 20:82.
282. J. H. M. Thornley and I. R. Johnson (1990): *Plant and Crop Modelling: A Mathematical Approach to Plant and Crop Physiology*, Oxford Univ. Press, Oxford.
283. J. Christophersen (1967): *Molecular Mechanisms of Temperature Adaptation* (C. L. Prosser, ed.), Am. Assoc. Adv. Sci., Washington, DC, p. 327.
284. R. de Visser, C. J. T. Spitters, and T. J. Bouma (1992): *Molecular, Biochemical and Physiological Aspects of Plant Respiration* (H. Lambers and L. H. W. Van der Plas, eds.), SPB Academic Publishing, The Hague, p. 493.
285. C. Schäfer, H. Simper, and B. Hofmann (1992): *Plant Cell Environ.*, 15:343.
286. J. S. Amthor (1994): *Advances in Carbon Dioxide Research* (L. H. Allen Jr., ed.), American Society of Agronomy, Madison, WI, in press.
287. J. Reuveni, J. Gale, and A. M. Mayer (1993): *Ann. Bot.*, 72:129.
288. N. E. Grulke, G. H. Riechers, W. C. Oechel, U. Hjelm, and C. Jaeger (1990): *Oecologia*, 83:485.
289. D. E. Collier, F. Ackermann, D. J. Somers, W. R. Cummins, and O. K. Atkin (1993): *Physiol. Plant.*, 87:447.

16

Barriers in the Wheat Leaf Rust Preinfection Phase

Robert E. Wilkinson

University of Georgia
Griffin, Georgia

John J. Roberts

Georgia Experiment Station, ARS, USDA
Griffin, Georgia

INTRODUCTION

Wheat (*Triticum aestivum* L. em Thell.) is a major food crop throughout the world (1). Pathogens develop on wheat that are specific to the species (2,3), and some of these pathogens are relatively specific to individual tissues (2,3). Loss of wheat production due to rusts has exceeded 10% of the crop in the United States in a mild rust year (4,5). Thus, study of the rust-wheat microcosm has some validity in terms of food and economics as well as to improve our understanding of plant response to stress.

Wheat leaf rust is caused by *Puccinia recondita* Rob. ex Desm., which is a pathogen whose infestation is essentially but not completely restricted to leaves (2). The spread of this pathogen is accomplished by the windborne dissemination of urediospores (3) containing a limited supply of triglycerides which must serve as (a) a metabolic energy source and (b) lipid substrate for the formation of urediospore germtubes (2,3).

When the urediospore lands on a wheat leaf in the presence of adequate moisture (dew) and temperature, the urediospore develops a germtube which grows laterally across the leaf until it encounters a stoma or other break in the cuticular layer. At this point, infection structures (i.e., appressorium, infection peg, vesicle, etc.) develop and the rust enters the leaf mesophyll (2–4). Obviously, growing conditions for the germtube on the leaf surface and the mycelia in the leaf intracellular mesophyll cavity are different. And, equally obviously, preclusion of germtube arrival at a stoma and/or development of infection structures at the stoma would decrease rust infection. Therefore, the preinfection phase of rust development is an extremely important stage in the rust infection process, and the microcosm leaf surface conditions affecting the germtube may greatly influence rust infectivity. Therefore, several questions arise: (1) Do leaf surfaces vary sufficiently to alter infectivity? (2) How do the leaf surfaces differ? (3) How do these differences influence rust germtube growth? (4) What conditions induce modifications in the leaf surface? and (5) How is the development of infection structures triggered?

RUST INFECTION VS. POSITION ON FLAG LEAF

Flag leaves were collected in the field from six wheat varieties. Pustules/cm^2 were determined microscopically on the distal, mid-, and basal fifths portions of the flag leaves. In the slow-rusting wheats (Ble Tendre, Menkemen, Lerma 50, and Veadrio) the numbers of pustules/cm^2 was highest on the basal fifth portion of the flag leaf, medium concentrated on the mid-fifth portion, and least concentrated on the distal fifth portion (Table 1). In the highly susceptible varieties (M1 and Red Bobs) the position on the leaf surface did not influence the development of pustules/cm^2.

Development of pustules on flag leaves depends on the period of time the leaves have been exposed to windborne urediospores. On a time series, the highly susceptible Red Bobs had pustule counts (pustules/cm^2) on the basal fifth portions of 2.9, 26.4, 122.5, and >200 at weekly intervals. The flag leaf distal fifth portion had pustule/cm^2 equivalent to the

Table 1 Slow-Rusting Pattern Infecting of Leaf Rust of Spring Wheats—1970

| | Pustules/cm^2 | | |
| | Distal | Mid- | Basal |
Variety	fifth	fifth	fifth
Menkemen	12c*	50b	80a
Ble Tendre	25c	123b	163a
Lerma 50	1c	6b	11a
Veadiro	9c	25b	104a
M1	329a	346a	309a
Red Bobs	250a	296a	254a

* Values on a line followed by the same letter are not significantly different at the 5% level.
Source: Data from Ref. 5.

base. But in the slow-rusting variety Mentana, the basal fifth had weekly pustule/cm^2 counts of 0.4, 0.5, 1.0, 6.2, and 24.3 while the distal fifth pustule counts were 0.3, 0.8, 1.4, 2.3, and 9.0. Thus, Red Bobs developed >30 times the pustules/cm^2 than Mentana in an equivalent time period when both varieties were exposed to the same population of windborne urediospores.

SCANNING ELECTRON MICROSCOPE VIEW OF WHEAT LEAVES

Grass leaf cells are produced by an intercalary meristem at the base of the blade (6). Stoma are produced in linear series that are offset by lateral distance and time of development. Therefore, if a germtube misses one line of stoma, continued germtube tip growth will eventually result in the arrival of the germtube tip at a stoma. This stoma offset condition is shown in Fig. 1. The epidermal cells present a series of ridges and valleys across which the germtube tip must grow.

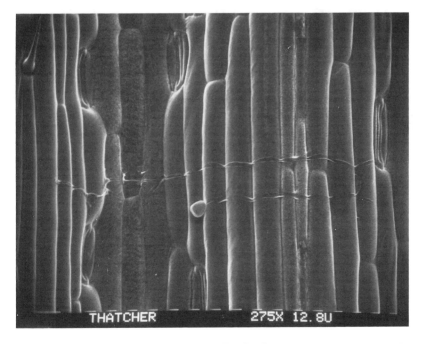

Figure 1 Scanning electron micrograph of wheat (*Triticum aestivum* L. cv Thatcher) leaf (×275) showing lateral urediospore germling growth and offset stoma lines.

Waxes are present on all leaf epidermal cell surfaces including the stoma guard cells (Fig. 2); but wax crystals are relatively rare on the trichomes (Fig. 2). The structure of the waxes varies from sheets to rods to plates (Fig. 3) and these structures have been ascribed to specific chemical compositions and modes of excretion (7). Germtubes adhere to the wax structures with sufficient force to remove the wax crystals from the leaf surface when the germtube dehydrates during preparation for electron microscopy (Fig. 4).

Trichomes are single or multicell extrusions of epidermal cells and are present on leaves in a genetically determined pattern. Germtube

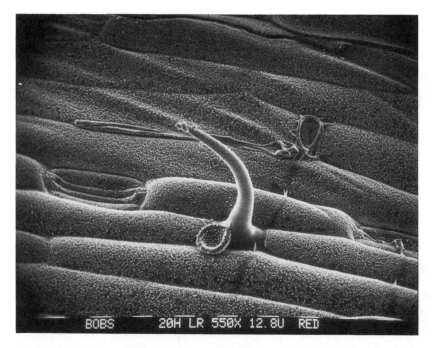

Figure 2   Scanning electron micrograph of wheat (*Triticum aestivum* L. cv Red Bobs) (550×) leaf showing wax presence on epidermal cells, stoma guard cells, and relative absence of wax on trichomes.

extension growth continues when a germtube touches the tip of a trichome (Fig. 5). But when the germtube tip touches the basal cell of a multicellular trichome, extension growth is reduced at the side touching the trichome basal cell and the germtube grows around the trichome (Fig. 6) (56). When the germtube tip grows up onto the germtube so that it no longer contacts the trichome basal cell, lateral growth across the leaf surface resumes (Fig. 6).

These electron micrographs present several problems. (1) Do the germtubes utilize the wax as substrates for energy and/or membrane synthesis? (2) How do the germtubes attach to the wax crystals? (3) What is the composition of the wheat leaf epicuticular wax? (4) Does the epicuticular wax vary between cultivars? (5) Does the epicuticular wax vary

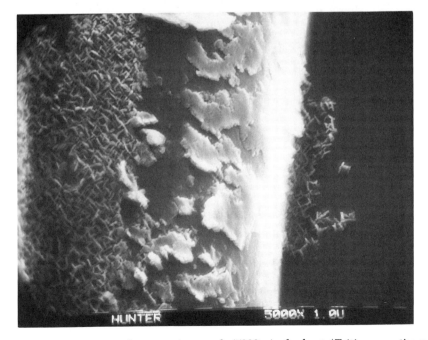

Figure 3 Scanning electron micrograph (5000×) of wheat (*Triticum aestivum* L. cv Hunter) leaf showing different types of epicuticular wax structure.

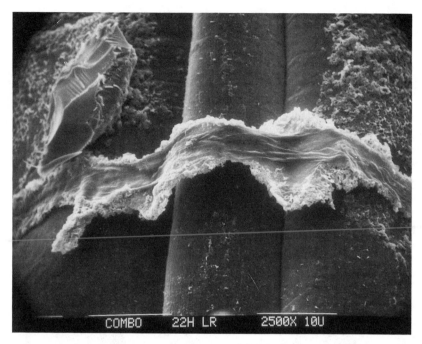

Figure 4 Scanning electron micrograph (2500×) of wheat (*Triticum aestivum* L. cv Combo) leaf showing epicuticular wax crystals attached to a collapsed urediospore germtube.

between portions of the leaf? and (6) If the wax composition varies from one portion of a leaf to another, what are the controlling mechanisms that produce different epicuticular wax products at different times?

EXOGENOUS LIPID UTILIZATION BY GERMLINGS

Urediospore germtubes utilized [1-^{14}C]linoleic acid to produce $^{14}CO_2$ (Fig. 7) (8). The production of $^{14}CO_2$ was significantly inhibited by the membrane impermeat sulfhydryl inactivator *p*-chloromercuribenzene-sulfonic acid (PCMBS) (Fig. 7). The inhibition of PCMBS was reversed

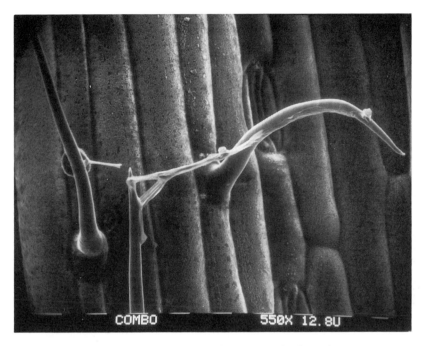

Figure 5 Scanning electron micrograph (550×) of wheat (*Triticum aestivum* L. cv Combo) leaf with urediospore germling growth after touching the tip of a trichome.

by a previous treatment with dithiothreitol (DTT) which prevents the inactivation of sulfhydryl groups by PCMBS (Fig. 7) (9). PCMBS is known to inactivate the membrane exofacial sulfhydryl group of H^+-ATPases which enable loading of [^{14}C]sucrose into phloem sieve tube members (9). Therefore, [1-^{14}C]linoleic acid movement through the germtube plasma membrane was inhibited by PCMBS and the attachment of the germtube to wax crystals (Fig. 4) is due to (at least in part) the attachment of the membrane bound H^+-ATPase to the wax crystal. Urediospores have triglycerides and are well known to have β-oxidation enzymes so that once the [1-^{14}C]linoleic acid is inside the germtube

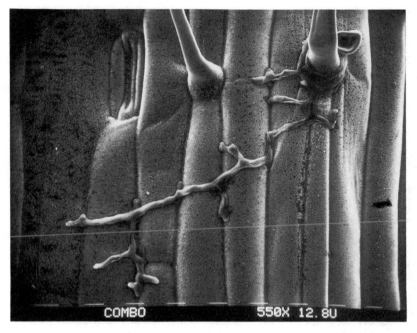

Figure 6 Scanning electron micrograph (550×) of wheat (*Triticum aestivum* L. cv Combo) leaf showing urediospore germling growth around the basal cell of a multicellular trichome.

plasma membrane, conversion of [1-^{14}C]linoleic acid to $^{14}CO_2$ occurs (10–14).

Wheat leaf rust urediospores were applied onto glass slides (937.5 mm^2) on which [1-^{14}C]palmitic acid (0.1 μCi) (60 μCi/mg) was spread evenly. Concomitantly, a concentration series of PCMBS (0, 0.01, 0.1, 1, or 10 mM) (50 μl/slide) was applied. After 16-hr germination in the dark at 28°C, average growth in length of five replications, as determined microscopically with an ocular micrometer at 125×, was inhibited by PCMBS (Table 2). Amounts of $^{14}CO_2$ (DPM) released decreased as PCMBS concentration increased (Table 2), and $^{14}CO_2$ production was correlated with germtube length (Fig. 8). These data show that the abil-

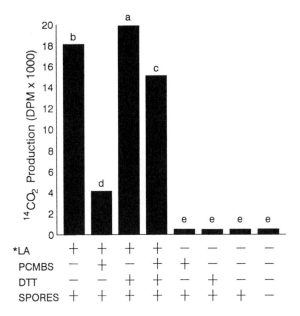

*LA	+	+	+	+	−	−	−	−
PCMBS	−	+	−	+	+	−	−	−
DTT	−	−	+	+	−	+	−	−
SPORES	+	+	+	+	+	+	+	−

Figure 7 Influence of p-chloromercuribenzensulfonic acid (PCMBS) (1 mM), a membrane-impermeant sulfhydryl group inactivator, and dithiothreitol (10 mM), a sulfhydrl group protector, on the utilization of [1-^{14}C]linoleic acid (*LA) to produce $^{14}CO_2$. Columns with the same letter are not significantly different at the 5% level as determined from analysis of variance on a randomized complete block design with means of five replications separated by the multiple LSD method.

Table 2 Influence of PCMBS on $^{14}CO_2$ Released by Wheat Leaf Rust Germ-tubes After 16-hr Exposure to [1-^{14}C]Palmitic Acid (0.1 μCi) (60 μCi/mg) on 937.5 mm^2 Glass Slides at 28°C (Values Are the Average of Five Replicates)

PCMBS (mM)	$^{14}CO_2$ Released (DPM)	Germtube length (μm)
0	10149a*	110a
0.01	9324b	72b
0.1	9362b	66b
1	4376c	52c
10	3431c	52c

*Values in a column followed by the same letter are not significantly different at the 5% level.

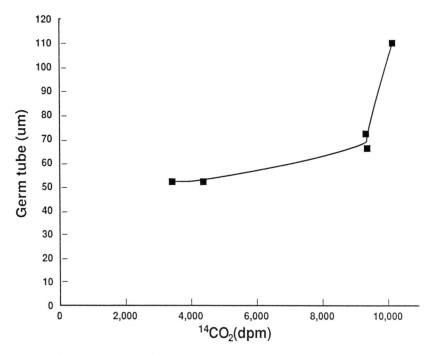

Figure 8   Correlation of $^{14}CO_2$ production from [$1\text{-}^{14}C$]palmitic acid by wheat leaf rust germtubes and length of germtube. Line = germtube length = 3161 − 0.957 (DPM) + 0.0000773 (DPM)^2. $R^2 = 0.813$.

ity of the germtube tip to grow is strongly correlated with the capacity of the germtube to obtain energy from an external lipid source. Additionally, germtube growth in length decreased markedly when absorption of the ^{14}C substrate was inhibited. Or, urediospore internal triglyceride concentrations could supply energy and membrane biosynthesis substrates for a very limited germtube extension ($\sim 50\,\mu M$) (Fig. 8). Growth beyond that limit requires support from lipids absorbed from an exogenous source.

Utilization of ^{14}C labeled fatty acids as an energy source has been evaluated in other rust urediospore germtubes. Urediospore germlings of *Puccinia graminis*, *P. substriata*, *P. sorghi*, *P. canaliculatus*, and *Uromyces appendiculatus* produced $^{14}CO_2$ from ^{14}C fatty acids, and PCMBS significantly inhibited the production of $^{14}CO_2$ (8). However, there were differences in susceptibility to PCMBS and DTT among the rust species (8).

Thus, rust germtubes utilized triglycerides and free fatty acids as an energy source. Free fatty acids are a component class of lipids found in leaf epicuticular waxes. But several other classes of lipids are also found in the epicuticular waxes. Do the rust germtubes use epicuticular wax constituents other than free fatty acids? If so, what are the relative rates of utilization of the different epicuticular wax constituent lipid classes?

UTILIZATION OF EPICUTICULAR WAX COMPONENTS

Wheat leaf epicuticular waxes are composed of alkanes, esters, β-diketones, hydroxy-β-diketones, free fatty acids, and free fatty alcohols (14). Occasionally, minor constituents (i.e., ketones, aldehydes) may accumulate which are intermediates in the biosynthesis of these major constituent classes (14). The biosynthetic mechanisms have been established (16). The β-diketones and hydroxy-β-diketones are found in only a few plant species, and chemically the β-diketones are represented by a very restricted number of compounds. As lipid wax classes, the fatty acids and fatty alcohols have constituents ranging from C_{14} to C_{34} and occasionally up to C_{38} (15,16). The lipid classes can be easily separated by thin layer chromatography. Separation of the individual C_n constituents within a single lipid class (i.e., alkanes, free fatty acids, etc.) requires careful analysis by gas–liquid chromatography. When plants are exposed to $^{14}CO_2$, the waxes contain ^{14}C and after separation by thin layer chromatography the lipid class can be used as substrates for urediospore germtubes.

Carbon-14 waxes from four wheat cultivars were prepared and used as substrates for wheat leaf rust urediospore germtubes. Percentages of the applied ^{14}C lipid classes recovered as $^{14}CO_2$ are shown in Table 3. Alkanes and esters were very poorly utilized as energy sources while the remaining lipid classes were utilized very readily (Table 3). The quantities of β-diketone constituents were analyzed spectrophotometrically on Coker 983 wheat flag leaves. β-Diketone content (mg/g FW) was least in the tip third of the leaf, intermediate in the mid-third of the leaf, and highest on the basal third of the leaf (Table 4). Major wax lipid constituent classes varied on the different flag leaf sections (Table 5). Basically, alkane plus esters represented a high percentage of the waxes on the leaf tip while the lipid classes that were most readily utilized as sources to produce $^{14}CO_2$ were present in greatest quantity on the basal third of the flag leaf (Table 5). When pustule counts (number/cm^2) on the individual third portions of flag leaves were plotted against wax quantity (mg/cm^3) [(β-diketones + hydroxy-β-diketones + free fatty acids + free fatty alcohols) minus (alkanes + esters)], a correlation coefficient of 0.8843 (P

Table 3 Utilization of ^{14}C Lipid Classes to Produce $^{14}CO_2$ by Wheat Leaf Rust
Spore Germtubes

Lipids	Coker 983	Florida 301	Red Bobs	Hunter
		% ^{14}C Applied		
Alkanes	38.9a*	1.0c	0.3c	23.8b
Esters	30.8a	4.5c	6.0c	13.6b
β-Diketones	70.8a	62.8b	21.8c	72.0a
Free fatty alcohols	98.0a	99.0a	99.0a	99.0a
HO-β-diketones	98.0a	99.0a	99.0a	98.0a
Free fatty acids	99.0a	98.0a	76.9b	99.0a

*Total 2000 DPM (after subfraction of background) applied/slide for each of four
replicates/lipid class. Values on a line followed by the same letter are not significantly
different at the 5% level as measured by the LSD multiple range test.

Table 4 β-Diketone Content of Coker 983 Flag Leaves

	Leaf (g)	Wax (mg/g FW)		
Segment	Fresh weight	Total	β-Diketone	Hydroxy-β-diketones
Tip	8.28	6.0	1.4c*	0.7b
Mid	17.13	5.3	3.1b	0.4c
Base	19.36	9.4	5.4a	1.0a

*Values in a column followed by the same letter are not significantly different at the 5%
level. Means separated by the LSD multiple means test.

Table 5 Coker 983 Flag Leaf Epicuticular Wax Contents (wt %)

Lipids	Leaf base	Mid-leaf	Leaf tip
Alkanes	3.10b*	6.74a	4.11b
Esters	13.83b	12.81b	46.53a
β-Diketones	57.42a	57.43a	23.28b
Free fatty alcohols	10.53b	14.06a	9.50b
HO-β-Diketones	10.99a	7.93b	11.00a
Free fatty acids	4.13a	1.03b	5.58a
Total (μg/g FW)	9376.5a	5349.4b	6004.4b

*Values on a line followed by the same letter are not significantly different at the 5% level
according to the LSD multiple means test.

= 0.01) was obtained [line = pustules = 83.67 + 46.55 [wax (mg/cm^2)]] (Fig. 9).

Therefore, urediospore germtubes utilized individual epicuticular wax constituent classes with differential facility; the lipid classes most easily utilized to produce $^{14}CO_2$ were present in greatest quantity on the basal third of the flag leaf; and pustule formation on the individual portions of the flag leaf was linearly correlated with the quantity of "favored" substrate.

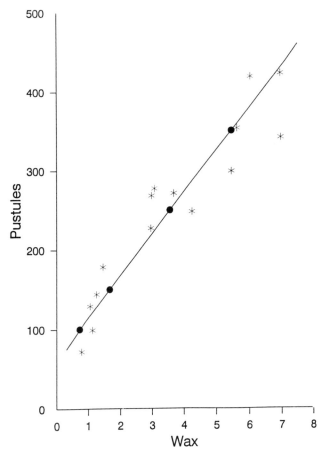

Figure 9. Regression of pustules (number/cm^2) vs. wax [(β-diketones + hydroxy-β-diketones + free fatty acids + free fatty alcohols) minus (alkanes + esters)] (mg/cm^2). R^2 = 0.8843, line = pustules = 83.67 + 46.55 [wax (mg/cm^2)]. Significance 0.01 %.

When stem rust is induced by *Puccinia graminis* L. which infects the leaf sheath more readily than the leaf blade. And wheat leaf rust infects leaf blades more readily than leaf sheaths. Some wheat cultivars are "resistant" to rust while other cultivars are "susceptible". Could a *portion* of these differential susceptibility responses be due to variation between (a) cultivars, (b) leaf blade vs. leaf sheath, or (c) distal vs. basal portions of a single flag leaf in epicuticular wax quantity and composition on the surface of the cuticle?

EPICUTICULAR WAXES ON WHEAT LEAVES

The plant chemical taxonomy literature is replete with epicuticular wax analyses of samples taken at one stage of leaf development (14,16). Differences between cultivars in epicuticular wax quantity and composition exist (17–20). but, data showing differences in epicuticular wax quantity and composition between different stages of growth are scarce.

In *Opuntia* (21) an increased quantity and changed composition of alkanes (mg/dm^2) on cladophylls of different ages (Fig. 10) could be explicable as a slow accumulation of higher C_n compounds accompanied by a cessation of synthesis of short C_n compounds. Or, these components were inert and excreted to the surface of the cladophyll in deposition regimes that were altered over the life of the cladophyll.

On tobacco (*Nicotiana tabacum* L.) leaves of the same age but grown under different photoperiod, temperature, and light intensity regimes, alkanes, fatty acid, and fatty alcohol contents were significantly different (22). These responses mirrored the amino acid content of the leaves (23). Since tobacco epicuticular waxes are characterized by the presence of branched chain alkanes, fatty acids, and fatty alcohols which are synthesized by the elongation of deaminated branched chain amino acids (i.e., valine, isoleucine), the epicuticular wax composition on the surface of the leaf was determined by the metabolic processes inside the leaf. Those waxes might be inert excretion products but the wax composition excreted was determined by the internal metabolism of the leaf.

Epicuticular wax quantity and composition of three wheat cultivars was determined when leaf samples were collected at eight different dates after planting (DAP) (17). Quantities of esters (mg/g DW) did not change appreciably from 24 to 100 DAP (Fig. 11) (17). Free fatty alcohol content decreased as DAP increased (Fig. 11) (17). Alkane ($\mu g/g$ DW) content increased as DAP increased on one cultivar but remained relatively uniform in the other two cultivars (Fig. 12) (17). The quantity of β-diketone present on the leaves increased markedly to a maximum at 66 DAP *and then decreased* in all three cultivars (Fig. 13) (17). The β-diketones are long-chain carbon compounds and are rather nonvolatile. If inert, non-

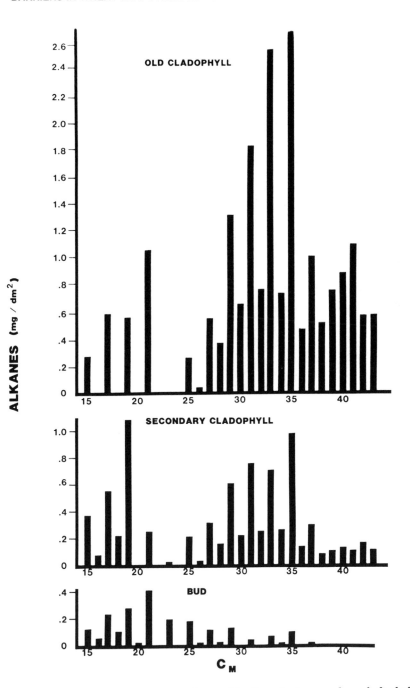

Figure 10 Alkane composition and quantities from bud, secondary cladophylls, and old cladophylls of *Opuntia engelmanni* var. *Texana* (Gr.) Weniger (21).

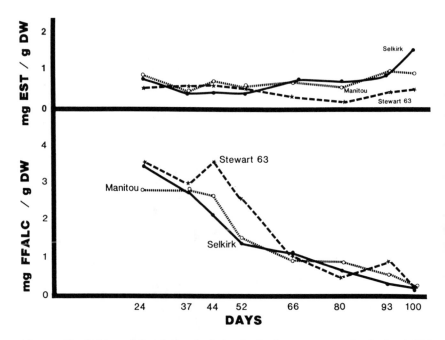

Figure 11 Ester and free fatty alcohol epicuticular wax contents of wheat (*Triticum aestivum* L. cv Selkerk, Manitou, and Stewart (63)) flat leaves as a function of days after planting (DAP) (17).

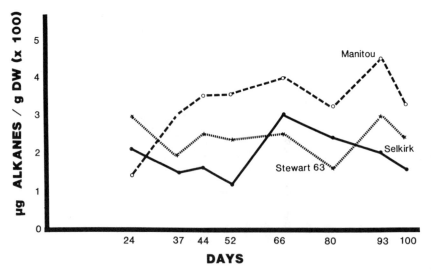

Figure 12 Flag leaf alkane content of wheat (*Triticum aestivum* L. cv Manitou, Selkirk, and Stewart 63) at various days after planting (DAP) (17).

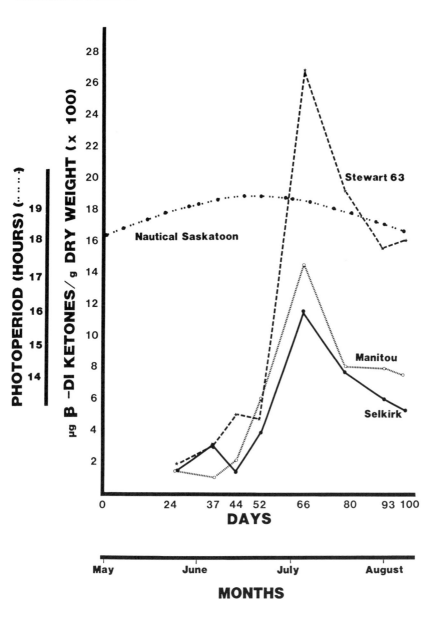

Figure 13 Flag leaf epicuticular wax β-diketone contents of wheat (*Triticum aestivum* L. cv Stewart 63, Manitou, and Selkirk) at various days after planting (DAP) (17).

volatile substances were excreted to the surface of the leaf and the leaf was no longer expanding rapidly, the quantity (μg/g DW) of that inert substance should remain relatively uniform on the leaf. β-Diketone content (μg/g DW) *decreased* by about 40% (Fig. 13) (17) while alkane and ester contents remained uniform and fatty alcohol contents decreased slightly (Figs. 11 and 12). Thus, rust infection is different at three positions on the flag leaf. The waxes adhere to the surface of the germtube and that adherence is at least partially explained by the attachment of membrane bound enzymes to the wax. The waxes can be absorbed into the germtube where constituents bearing ^{14}C are converted to ^{14}CO$_2$; and, the lipid constituent classes are utilized to produce ^{14}CO$_2$ at very different rates. The quantity of the various lipid classes in the surface epicuticular wax can change with DAP (i.e., photoperiod, temperature, etc.). And the composition of the epicuticular wax is a reflection of the metabolic state of the epidermal cells. So how does altered environment change wheat cellular metabolism? Does the altered metabolism induce a changed epicuticular wax composition?

EPICUTICULAR WAX RESORPTION

After ovule fertilization, the physiology of seed-bearing plants is altered dramatically. Vegetative cell enlargement ceases, seed filling commences, and vegetatively stored food reserves are metabolized to translocatable substances, moved to the seed, and deposited as seed food reserves.

Vegetative cell enlargement depends on a supply of membrane phospholipids. Phospholipid biosynthesis is a very complex system which includes fatty acid synthesis in plastids. The major product fatty acids synthesized by the plastids are monoenoic [i.e., palmitoleic acid ($C_{16:1}$) or linoleic acid ($C_{18:1}$)] (24). The highly hydrophobic monoenoic fatty acids are translocated by carrier proteins from plastids to microsomes where the fatty acids are dehydrogenated to dioenoic fatty acids. Further desaturation occurs in the plastids after transfer via carrier proteins back to the plastids. Microsomes are a cytosolic location of glyoxylate metabolism of fatty acids (10–13). Therefore, an active microsomal glyoxylate system could remove fatty acids from the lipid pool requisite to produce phospholipids. If a sufficient quantity of phospholipids are not available to maintain membrane integrity by replacing lipid units made inoperative by various causes *and* supply the phospholipids requisite to membrane enlargement, vegetative cell enlargement ceases. Thus, glyoxylate metabolism of fatty acids in the microsomes during seed filling would accomplish two major items: (a) vegetative cell enlargement would cease and (b) the products of glyoxylate metabolism would be converted to sucrose which would be translocated to the seed where the carbon would be stored as starch or triglyceride.

Synthesis of both α-amylase and the glyoxylate system are influenced by gibberellic acid (10–13,25–33). Increased α-amylase activity mobilizes vegetatively stored starch causing a decrease in stem-stored starch (34). But the only major vegetative repository for excess lipids is in epicuticular waxes, which are supposedly inert excretion products outside the symplast. The epicuticular waxes are basically derived from long-chain fatty acids which are extruded through the plasma membrane (15). If that extrusion process were reversed, certain constituents in the epicuticular waxes might be resorbed. And exogenously applied ^{14}C fatty acids have been recovered from complex membrane lipids inside the cell (35).

When [UL-^{14}C]palmitic acid was applied to wheat flag leaves, ^{14}C was extracted from the stem and head with boiling ethanol (Table 6). Fatty acids were not translocated 20 cm above the flag leaf to the head, and starch was not extracted in the boiling ethanol. The extracted ^{14}C substances cochromatographed with [^{14}C-UL]sucrose on TLC and only the sucrose not yet converted into starch was extracted. When wheat leaf isocitric lyase (IL) content was assayed, up to 16 nM phenylhydrazone produce/g FW/min was found. Since IL is a requisite enzyme in the glyoxylate system, these data indicate that [UL-^{14}C]palmitic acid was converted to sucrose via the glyoxylate system and the [^{14}C]sucrose was translocated to the head. Thus, a potential for resorption and metabolic utilization of some epicuticular wax constituents exists. Therefore the epicuticular wax composition would be altered during specific metabolic stages of plant development and the altered wax composition and/or quantity would alter rust infection as shown in Figure 9.

Photoperiod influences GA biosynthesis (36). GA influences IL synthesis and activity in cereal seeds (10–13). In wheat grown under 12-hr days at 20°C, GA_3 induced increased growth in length and fresh weight (Table 7) but a decreased rate of deposition in total epicuticular wax (μg/plant) and β-diketone (% total wax) contents (Table 7) was found.

Table 6   [14]C Recovered in Boiling Ethanol from Stem and Head of Hunter Wheat After Application of [UL-^{14}C]Palmitic Acid to the Flag Leaf (Average of Five Replications; dpm/mg FW)

Hours	Stem	Head
0	0	0
4	28596	758

Table 7   Influence of GA$_3$ on the Synthesis of Epicuticular Wax on Wheat (*Triticum aestivum* L. cv Stacy) Leaves 14 Days After Spraying (Each Value is the Average of Five Replications)

GA$_3$ (μM)	Height (ΔCM)	Fresh weight (mg/plant)	Total wax ($\Delta\mu$g/plant)	β-Diketones (Δ% of total wax)
0*	+3.0b**,†	769c	+855a	+0.99a
1	+6.3a	878b	+725b	+0.73b
10	+7.3a	934a	+734b	+0.73b

*Values presented are the *changes* (Δ) from the original untreated material.
**Values in a column followed by the same letter are not significantly different at the 5% level according to the LSD multiple range test.
† At treatment: height = 30.7 cm; total wax = 295.1 mg/plant; and β-diketones = 3.7% of total wax.

The rate of epicuticular wax deposition decreased when GA$_3$ was applied. When GA$_3$–methyl ester was applied the same relationships were found and IL quantity and activity were increased (Table 8). These relationships are in accord with currently accepted theories of GA-induced responses in cereal seeds and coleoptiles.

Certain herbicides inhibit GA biosynthesis and inhibit length growth (37). When GA$_3$ is applied to the sand potting medium the herbicide-induced growth in length inhibition was reversed (37). When GA$_3$ was solubilized with 0.5 N KOH and incorporated into the sand, growth in leaf length was not significantly altered while the rate of increase in fresh weight (Δ mg/plant) was significantly decreased (Table 9). Total epicu-

Table 8   Influence of Gibberellic Acid–Methyl Ester (GA$_3$) on the Isocitric Lyase Content of Wheat (*Triticum aestivum* L. cv. Stacy) Leaves at 7 Days After GA$_3$ Application (Each Value Is the Average of Five Replications with Triplicate Assays per Replication)

Isocitric lyase	GA$_3$–methyl ester (mM)			
	0	0.1	1.2	5.0
(*nM Hydrazone product* min/mg protein)	7.5c*	17.1a	15.5b	15.7b
(*nM Hydrazone product* min/g fresh weight)	2.6b	3.8a	3.7a	3.7a

*Values on a line followed by the same letter are not significantly different at the 5% level according to the LSD multiple comparison test.

Table 9 Influence of a GA_3 Degradation Product (TCCTHDA) on the Growth and Epicuticular Wax Quantity and Composition at 6 Days After Treatment of 28-Day Old Wheat (*Triticum aestivum* L. cv Stacy) Grown in Nutrient Solution (Each Value Is the Average of Five Replications)

		TCCTHDA (μM)			
	Days	0	1	10	100
Height (Δcm)	6	8.4a*	6.2a	7.0a	7.0a
Fresh weight (Δmg/plant)	6	483a	373b	340b	319b
Total wax (Δmg/plant)	6	1531a	1005b	1001b	961b
β-Diketone (Δmg/plant)	6	51a	6b	3b	1b

*Values on a line followed by the same letter are not significantly different at the 5% level according to the LSD multiple range test.

ticular wax deposition (Δ mg/plant) was decreased and β-diketone synthesis (Δ mg/plant) virtually stopped (Table 9). Since growth in length and fresh weight were not modified by this exogenous chemical, the compound was not GA_3. Treatment of GA_3 with KOH causes a rupture of the lactone ring (37,38) with the quantitative production of TCCTHDA (Fig. 14). Therefore, a non–growth-producing gibberellin carbon skeleton had induced a decreased total non-β-diketone wax content and virtually eliminated the synthesis of β-diketone (Table 9). Topical application of gibberic acid–methyl ester (Fig. 14) did not induce altered growth in length or fresh weight (Table 10). But total non-β-diketone epicuticular wax contents (Δ μg/plant) were increased threefold while β-diketone contents (% of wax) were decreased (Table 10). Thus, the rate of β-diketone biosynthesis was decreased but the increase in total non-β-diketone epicuticular wax was greater than could be accounted for as a decreased β-diketone content.

When gibberic acid–methyl ester (G-ME) (5 μM) (Fig. 14) and GA_3-methyl·ester (GA_3-ME) (10 μM) were topically applied to wheat separately and concomitantly, the rate of increase in leaf fresh weight was decreased by G-ME, increased by GA_3-ME, and unchanged by the mixture (Table 11). Or G-ME induced a decreased growth rate which was reversed by GA_3-ME. β-Diketone contents (μg/g FW) were significantly decreased by G-ME and GA_3-ME (Table 11). Total epicuticular wax content (μg/g FW) was decreased by GA_3-ME, increased by G-ME, and plants treated with the mixture had wax content equal to that of the untreated plants (Table 11). Or G-ME induced an increased wax synthesis while GA_3-ME induced a decrease in wax synthesis. Both G-ME and GA_3-ME induced increased protein synthesis and IL activity and

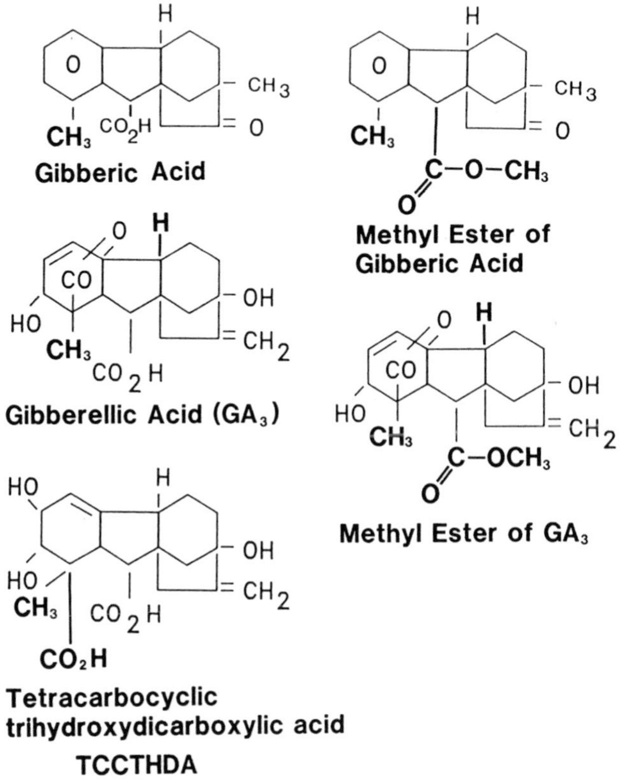

Figure 14 Chemical structures of gibberic acid, gibberic acid–methyl ester, gibberellic acid (GA_3), GA_3–methyl ester, and tetracarboxylic trihydroxydicarboxylic acid (TCCTHDA).

Table 10 Influence of Gibberic Acid–Methyl Ester on the Epicuticular Wax Composition of Wheat (*Triticum aestivum* L. cv Stacy) Leaves 7 Days After Application

	Gibberic acid–methyl ester conc. (μM)			
Constituent	0	0.1	1.2	5.0
Height (ΔCM)	+4a	+5a	+5a	+5a
Fresh weight (Δmg/plant)	+62.8a	+180.5a	+170.8a	+192.6a
Total non-β-diketone wax ($\Delta\mu$g/plant)	+1430d*	+2490c	+3620b	+4070a
β-diketones (% of total wax)	+25a	+17b	+16b	+11c

*Each value is the average of five replications. Values on a line followed by the same letter are not significantly different at the 5% level according to the LSD multiple range test.

Table 11 Influence of Gibberic Acid–Methyl Ester (G-ME) and Gibberellic Acid–Methyl Ester (GA^3-ME) on the Growth, β-Diketone Content, Total Epicuticular Wax, and Isocitric Lyase Content of Wheat (*Triticum aestivum* L. cv. Hunter) at 7 Days After Spraying

Treatment (μM)		Δmg Leaf fresh weight (%)	$\Delta\mu$g β-diketone g FW (%)	Δmg Total wax g FW (%)	$\Delta\mu$g Protein ml (%)	μM Hydrazone product/min			
G-ME	GA_3-ME					ml Enz (%)	plant (%)	mg protein (%)	g FW (%)
0	0	+573b* (100)	+3.9a (100)	−0.48b (100)	149.2c (100)	2.23c (100)	2.41c (100)	14.22c (100)	10.53c (100)
5	—	+327c (57)	+2.0b (53)	+1.99a (514)	189.8b (127)	8.85ab (397)	9.26ab (348)	52.32a (368)	31.04ab (295)
—	10	+719a (125)	+1.8b (46)	−277c (−477)	249.8a (167)	7.44b (334)	7.80b (323)	29.62b (208)	25.60b (243)
5	10	+481b (84)	+3.9a (100)	−0.65b (65)	269.1a (180)	10.17a (457)	11.28a (467)	38.38a (270)	36.15a (343)

*Values in a column followed by the same letter are not significantly different at the 5 % level as separated by the LSD multiple comparison test.

synthesis (Tables 8 and 10). These activities by GA_3 are totally within the currently published parameters of activity for GA_3. The activities of TCCTHDA and gibberic acid are real but totally anomalous in current theory of GA activity in plants. All these data really show is that epicuticular wax synthesis is controlled by cytosol enzymology, which in turn is controlled by environmental parameters; and that non–growth-promoting gibberellin carbon skeletons may modify selected enzymology in the cytosol. Additionally, epicuticular waxes are not inert excretion products on the outside of the cell which are never resorbed, metabolized, or utilized. That resorption and utilization may be slow and/or limited to specific lipid classes, but it exists. Therefore, altered photoperiod and temperature induce change plant growth regulator (PGR) content, which alters the synthesis and activity of specific enzymes which in turn alter the quantity and composition of epicuticular waxes. These, in turn, influence growth of rust germtubes and alter rust infectivity.

WHEAT LEAF RUST INFECTION STRUCTURE FORMATION

Wheat leaf rust germtubes laterally traverse the leaf surface until the germtube tip touches a stoma or other break in the cuticle (2,3) whereupon very specific structures are developed (2,3). The time sequence, enzymology, nucleation, and other metabolic processes have been brilliantly described (2,3,40). But the triggering mechanism(s) for the conversion of a germtube to infection structures are not understood. The time sequence of events includes (a) cessation of germtube lateral elongation, (b) thickening of the germtube tip, (c) development of appressorium, infection peg, and vesicle. Accompanying the development of these structures are (a) changes in nuclear division and (b) large changes in protein, RNA, DNA, and enzyme synthesis (2,3). These structural and metabolic changes have been correlated with stoma ridges (43–45) and ion concentration (41,42,46–50). The structural and metabolic changes in the germtube at an opening in the leaf are strongly reminiscent of plant response to PGRs. If a PGR were involved in this process, the PGR would have to (a) be volatile, (b) be produced by the host plant, (c) induce cessation of lateral elongation, (d) induce the initiation of protein synthesis, and (e) induce nuclear division. Ethylene is a volatile naturally occurring PGR that is produced by all parts of plants at differing evolution rates, and ethylene induces cessation of lateral growth, initiation of vertical growth, nuclear division, and protein synthesis (51–54).

When wheat leaf rust urediospore germtubes were exposed to ethylene, appressorium development was initiated (Fig. 15). The rate of appressorium initiation was strongly influenced by ethylene concentra-

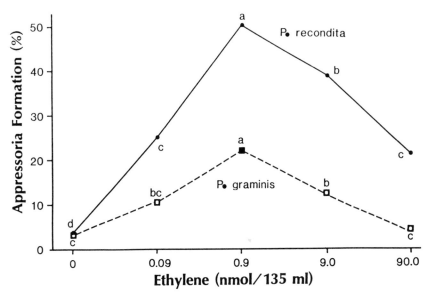

Figure 15 Influence of ethylene on the formation of appressoria by wheat leaf rust (*Puccinia recondita* Rob. ex. Desm.) and wheat stem rust (*Puccinia graminis* Pers. f. sp. *tritici*) germlings. Points on a line followed by the same letter are not significantly different at the 5% level.

tion (Fig. 15), and excessive ethylene concentrations (>100 nmol/135 ml) inhibited germtube elongation without concomitant infection structure development (data not presented). Wheat stem rust urediospore germtubes responded to ethylene at the same concentration but with a decreased rate of appressorium development (Fig. 15).

The time sequence of infection structure development has been shown to be rather strict (2,3,40). Appressorium formation is followed by peg formation within a relatively uniform time frame. It can be hypothesized that the entire infection structure development is a series of interdependent sequential processes. First, the germtube grows across a stoma. Horizontal elongation ceases but germtube wall formation continues, forming a thick, strong appressorium wall. The increased concentration of H^+-ATPases in the germtube membrane, now the appressorium, attach to the waxes on the stoma guard cells. When germtubes were grown on leaves from waxless mutant host plants, the germtubes were totally disoriented and infection structure formation was random (54). Potassium from the surface of the closed guard cell (55) stimulates H^+-ATPase activity which increases metabolic energy which in turn stimulates the capacity for growth. But the appressorium is by this time strongly attached to the guard cells and the least strenuous direction for growth is vertically into the stoma chamber (40). Inside the mesophyll activity, environmental conditions are totally different from that presented to the germtube on the leaf surface but the vesicle and hyphae present in the mesophyll have different nuclear and enzymatic capacities from those found in the germtube.

Ethylene concentration and production varies between tissues and species (51). Smuts have cutinases and produce appressoria on cuticles without benefit of any (presently) recognized triggering device. Certainly wheat stem and leaf rusts respond to ethylene differently (Fig. 15). And it should be axiomatic that each particular genome will respond to stress (stimulus) in a pattern determined by growth stage, genome, individual stress, and so forth.

SUMMARY

The literature reviewed and data presented herein have shown that wheat leaf rust infects portions of flag leaves with different concentrations of pustules. Urediospore germtubes utilize the epicuticular waxes as energy sources and the germtube is strongly attached to the wax. Germtube growth is correlated with the ability of the germtube to obtain lipids from outside the germtube. The composition of wheat leaf epicuticular waxes is not uniform across leaf sections, individual leaves, or cultivars. Rust infectivity is highly correlated with leaf epicuticular wax composition and quantity. Therefore, at least a portion of natural rust "resistance" and "susceptibility" is determined by the quantity and composition of epicuticular waxes available to the germtube for use as an energy source. If epicuticular wax synthesis and deposition on the surface of individual cells varies (Fig. 3) so that the waxes present at the ridge differ somewhat from the waxes in the valleys, then wax composition may be a triggering mechanism for the growth of germtubes in different directions (Fig. 6). Wax quantity and composition is not static. Epicuticular wax components are not inert excretion products but can be resorbed and utilized. The composition and quantity of epicuticular waxes are greatly controlled by the leaf cell metabolism. Thus, as the environmental exposure and growth stage of the host plant changes, metabolic modification in the cells results in altered wax quantity and composition available to the rust germtube. Urediospore germtubes respond to the gaseous plant growth regulator, ethylene, so that the capacity to produce infection structures at the stoma can be greatly influenced by the gaseous environment at the stoma. In short, each organism at each stage of development responds to stress (stimulus) in modes that are determined by genetics, growth stage, enzymology, other stimuli, and previous history of that particular response unit. The responses presented here in wheat leaf rust interactions are only representative examples of the responses of all plants to environmental stress. Wheat rusts operate in a very restricted environment but "big" plants show similar responses.

ACKNOWLEDGMENTS

This work was funded by the U.S. Department of Agriculture, Agricultural Research Service and University of Georgia, College of Agriculture Experiment Stations (State-Hatch Project 1451). Electron photomicrographs in Figures 1–6 were provided through the deeply appreciated assistance of the Electron Optical Facility of the University of Minnesota Agricultural Experiment Station.

REFERENCES

1. D. L. Long (1991): Estimated small grain losses from rust in 1990. USDA-ARS Cereal Rust Lab.
2. W. R. Bushnell and A. P. Roelfs (eds.) (1984): *The Cereal Rusts, Vol. I, Origins, Specificity, Structure, and Physiology.* Academic Press, New York.
3. A. P. Roelfs and W. R. Bushnell (eds.) (1985): *The Cereal Rusts, Vol. II, Diseases, Distributions, Epidemiology, and Control.* Academic Press, New York.
4. C. L. Lennox and F. H. J. Rijkenberg (1989): *Plant Pathol.*, 38:547.
5. R. M. Caldwell, J. J. Roberts, and Z. Eyal (1970): *Phytopathology*, 60:1287.
6. K. Esau (1953): *Plant Anatomy.* John Wiley and Sons, New York.
7. D. Simpson and P. von Wettstein-Knowles (1980): *Carlsberg Res. Commun.*, 45:465.
8. R. E. Wilkinson, J. J. Roberts, and J. W. Johnson (1991): *Plant Lipid Biochemistry, Structure, and Utilization* (P. J. Quinn and J. L. Harwood, eds.), Portland Press, London, pp. 370–372.
9. R. Giaquinta (1976): *Plant Physiol.*, 57:872.
10. H. Beevers (1980): *The Biochemistry of Plants, Vol. 4, Lipids: Structure and Function* (P. K. Stumpf, ed.), Academic Press, New York.
11. D. L. Doig, A. J. Colborne, G. Morris, and D. L. Laidman (1975): *J. Exp. Bot.*, 26:387.
12. H. L. Kornberg and H. Beevers (1957): *Biochim. Biophys. Acta*, 26:531.
13. M. Ricchti and F. Widmer (1986): *J. Exp. Bot.*, 37:1685.
14. A. P. Tullock (1976): *Chemistry and Biochemistry of Natural Waxes* (P. E. Kolattukudy, ed.), Elsevier, New York, pp. 235–250.
15. P. E. Kolattukudy, R. Crotean, and J. S. Buckner (1976): *Chemistry and Biochemistry of Natural Waxes* (P. E. Kolattutudy, ed.), Elsevier, New York, pp. 290–349.

16. T. Swain (1963): *Chemical Plant Taxonomy*. Academic Press, New York.
17. A. P. Tulloch and A. P. Hoffman (1971): *Phytochemistry*, *12*:2225.
18. G. Bianchi and C. Murelli (1988): *Genet. Agr.*, *42*:403.
19. G. Bianchi (1987): *Gaz. Chim. Ital.*, *117*:707.
20. G. Bianchi (1985): *Genet. Agr.*, *39*:471.
21. R. E. Wilkinson and H. S. Mayeux, Jr. (1990): *Bot. Gaz.*, *151*:342.
22. R. E. Wilkinson and M. J. Kasperbauer (1972): *Phytochemistry*, *11*:2439.
23. R. E. Wilkinson, M. J. Kasperbauer, and C. T. Young (1981): *J. Agr. Food Chem.*, *29*:658.
24. G. Roughan and R. Slack (1980): *Biogenesis and Function of Plant Lipids* (P. Mazliak, P. Benveniste, C. Costes, and R. Douce, eds.), Elsevier, Amsterdam, pp. 11–18.
25. T. Murata, T. Akazawa, and S. Fukuchi (1968): *Plant Physiol.*, *43*:1899.
26. T. Nomura, Y. Kono, and T. Akazawa (1969): *Plant Physiol.*, *44*:765.
27. Y. Tanaka, T. Ito, and T. Akazawa (1970): *Plant Physiol.*, *46*:650.
28. L. Paleg (1960): *Plant Physiol.*, *35*:293.
29. J. E. Varner (1964): *Plant Physiol.*, *39*:413.
30. H. Yomo (1960): *Hakko Kyokaishi*, *18*:600.
31. M. J. Chrispeels and J. E. Varner (1967): *Plant Physiol.*, *42*:398.
32. R. I. Doig, A. J. Colborne, G. Morris, and D. L. Laidman (1975): *J. Exp. Bot.*, *26*:387.
33. J. V. Jacobsen and T. J. V. Higgins (1982): *Plant Physiol.*, *70*:1647.
34. I. F. Wardlaw (1980): *Biology of Crop Productivity* (P. S. Carlson, ed.), Academic Press, New York, pp. 297–339.
35. P. G. Roughan, G. A. Thompson, Jr., and S. H. Cho (1987): *Arch. Biochem. Biophys.*, *259*:481.
36. S. J. Gilmore, J. A. D. Zeevaart, L. Schwenen, and J. E. Graebe (1986): *Plant Physiol.*, *82*:190.
37. R. E. Wilkinson (1989): *Crop Safeners for Herbicide Development, Uses, and Mechanism of Action* (K. K. Hatzios and R. E. Hoagland, eds.), Academic Press, New York, pp. 221–240.
38. B. E. Cross, J. F. Grove, P. McCloskey, J. MacMillan, J. S. Moffatt, and T. P. C. Mulholland (1961): *Adv. Chem.*, *28*:3.
39. J. F. Grove (1961): *Quart. Rev., Chem. Soc.*, *15*:56.
40. H. C. Hoch and R. C. Staples (1987): *Annu. Rev. Phytopathol.*, *25*:231.
41. H. J. Grambow and H. J. Reisener (1976): *Ber. Deutsch. Bot. Ges.*, *89*:555.
42. J. H. Grambow, G. Garden, F. Dallacker, and A. Lehman (1977): *Z. Pflanzenphysiol.*, *82*:62.
43. H. J. Grambow and D. Müller (1978): *Can. J. Bot.*, *56*:736.
44. W. K. Wynn (1976): *Phytopathology*, *66*:136.
45. R. C. Staples, H. J. Grambow, H. C. Hoch, and W. K. Wynn (1983): *Phytopathology*, *73*:1436.
46. J. H. Grambow and G. E. Grambow (1978): *Z. Pflanzenphysiol.*, *90*:1.
47. H. J. Grambow (1978): *Z. Pflanzenphysiol.*, *88*:369.
48. S. G. W. Kaminsky and A. W. Day (1984): *Exp. Mycol.*, *8*:63.
49. R. C. Staples, H. J. Grambow, and H. C. Hoch (1983): *Exp. Mycol.*, *7*:40.

50. H. J. Grambow (1977): *Z. Pflanzenphiol.*, *85*:361.

51. E. M. Meyer, P. W. Morgan, and S. F. Yang (1984): *Advanced Plant Physiology* (M. B. Wilkins, eds.), Pitman, London, pp. 111–126.

52. B. S. M. Ingemarsson, L. Eklund, and L. Eliasson (1991): *Physiol. Plant.*, *82*:219.

53. J. I. Sarques, W. R. Jordan, and P. W. Morgan (1991): *Plant Physiol.*, *96*:1171.

54. W. K. Wynn and R. C. Staples (1981): *Plant Disease Control: Resistance and Susceptibility* (R. C. Staples and G. A. Toenilssen, eds.), John Wiley and Sons, New York, pp. 45–69.

55. R. Hedrich and J. I. Schroeder (1989): *Annu. Rev. Plant Physiol.*, *40*:539.

56. J. J. Roberts, D. L. Long, R. E. Wilkinson, and G. G. Ahlstrand (1990): *Abstr. Phytopathology*, *80*:1046.

Index

Abscisic acid, 12, 26, 183
 chilling, 403
 flooding, 301
 ozone, 370
 photoperiod, 55
 stoma, 331
 temperature, 61
Absorption, 163
Acclimation, 393
 temperature, 522–525
Acetate kinase, 154
Acetylene reduction, 250
Acetyl thiokinase, 154
Acid phosphatase, 151, 164
Acid soil stress, 125–148
 sensitivity, 126, 185
 tolerance, 126, 185
Aconitase, 169, 170
Adaptation respiration
 temperature, 522–525
Adenosine
 diphosphate, 151

[Adenosine, *continued*]
 triphosphate, 151
Adenosuccinate, 154
S-adenosylmethionine, 162
Advective freezing, 392
Aeration
 soil, 277
Agmatine, 154
Agravotropism, 68
Agromanagement
 salinity, 219
Air pollution, 125
 climate, 300
 concepts, 364
 definition, 357
 direct effects, 366
 edaphic, 360
 factors influencing, 360
 history, 358
 hydrogen fluoride, 369
 meteorology, 360
 peroxyacetylnitrate, 360

585

[Air pollution, *continued*]
   plant susceptibility, 360–364
   salinity, 219
   topography, 360
Alcohol dehydrogenase, 172
Aleurone, 128
Alkali soils, 201
Alternative pathway
   respiration, 510
Aluminum
   accumulation, 136
   oxygen donor ligands, 137
   respiration, 540–543
   toleration, 9
   toxicity, 9, 22, 24
Amine oxidase, 166
Ammonium
   uptake, 3
α-amylase, 128
Amyloplast, 65
Antidote
   herbicide, 194
Antiports, 209
Apoplast, 202
Apparent growth
   efficiency, 511–516
Arginine decarboxylase, 12
Ascorbic acid oxidase, 166
ATPase, 208
ATP sulfurylase, 162
Autumn syndrome, 60
Auxin
   calcium, 157
   flooding, 301
   sensitivity, 71–74
Avenic acid, 7

Bacteroid
   definition, 250
   nodule, 250
Betaine, 13
Biotin, 162

Blackheart, 398
Boron, 173–178
   accumulation, 10
   carbohydrate synthesis, 174
   critical level, 174
   deficiency, 174
   distribution, 10
   efficiency, 22
   nutrient interactions, 176
   plant morphology, 176
   protein synthesis, 174
   toxicity, 10, 174
   translocation, 176
   uptake, 173
Budkill, 398

C3 photosynthesis, 413–499
Caffeic acid, 168
Calcium, 157–160
   auxin, 157
   calmodulin, 158
   deficiency, 144
   efficiency, 11, 157
   interactions, 159
   nutrient enzymes, 157
   translocation, 159
   uptake, 157
   utilization, 11
Calmodulin, 68, 157
   NAD kinase, 158
Calorigen, 55
Calvin cycle, 478–482
Carbamoyl phosphate synthetase, 154
Carbohydrate synthesis
   copper, 167
   iron, 169
   magnesium, 161
   manganese, 164
   phosphorus, 151
   potassium, 154
   sulfur, 162
   zinc, 172

Carbon balance, 501–554
Carbon dioxide
  respiration, 527–536
    direct effects, 530–536
    indirect effects, 528–530
  stomata, 329
Carbonic anhydrase, 172
γ-carotene, 190
β-carotene, 183–198
Casparian strip, 202
Catalase, 169
Cation exchange capacity, 2
Cell wall
  solute
    root, 207
Chilling
  abscisic acid, 403
  chlorophyll, 402
  injury, 391, 394
  membrane damage, 399
  photoinhibition, 402
  photosynthesis, 401
  porphyrin, 402
  protection, 403
  respiration, 400
  root, 402
  sensitive, 394
  tolerance, 400
  xanthophyll, 402
  zeaxanthin, 402
Chill stress, 391
Chloride
  exclusion, 23
  guard cell, 327
Cholesterol, 208
Chlorophyll photooxidation, 402
p-Chloromercuribenzoic acid, 143
Citrate isomerase, 169
Citric acid, 9, 137
Climate
  ozone, 374

Coefficient
  respiration
    growth, 511–516
Cold
  hardiness, 393
  temperature, 391–411
Compaction
  soil, 263–287
    coarse-texture, 281
    measurement, 264
Compact layers
  soil, 266
Companion cell, 214
Cone index
  soil, 265
Copper, 7, 165–168
  carbohydrate synthesis, 167
  content, 165
  deficiency, 167
  efficiency, 22
  nutrient,
    enzymes, 166
    interactions, 167
  plant morphology, 167
  protein synthesis, 167
  tolerance, 10, 166
  toxicity, 10, 166
  uptake, 165
Coprogen, 8
Critical level
  boron, 174
Cysteine, 161
Cytochrome, 57
Cytochrome c, 14
  oxidase, 166
Cytokinin, 12, 26
  flooding, 301
  freezing, 404
  photoperiod, 53

Day-neutral plants, 85
Deacclimation, 393

Deficiency
  boron, 174
  copper, 167
  iron, 204
  magnesium, 160
  manganese, 165
  molybdenum, 176
  potassium, 154
  sulfur, 162
  zinc, 171, 173
Dehydration
  avoidance, 13
  tolerance, 13
Desiccation
  nitrogen fixation, 246
Diagravitropism, 64
Diamine oxidase, 166
2,4-dichlorophenoxyacetic acid, 126
Dieback, 398
Diltiazem, 143
Drought, 1
  adaptation, 14
  escape, 14
  plant hormones, 62
  stress, 13
  tolerance, 12, 24

Edaphic environments
  plant adaptations, 1
EDTA, 7
Efficiency
  calcium, 157
  phosphorous, 150
Electrolyte leakage, 398
Electron transfer, 134
  thylakoid, 484–486
Endodermis, 132, 202
Energy currency, 505
Enolase, 160
Environment photosynthesis, 414
Enzymes
  phosphorus, 151
  potassium, 153

Epicuticular wax
  resorption, 572–580
  utilization, 560–568
  wheat, 568–578
Epidermal
  conductance, 15
  transpiration, 15
Epidermis
  plasma membrane, 141
EPTC, 190
Escape mechanisms, 144
Ethylene, 250, 578
  flowing, 292
  photoperiod, 55
Evapotranspiration, 201–244
Extracellular ice, 397
Extraorgan freezing, 405

Farnesol, 14
Farnesyl pyrophosphate, 184
Fatty acid resorption, 573
Ferrichrome, 8
Ferridoxin, 162
Ferrioxamine, 8
Fertilizer utilization, 1
Flooding, 289–321
  abscisic acid, 301
  auxin, 301
  chlorosis, 308
  cytokinin, 301
  epinasty, 307
  ethylene, 292
  gibberellic acid, 301
  growth regulators, 301
  hypertrophy, 308
  leaf dehydration, 291
  nutrition, 300
  photosynthesis, 293
  plant growth, 303–307
  root permeability, 291
  senescence, 308
  stomata, 292

[Flooding, *continued*]
  sulfide, 298
  tolerance, 308–312
    anatomy, 309
    metabolism, 310
    morphology, 309
    water relations, 291
    wilting, 293
Florigen, 52
Freeze-thaw
  soil, 405
  stress, 396
Freezing
  extraorgan, 405
  injury, 395
  tolerance
    cytokinin, 404
    protein, 404
Frost
  cracks, 398
  injury, 395
  resistance, 395–399

Galactooxidase, 166
Gallic acid, 75
Genetic manipulation, 1
Geranylgeraniol, 184
Geranylgeranyl pyrophosphate, 184
  synthase, 187
GGPP synthase, 187
Gibberellic acid, 26, 128, 183–198, 573
  flooding, 301
  precursor, 191
  photoperiod, 54, 108, 576
Gibberic acid, 576
Gliadin, 163
Glucan synthase, 157
Glucose-6-phosphate
  dehydrogenase, 162
Glutamate, 170
Glutanic acid dehydrogenase, 172
Glycerol, 207

Glycinebetaine, 11, 207
Glycolipid, 210
Glycolysis, 504
Glycophytes, 11
Glyoxylate, 136
Gravitropism, 64–74
  coleoptiles, 68–74
  memory, 71
  negative, 68
  perception, 69
  plant hormone, 64–74
  positive, 64
  root, 64–68
  stem, 68–74
  transduction, 66–74
    auxin, 71
    inhibitor, 66–68
Growth efficiency, 511–516
Growth regulators
  flooding, 301
Growth respiration, 511–516
  coefficient, 511–516
Guard cell, 324
  biochemistry, 326
  chloride, 327
  malate, 327
  pH, 327
  phosphoenolpyruvate carboxylase, 327
  potassium, 327
  proton extrusion, 327
  solutes, 326

Halophyte, 10, 202–244
Hardening, 391–411
Heat tolerance, 1
Heliotropic, 116
Herbicide
  antidote, 194
  resistance, 1
  safeners, 194
High humidity
  salinity, 219

Hill reaction, 164
H+ excess, 126–134
H+ symports, 209
2-Hydroxybenzoic acid, 58
Hydraulic conductivity, 208
Hypoxia, 290–321

IAA oxidase, 164, 174
Ice-nucleation, 396
Indoleacetic acid, 26, 132
Infection structures
    wheat leaf rust, 518
Infection thread, 247
Influx
    phosphorus, 150
Intracellular freezing, 396
Ion
    absorption, 205
    accumulation, 206
    carriers, 208
    channels, 208
    chelation, 145
    concentration, 205
    discrimination, 207
    exclusion, 139, 206
    homostasis, 214
    metal, 9
    movement, 144
    pores, 140
    pumps, 140, 208
    toxicity, 9
    transport, 205
    uptake, 139, 144
Iron, 168–171
    acquisition, 7
    carbohydrate, synthesis, 169
    chlorosis, 6, 24
    deficiency, 6, 324, 204
    efficiency, 6, 22
    nutrient
        enzymes, 169
        interactions, 171

[Iron, continued]
    protein synthesis, 169
    translocation, 171
    transport, 7
    uptake, 168, 171
        efficiency, 7
Isocitric
    dehydrogenase, 164
    lyase, 136, 573
Isofloridoside, synthase, 158
Isopentenylpyrophosphate, 184
    isomerase, 187
Isoprenoid, 134
    biosynthesis, 183–198

Jasmonic acid, 57

ent-Kaurene synthesis, 136, 164
α-ketoglutaric acid, 170

Laccase, 166
Lactic acid dehydrogenase, 172
Leaching, 201
Leaf
    blade, 11
    sheath, 11
    wax, 556
Light
    absorption, 115
    duration, 50–57
    guard cell, 328
    intensity
        salinity, 219
    plant development, 83–121
    quality, 50
        photoperiod, 94
        seed germination, 45–49
        stem elongation, 49
    reflection, 110
    transmission, 110
Lipoic acid, 162
Lycopene, 190

Long-day plants, 86–121

Macropore
  soil, 263
Magnesium, 160
  carbohydrate synthesis, 161
  deficiency, 160
  nutrient enzymes, 160
  protein synthesis, 161
  translocation, 161
  uptake, 160
Malate
  dehydrogenase, 162, 164
  guard cell, 327
Malic acid, 9, 63
Malic enzyme, 164
Manganese, 7, 163–165
  absorption, 9
  carbohydrate synthesis, 164
  chelation, 136
  complexation, 135
  deficiency, 134, 165
  entrapment, 9
  nutrient enzymes, 164
  precipitation, 136
  protein synthesis, 164
  sequestration, 135
  toxicity, 10, 23, 134, 164
  translocation, 9, 135, 162, 165
  uptake, 135, 163
Mannitol, 207
Mechanical impedance
  soil, 265
Membrane
  damage
    chilling, 399
  fluidity, 395
  semipermeability, 395
Metabolic messengers, 39
Metal detoxification, 9
Metal immobilization, 9
Methionine, 161

Metolachlor, 193
Mevalonic acid, 184
Mevalonic kinase, 136
Mineral
  deficiency, 2
  stress, 1
  toxicity, 2
Mitochondrial P/O, 507
Molybdenum, 176
  deficiency, 176
  nutrient
    enzymes, 177
    interactions, 177
  -pterin, 3
  uptake, 176
o-Monooxygenase, 166
Mucigel, 138
  aluminum, 138
Mucilage, 9, 269
Mugineic acid, 7
Mulch
  reflected light, 118

NADH
  nitrate reductase, 158
  oxidase, 158
NAD
  kinase, 158
    calcium, 158
NADP$^+$/NADPH, 507
Neighboring cells
  stomata, 324
Nicotinamine, 7
Nifedipine, 143
Nitrate
  reduction, 2, 163
  uptake, 2
Nitrate reductase, 3, 8, 22, 177
  phosphorus, 151
Nitrogen
  accumulation, 3
  efficiency, 5

[Nitrogen, *continued*]
  fixation, 245–262
    dessication, 246
    salinity, 246
    symbiotic, 245–262
    temperature, 246
    transpiration, 254
  remobilization, 3
  respiration, 540–543
  use efficiency, 3
Nitrogenase, 177, 250
  relative efficiency, 253
Nodule
  development, 247
  diffusion, barrier, 251
  function
    mechanism, 256
    oxygen, 255
    relative efficiency, 250
    temperature
      diurnal, 250
      high, 252
      low, 251
    water stress, 253
      carbohydrate, 254
  gas permeability, 251
  structure, 247
  water transport, 250
Nonsaline soils, 201
Norflurazon, 190
Nutrient
  availability, 2
  enzymes
    calcium, 157
    copper, 166
    iron, 169
    magnesium, 160, 164, 165
    manganese, 165
    molybdenum, 177
    sulfur, 161
    zinc, 172
  interactions

[Nutrient, *continued*]
  boron, 176
  copper, 167
  iron, 171
  molybdenum, 177
  phosphorus, 152
  zinc, 173
  soil content, 2
  uptake, 2

Ornithine decarboxylase, 12
Orthogravitropism, 68
Orthophosphate dikinase, 154
Osmotic adjustment
  solute
    root, 206
Osmotic potential, 207
  soil, 10
Osmotin synthesis, 12
Oxidative phosphorylation, 504
Ozone, 357–389
  abscisic acid, 370
  climate, 374
  crop, 374
    economics, 379
    losses, 376
  forests, 380
  fruit, 370
  leaf production, 372
  leaf senescense, 372
  metabolism, 369
  mycorrhizae, 373
  organismic response, 371
  photosynthesis, 370
  pollen, 370
  roots, 373
  salinity, 220

Pathogen
  encroachment, 1
  resistance, 1
PCMBS, 143

Penetrometer, 265
Pericycle, 202
Peroxidase, 169
Peroxyacetylnitrate, 360
Pest
  encroachment, 1
  resistance, 1
pH, 2
  root growth, 131
  shoot growth, 131
Phloem, 141
Phosphate
  deficiency, 144
  phytic-, 8
  regeneration photosynthesis, 436
Phosphoenolpyruvate carboxylase,
    327
  guard cell, 327
Phospholipid, 208
Phosphorus
  absorption, 5
  acid phosphatase, 14
  assimilation, 22
  carbohydrate synthesis, 151
  deficiency, 5
  efficiency, 5, 22, 24, 150
  enzymes, 151
  influx, 150
  nitrate reductase, 154
  nutrient interactions, 152
  pH gradient, 150
  plant morphology, 152
  protein synthesis, 151
  redistribution, 5
  stress, 152
  translocation, 152, 156
  uptake, 2, 22, 150
  utilization, 5
Photoperiod, 573
  adaptive response, 414
Photorespiration, 431, 481

Photosynthesis
  acclimation, 455
  analysis
    biochemical, 420
    fluorescense, 419
    metabolite, 423
  atmospheric carbon dioxide, 459–
    463
  C3 plants, 413–499
  Calvin cycle, 478–482
  carbon dioxide, 441–448
    long-term, 439
    modeling, 437
  chilling, 401
  control
    analysis, 418–423
    gas exchange, 419
  environment
    long-term responses, 414
    physical-chemical, 416
    sensory perception, 417
    short-term responses, 414
    substrate, 415
  flooding, 293
  humidity, 441, 470–472
  light, 441
    changing, 457
    long-term, 453
    low, 453
  limitations, 423–441
    boundary layer, 429
    diffusion, 424–444
    phosphate regeneration, 436
    stoma, 425–429
  mesophyll, 430
  mineral availability, 416
  nitrogen, 441, 475–478
  ozone, 370
  photoinhibition, 450-
  postillumination, 457
  rubisco, 432, 434
    oxygenation, 431

[Photosynthesis, *continued*]
  RuBP, 432
  starch synthesis, 452, 478–482
  sucrose synthesis, 452, 478–482
  temperature, 441, 464–470
  thylakoid limited, 435
  thylakoid regulation, 448
  water, 441, 472–475
Phytochrome, 152
Phytoene, 190
  synthetase, 184, 190
Phytoferrin, 170
Phytofluene, 190
Phytol, 183
Photomorphogenesis, 84–121
Photoperiod
  abscisic acid, 55
  alkanes, 109
  biological clock, 51
  bud dormancy, 51
  chlorophyll a, 106
  chlorophyll b, 106
  cytokinin, 53
  discovery, 84–89
  ethylene, 55
  fatty acids, 109
  fatty alcohols, 109
  flowering, 51–56, 91
  free amino acids, 109
  gibberellin, 54, 108
  growth retardants, 54
  leaf, 104
  light quality, 94
  night break, 104
  photosynthetic efficiency, 110
  reflected light
    plant, 110
    soil, 110
  reversible, 95
  seed germination, 51, 96
  system elongation, 51, 104
  storage organs, 56
  tuber formation, 56
  vegetative reproduction, 51

Photosystem
  I, 170
  II, 134, 170, 214
Phytic phosphate, 6
Phytochrome, 40, 45–50, 90
Phytoene synthetase, 136
Pinitol, 12
Plant development
  light, 83–121
  growth regulators
    definition, 40
  growth substances, 40
  hormones, 39–75
    concentration, 44–75
    cytokinin, 62
    definition, 40
    drought, 62
    gibberellic, 62
    gravity, 64–74
    perception, 43–75
    seed germination, 45–49
    sensitivity, 39–75
    temperature, 60–62
    translocation, 39–75
  morphology
    boron, 176
    copper, 167
    phosphorus, 152
    potassium, 156
    sulfur, 163
  nutrition, 149–182
Plasmadesmata, 141, 202
Plasma membrane, 207
Plastocyanin, 166
Plowpan, 267
Plowsole, 267
Poiukilotherm, 392
Pollen
  ozone, 370
Pollutant
  concentration, 360
  type, 360
  uptake, 367

Pollution
  air, 357–389
Polyphenol oxidase, 166
Potassium, 6
  accumulation, 6
  carbohydrate synthesis, 154
  cell extension, 156
  deficiency, 154
  enzymes, 153
  guard cell, 327
  plant morphology, 156
  protein synthesis, 154
  root growth, 6
  root morphology, 6
  stoma, 325
  translocation, 6
  uptake, 6, 153
  utilization, 6
Proline, 11, 207
Protein synthesis
  boron, 174
  copper, 167
  iron, 169
  magnesium, 161
  manganese, 164
  phosphorus, 151
  potassium, 154
  sulfur, 162
  zinc, 172
Protocatechyic acid, 75
Proton extrusion
  guard cell, 327
Psiderophore, 168
Putrescine, 12, 154
Pyrophosphatase, 208
Pyruvate kinase, 154

Quinones, 183–198

Radiation freezes, 392
Rainfall, 1
Reflected light
  mulch, 118

[Reflected light, *continued*]
  plant, 110
  soil, 110–118
Relative humidity, 329
Respiration, 501–554
  alternative pathway, 510
  aluminum, 540–547
  apparent, 511–516
  carbon dioxide, 527–536
    direct, 530–536
    indirect, 528–530
  chilling, 400
  control
    biochemical 504–511
    physiological, 511–516
    pull, 505–508
    push, 508–511
  drought, 540–543
  growth, 511–516
    coefficient, 511–516
    rate, 511–518
  ion transport, 519
  light
    low, 539
  maintenance, 511–520
    coefficient, 511–516
    rate, 511–516
  nitrogen, 540–543
  nutrient deficiency, 540–543
  ozone, 537
  products, 504
  protein turnover, 519, 540–543
  repair, 518–520
  resource limitation, 543
  salinity, 540–543
  stress, 536–543
    aerial, 537–540
    soil, 540–543
  substrate, 504
  sulfur dioxide, 537
  temperature, 521–527
    acclimation, 522–525
    adaptation, 522–525

[Respiration, *continued*]
  high, 525–527
  low, 525–527
  toxins, 544
Rest, 393
Rhizobia
  polysaccharide, 247
  salinity, 246
  temperature, 246
  water stress, 246
Rhodotorulic acid, 8
Riboflavin, 169
Ribulose biphosphate carboxylase,
    161
Root
  anatomy, 268
  cap, 269
  chilling, 400
  cortex, 202, 269
  diameter, 270
  elongation, 141
  growth, 267–279
    pH, 131
    soil pores, 273
  hairs, 270
  lateral, 270
  mechanical stress, 273
  nodal, 271
  ozone, 373
  proliferation, 3
  relative length, 270
  seminal, 271
  solute, 202–211
  stele, 202
  tap, 271
  water movement, 202–211
Rubisco limitation, 432

Safeners
  herbicide, 194
Salicylic acid, 55
Saline-sodic soils, 201

Saline soil
  environment, 200–200
  stress, 12
  tolerance, 1
Salinity, 10, 199–244
  aerial environment, 218–220
  agromanagement, 219
  high humidity, 219
  light intensity, 219
  natural environment, 218–200
  nitrogen fixation, 246
  nodule performance, 246
  pollutants, 219
  respiration, 540–543
  rhizobia, 246
  stress, 207
  temperature, 219
  tolerance, 12
Salt
  glands, 204, 213, 218
  sieving, 208
  stress, 202
  tolerance, 12, 24, 200, 202,
    220–227
    cereal, 222
    forage, 223
    ornamental, 226
    tree, 224
    vegetable, 221
    vine, 224
  toxicity, 23
Salt exclusion, 204
Scandium, 138
Scutellium, 128
Seed germination
  photoperiod, 96
Sesquioxide content, 2
Shoot growth
  pH, 131
Short-day plants, 86–121
Siderophytes, 7
  microbial, 8
  phyto-, 7

Siroheme, 3
Sitosterol, 210
Smuts, 580
Sodic soil, 201
    tolerance, 1
Sodium chloride, 10
Soil
    aeration, 277
    aggregation, 278
    alkaline, 201
    available water, 278
    base saturation, 2
    bulk density, 264
    bulk volume, 264
    coarse
        compaction, 281
    compaction, 263–287
    definition, 263
    fine-texture, 283
    yield, 279
    compact layers, 266
    cone index, 265
    fissures, 265
    flooding, 290
    freeze-thaw, 405
    impedance
        root growth, 275
    macropore, 263
    matrix potential, 10
    mechanical impedance, 265
    moisture capacity, 2
    moisture retention, 2
    nonsaline, 201
    organic matter, 2
    oxygen levels, 2
    permeability, 2
    pores
        root growth, 273
    porosity, 264
    reflected light, 118
    saline-sodic, 201
    salinity, 200

[Soil, continued]
    sodic, 201
    strength, 2
    temperature, 2
    transmission pores, 265
    waterlogged, 202
Solanesyl-45-pyrophosphate, 184
Solute
    root, 202–211
        anatomy, 202–204
        cell walls, 207
        ion exclusion, 205
        ion pumps, 208
        ion transport, 205
        lipids, 209
        morphology, 202–204
        organelles, 210
        osmotic adjustment, 206
        solute synthesis, 206
        water relations, 204
    shoot, 211–218
        anatomy, 213
        morphology, 213
        reproduction, 215
        /root, 216
        salt loading, 212
        succulence, 217
Sorbitol, 11, 12
Squalene, 189
    synthetase, 190
Starch synthesis, 452, 478–482
Statocytes, 65
Stele, 141
Sterol, 183–198
Stigmasterol, 210
Stomata, 323–355
    abscisic acid, 331
    canopy, 334
    carbon dioxide, 329
    complex, 324
    control, 328–334
    density, 15

[Stomata, *continued*]
  dioecious plants, 346
  guard cells, 324
  light, 328
  mechanisms, 325–328
  neighboring cells, 324
  Populus, 339–346
  relative humidity, 329
  scaling, 334
  subsidiary cells, 324
Stress
  environmental, 1
  respiration, 536–543
  salinity, 207
  salt, 202
  tolerance
    gene control, 25
    selection, 25
    traits, 25
Subsidiary cell, 324
Sucrose synthesis, 12, 207
  photosynthesis, 452, 478–482
Sulfur, 161–163
  absorption, 163
  carbohydrate synthesis, 162
  deficiency, 162
  nutrient enzymes, 161
  plant morphology, 163
  protein synthesis, 162
  translocation, 163
  uptake, 161
Sunscald, 398
Supercooling, 396
Superoxide dismutase, 164
Symplast, 202

TCA cycle, 504
Temperature
  abscisic acid, 61
  nitrogen fixation, 246
  plant hormone, 60–62

[Temperature, *continued*]
  respiration, 521–527
    acclimation, 522–525
    adaptation, 522–525
  rhizobia, 246
  salinity, 219
Tetcyclasis, 60–62
Tetraethylammonium, 154
Theichodes, 15
Thiaminepyrophosphate, 162
Thigmonasty, 74
Thylakoid membrane
  electron transport, 483–485
Tillage pan, 267
Tocopherol, 184
Tocopherylquinone, 184
Tolerance
  salt, 202
Toxicity
  boron, 174
  copper, 166
  manganese, 164
  zinc, 171
Traffic sole, 267
Transaconitic acid, 9
Translocation
  boron, 176
  calcium, 159
  iron, 171
  magnesium, 161
  manganese, 162, 165
  phosphorus, 152
  potassium, 156
  sulfur, 163
  zinc, 172
Transport
  nodule
    water, 250
Trichomes, 558
Tuber formation
  photoperiod, 56
Turgor, 325

Turgorins, 75
Tyrosinase, 166

Uptake
  boron, 173
  calcium, 157
  copper, 165
  iron, 168, 171
  magnesium, 160
  manganese, 163
  molybdenum, 176
  phosphorus, 150, 153
  potassium, 153
  sulfur, 161
  zinc, 171
Uracil, 174
Urediospore, 555

Vascular stele, 269
Verapamil, 163
Violaxanthin, 196
Vitamin K, 184

Water potential
  soil, 10
Water utilization, 1
Wax
  epicuticular, 558
    resorption, 572–580

Wheat
  epicuticular wax, 568–578
  flag leaf, 555–583
    wax, 555–583
  leaf rust, 555–583
    germling, 555–583
    infection structures, 578
  leaf surface, 557–560
  stem rust, 579
    ethylene, 579
Winter burn, 398

Xanthophyll, 183
  chilling, 402
Xanthoxin, 59
Xylem, 141
  parenchyma, 141, 214

Zeaxanthin, 17
  chilling, 402
Zinc
  carbohydrate synthesis, 172
  deficiency, 171, 173
  nutrient, enzymes, 172
  nutrient interactions, 173
  protein synthesis, 172
  tolerance, 10
  toxicity, 10, 171
  translocation, 172
  uptake, 171